MECHANISMS OF MORPHOGENESIS

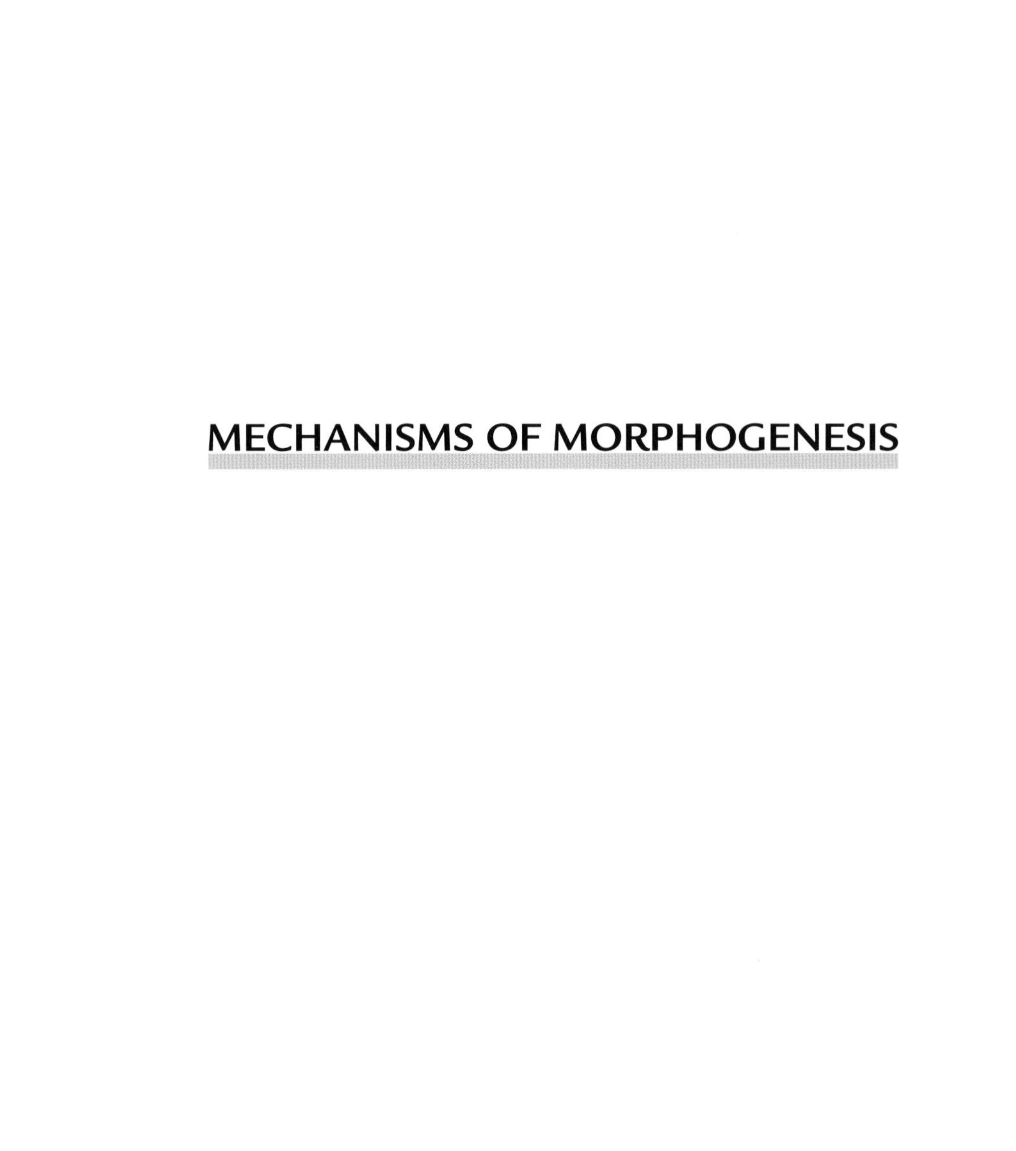

MECHANISMS OF MORPHOGENESIS

JAMIE A. DAVIES

Morphogenesis Laboratories
University of Edinburgh College of Medicine
Edinburgh, UK

ELSEVIER
ACADEMIC
PRESS

AMSTERDAM • BOSTON • HEIDELBERG • LONDON • NEW YORK • OXFORD
PARIS • SAN DIEGO • SAN FRANCISCO • SINGAPORE • SYDNEY • TOKYO

Elsevier Academic Press

30 Corporate Drive, Suite 400, Burlington, MA 01803, USA
525 B Street, Suite 1900, San Diego, California 92101-4495, USA
84 Theobald's Road, London WC1X 8RR, UK

This book is printed on acid-free paper. ♾

Library of Congress Cataloging-in-Publication Data
Application submitted.

British Library Cataloguing in Publication Data
A catalogue record for this book is available from the British Library

ISBN 13: 978-0-12-204651-3
ISBN 10: 0-12-204651-x

For all information on all Elsevier Academic Press publications visit our Web site at www.books.elsevier.com

Printed in China
05 06 07 08 09 10 9 8 7 6 5 4 3 2 1

To Katie

CONTENTS

A NOTE ON THE REFERENCES ix

1 Introductory Section 1

1.1 INTRODUCTION: The Aims and Structure of This Book 3

1.2 KEY CONCEPTS IN MORPHOGENESIS 7

1.3 THE POWER AND LIMITATIONS OF MACROMOLECULAR SELF-ASSEMBLY 17

2 Cell Shape and the Cell Morphogenesis 31

2.1 MORPHOGENESIS OF INDIVIDUAL CELLS: A Brief Overview 33

2.2 THE SHAPES OF ANIMAL CELLS: Tensegrity 41

2.3 CELLULAR MORPHOGENESIS IN ANIMALS 59

2.4 CELLULAR MORPHOGENESIS IN PLANTS 77

3 Cell Migration 93

3.1 CELL MIGRATION IN DEVELOPMENT: A Brief Overview 95

3.2 CELL MIGRATION: The Nano-Machinery of Locomotion 103

3.3 CELL MIGRATION: Navigation by Chemotaxis 119

3.4 CELL MIGRATION: Galvanotaxis 137

3.5 CELL MIGRATION: Navigation by Contact 149

3.6 CELL MIGRATION: Waypoint Navigation in the Embryo 167

3.7 CELL MIGRATION: Condensation 185

4 Epithelial Morphogenesis 197

4.1 THE EPITHELIAL STATE: A Brief Overview 199

4.2 EPITHELIAL MORPHOGENESIS: Neighbour Exchange and Convergent Extension 213

4.3 EPITHELIAL MORPHOGENESIS: Closure of Holes 223

4.4 INVAGINATION AND EVAGINATION: The Making and Shaping of Folds and Tubes 231

4.5 EPITHELIAL MORPHOGENESIS: Fusion of Epithelia 247

4.6 EPITHELIAL MORPHOGENESIS: Branching 259

5 Morphogenesis by Cell Proliferation and Death 289

5.1 GROWTH, PROLIFERATION, AND DEATH: A Brief Overview 291

5.2 MORPHOGENESIS BY ORIENTATED CELL DIVISION 311

5.3 MORPHOGENESIS BY ELECTIVE CELL DEATH 329

6 Conclusions and Perspectives 347

6.1 CONCLUSIONS AND PERSPECTIVES 349

INDEX *361*

A NOTE ON REFERENCES

The purpose of the references in this book is to allow readers to explore aspects of morphogenesis in more detail and to judge key experiments for themselves. Their emphasis is therefore on recent reports and reviews that illustrate current knowledge and thought about particular problems. The seminal papers of important fields are acknowledged, but a comprehensive listing of all publications pertaining to each subject discussed would be impossible; the book already contains well over a thousand references and it would need tens of thousands if full justice were to be done to all who have contributed to the stories herein. Nevertheless, I apologize to those whose valuable contributions have not been listed explicitly.

References are generally given as superscripts like this.[1] Where a superscript of this type could be confused with a power, because it applies to a formula or units of measurement, references are tagged explicitly like this.[ref2]

1

Introductory Section

CHAPTER 1.1

INTRODUCTION: The Aims and Structure of This Book

This book is about the shapes of living tissues and bodies. More specifically, it is about how these shapes are created as organisms develop from simple fertilized eggs into complex animals and plants. This production of shape is usually called "morphogenesis."♣

The processes that are responsible for the development of animals and plants can be divided, for convenience, into three main categories: communication, differentiation, and morphogenesis.

Morphogenesis is the only one of these activities that is obvious to even a casual observer of embryos, and was therefore the first to be studied. Indeed, the study of morphogenesis is one of the oldest of all the sciences, dating back to ancient Greece where Aristotle (384–323 BCE) described the broad features of morphogenesis in birds, fish, and cephalopods. Even at this early stage of scientific thinking, he understood that an animal's egg had the "potential" for its final form but that it did not contain a miniature version of the form itself.[2] Since those days, the deepest questions of animal and plant development have continued to concern the creation of shape and form. Until very recently, however, the molecular revolution in biology has almost ignored this most fundamental aspect of animal and plant development and has instead concentrated on differentiation and communication.

It is easy to understand why. The immunologist and philosopher Peter Medawar once defined science as "the art of the soluble"[3] and later wrote an entertaining series of essays under that title that pointed out the futility of researchers devoting their lives to tackling, head-on, problems that were simply too difficult to solve by brute force when an intelligently thought-out series of studies on related but soluble problems might provide an alternative line of attack on the original issue.[4] The morphogenesis of a tissue is a formidably difficult problem, being four-dimensional (three space, one time) and involving the interaction of thousands of different molecules. This makes it much more complicated to study at a mechanistic level than either differentiation or communication.

♣From the Greek roots *morph*, meaning "form," and *genesis*, meaning "creation." As C.H. Waddington explained almost half a century ago,[1] "*The word 'morphogenesis' is often used in a broad sense to refer to many aspects of development, but when used strictly it should mean the molding of cells and tissues into definite shapes.*" This book uses the term strictly.

Differentiation can be studied using simple assays for gene expression such as Northern blots, PCR bands, or microarrays, and the control of gene expression can be investigated by mutating a one-dimensional length of DNA and monitoring the effects; much of this work can even be done in cell-free systems. Communication can be studied by grafting pieces of embryo to identify the presence of signalling events and then testing the ability of candidate signalling molecules to drive developmental events by removing the candidates or applying them ectopically. Even these experiments, which use limited numbers of components and usually simple assay systems, can be difficult and time consuming, but, when performed well, they usually generate clear and unambiguous results. For this reason, study of differentiation and communication has dominated developmental biology in the molecular era.

The result of this concentration on differentiation and communication has been a vastly improved knowledge of how plants and animals develop, and also a new understanding of the molecular bases of human congenital disease. The best textbooks of the pre-molecular era were dominated by descriptive embryology and by speculation about mechanisms, this speculation being based mainly on extrapolation from unusually tractable but dangerously distant models (such as gene control in viruses). These books, which provided the foundations on which programs of molecular biology could stand, have been succeeded in recent years by large and solid works full of well-established mechanisms of signalling, pattern formation, and gene control.[5,6,7,8] Readers under the age of about 35 will no doubt have been guided into developmental biology by one of these excellent volumes.

What has been gained has been a triumph, but it has created a temporary imbalance in the field of developmental biology. The spotlight of developmental textbooks and review volumes is usually focused on the molecular biology of gene control and pattern formation. Morphogenesis, the deep developmental question that held the centre stage of embryological thought for over two millennia, has been somewhat eclipsed. This shows up in several ways. Most obviously, it shows up in the proportion of pages that current texts devote to molecular mechanisms of morphogenesis compared with those devoted to other aspects of developmental biology; in the currently popular and excellent general textbooks of developmental biology, this proportion is generally less than 10 percent and is sometimes close to zero. The taking for granted of morphogenetic mechanisms also shows up, more subtly, in the "blob and arrow" diagrams that are usually used to summarize developmental pathways. These tend to show numerous arrows between genes and promoters and signalling pathways that set up the control logic for a developmental event. These arrows then point to "black boxes" in which the actual morphogenesis—the point of all that lies above them in the diagram—actually takes place (see Figure 1.1.1). The reason for the black boxes is obvious; for the most part, morphogenesis still *is* a "black box," the inner workings of which remain mysterious. And, lest I have offended anyone by the comments in this paragraph, I stress that many of my own reviews on organ development[9,10,11,12,13,14] have been as full of black boxes that stand for morphogenesis as anyone else's have been, for the same reason.

The approach to "insoluble" problems that was recommended by Medawar is, however, starting to pay off. When molecular biological techniques were first applied to embryonic development, it would have been impossible to solve the problems of morphogenesis directly and, even where the reasonableness of hypothetical mechanisms could be confirmed by computer modelling, it was generally not possible to verify them in real developing systems. The great advances in our understanding of the molecular

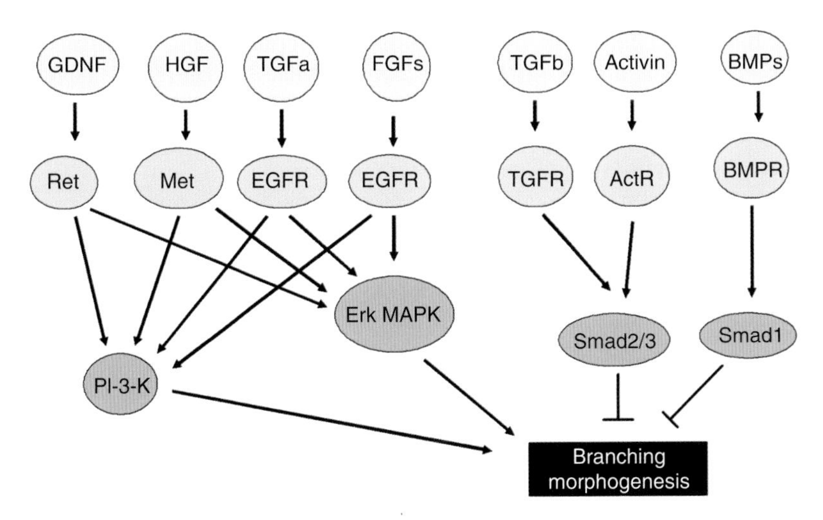

FIGURE 1.1.1 A typical developmental "blob and arrow" diagram, in which signalling events are depicted in detail but morphogenesis is treated as a black box. This particular diagram is adapted from one of my own summaries of processes that control renal development.[10]

genetics of development and of the biochemistry of cell signalling may have unbalanced developmental biology temporarily, but they have allowed researchers to create tools that are powerful enough to be applied fruitfully to morphogenesis itself. Very recently, mainly in the 21st century, the application of these tools has begun to transform our understanding of how cells can translate a "command" to make a shape into that shape itself. The long-sought molecular mechanisms of morphogenesis are now being glimpsed for the first time.

The purpose of this book is to bring together in one place some of the most significant advances that have been made in identifying mechanisms of morphogenesis from a variety of species, systems, and scales. The field is at too young a stage for the book to be able to present a complete, rounded story. Rather, it provides examples of the mechanisms that have been found, or are strongly suspected, to drive the types of morphogenetic event most common in animal and plant development. The order of chapters in the book does not follow the order of events in the development of any particular organism; instead, the book is organized approximately by scale, beginning with the assembly of supramolecular complexes, then the morphogenesis of individual cells, and working mainly upwards in size toward the scale of tissues. This order has been chosen carefully because the morphogenetic mechanisms described in later chapters tend to invoke, and "take for granted," the fine-scale mechanisms described in earlier chapters. Cross-references between chapters have been provided, wherever possible, for readers who are "dipping" rather than reading cover to cover.

As well as concentrating on the molecular biology of morphogenetic mechanisms, this book lays some stress on the more abstract principles that seem to be emerging from them. There are two reasons for this. The first is that it is easier for an author to present, and for a reader to understand, detailed molecular information when there is a framework on which it can be organized. The second is a stubborn belief on the part of the author that there must be more to understanding biology than a lot of diagrams with blobs and squiggly arrows. The issue of "understanding" morphogenetic mechanisms is discussed further in Chapter 6.1. Some of the principles that run most pervasively through this book are described briefly in Chapter 1.2; readers intending only to dip into the sections of this book that are most relevant to them are encouraged to look at Chapter 1.2 first because it defines terms used in later chapters.

This book is absolutely not intended to compete with the general texts on development mentioned above. Whereas they set out the processes of differentiation and signalling and pattern formation in great detail and take morphogenesis somewhat for granted, this book concentrates solely on mechanisms of actual morphogenesis and takes all of the differentiation and patterning mechanisms that control it for granted. The decision to focus only on morphogenesis itself was made to keep the length of this volume reasonable and its price down,♠ and because other authors have anyway discussed the other aspects of development far better than I could. Most readers will already have a sophisticated understanding of the genes and signalling aspects of modern developmental biology anyway; it is my hope that this book will help them in their efforts to link these aspects to the mechanisms that actually generate shape.

Reference List

1. Waddington, CH (1956) Principles of embryology. (George Allen & Unwin)
2. Ronan, CA (1983) Cambridge history of the world's science. (Cambridge University Press)
3. Medawar, PB (1964) Act of creation. *New Statesman*
4. Medawar, PB (1967) The art of the soluble. (Methuen)
5. Gilbert, SF (2003) Developmental biology. **7th**: (Sinauer)
6. Wolpert, L (2002) Principles of development. (Oxford University Press)
7. Alberts, B, Bray, D, Lewis, J, Raff, M, Roberts, K, and Watson, JD (1994) Molecular biology of the cell. **3rd**: (Garland)
8. Slack, JMW (2001) Essential developmental biology. (Blackwell)
9. Davies, J (2001) Intracellular and extracellular regulation of ureteric bud morphogenesis. *J. Anat.* **198**: 257–264
10. Davies, JA (2002) Do different branching epithelia use a conserved developmental mechanism? *Bioessays.* **24**: 937–948
11. Davies, JA and Bard, JB (1998) The development of the kidney. *Curr. Top. Dev. Biol.* **39**: 245–301
12. Davies, JA and Fisher, CE (2002) Genes and proteins in renal development. *Exp. Nephrol.* **10**: 102–113
13. Davies, JA (2004) Branching morphogenesis. (Landes Biomedical)
14. Barnett, MW, Fisher, CE, Perona-Wright, G, and Davies, JA (1-12-2002) Signalling by glial cell line-derived neurotrophic factor (GDNF) requires heparan sulphate glycosaminoglycan. *J. Cell Sci.* **115**: 4495–4503

♠A wish to make this book affordable is also the reason that it is illustrated by drawings rather than by full-colour micrographs.

CHAPTER 1.2

KEY CONCEPTS IN MORPHOGENESIS

The advancement of developmental biology depends on facts about real organisms, acquired by painstaking experiment and careful observation. No science, however, consists of facts alone; only when hard-won knowledge is synthesized into principles can researchers hope to reach something that deserves to be called "understanding." Principles serve several functions: they unite a body of factual knowledge into something concise enough to grasp, they assist with the design of insightful experiments, they help provide a framework for the presentation and evaluation of new discoveries, and, when the principles are not too vague, they can be used to make testable predictions.

There are two broad theories about how principles arise in science. In one view, traditionally associated with Francis Bacon but more fairly attributable to a combination of William of Ockham and John Stuart Mill,[1] experiments come first and principles emerge later from consideration of their results. In the other view, traditionally associated with Karl Popper, hypothetical principles come first and the role of experiments is to attempt to disprove them.[2,3] Biology operates in both Baconian and Popperian modes[4] and can switch between the two many times even during the course of a single project. It is common, for example, for the phenotype of a random mutant to suggest a new principle, in the Baconian manner, and then for that principle to be tested by experiment in the Popperian manner. This dual mode of thinking, used by almost all biologists whether they are explicitly aware of it or not, poses a dilemma for the author of a textbook: should the presentation of a conceptual framework follow the presentation of experimental data, or vice versa? Discussing concepts only after the data have been presented would enable a reader to assess their meaning with an unbiased mind rather than the filter of principles that may turn out to be fallacious. Experimental results are, however, much easier to present and to read in the context of a conceptual framework, if only because such a framework can help to explain why there is a point to discussing a particular experiment in the first place. This book therefore attempts a compromise between the two strategies. This chapter and the next will set out some key concepts of morphogenesis in the broadest terms because these concepts set the agenda for, and are used many times by, the chapters that constitute the experimental core of the book. Chapter 6.1, at the end of the book, will then discuss the principles in more detail, in light of the data described.

The key concepts introduced briefly in this chapter are the idea of a "mechanism," the principle of emergence, the use of feedback, and the differences between self-assembly and adaptive self-organization.

THE IDEA OF "MECHANISM"

The word "mechanism," as used throughout this book, connotes a sequence of events that takes place at a molecular level and that can be explained by interactions of molecules that follow the ordinary laws of physics and chemistry. Mechanism, in this sense, is the antithesis of vitalism. There is no "bio" prefix to the words "physics" and "chemistry" in the first sentence because there is no need for one; the biological character of morphogenetic mechanisms lies not in the individual interactions of their molecular components but rather in the elaborate systems of feedback that regulate where and when those interactions take place. This emergence of biological character from simple physics is discussed further, later in this chapter.

One of the key requirements of a molecular mechanism, as the term is used in this book, is that it must involve only local interactions, for molecules can be affected only by events that take place at their surfaces and they cannot, generally,♣ be affected by action-at-a-distance. This point is stressed here because traditional embryology has often used the concept of gradients within a "morphogenetic field" to explain particular aspects of development. The idea of morphogenetic gradients and fields[5,6,7,8] is very useful for providing a high-level view of events, but it cannot be a part of a molecular-level explanation because a gradient is, by definition, non-local and cannot be sensed directly by a single molecule. At a given time, a single receptor molecule will either be binding one of the molecules involved in the gradient or will not; that is all. A gradient may be able to be sensed by a *system* of interacting molecules, but to explain the mechanism of such a system, one has to focus on the purely local influences within it. These are, after all, all that a molecule can sense. For this reason, high-level concepts such as classical morphogenetic fields are useful for considering large-scale organization in the embryo, but they do not belong in a molecular-level discussion. The more mystical definitions of morphogenetic fields that have been offered by writers such as Rupert Sheldrake are even less relevant to the mechanistic explanations being sought here.[9]

The quest for "mechanisms" of morphogenesis therefore aims to account for the shape changes at the scales of cells and tissues in terms of events that take place only at the scale of individual molecules. A simple listing of molecular interactions is not enough, though, because an understanding of mechanism can be obtained only when the pathways of control that regulate the behaviour of the components have also been identified and characterized. This book therefore uses the word "mechanism" to refer to a combination of the molecular interactions that directly result in shape, of other molecular interactions that control them, and of the principles of control that are involved.

♣Electric fields are an exception to this generalization that may be important in morphogenesis; their role in development is discussed in Chapter 3.4. Gravitational and magnetic fields to which embryos are normally subjected are very weak and have a negligible effect on the behaviour of individual molecules.

EMERGENCE

Emergence is one of the most important concepts in the study of morphogenesis, and indeed in developmental biology as a whole. Emergence, which crops up in fields as diverse as cell biology, neuroscience, information theory, robotics, ecology, game theory, economics, and town planning,[10,11] does not yet have one single rigorous definition. The various definitions that have been offered are all connected with the concept of something being "more than the sum of its parts." The aspect of emergence that is most relevant to understanding embryonic development is that very complex behaviours can emerge from the action of very simple operations, and, by extension, *very complex forms can emerge from the action of comparatively simple machines.*

An entertaining illustration of these aspects of emergence, one which is accessible to anyone with a computer, is provided by John Conway's "Game of Life." The Game of Life is a computer simulation of a cellular automaton. The "playing area" of the game consists of a square grid of locations, each of which can be empty or occupied by one living cell.♣ The initial pattern of occupancy is set up manually, and the game then begins. Time, in the world of the game, is divided into intervals that are separated by the ticks of a master clock. At each tick, the pattern of occupancy for the next time interval is determined by the pattern of occupancy in the previous interval, according to the following simple rules:

1. A cell that has fewer than two neighbours dies from lack of trophic support.
2. A cell that has four or more neighbours dies from overcrowding (for example, through build-up of toxins).
3. If exactly three cells are neighbours of an empty location, one of them divides so that one daughter stays where the mother was and the other occupies the previously empty location. (Since mothers and daughters are instantly equivalent, it makes no difference which of the cells is considered to have divided.)

In each of these rules, a "neighbour" must be in any of the eight locations that border a given location (see Figure 1.2.1).

The best way to gain a feeling for the game is to play it, using one of the many programs and applets that can be found by searching for "Game of Life" on the Web.

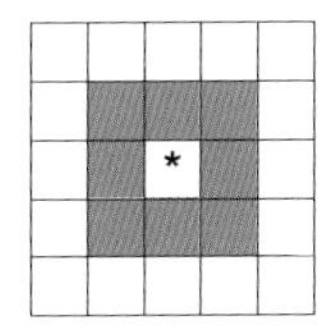

FIGURE 1.2.1 The concept of "neighbours" in the Game of Life. A cell in the location marked "*" is surrounded by eight locations, marked in blue, each of which might contain a cell.

♣Conway's original description used the word "cell" to refer to a location on the grid rather than to its contents (*cf.* "prison cell"). For consistency with the rest of this book, I have used "cell" to refer to the living thing and "location" to refer to place when describing the rules of the game, as many websites devoted to the subject also now do. To describe the game, I have also translated the rules into language that is as "biological" as possible, but have not changed their action at all.

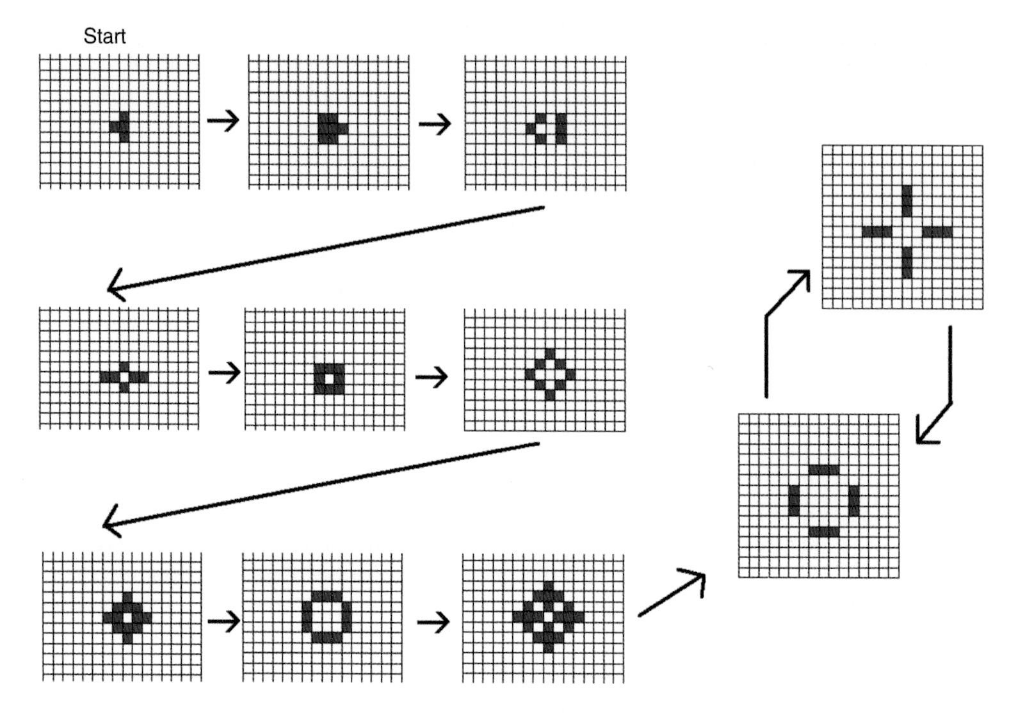

FIGURE 1.2.2 The development of a simple four-cell starting arrangement in the Game of Life. This particular arrangement ends in an oscillating pattern rather than a stable state.

What is immediately apparent is that the three simple rules of the game give rise to very rich behaviour that, while predictable formally, is far from predictable intuitively. One very simple example of the development of a simple four-cell pattern is shown in Figure 1.2.2: the initially asymmetrical pattern acquires symmetry and goes on to develop, via a sequence of intermediate stages, into an oscillating arrangement of 12 cells. Different starting conditions result in different behaviour of the system. Some initial patterns die out, some settle down quickly into stability, whereas others keep changing for as long as one continues to watch. Fanatics of the game spend much time discovering patterns whose behaviour is particularly interesting. One example of interesting behaviour is shown by "gliders," patterns that recreate their starting configuration at a location a little displaced from the starting position. Gliders can therefore move across the playing area forever unless they meet a boundary or collide with other cells (see Figure 1.2.3). Patterns have even been found that eject streams of gliders, one after the other—such patterns are called "glider guns."

The Game of Life is obviously not meant to be a literal simulation of any particular living system. Nobody, not even in 1970 when the game was invented, thought that cells really behaved according to such simple rules. The game does, however, share two important characteristics with most morphogenetic processes in real biology. The first is that the operation of the "rules" is strictly local and requires no knowledge of the state of the system as a whole; each cell senses only the conditions in its immediate neighbourhood. Biological molecules are affected only by the other molecules with which they make contact, and the rules that govern them are therefore local to the nanometer scale.

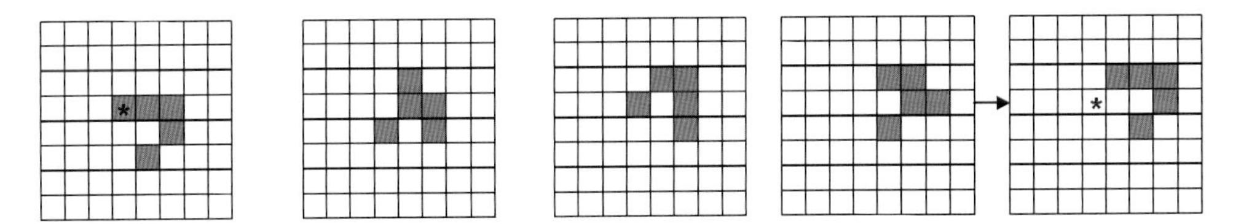

FIGURE 1.2.3 A "glider": The arrangement of five cells appears again, four ticks later, displaced one location up and one location to the right (for this orientation of the glider). In the figure, the asterisk marks the same location in the first and last diagrams to emphasize the displacement of the glider pattern.

It is therefore encouraging to see that strictly local rules can generate rich behaviour even in artificial systems that are simple enough to understand completely. The second characteristic is that all cells are equal—in the Game of Life, there is no hierarchy and no one cell is in command over the others. This, again, is similar to real life; in many systems described in this book, no one cell takes command, and even in the exceptions, once one drops down to a molecular level, there comes a point when all agents of a given type—for example, molecules of a myosin light chain—must be considered as equal in principle so that any special nature of any one molecule emerges from the behaviour of the system as a whole rather than an in-built privilege. It is precisely because it shares these characteristics with real biology that the game is such a useful illustration of the power of simple, local processes to create highly complex behaviour. The rules of the game, for example, say nothing about the translocation of anything across the field, yet with the correct starting conditions, the "glider" in Figure 1.2.3 moves steadily and forever (or at least until it reaches a boundary or another group of cells). These observations, which can be made by anyone who plays with the game for even a few minutes, provide both hope and warning to researchers who study the development of form in real biological systems. The hope is that, in real biology as in the Game of Life, the rich variety of shapes may result from mechanisms that are simple enough to understand. The warning is that underlying mechanisms may have a structure completely different from that of their result, and identifying these mechanisms therefore requires both persistence and imagination.

EMERGENCE, TRAP-DOOR PROCESSES, AND THE DANGERS OF POST-HOC REASONING

Emergence often involves processes that are analogous to the "trap-door" algorithms beloved of cryptographers. Trap-door algorithms are those that cannot be reversed unambiguously. This concept can be illustrated with a simple arithmetic example: if one is told that the multiplication product of two numbers is 12, one cannot deduce what those numbers were. They may have been 1 and 12, or 2 and 6, or 3 and 4, or –3 and –4, or 1/3 and 36, or an infinite number of other possibilities.♣ Multiplication is a "trap door" because, once a pair of numbers has fallen through the process, some information is lost. Rule-based emergent systems such as the Game of Life also involve trap-door

♣Infinite because, for any integer, n, multiplying 12n by 1/n will yield the result 12.

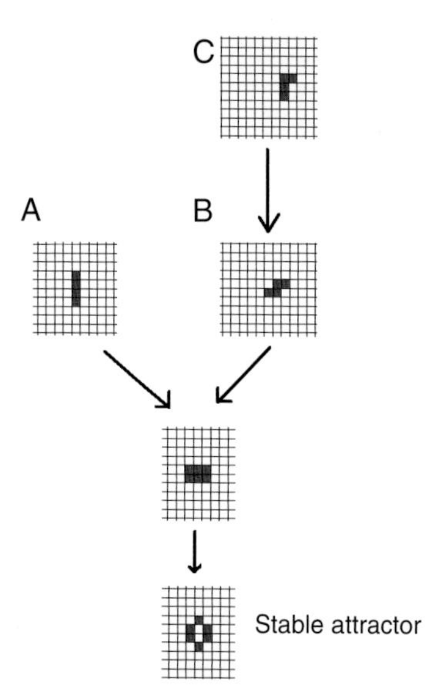

FIGURE 1.2.4 Trap-door functions in the Game of Life; knowing the arrangement of cells in the stable attractor in which the system ends up does not indicate how it got there. The initial conditions may have been the straight line A, the staggered line B, or the "knight's move" C, which generates B, or a host of other possible starting conditions not illustrated. The final state is called an "attractor" because several other states converge to it.

functions: knowing that the result of running a game is a particular arrangement of cells gives no indication as to the original layout of cells (see Figure 1.2.4). This illustrates a very important point in the vastly more complex emergent systems that constitute real developing embryos: *morphogenetic processes cannot be deduced from final form.* Of course, hypotheses can be proposed, especially when a related developmental system is already understood, but these hypotheses must always be tested by observing and experimenting on the system as it develops. A concrete example of this is provided by the elongation of tissues, which is described in detail in later chapters of this book. Some elongation events are driven directly by elongation of cells as the tissue elongates, whereas others are driven by the rearrangement of cells without any elongation of the cells themselves, and still others involve flattening of cells before tissue elongation takes place, followed by a return of cells to their normal shape as the tissue elongates. Simply observing that a tissue becomes longer does not, therefore, lead one directly to the mechanism. Similar ambiguities are seen at the molecular level.

This point is stressed because drawing conclusions from post-hoc reasoning has been a long and dangerous tradition in morphogenetic research. Since the early decades of the last century, long before the advent of computers, attempts have been made to model possible mechanisms of development mathematically. If the model produced the correct shapes, the modeller tended to accept the reality of the proposed mechanism. This tendency has become more marked as computers have become more available. Used properly, in conjunction with experimental observation, computer modelling is a very important tool for investigation of morphogenesis (this point is discussed in more detail in Chapter 6.1). The fact that a mechanism works on a computer is not, however, itself strong evidence that it works in life; usually, many possible mechanisms will produce the "correct" result, and only observation of the real embryo will indicate which is used. Even now, there seem to be far too many mathematically inclined researchers, usually entering biology from outside, who miss or ignore this point.

FEEDBACK, SELF-ASSEMBLY, AND ADAPTIVE SELF-ORGANIZATION

The development of complex structures is not unique to the biological world. Purely physical systems can generate rich patterns that range from the sub-millimeter scales of crystals, through the centimeter scale of snowflakes, to the meter scale of sand dunes, and up to, and possibly beyond, the hundreds-of-light-years scale of galaxies. Morphogenesis of living things is usually recognizably different from that of the physical world, though, so it is worth taking a moment to consider what the main differences are, in the hope that these differences will highlight the key features that enable a living organism to use physical processes in a biological way.

When asked what makes something "biological" rather than purely physical, most biologists will instinctively, and correctly, focus on the Darwin-Wallace theory of evolution: if the workings of a process are the result of, and can be modified by, natural selection, the process is likely to be biological. Otherwise, it is not. That answer begs the question: what makes a process open to modification by natural selection? Essential attributes of such a process are flexibility and function.

Flexibility tends to be in short supply in purely physical morphogenesis. The simplest way in which physical structures are built is self-assembly. Self-assembly is discussed in detail in Chapter 1.3 because it is a core component of biological morphogenesis too. All that needs to be said here is that self-assembly is the coming together of subunits to make a structure, because their association is energetically favourable and their association reasonably probable (these conditions are explained further in Chapter 1.3). Pure self-assembly tends to be relatively inflexible in terms of the range of structures that can be produced. Most elements and compounds that can form crystals, for example, can form only one possible structure. There is therefore no possibility for their modification by selection, even though crystals can "reproduce" in a limited way (by being broken off and seeding further crystal growth). Even where the same atoms or molecules can form more than one type of crystal—for example, the way that carbon can form graphite or diamond and DNA can form A-, B-, and Z-type crystals—the form taken depends purely on the physical conditions (such as pressure) during which crystallization takes place and has nothing to do with fitness for any purpose.

Some physical processes, particularly ones that involve flows of energy, can produce a variety of structures and do, therefore, have some degree of flexibility. An example of such a physical process is provided by the formation of sand dunes where wind sweeps across a desert. Any slight hillock in the sand, caused by random chance if nothing else, creates a wind shadow to its leeward side. As air movement slows in this wind shadow, blown sand grains carried in the air fall, and the hillock grows, its peak moving a little to leeward as it grows in height as the process repeats. The air in the lee of such a hillock of sand carries few sand grains, not enough to amplify a hillock that is very close by, so only at some distance from a growing dune may another dune begin to grow and thrive. The result of these constraints is that windswept sand is sculpted into a series of spaced-out dunes, all moving slowly to leeward across the desert. The height and spacing of the dunes depends on the sizes of the grains, the speed of the wind, and the history of the system. As long as wind energy continues to be available, morphogenesis of the system continues. The system even involves some degree of internal feedback, as the presence of an existing dune alters the probability of sand deposition. Sand dune "morphogenesis" is still intuitively different from biological morphogenesis, though, because the form of the sand is dictated simply by the physical attributes of the components involved and not by any feedback from how well the form is adapted to function, because there is no function.♣

Morphogenetic mechanisms of biology generally have another "layer" to them that provides negative feedback and adjusts morphogenetic processes to optimize them for a specific function. For example, the morphogenesis of the cytoskeleton of an animal cell (Chapters 2.1, 2.2, and 3.2) depends on the association of subunits by simple self-assembly, but the probabilities of self-assembly and of disassembly at any place in the cell are modulated by whether the structure at that place is fulfilling a useful structural role (for example, on whether it is carrying mechanical tension or is flapping about doing nothing). Feedback of this type arises when the output of a process (for example, the shape and location of a structure) is used to control the processes that produce that output (for example, self-assembly and enzymatic destruction). It is the addition of this type of feedback to processes of physical self-assembly that results in a truly "biological" system, and when reduced to the most basic level of description, many morphogenetic processes have the form of Figure 1.2.5. This figure will be discussed in more detail in Chapter 6.1.

Of course, Figure 1.2.5 is oversimplified, and multiple layers of negative feedback are common. For example, the innervation of a particular muscle by a particular neuron may depend at the finest level on the self-assembly, in the neuron, of cytoskeletal structures characteristic of axon extension (see Chapter 3.2). Fast, highly localized (10-nm-scale) feedback will modulate the assembly and disassembly of these structures so that they are optimized for the advancement of the plasma membrane at the leading edge of the extending axon. Feedback on the 100- to 1,000-nm scale will further modulate the self-assembly and disassembly of motile cytoskeleton to ensure that the axon extends along a trail of some guidance molecule rather than in a random direction (see Chapter 3.5). A cell-scale feedback system may then be used to switch off the axon extension mechanism completely once the cell has reached its target, or it may be used to kill the cell if it has reached the wrong target. And, in any biological system, there is always the slow feedback of natural selection that will favour the reproduction of those organisms whose genomes encode morphogenetic mechanisms that build organisms most able to thrive and reproduce.

The addition of feedback to a morphogenetic system gives it a property far more powerful than mere self-assembly. In this book, that property is referred to as "adaptive self-organization."♠ Systems that show adaptive self-organization can arrange their structures in ways not simply dictated by the properties of the structures' subunits, but also according to the (unpredictable) environment in which they find themselves. Using

♣This discussion assumes a desert devoid of life, for example in the Martian desert. By stabilizing surface sand, plants can modify sand dune morphogenesis to promote the development of dune systems optimized for their needs.

♠Adaptive self-organization is also known by the terms "swarm intelligence," "hive intelligence," "distributed optimization," "adaptive routing," and other similar phrases, depending on context.

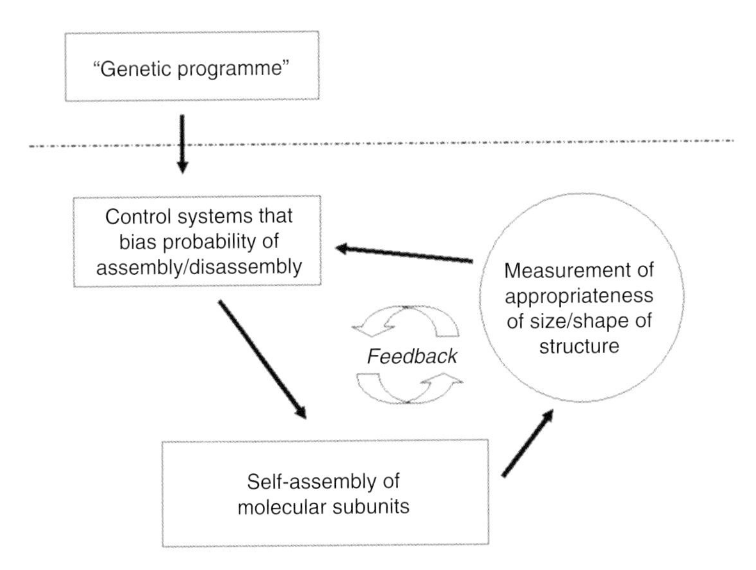

FIGURE 1.2.5 A generalized view of a morphogenetic mechanism typical of an embryo. As explained in the text, feedback usually operates on many layers, and the picture is not as simple as this diagram implies. This book concentrates mainly on events below the dotted line; events above the line are covered very well by other texts.

adaptive self-organization, a mitotic spindle can locate kinetochores that are as hidden in the vast complexity of a living cell as a needle would be in a haystack (see Chapter 2.2); two cells that meet can adhere and align their junctions and cytoskeletons precisely (see Chapter 4.1); a wound that was never "expected" in a developmental program can be closed automatically (see Chapter 4.3); the different structures of the body can form in perfect proportion to one another (see Chapter 5.1) and at a spatial resolution far finer than the "patterning" signals of a growing embryo. Adaptive self-organization will be a recurring theme throughout this book. It has to be; it is one of the defining features of life itself.

Reference List

1. Harre, R (1985) The philosophies of science. (Oxford University Press)
2. Popper, K (1959) Logic of scientific discovery. (Hutchison)
3. Popper, K (1963) Conjectures and refutations. (Routledge & Kegan Paul)
4. Mayr, E (1997) This is biology. (Belknap/Harvard)
5. Weiss, P (2004) Principles of development. (Holt)
6. Raftery, LA and Sutherland, DJ (2003) Gradients and thresholds: BMP response gradients unveiled in *Drosophila* embryos. *Trends Genet.* **19**: 701–708
7. Green, J (2002) Morphogen gradients, positional information, and Xenopus: interplay of theory and experiment. *Dev. Dyn.* **225**: 392–408
8. Wolpert, L (1969) Positional information and the spatial pattern of cellular differentiation. *J. Theor. Biol.* **25**: 1–47
9. Sheldrake, R (1981) A new science of life: The hypothesis of formative causation. (Anthony Blond)
10. Johnson, S (2001) Emergence. (Penguin)
11. Holland, JH (1998) Emergence: From chaos to order. (Oxford University Press)

CHAPTER 1.3

THE POWER AND LIMITATIONS OF MACROMOLECULAR SELF-ASSEMBLY

INTRODUCTION TO SELF-ASSEMBLY

Self-assembly is one of the simplest possible mechanisms for morphogenesis. Its defining feature is that the components that assemble together contain, in their shapes and binding characteristics, sufficient information to determine the structure they produce: there is no need for any external regulation and no need for any special prior spatial arrangements or special timing. For some highly specialized biological systems, such as viruses, self-assembly can be the sole mechanism of morphogenesis, although most biological systems are more complex. Nevertheless, it is important that researchers into morphogenesis are familiar with the characteristics of self-assembling systems for two main reasons. First, even complex morphogenetic systems may use self-assembly many times in the construction of sub-assemblies. Second, an understanding of the power and limitations of simple self-assembly will, by providing a contrast, assist one in understanding the more complex mechanisms covered later in this book.

Self-assembly of macromolecular complexes—whether by (non-covalent) "polymerization" of a series of identical monomers or by coalescence of a structure composed of several different constituents—is driven by the ability of subunits to bind to each other in a specific way. As in all chemistry, this binding is determined by the difference in free energy between the system (consisting of the subunits, the surrounding water, and so on) in the separated state and in the assembled state. Generally, the interactions between subunits depend on a multitude of individually weak, non-covalent bonds such as dipolar forces, hydrogen bonds, and van der Waals interactions. Stable associations form when the free energy change caused by bonding is substantially larger than the energies of thermal collisions that would tend to knock a complex apart. The use of these weak bonds, rather than covalent ones, has consequences to the abilities of self-assembling structures to perform quality-control functions, as will be discussed below.

Reliable self-assembly requires that the coming together of components in the right way is not too improbable. This partly reflects the energy requirements of the process (particularly if the transition of a subunit from an unbound state to a bound one involves the crossing of an energy barrier) and partly reflects the proportion of phase space that is occupied by the structure to be assembled. Phase space is an abstraction that represents, as a single location in multidimensional space, the location and momentum of every subunit in the system at any one time.♣ Neglecting momentum for a moment and concentrating on location, the state space for a single particle can be represented in three-dimensional space simply by the x, y, z coordinates of that particle. The locations of two particles can be represented together as a unique location in six-dimensional space, specified by the three coordinates of each particle, and so on. The location of **n** particles can be represented simultaneously by a unique location in (3**n**)-dimensional space [(6n)-dimensional when momenta are included]. A system will tend to "explore" its multidimensional state space, the movement of the unique point in state space being dictated by the interactions between the particles it contains. Some locations in state space will represent states in which all particles are unbound, some will represent partially completed assemblies, and some will represent a complete assembly. If the volume of state space that represents an acceptable assembly is very small in comparison to the volume of state space itself, it is highly unlikely that the system will ever visit this volume, and assembly is therefore unlikely in a reasonable amount of time. This would be the case if, for example, subunits had to come together in just one precise order and orientation in space for a structure to be formed. If, on the other hand, the volume of state space that represents an acceptable assembly occupies a reasonable fraction of state space, the system is much more likely to enter this volume. This would be the case, for example, if all subunits were identical and could come together in any order to make a filament that would be acceptable whatever its orientation in space.

The abstraction in the last paragraph boils down to a common-sense conclusion: self-assembly works well for uncomplicated structures but is less useful for ones that are complex or have to be orientated very precisely in space.

An intuitive feel for the process of self-assembly can be provided by macroscopic models of the process, one of the best known of which consists of small, light, magnetic tiles that are floated onto water in a Petri dish (see Figure 1.3.1). The dish is subjected to gentle agitation, to provide a macroscopic analog of the Brownian motion that keeps the molecular constituents of biological self-assembling systems in motion, and the behaviour of the tiles is observed. The random motion of the tiles brings them into collision, and if the surfaces that meet are magnetically compatible, they may stick. This random meeting and sticking causes the tiles to form larger and larger aggregates, "tiling" part of the surface of the water. The "binding energy" of the magnets is not enough to make binding an irreversible step, and tiles are sometimes lost from the aggregate. More strikingly, when two small aggregates combine, their association tends to be unstable if their edges do not fit together well but stable if there is a good fit with no gaps. The final result is the spontaneous construction of a large area of regular tiling—a movie of the process can be found at http://mrsec.wisc.edu/edetc/cineplex/self/self02.html.

SELF-ASSEMBLY OF BILAYERED MEMBRANES

A more biologically relevant self-assembling system, but one still capable of producing structures that can be viewed with a simple light microscope, is provided by

♣A clear and accessible introduction to phase space appears in Chapter 5 of R. Penrose's *The Emperor's New Mind.*[1]

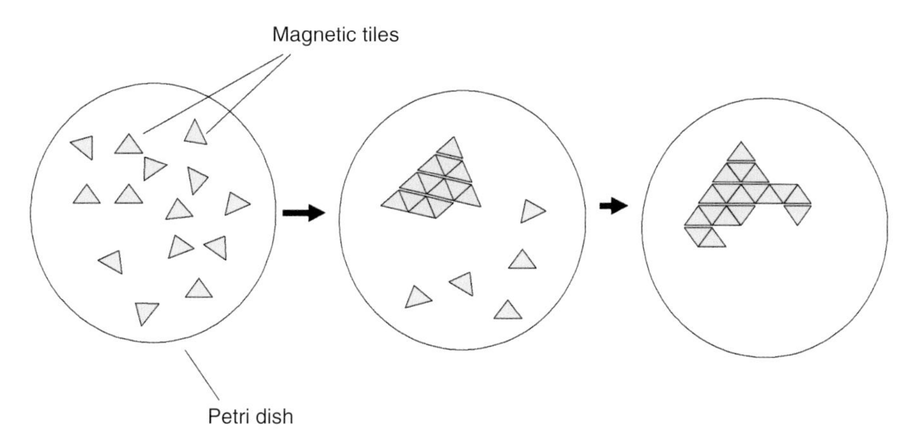

FIGURE 1.3.1 Self-assembly of floating magnetic tiles into arrays. This drawing is based on images in a movie that can be found at http://mrsec.wisc.edu/edetc/cineplex/self/self02.html.

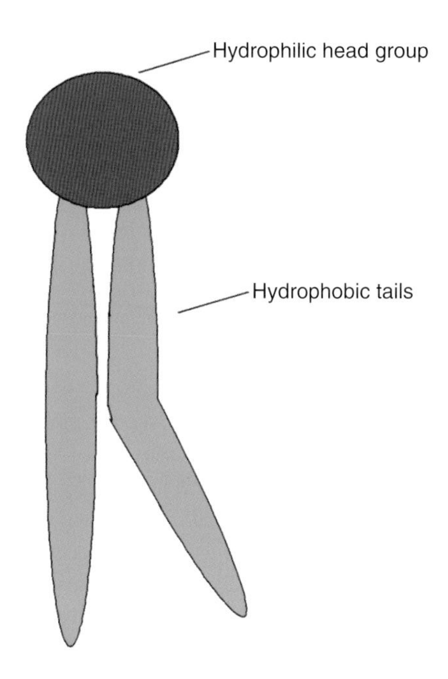

FIGURE 1.3.2 The general structure of a membrane phospholipid.

phospholipids, which are the main constituent of cell membranes. Phospholipids are amphipathic; they have both hydrophilic and hydrophobic characters that are separated into different domains of the molecule; the "head" is hydrophilic and the "tail(s)" hydrophobic (see Figure 1.3.2). The hydrophilic head can form energetically favourable hydrogen bonds with water, but the tail cannot; the effect of immersing the tail region in water would be to disrupt the hydrogen bonds between many water molecules, which would be energetically unfavourable from the point of view

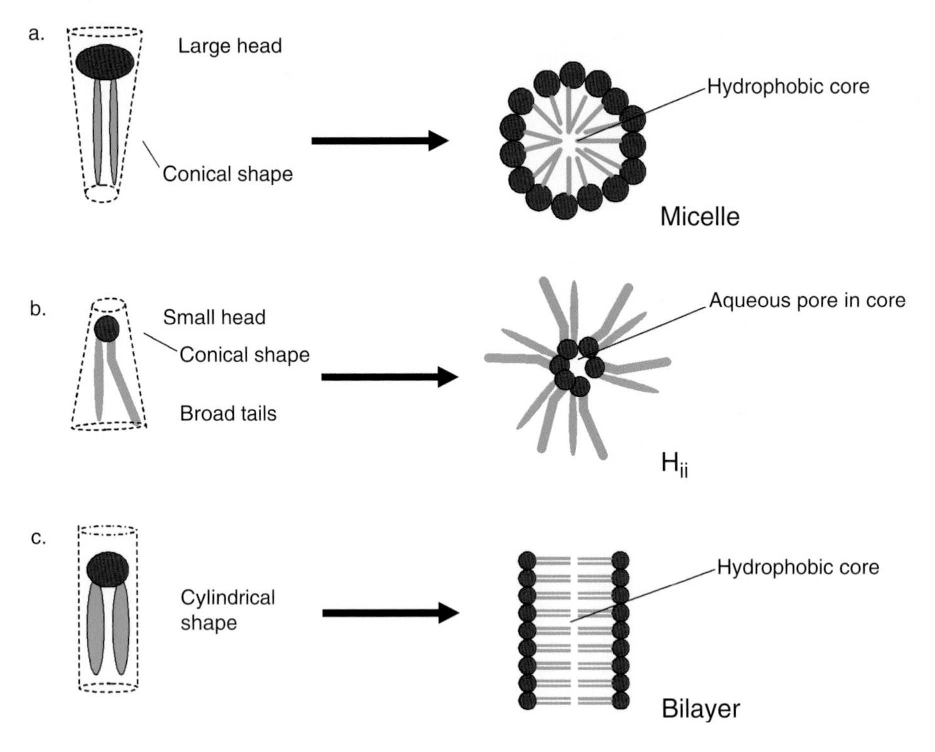

FIGURE 1.3.3 Phospholipids and glycolipids with different geometries self-assemble into different structures when suspended in water. The bilayer in part c extends upwards and downwards from the short section illustrated.

of the complete solute-plus-solvent system. Phospholipids therefore combine to form structures in which the net free energy change, including disruption of water–water interactions, is as favourable as possible.

The structure formed by a given type of phospholipid is determined largely by the gross shape of the monomer, particularly the ratio of the widths of the head region and the tail regions.[2] Some phospholipids, such as gangliosides, have large head groups, giving the molecules an approximately conical shape (see Figure 1.3.3a). When mixed with water, such wide-head conical molecules tend to associate together in spherical monolayered shells called micelles, in which all of the heads face outwards and can associate with water and the tails all face in toward the water-free core. Some lipids, such as the monogalactosyldiglycerides found in chloroplasts, are conical the other way around and have a small head and broad tails. For these molecules, a micellar aggregate would be unstable because of the difficulty of packing the thick tails in the cramped interior of the micelle, and the shape of the molecules forces them to pack the other way around. The most common resulting structure, called the H_{ii} phase, consists of a hexagonal array with heads oriented inwards to form a narrow aqueous channel and the tails oriented outwards[3] (see Figure 1.3.3b).

Phospholipids whose heads and tails are of approximately equal effective diameter, and which therefore have cylindrical rather than conical shapes, associate most stably in double-layered sheets in which the hydrophobic tails of one layer face the

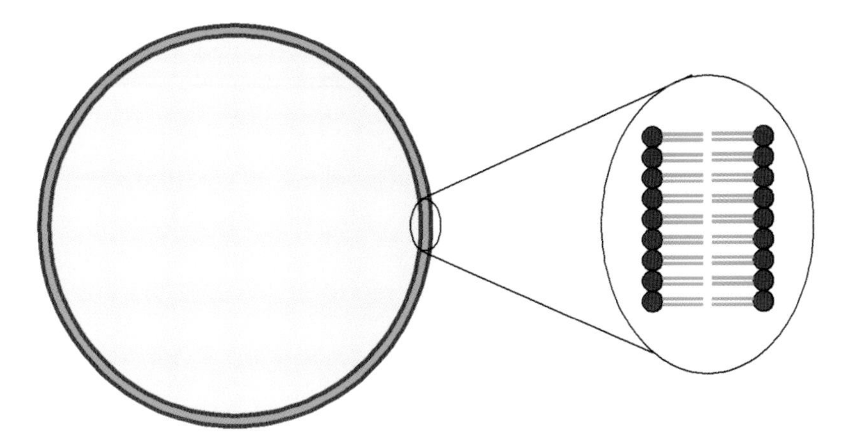

FIGURE 1.3.4 The structure of a single-layered phospholipid vesicle.

tails of the other layer and the hydrophilic heads face the water (see Figure 1.3.3c). An elegant demonstration of the importance of phospholipid geometry to the shape of the supramolecular structure has been provided by mixtures of different molecules. The phospholipid lysolecithin is conical, narrowing toward the tail, and it tends to form micelles. If it is mixed with cholesterol, the cholesterol associates with the tail region and fills in "space" to make the dimer more cylindrical; under these conditions, flat sheets form in preference to micelles.[4] A simple flat sheet of such a bilayer would have the problem of unstable edges, at which the radius of curvature of the sheets would have to be very short. A more stable configuration is a closed, and therefore edgeless, surface such as a vesicle (see Figure 1.3.4).

The production of spherical vesicles, in which all parts are alike, is a common result of mixing approximately cylindrical phospholipids with water, but other geometries are also possible. Closed cylinders, for example, are common, and they are stable enough for measurements to be made on their mechanical properties.[5] With the addition of divalent cations that can promote association between membranes, helical morphologies are also possible.[6] Filaments can also form, either when a hot isotropic solution of certain (but not all) lecithin derivatives is slowly cooled[7] or when water is added to a smear of lecithin dried on a microscope slide. The formation of these filaments is a dynamic process; the structures shown in the still micrograph in Figure 1.3.5, taken at the interface between a drop of water and a smear of lecithin (= phosphatidylcholine), were in constant motion and tumult for many minutes and, indeed, when I was obtaining this image,♣ colleagues glancing at the video monitor assumed that they were seeing a complex living system.

The dynamism of vesicles has provided the basis of what is so far the most advanced attempt to construct a self-reproducing "cell" from non-living components. The starting material is not a phospholipid but a smaller amphipathic molecule, sodium oleate (Figure 1.3.6). At high concentrations, sodium oleate forms micelles in water. If a

♣Dry a drop of lecithin, dissolved in chloroform, on a microscope slide. When it is fully dry, place a drop of water so it touches the edge of the lecithin smear, and observe the interface by phase contrast. Interesting morphologies will result in minutes. This can make an interesting lecture theatre demonstration if a projection microscope is available.

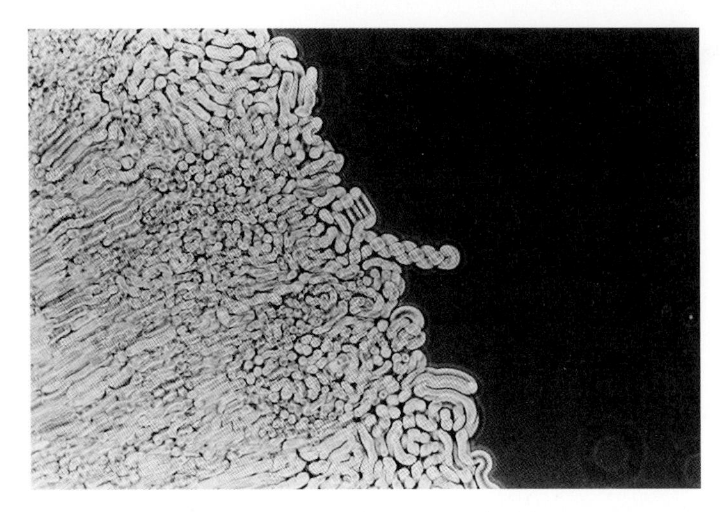

FIGURE 1.3.5 Bending and twisting tubes formed when water is added to a smear of dried lecithin on a microscope slide.

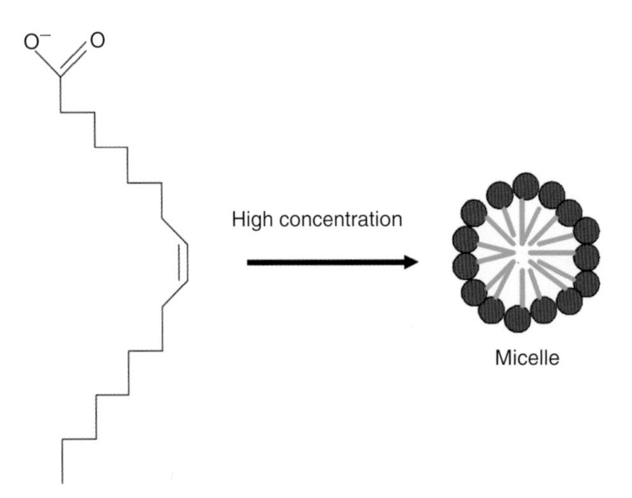

FIGURE 1.3.6 The formation of micelles by sodium oleate at high concentration.

suspension of such micelles is injected into a large volume of pH 8.8 buffer, the oleate rearranges to form vesicles. The kinetics of this rearrangement show a sigmoidal curve[8] (see Figure 1.3.7). This curve, with an initial lag followed by a marked acceleration of vesicle formation once some vesicles had already formed, indicates an autocatalytic process in which the presence of existing vesicles encourages the conversion of micelles to vesicles. Furthermore, if pre-existing vesicles are added to the system very early in its "development," the lag phase is markedly reduced, and fast conversion of micelles to vesicles takes off immediately. Vesicles therefore catalyze their own growth by "feeding on" the micellar stocks of oleate (see Figure 1.3.8).

The vesicles of this system are capable of more than simple growth, however: they can also reproduce. The first evidence for this was the observation that, if vesicles

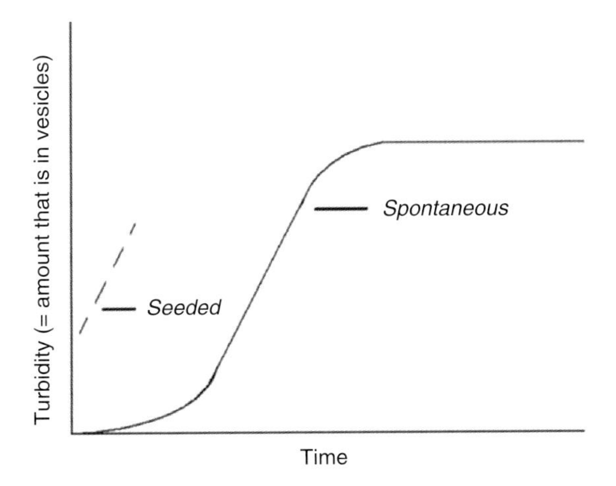

FIGURE 1.3.7 The kinetics of vesicle formation by sodium oleate. Spontaneous formation of vesicles has an appreciable lag, whereas vesicle formation takes off at full speed if the solution is seeded with pre-assembled vesicles. Vesicles therefore catalyze their own growth, which is the basis of their "reproduction." Figure based on data in Ref. 8.

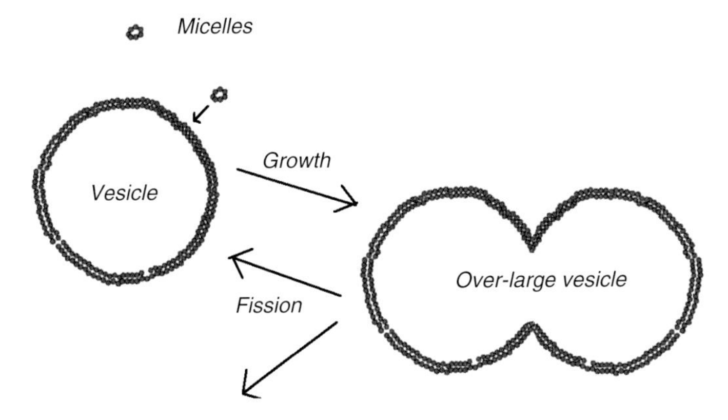

FIGURE 1.3.8 The "reproductive cycle" of vesicles of sodium oleate. The precise details have not yet been elucidated, but it appears that vesicles grow by accretion of material in micelles, and once they become large enough, they undergo fission.

of 100 nm diameter are used to "seed" the solution, the result is a larger number of vesicles of about 100 nm rather than the simple enlargement of "seed vesicles." This suggests that the vesicles must be using new material to produce further vesicles of their own average size, perhaps by a cycle of growth and fission. Visual evidence suggestive of fission comes from cryo-electron microscopy, which reveals the presence of paired vesicles.[8] Biochemical evidence comes from experiments in which the "seed" vesicles are tagged with ferritin, an electron-opaque marker.[9] As vesicle reproduction proceeds, the ferritin tag becomes shared between old and new vesicles, implying that formation of new vesicles occurs by fission of existing ones rather than by catalyzed production of completely separate new ones.[10] (This method of analysis, and its interpretation, is analogous to the classical experiments of Meselson and Stahl that proved that DNA replicates semi-conservatively.) The precise method of vesicle reproduction is not yet understood, but the ability of a simple molecule to produce replicating vesicles is nevertheless a striking example of the power of self-assembly, and one that may be of great relevance to the problem of the origin of life.

Artificial phospholipid bilayers are frequently used as models for study of the properties of real cell membranes, with which they share many properties and, in the case of vesicles, a shape very common in biology. It is important to note, however, that the process of self-assembly seen in artificial systems is not used to produce phospholipid bilayers *in vivo*. Real membranes are stabilized by the same intermolecular forces, and membrane curvature may be determined in part by the mix of phospholipid shapes present, but in living cells, new membranes form from existing ones and never form spontaneously from their separated components. In eukaryotic cells, for example, new phospholipids are inserted into the membrane of the endoplasmic reticulum by enzymes that are already associated with that membrane, and the new components reach other parts of the cell as part of the bilayer of trafficking vesicles.[11,12] This is another illustration of the point made in Chapter 1.2 that a given form may be produced in several ways so that a researcher cannot deduce a morphogenetic mechanism solely from its result.

ONE-DIMENSIONAL SELF-ASSEMBLY: Actin

Actin is the chief constituent of microfilaments, an important cytoskeletal system that will be described in more detail in Chapters 2.2, 3.2, and others. Actin monomers are proteins that consist of a single peptide chain, folded into two lobes, between which there is a deep cleft occupied by a single molecule of ADP or ATP. The monomers can polymerize to form a twisting filament two subunits wide, in which each subunit makes contact with four other subunits. A solution of actin monomers in an aqueous, Mg^{2+}-containing buffer will produce filaments by spontaneous self-assembly as long as the monomer concentration is above the critical concentration (about 0.1 μM for ATP-actin). The dynamics of the process are not simple, however, and follow a sigmoidal plot in which polymerization becomes rapid only after a time lag. This lag reflects the fact that actin polymerization is to some extent "autocatalytic": the probability of monomers being added to an existing filament is significantly higher than the formation of filaments *de novo*. Actin dimers, the smallest possible multimers, are highly unstable and tend to fall apart before additional monomers can be added. Trimers, when they do eventually occur, are much more stable and are therefore able to nucleate the formation of stable filaments, which grow as new monomers locate their ends. The lag period *in vitro* is a result of the low probability of producing trimers.

Cells, and even some infectious bacteria that grow within them, use the ability of actin to self-assemble into filaments but control it using (1) proteins that sequester monomers to prevent inappropriate polymerization, (2) proteins that encourage nucleation when the formation of new filaments is wanted, and (3) proteins that bind to the ends of filaments to block their growth or to block their destruction. This is typical of how self-assembly is used in real morphogenesis: a self-assembling process is embedded in the wider context of regulatory proteins that control the self-assembly in response to other factors and therefore direct it to serve higher levels of organization.

ONE-DIMENSIONAL SELF-ASSEMBLY: Collagen

Another example of biological self-assembly that has been the subject of much attention over the last 30 years or so is that of the extracellular matrix protein collagen. Collagen proteins assemble by associating into fibrils. Many other molecules associate with collagen and are required for its function, but they are not required for the specific act of fibril assembly. There are at least 20 types of collagen in mammals; only five of these (types I, II, III, V, and XI) seem to be capable of self-assembly into fibrils, but these types constitute most of the total mass of collagen in an individual animal. The basic unit of fibril assembly, a collagen protein, itself consists of three peptides that are, for most of their length, wrapped around each other in a tight triple-helix that forms a rod about 300 nm long and just 1.5 nm wide. Cells secrete procollagen molecules in which the triple-helical rod is flanked at each end by large non-helical peptides; these are cleaved by specific extracellular proteases to leave the helical rod domain flanked just by short telopeptides that together amount to no more than about 2 percent of the length of the molecule (see Figure 1.3.9). It is possible to perform this enzyme-driven procollagen–collagen transition *in vitro*, and therefore to study the process of fibril self-assembly.[13] Alternatively, collagen proteins purified from tissues under conditions in which existing fibrils disassemble (for example, dilute acetic acid) will self-assemble into fibrils when their solution is neutralized.[14]

In most cases, the initial step of self-assembly is the formation of a staggered dimer in which two collagen proteins lie next to each other, facing the same direction, but with one "in the lead." This dimerization depends critically on the presence of the telopeptide sequences.[15] There is some evidence that subsequent addition of a monomer to form a staggered trimer is a critical step before the assembly of larger structures can proceed[16] (the "lag" phase ends as these structures appear). Over some hours, the trimers associate with each other to form higher-order multimers that increase in both girth and length to become true fibrils. Again, there seems to be a critical transition in this process, the rate

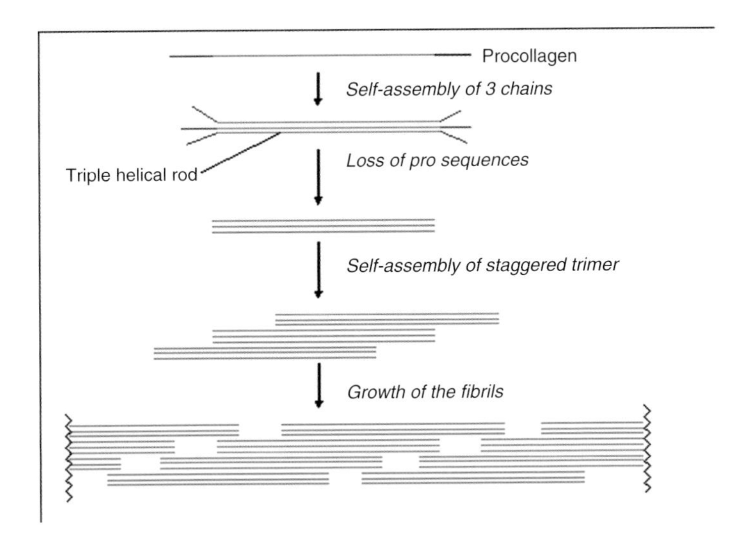

FIGURE 1.3.9 The process of collagen assembly.

of growth up to the size of about five trimers being modest but much faster thereafter.[17] This may help to ensure that collagen polymerization tends to make a few large fibrils rather than many tiny ones. The staggered binding seen in the dimer and trimer is also seen in the fibrils. The stagger creates zones of the fibril that include "gaps" capable of trapping electron-microscopic stains such as phosphotungstate and regions with no gaps that stain only lightly. The overall appearance of a fibril stained in this way is therefore one of alternating light and dark bands, the periodicity of which is a little under 70 nm.

When samples of a collagen solution are taken throughout fibril self-assembly, it seems that maximum fibril diameter is reached when the fibril has only reached about 20 percent of its final size, suggesting that subsequent addition of collagen is at the fibril's ends. The reason for this is not yet clear, but it suggests a subtle property of the self-assembling system that is sensitive to the diameter of the complete fibre. The ends of growing fibrils tend to be tapered, and it is possible that the structure of these ends guides the arrival of new components into the correct "stagger." In most cases *in vitro* and *in vivo*, collagen proteins self-assemble in a staggered side-by-side arrangement in which all chains lie in the same direction in terms of their $NH_2 \rightarrow COOH$ polarity. Such fibrils are described as unipolar. (Bipolar fibrils can occur in some *in vitro* systems[18] and probably *in vivo,* too,[19] but they will not be discussed further here.)

It is important to note that the fibril diameters of collagens assembled *in vitro* from purified collagens isolated from tissues generally differ from their native diameters in those tissues. Therefore, while collagen fibrils clearly self-assemble, in living systems this self-assembly must be subject to additional regulation. This is a general pattern: self-assembly of components is a vital part of most morphogenetic processes, but it is regulated to a greater or lesser extent by other factors and, when the extent of regulation is particularly great—especially when feedback is involved—the term "self-assembly" ceases to be used at all, and other terms such as self-organization come to be used instead. It is also important to note that the assembly of fibrils goes only part of the way to assembling biologically useful collagen structures. In real extracellular matrices, collagens associate with a large number of other matrix components and with cellular receptors. Together, these organize them into an appropriate spatial arrangement—for example, a precise, optically transparent orthogonal array of fine fibrils in the cornea of the eye or a dense, parallel array of thick fibrils in the Achilles tendon. These arrangements certainly do not arise by simple self-assembly of their components, and they are subject to very fine spatial regulation by the cells that lay them down.

THREE-DIMENSIONAL SELF-ASSEMBLY: Simple Viruses

The only complete organisms capable of building themselves entirely by simple self-assembly are small viruses. A striking example is provided by Tobacco Mosaic Virus (TMV). The genome of this virus consists of a single-stranded RNA molecule of 6,395 nucleotides that has an inverted 7-methyl-G cap at the 5′ end.[20,21] In the mature virus particle, this RNA is packaged in 2,131 molecules of coat protein, each protein unit making contact with three nucleotides of the RNA. The wedge-shaped protein units are arranged as a hollow-centred, right-handed spiral with $16\frac{1}{3}$ proteins per turn and 131 turns in all, the complete virus measuring 300 nm in length and 18 nm in diameter. The RNA is also arranged in a spiral, about 4 nm from the axis and therefore just within the

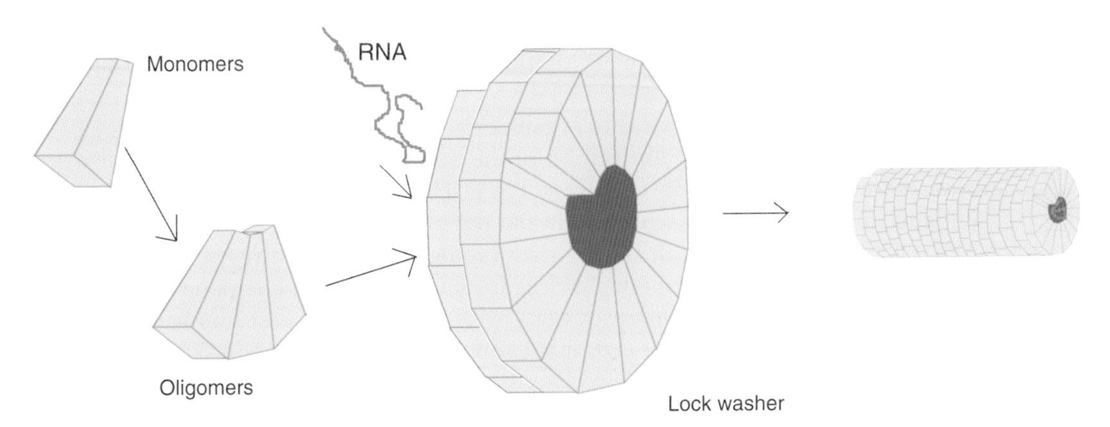

FIGURE 1.3.10 Assembly of the Tobacco Mosaic Virus (TMV). Protein subunits (green) form di- and trimers spontaneously; they also form a few discs (not shown). In the presence of TMV RNA, one of these discs flips to a lock-washer shape. These lock-washers become two-layered and, once the RNA/lock-washer complex has formed, further protein subunits are recruited, generally in the form of discs, until the complete virus is assembled.

protein's space, and a hole about 2 nm across runs down the axis of the whole structure (see Figure 1.3.10).

If purified TMV coat protein and RNA are mixed together *in vitro*, they self-assemble to form morphologically normal and infective virus particles.[22] The proteins alone, however, will not self-assemble to form a complete coat structure (at least, not at physiological pH), but instead form smaller oligomers, mainly di- and trimers, and some one-layered non-helical discs of 17 subunits and two-layered non-helical discs of 34 subunits. Promotion of the rest of the assembly requires association of one of these discs with RNA. The association is specific and requires an unusual stemmed-loop structure that is present around nucleotide 5476 of TMV's genome (the figure is correct for the *vulgare* strain of the virus). The requirement for this stemmed loop ensures that TMV coat protein is used to package only TMV RNA and is not wasted on general RNA from the host cell.

During normal assembly, the stemmed loop enters the hole in the middle of a two-layered disc, so that both free ends of the RNA dangle out from the same side. The strength of the interaction between the protein and the stemmed loop causes the RNA-RNA duplex to melt and to generate more single-stranded RNA. This associates with more of the protein in the centre of the hole, until a complete turn binds. The total energy change involved in this causes the disc to flip to its helical lock-washer shape, an action that traps the RNA and that also enables the structure to initiate the addition of more coat protein. Extra protein seems to attach mainly in the form of discs—a fact that can be demonstrated by running gels of partially completed virus particles. The result is a discrete ladder representing successive addition of discs rather than the smear that would result if addition were monomer by monomer. As new discs are added, the 5′ end of the RNA is pulled up through the middle of the existing structure until it has all been packaged. The remaining projecting length of the 3′ end is then covered. Stable attachment of new discs requires the energy change produced by binding RNA so, once the complete length of TMV RNA has been packaged, further growth ceases. The RNA therefore acts as a "molecular ruler" that determines the precise length of the finished virus.

The ability of an entire virus to organize itself *in vitro* from a simple mixture of its components is a remarkable illustration of the power of self-assembly for small structures. Indeed, the first demonstration of TMV self-assembly, by Fraenkel-Conrat and Williams in 1955, caused a sensation in the popular press at the time, being heralded as the creation of "life in a test tube." Most biologists would, however, distance themselves from the idea that assembling a virus from its constituent components is creation of "life," for the virus can only propagate with the help of a vast number of systems in a host cell, and the host cell certainly cannot be created by mixing its components *in vitro*.

QUALITY CONTROL IN SELF-ASSEMBLING STRUCTURES

Although they are simple, self-assembling structures are nevertheless able to perform limited error correction and quality control. Their ability to do this comes from two main sources: the relative strengths of their chemical bonds and the use of binding sites to which more than one monomer contributes.

The relatively weak bonds that connect the molecules of biological self-assembling structures are based on hydrogen bonding and dipolar and van der Waals interactions. These are orders of magnitude weaker than the covalent linkages used within the constituent molecules themselves; compare, for example, the mean bond dissociation enthalpy of a C–C covalent bond, 348 kJ/mol, with that of a hydrogen bond (10–30 kJ/mol).[23,24] All chemical bonds are equilibria, but for most covalent bonds within biological molecules, the equilibrium is located so far toward the bonded state that the fact of the equilibrium can almost be disregarded and the covalent bonds can be regarded as essentially permanent. There are, of course, exceptions, particularly for the small molecules of intermediary metabolism and for regulatory groups such as phosphates, but the foregoing is true for most bonds *within* most structural molecules. The weakness of *inter*molecular bonding, however, means that the fact of the equilibrium between molecules in the bound and unbound states is significant, and even if the average direction of progress is toward polymerization, the addition of any particular subunit is reversible. In the case of misfit subunits, for which the bonding energy is even lower (because of the misfit), it is especially likely that the subunit will dissociate again and that the error will therefore be corrected.[25]

The second source of error correction, common in self-assembling structures but by no means universal, is the use of binding sites for a new component that are created or exposed only when earlier components in the sequence of assembly are bound correctly. For example, bacteriophages such as T4 assemble their many different types of components in a strict order, and the binding site for a late-adding component is usually created by the binding of its predecessor.[26,27] The use of this strategy has two consequences: it forces the associations between components to take place in a specific order, and it prevents further construction of an assembly that has just incorporated a misaligned or wrong component until that erroneous component has dissociated (in the thermodynamic equilibrium described above) and been replaced with a correct one.

LIMITATIONS TO SELF-ASSEMBLY

Self-assembly is a powerful mechanism for bringing subunits together to form a macromolecular complex, and it is used extensively for small-scale structures throughout biology. It does, however, suffer from a number of limitations that prevent its use for large-scale and complex structures. The most serious limitation of pure self-assembly is its inflexibility. Small, standalone structures such as simple viruses are self-contained and do not have to be aligned precisely with anything else. There is therefore no need for them to be able to be flexible enough to adapt their forms according to circumstances. Even where some degree in adaptation is available—as for example in TMV, in which RNA that is shorter or longer than wild type will be packaged using fewer or more than the normal number of proteins (within limits)—the complete information for assembly resides within the components involved, and no notice need be taken of anything external. Where structures form part of larger entities such as cells and embryos, information carried in subunits is not enough because the precise layout of the larger entity is unpredictable, either because of random noise and error in its internal workings or because it exists in an unpredictable environment. For these structures, self-assembly is not enough, and the feedback systems described briefly in Chapter 1.2 are added to turn mere self-assembly into adaptive self-organization.

Reference List

1. Penrose, R (1989) The emperor's new mind. (Oxford University Press)
2. de Kruijff, B (15-10-1987) Polymorphic regulation of membrane lipid composition. *Nature.* **329**: 587–588
3. Sen, A, Williams, WP, and Quinn, PJ (23-2-1981) The structure and thermotropic properties of pure 1,2-diacylgalactosylglycerols in aqueous systems. *Biochim. Biophys. Acta.* **663**: 380–389
4. Rand, RP, Pangborn, WA, Purdon, AD, and Tinker, DO (1975) Lysolecithin and cholesterol interact stoichiometrically forming bimolecular lamellar structures in the presence of excess water, of lysolecithin or cholesterol. *Can. J. Biochem.* **53**: 189–195
5. Schneider, MB, Jenkins, JT, and Webb, WW (1984) Thermal fluctuations of large cylindrical phospholipid vesicles. *Biophys. J.* **45**: 891–899
6. Lin, KC, Weis, RM, and McConnell, HM (11-3-1982) Induction of helical liposomes by $Ca2^{+}$-mediated intermembrane binding. *Nature.* **296**: 164–165
7. Rudolph, AS, Ratna, BR, and Kahn, B (4-7-1991) Self-assembling phospholipid filaments. *Nature.* **352**: 52–55
8. Blöchliger E, Blocher M, Walde, P, and Luisi, P (1-11-1998) Matrix effect in the size distribtion of fatty acid vesicles. *J. Phys. Chem. B.* **102**: 10383–10390
9. Berclaz, N, Blochliger, E, Muler, M, and Luisi, PL (2001) Matrix effect of vesicle formation as investigated by cryotransmission electron microscopy. *J. Phys. Chem. B.* **105**: 1065–1071
10. Berclaz, N, Muller, M, Walde, P, and Luisi, PL (2001) Growth and transformation of vesicles studied by ferritin labeling and cryotransmission electron microscopy. *J. Phys. Chem. B.* **105**: 1056–1064
11. Mayer, A (2002) Membrane fusion in eukaryotic cells. *Annu. Rev. Cell Dev. Biol.* **18**: 289–314
12. Huijbregts, RP, Topalof, L, and Bankaitis, VA (2000) Lipid metabolism and regulation of membrane trafficking. *Traffic.* **1**: 195–202
13. Miyahara, M, Hayashi, K, Berger, J, Tanzawa, K, Njieha, FK, Trelstad, RL, and Prockop, DJ (10-8-1984) Formation of collagen fibrils by enzymic cleavage of precursors of type I collagen in vitro. *J. Biol. Chem.* **259**: 9891–9898
14. Bard, JB and Chapman, JA (21-11-1973) Diameters of collagen fibrils grown in vitro. *Nat. New Biol.* **246**: 83–84

15. Capaldi, MJ and Chapman, JA (1982) The C-terminal extrahelical peptide of type I collagen and its role in fibrillogenesis in vitro. *Biopolymers.* **21**: 2291–2313
16. Silver, FH (25-5-1981) Type I collagen fibrillogenesis in vitro. Additional evidence for the assembly mechanism. *J. Biol. Chem.* **256**: 4973–4977
17. Chen, JJ, Silver, DP, Walpita, D, Cantor, SB, Gazdar, AF, Tomlinson, G, Couch, FJ, Weber, BL, Ashley, T, Livingston, DM, and Scully, R (1998) Stable interaction between the products of the BRCA1 and BRCA2 tumor suppressor genes in mitotic and meiotic cells. *Mol. Cell.* **2**: 317–328
18. Kadler, KE, Hojima, Y, and Prockop, DJ (1-6-1990) Collagen fibrils in vitro grow from pointed tips in the C- to N-terminal direction. *Biochem. J.* **268**: 339–343
19. Holmes, DF, Lowe, MP, and Chapman, JA (7-1-1994) Vertebrate (chick) collagen fibrils formed in vivo can exhibit a reversal in molecular polarity. *J. Mol. Biol.* **235**: 80–83
20. Goelet, P, Lomonossoff, GP, Butler, PJ, Akam, ME, Gait, MJ, and Karn, J (1982) Nucleotide sequence of Tobacco Mosaic Virus RNA. *Proc. Natl. Acad. Sci. U.S.A.* **79**: 5818–5822
21. Zimmern, D (1975) The 5′ end group of tobacco mosaic virus RNA is m7G5′ppp5′Gp. *Nucleic Acids Res.* **2**: 1189–1201
22. Butler, PJ (29-3-1999) Self-assembly of Tobacco Mosaic Virus: the role of an intermediate aggregate in generating both specificity and speed. *Philos. Trans. R. Soc. Lond. B Biol. Sci.* **354**: 537–550
23. Pauling, L (1960) The nature of the chemical bond. **3rd**: (Cornell University Press)
24. Suresh, SJ and Naik, VM (2000) Hydrogen bond thermodynamic properties of water from dielectric constant data. *J. Chem. Phys.* **113**: 9727–9732
25. Lindoy, LF and Atkinson, IM (2000) Self-assembly in supramolecular systems. (Royal Society of Chemistry)
26. Wood, WB, Edgar, RS, King, J, Lielausis, I, and Henninger, M (1968) Bacteriophage assembly. *Fed. Proc.* **27**: 1160–1166
27. Wood, WB and Edgar, RS (1967) Building a bacterial virus. *Sci. Am.* **217**: 61–66

2

Cell Shape and the Cell Morphogenesis

CHAPTER 2.1

MORPHOGENESIS OF INDIVIDUAL CELLS: A Brief Overview

Some of the most spectacular examples of morphogenesis that occur during development and adult life take place not at the scale of tissues and organs but rather at the scale of individual cells. The shapes of cells vary, according to differentiation state and environment, from simple spheres to cubes, rods, discs, branched trees, and even hollow tubes. In some cases, a specific cell shape is dictated mainly by the physiological function of the cell, whereas in other cases, changes of cell shape are required to change the shape of a developing tissue. The purpose of this brief chapter is to give an overview of the types and developmental roles of morphogenetic changes in individual cells; readers already familiar with the histology of both animals and plants should feel free to skip it. The chapters that follow will delve more deeply into the molecular mechanisms involved.

FLATTENING AND ELONGATION OF CELLS

Drawings of a "typical" animal cell, of the kind that appear near the beginning of most biology textbooks,[1,2,3,4] usually depict a spherical or cubic entity, all axes of which are equal. Many cells really are like this; for example, lymphocytes in peripheral mammalian blood are approximately spherical,[5] and epithelial cells in the distal tubule of the mammalian kidney are approximately cuboidal[1] (see Figure 2.1.1a). A change in the ratio of length to width is a simple and common example of change of cell shape. For example, epithelial cells may flatten considerably in their apico-basal axis (see Figure 2.1.1.b); this flattening is often seen where substances have to travel across the epithelium, as in the alveoli of the lung, and is presumably an adaptation to minimize the diffusion path. Alternatively, cells can elongate in the apico-basal axis to become long and thin, as they do in the placodes of vertebrate embryos[6] (see Figure 2.1.1c).

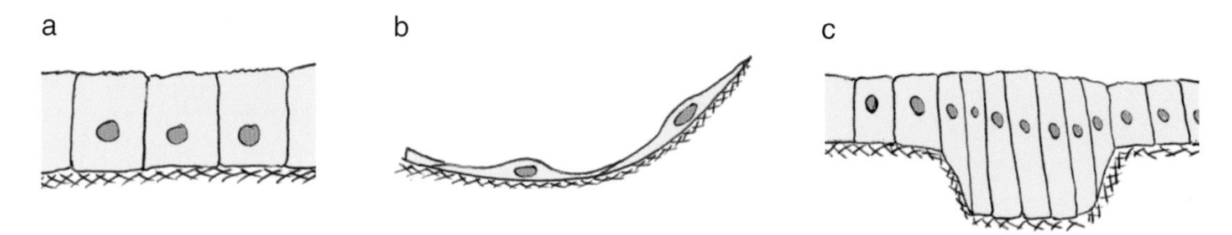

FIGURE 2.1.1 The flattening and elongation of epithelial cells. (a) Epithelial cells of cuboidal form, for example in the distal tubule of kidneys. (b) Flattened epithelial cells, for example in the alveoli of the lung. (c) Elongated epithelial cells, for example in an embryonic placode (see Chapter 4.4). The crosshatched lines depict a basement membrane.

PRODUCTION OF CELL PROCESSES

One of the most common types of cell morphogenesis is the formation of processes. The reasons for this include increasing surface area, connecting with other cells, exploring the environment, and sensing vibration. The production of microvilli by epithelial cells of the small intestine or of the renal proximal tubule is an example of process formation to increase the surface area available for substance exchange (see Figure 2.1.2a). Cell processes that are produced for the purposes of connecting cells together can be temporary or permanent. Temporary processes are often seen when cells that have met are about to adhere (see Figure 2.1.2b); the processes from each cell push against the membrane of the other, and their interdigitation provides a large surface area for the establishment of adhesions to form (see Chapter 4.3), although as the adhesions mature, the processes disappear. A morphologically similar interdigitation of processes—but one that is permanent—is used by glomerular podocytes to create the main filtration element of mammalian kidneys; podocytes produce many processes that lie parallel to those of other cells, and a fine filter develops between them where their membranes fuse[7] (see Figure 2.1.2c).

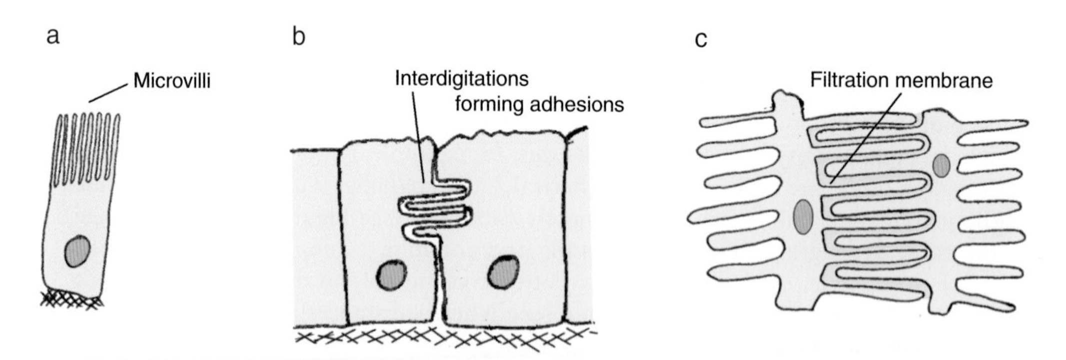

FIGURE 2.1.2 Examples of cell morphogenesis by the production of processes (part 1). (a) An absorptive cell typical of gut and renal epithelia. (b) Two epithelial cells meeting and forming adhesions, although the scale of the processes involved has been exaggerated for clarity: in reality there are many more, much finer processes. (c) The interdigitating processes of adjacent podocytes in the glomerulus of a mammalian kidney.

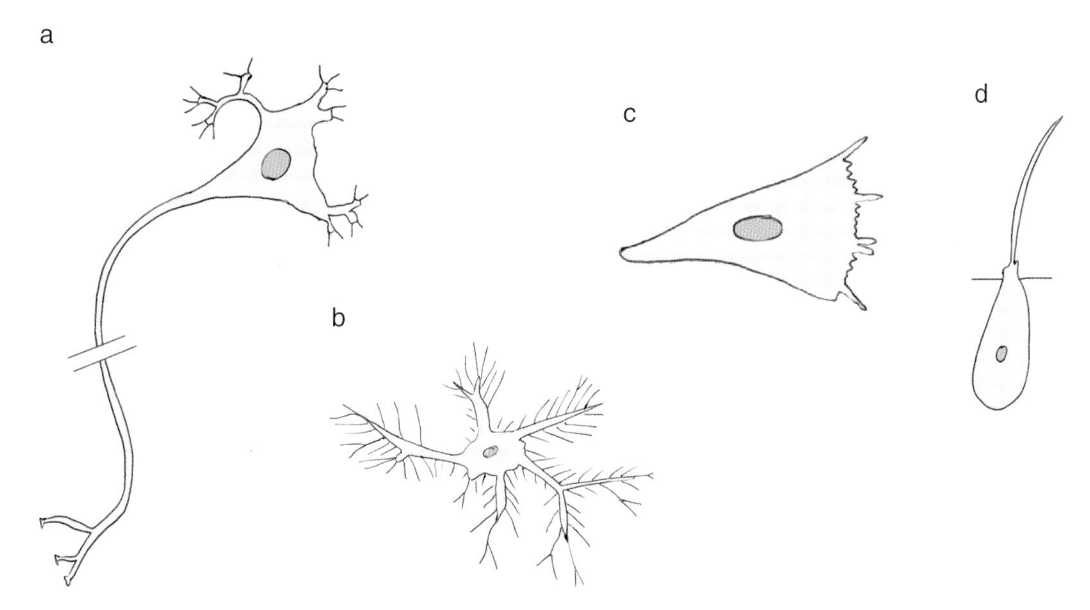

FIGURE 2.1.3 Examples of cell morphogenesis by the production of processes (part 2). (a) A neuron, the long axon of which (shown interrupted) can reach lengths of several meters in the largest animals. (b) A follicular dendritic cell of a mammalian lymph node. (c) A motile cell that shows a leading edge and filopodia. (d) A mechanosensory cell of the ear's cochlea.

Cells can also produce processes for interacting with cells of a different type. For example, the follicular dendritic cells on the lymph nodes of the vertebrate immune system produce large numbers of branched processes (see Figure 2.1.3b). These processes make contact with the lymphocytes around them and stimulate their survival and proliferation. They do this either because they present specific processed antigens to them and stimulate only cells that recognize those antigens (the traditional view[8,9]) or because they present non-specific survival and proliferation signals to lymphocytes that can be used only by cells that are already binding soluble antigen (an emerging view[10,11]). The connectivity of the nervous system relies on the production of cell processes that connect neurons to neurons or connect neurons to effector cells such as muscles and glands. Neurons can bear very large numbers of small processes, dendrites, and also sprout one axon that can reach lengths of several meters in large mammals (see Figure 2.1.3a). In the case of neural connections, processes have to locate their targets very precisely, which normally requires exploration of the environment provided by other cells (see Chapter 3.5). This exploration usually involves cell processes—filopodia (see Figure 2.1.3c)—which project forward from the leading edge of a moving cell and sample the environment (see Chapter 3.5).

Processes can also be produced for sensing mechanical movement, including the high-frequency movements of sound waves or the slow and subtle movements of otoliths in the balance organs of the vertebrate ear. The processes are often modifications of microvilli (see Chapter 2.3) that project into the gas or fluid in which the movement takes place (see Figure 2.1.3d).

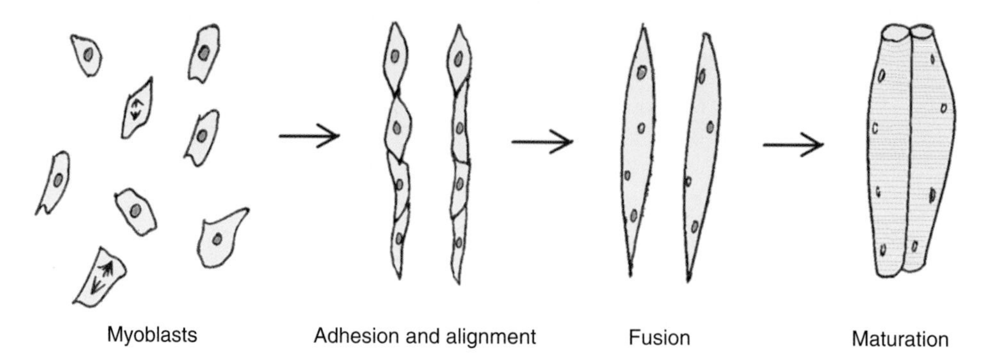

FIGURE 2.1.4 A schematic view of the cell alignment and fusion involved in the development of vertebrate skeletal muscle.

CELL FUSION

Cells can also alter their shapes by fusing with other cells, or by fusing with themselves. A closely studied example of fusion between cells is provided by the development of vertebrate skeletal muscle. The progenitor cells of muscle are fibroblastic in morphology. Once they are induced, by signalling molecules such as Wnts and Shh, to form muscle, they multiply and align with each other to form long chains of cells. The cells within each chain then fuse to produce a single multinucleate cell, the myotube, which matures into a long, thin muscle fibre (see Figure 2.1.4).[12] An essentially similar process takes place in the formation of somatic muscles in insects such as *Drosophila melanogaster*.[13]

Cell fusion can also take place between one part of a cell and another part of the same cell. This type of cell morphogenesis is used to create tubes (see Chapter 4.5).

CELL CAVITATION

The finest blood capillaries of the body are so narrow that their lumens run through individual cells rather than through tubes formed by multiple cells. The formation of lumens begins when the cells concerned form large intracellular vacuoles by pinocytosis (see Figure 2.1.5). These vacuoles coalesce with each other and with the plasma membrane so that a hollow lumen is formed,[14] and the cavitation is coordinated between neighbouring cells so that their lumens join up to form a very fine blood vessel.

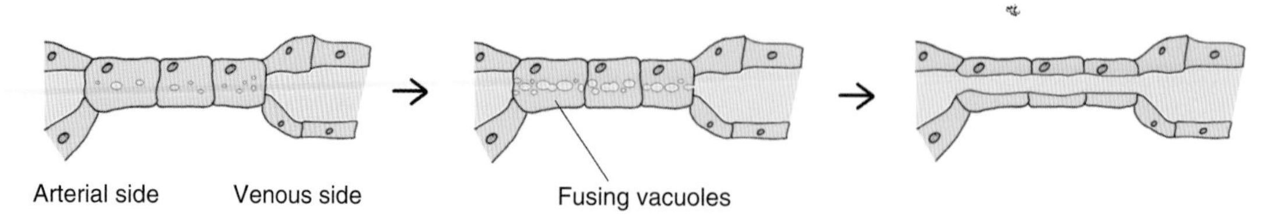

FIGURE 2.1.5 Cavitation of endothelial cells by fusion of vacuoles to create fine-vessel lumens.

CHANGES IN CELL SHAPE CAN DIRECTLY DRIVE MORPHOGENESIS OF TISSUES

All morphogenesis depends to some extent on changes of cell shape, but in some cases the change in shape of a large-scale structure is driven directly by a change in the shape of the cells that comprise it. These cases underline the more general importance of understanding the basis of cell shape, if the shapes of organisms are to be understood.

Direct relationships between changes of cell shape and morphogenesis of larger structures are shown most clearly by organisms in which cell movement is not possible: these organisms include some multicellular prokaryotes and all multicellular plants and fungi. Much of the gain of size of developing plant tissues is achieved by cell enlargement rather than cell multiplication. One of the simplest ways in which an expanding cell can change shape is by expanding more in one direction than others ("simplest" in terms of the shape change, not necessarily simplest in terms of the molecular mechanisms that underlie it). In a tissue composed of cells that cannot move with respect to one another, a single cell cannot expand in this way unless it is on the edge of a tissue and expands outwards, but if the behaviour of cells is regulated so that all expand together, the entire tissue will change shape (see Figure 2.1.6).

This system of simultaneous anisotropic cell expansion is used a great deal by growing plants. At the tip of a root, for example, there is a zone of cell proliferation immediately behind the protective root cap and, proximal to this, is a zone of cell expansion. In the zone of expansion, cell walls are allowed to yield, in a controlled manner, to osmotic pressure inside the cell (see Chapter 2.4). If this cell expansion were isotropic, a broad and relatively short root would result (see Figure 2.1.7). In reality, the cell expansion is anisotropic so that cells enlarge a great deal along the axis of the root and little in girth. Since cells cannot move past each other, this section of the root is therefore forced to elongate as a whole, and a long, thin root is the result (see Figure 2.1.7). Similar anisotropic cell expansions are seen in shoots and leaves.

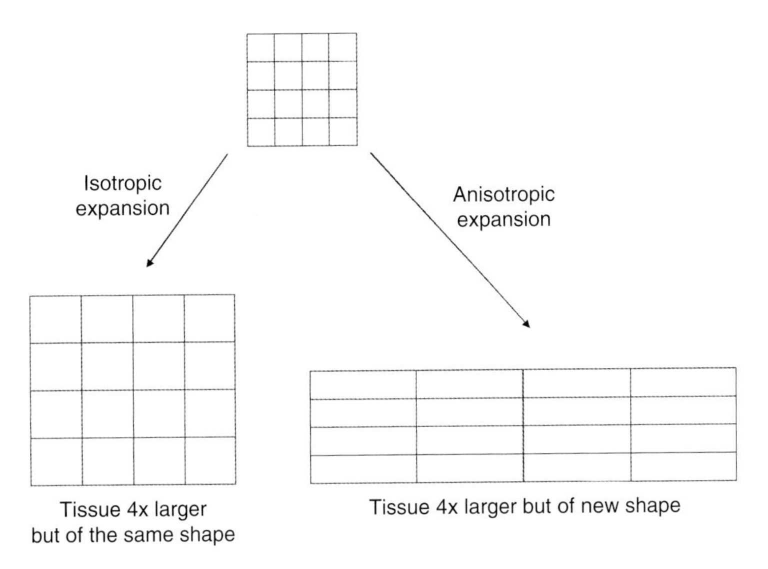

FIGURE 2.1.6 Simultaneous anisotropic expansion by cells can reshape an entire tissue.

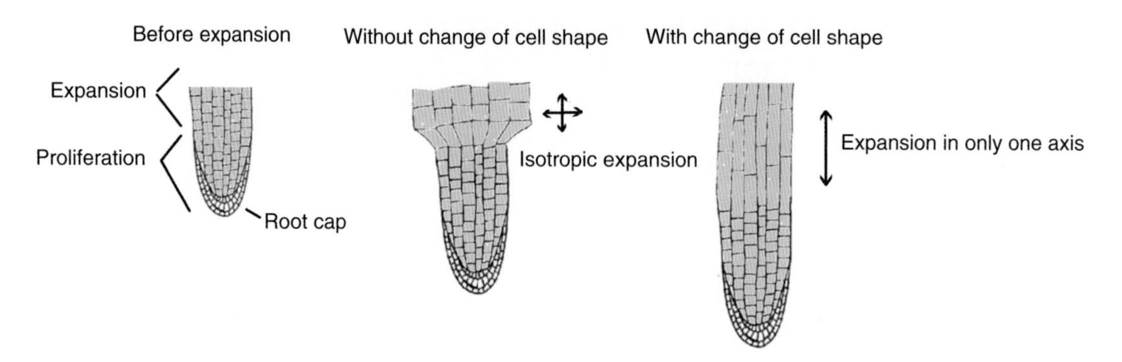

FIGURE 2.1.7 The consequences of cell expansion proximal to the zone of proliferation in a developing root of a flowering plant. If cell expansion were to be isotropic, the root would grow in girth as well as length, whereas a change in cell shape so that cells expand only along the root axis elongates the root as a whole without broadening it.

Where cell expansion takes place more on one side of a shoot than on the other, it causes the shoot to bend: this is an important mechanism for organism-scale morphogenesis in plants.

Anisotropic expansion of cells can be achieved either by distributing the expansion throughout the cell walls along the elongating axis of the cell ("diffuse expansion") or by concentrating new cell growth at one end of the cell ("focused" or "tip" growth). Flowering plants use tip growth as well as diffuse expansion; tip growth is, for example, responsible for the production of absorptive "root hairs" from the sides of roots just proximal to the zone of cell elongation described above (see Figure 2.1.8). Tip growth is

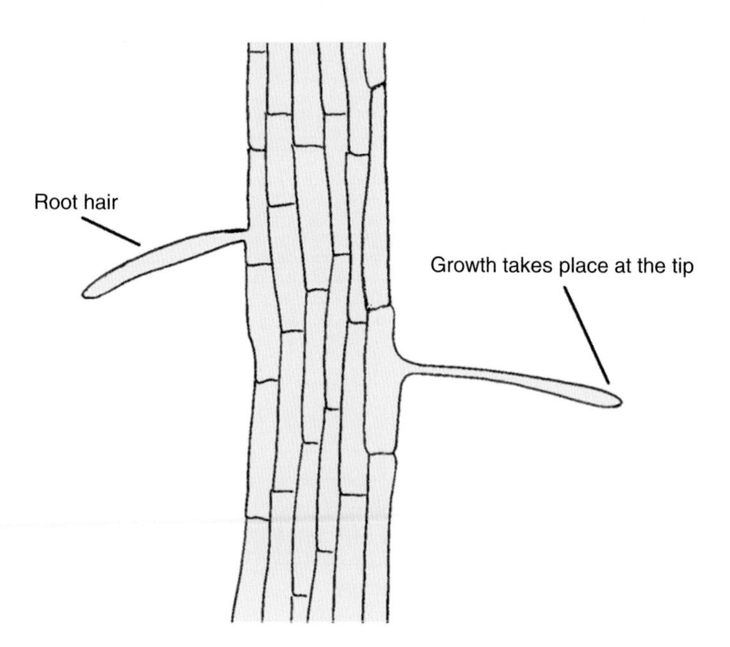

FIGURE 2.1.8 Root hairs of flowering plants grow from their tips.

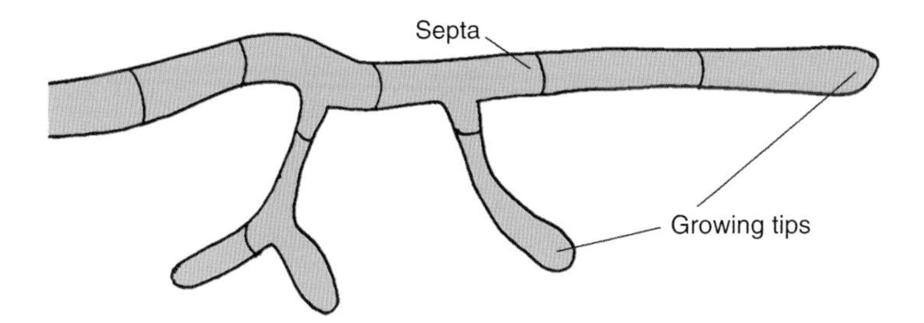

FIGURE 2.1.9 Tip growth of the hyphae of a basidiomycete fungus.

also responsible for the morphogenesis of the largest organisms discovered on Earth—multicellular fungi.

Multicellular fungi (most of which are modest in size, but some of which span square kilometers of forest floor[15]) consist mainly of a network of branched hyphae that ramifies through the soil or other substrate. Hyphae are either coenocytic or separated into cells by transverse septa, and all of their growth takes place at the hyphal tips. Occasionally, patches on the side walls of hyphal cells can develop a "tip" character themselves, and this tip then begins to extend as a branch; in this way, a network can be built up[16] (see Figure 2.1.9).

This chapter has presented only the briefest outline of cellular morphogenesis, with the aim of illustrating typical changes of shape and of sketching some of the ways in which the morphogenesis of cells can contribute directly to the form of the organism. The variety of cell shapes is far wider than those listed above, and the shapes themselves can become fantastically complicated, especially among unicellular plankton. The examples on which this chapter has focused have been chosen because they represent the types of cell morphogenesis that have been studied most closely. The molecular mechanisms that drive cell morphogenesis are the focus of the next chapters in this section.

Reference List

1. Cormack, DH (1984) Introduction to histology. (Lippincott)
2. Marieb, EN (2004) Human anatomy and physiology. (Pearson International)
3. Starr, C and Taggart, R (2004) Biology: The unity and diversity of life. (Thomson)
4. Pollard, TD and Earnshaw, WC (2002) Cell biology. 725–765 (Saunders)
5. Schmaier, AH and Petruzzelli, LM (2003) Hematology. (Lippincott Willimas and Wilkins)
6. Mendoza, AS, Breipohl, W, and Miragall, F (1982) Cell migration from the chick olfactory placode: A light and electron microscopic study. *J. Embryol. Exp. Morphol.* **69**: 47–59
7. Kobayashi, N, Gao, SY, Chen, J, Saito, K, Miyawaki, K, Li, CY, Pan, L, Saito, S, Terashita, T, and Matsuda, S (2004) Process formation of the renal glomerular podocyte: Is there common molecular machinery for processes of podocytes and neurons? *Anat. Sci. Int.* **79**: 1–10
8. Klaus, GG, Humphrey, JH, Kunkl, A, and Dongworth, DW (1980) The follicular dendritic cell: Its role in antigen presentation in the generation of immunological memory. *Immunol. Rev.* **53**: 3–28
9. Tew, JG, Wu, J, Qin, D, Helm, S, Burton, GF, and Szakal, AK (1997) Follicular dendritic cells and presentation of antigen and costimulatory signals to B cells. *Immunol. Rev.* **156**: 39–52

10. Haberman, AM and Shlomchik, MJ (2003) Reassessing the function of immune-complex retention by follicular dendritic cells. *Nat. Rev. Immunol.* 3: 757–764
11. Kosco-Vilbois, MH (2003) Are follicular dendritic cells really good for nothing? *Nat. Rev. Immunol.* 3: 764–769
12. Abmayr, SM, Balagopalan, L, Galletta, BJ, and Hong, SJ (2003) Cell and molecular biology of myoblast fusion. *Int. Rev. Cytol.* **225**: 33–89
13. Taylor, MV (16-12-2003) Muscle differentiation: Signalling cell fusion. *Curr. Biol.* **13**: R964–R966
14. Davis, GE and Camarillo, CW (10-4-1996) An alpha 2 beta 1 integrin-dependent pinocytic mechanism involving intracellular vacuole formation and coalescence regulates capillary lumen and tube formation in three-dimensional collagen matrix. *Exp. Cell Res.* **224**: 39–51
15. Ferguson, BA, Dreisbach, TA, Parks, CG, Filip, TM, and Schmidt, CL (2004) Coarse-scale population structure of pathogenic *Armillaria* species in a mixed-conifer forest in the Blue Mountains of northeast Oregon. *Can. J. For. Res.* **33**: 612–623.
16. Moore, D (1998) Fungal morphogenesis. (Cambridge University Press)

CHAPTER 2.2

THE SHAPES OF ANIMAL CELLS: Tensegrity

The ability of animal cells to acquire specialized shapes, to produce processes, to become flat and wide or tall and thin, and to move depends on their ability to modify the mechanisms that control cell shape even in "ordinary" cells that are not doing anything special. It is therefore important to become familiar with the ordinary regulation of cell shape before considering unusual deviations from it.

The shape of any flexible body is governed by the second law of thermodynamics, which favours an arrangement that minimizes the free energy of the system over any other possible arrangements (see Chapter 1.3). If the plasma membrane were a simple phospholipid bilayer, then an isolated cell would minimize its energy by minimizing its surface area. It would therefore become spherical (neglecting a small distortion due to gravity), and groups of adhering cells would take on the shapes associated with groups of adhering soap bubbles. Indeed, early mathematical treatments of cell shape modelled cells precisely according to these simple laws of surface tension,[1] and the shapes of non-adherent cells, or of cultured cells that have just been trypsinized for passage and have therefore lost their junctions with the matrix and so on, are usually bubble-like (see Figure 2.2.1). Not all cells are like this, however, even in suspension; the biconcave disc of a mammalian erythrocyte is an exception that is familiar to most biologists, and swimming protozoa come in a vast variety of shapes that seem to defy the principle of minimizing surface area (see Figure 2.2.1). It is clear, therefore, that other forces must be at work even for cells in suspension, and certainly for those that associate with each other.

While surface tension at the plasma membrane is no doubt important, the dominant influence on the shapes of animal cells is the cytoskeleton. This modifies membrane shape both by forming a contractile network in the cortical region, just inside the plasma membrane, and by forming systems of long filaments that run across the cell and can push or pull on the membrane. Of the three basic types of cytoskeleton (microfilaments, intermediate filaments, and microtubules), microfilaments and microtubules play the largest role in controlling the shape of most cells. This point is illustrated by the relatively mild phenotype of the vimentin knockout mouse,[2,3] in which cells that ought to express vimentin and that have no alternative intermediate filament system are still able to play their normal roles in development.

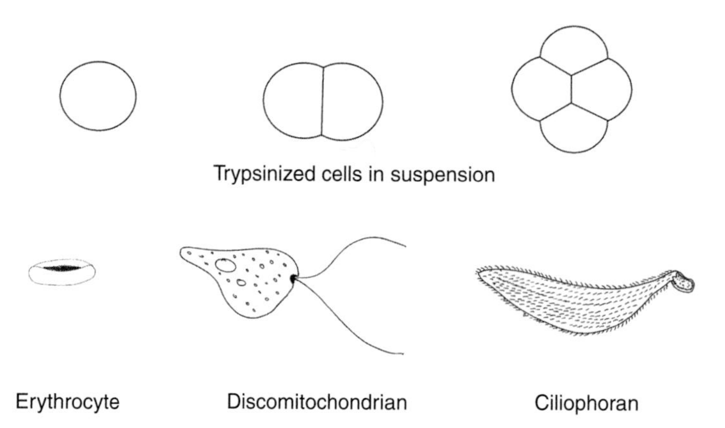

FIGURE 2.2.1 The shapes of some cells in suspension. Mammalian cells that normally grow attached to a substrate and to each other typically round up and have soap-bubble-like morphologies, minimizing their surface area, when released from their substrate by trypsinization (the drawings in the top row are of mIMCD3 renal epithelial cells during passage). Erythrocytes, on the other hand, normally exist in suspension but nevertheless maintain a biconcave discoid shape and do not round up into spheres. Protozoa can maintain very complex shapes.

The basic structures of microfilaments have been described in Chapter 1.3; microtubules are also formed by polymerization of monomers, as will be described in more detail below. The arrangement of these cytoskeletons in cells is so complex that we still do not have a clear picture of all of the morphogenetic forces that act on even simple cells, but there are at least theories about how the cytoskeleton-membrane system works in principle. One of the most comprehensive models[4,5] treats the cell as an assembly of tension and compression elements. This model views the cell as analogous to buildings and sculptures that gain their structural stability from tension rather than from the pure compression that stabilizes an arch. It is therefore referred to by the same term—tensegrity—that was coined by the architect R. Buckminster Fuller[6] to describe the "tensional integrity" of buildings and of the bar-and-wire sculptures by the artist Kenneth Snelson.[7] The principle of tensegrity can be illustrated by constructing a simple kite shape from two pencils and an elastic band (see Figure 2.2.2); tension in the band

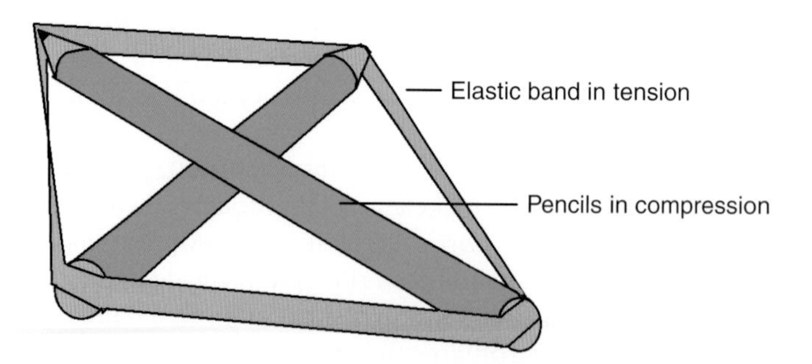

FIGURE 2.2.2 A desktop model of tensegrity, in which the balance of forces between compressed pencils and a tense elastic band stabilizes a simple structure. In a better version, the pencils would be filed down in the middle so that they do not quite make contact as they cross and there would be no possibility of frictional forces between them.

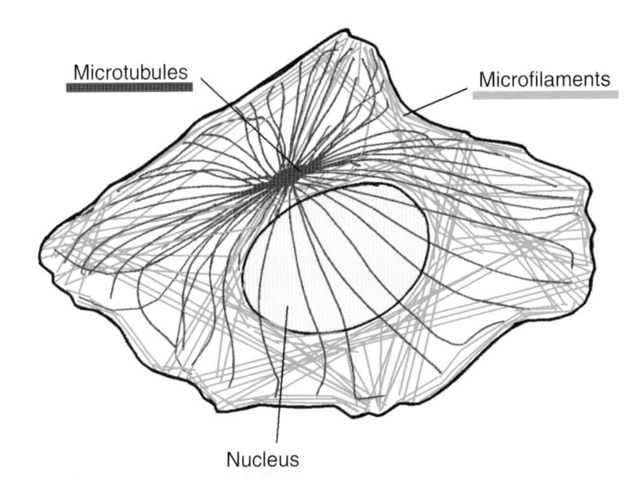

FIGURE 2.2.3 The principal tension and compression structures in a cell. Actin microfilaments (green) are generally in tension and microtubules (red) are generally in compression.

holds the pencils firmly while compression in the pencils prevents the tension in the band from being released.

In the current formulation of the tensegrity model, tension is borne and generated mainly by the microfilaments, especially in conjunction with myosin, and also by intermediate filaments and by the membrane itself as "surface tension." Compression is borne mainly by microtubules and by the extracellular matrix[5] (see Figure 2.2.3). These divisions of duty are not completely clear-cut, however, because actin bears compressive forces at the leading edge of a cell (see Chapter 3.2) and in protrusions such as microvilli (see below). The shapes of typical actin and tubulin cytoskeletons are compatible with the roles proposed for them in tensegrity structures; microfilaments are usually straight, which would be expected for a filament under tension, whereas microtubules are usually curved, which would be expected for a strut under compression. Evidence for this balance of forces is provided by experiments in which one component is removed. If, for example, spread cells are removed from their matrix (a culture dish), they tend to lose their shapes and become spherical, suggesting that having a substrate in compression and being able to resist the tensile forces from the inside of the cell was critical to their former shape. Similarly, if microtubules are depolymerized with drugs so that they are no longer able to share the burden of compressive loads, the force on the extracellular matrix is increased.[8]

According to the tensegrity model, the shape of a cell is governed mainly by the arrangement of, and forces within, microfilaments, microtubules, adhesions, and the matrix (assuming that the surface tension of the membrane is constant). Understanding cell shape therefore boils down mainly to understanding the placement of elements of the cytoskeleton.

The placement of cytoskeletal elements has to be done with an accuracy that is in the ± 50 nm range, particularly when the structures of adjacent cells have to line up and connect via junctions. The spatial resolutions offered by the patterning mechanisms of classical developmental biology are nowhere near as fine as this; differentiation state varies only cell by cell (spatial resolution in the ± 10 μm range), while gradients of morphogens such as retinoic acid can be read with an accuracy no greater than a few μm, for reasons explored in more detail in Chapter 3.3. The embryo solves this problem by using the "classical" mechanisms of developmental biology only to invoke the building of molecular machines at the right times and places. It then relies on the ability of these machines to build appropriate fine-scale structures autonomously. Thus, almost every aspect of the fine-scale structure of an embryo is the result of adaptive self-organization rather than specific "programmed" spatial information.

BUILDING AND PLACEMENT OF TENSILE MICROFILAMENTS

The actin microfilaments of animal cells are arranged in two main ways: as a network of fine filaments in the cortical region of the cytoplasm, just under the membrane, and as a network of tense filament bundles, often called "stress fibres," that connect the sites at which cells adhere to each other and to the basement membrane.

Branched networks of actin are characteristic of the cell cortex and of the motile leading edge described in Chapter 3.2. In most cells, the cortex immediately below the plasma membrane is occupied by a highly cross-linked network of microfilaments that forms a viscous gel. This gel is capable of resisting tangential tension forces and all compression forces, although, because this resistance is based on viscosity, short-term forces are resisted better than long-term ones. In morphologically simple cells such as adult mammalian erythrocytes, the cortex is less than 1 μm thick, but it plays a major role in controlling cell shape. Indeed it has to, for there is no other cytoskeletal system in these cells.[9] Transmembrane proteins such as the HCO_3^-/Cl^- exchanging Band III glycoprotein connect the membrane bilayer to cortical microfilaments via ankyrin, spectrin, and band 4.1 protein[10] (see Figure 2.2.4). The microfilaments are themselves linked together by cross-linking proteins such as filamin[11] and myosin[12], the latter of which can act on microfilaments to place them under tension (see below). Together, these proteins create a viscous, tense gel just inside the plasma membrane. Erythrocytes deficient in components of the cortical cytoskeleton lose their conventional biconcave shape, round up, and become fragile. Erythrocytes from which membrane lipids have been extracted, however, retain the basic shape of the skeleton, which confirms that the protein network is more important than the membrane in maintaining cell shape.[13] The comparative simplicity of the mammalian erythrocyte allows its shape to be modelled mathematically and the results of possible manipulations to be predicted. Other cells, however, have an elaborate cytoskeletal system internal to the cortex, in addition to the cortical gel, and this plays a major role in morphogenesis. It is this internal system that has attracted most attention in relation to the tensegrity model.

As was described in Chapter 1.3, actin polymerizes spontaneously *in vitro* when it is sufficiently concentrated, having to form first a dimer, then a trimer. Formation of the trimer involves an unfavourable reaction, but once the trimer is formed, polymerization is favourable. The two ends of the filament elongate at different rates, the "barbed" end elongating more rapidly than the "pointed" end (the terms "barbed" and "pointed" derive from the appearance of actin filaments decorated with myosin). In conditions typical of a living cytoplasm, the rate of spontaneous trimer formation is so low as to be insignificant, and the nucleation of new filaments depends on protein complexes that can

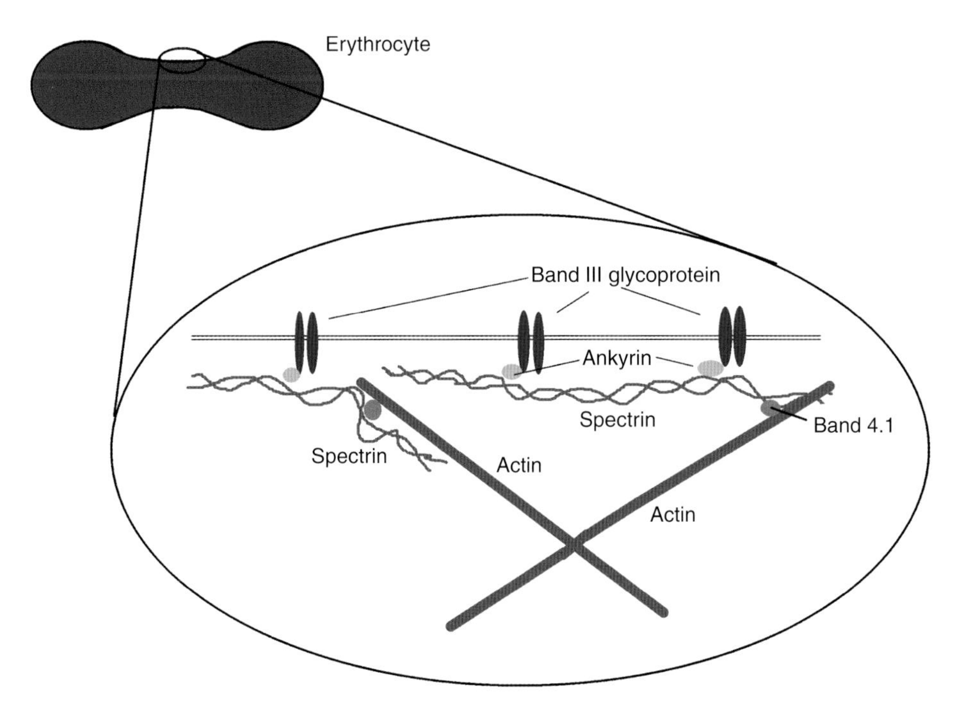

FIGURE 2.2.4 The cortical cytoskeleton of a mammalian erythrocyte (for clarity, the diagram shows only the most significant of the many components present).

stabilize or mimic trimers. These protein complexes include Arp2/3 and proteins of the Formin family. Each type of complex is used to nucleate specific types of microfilament, and mutation of each protein causes loss of a specific subset of microfilaments.[14,15,16] Broadly, Arp2/3 is used to nucleate branched networks of actin typical of the leading edges of motile cells (see Chapter 3.2) and of the cortical actin gel network, whereas Formins are used to nucleate straight filaments.[17,18]

The formation of cell-cell and cell-matrix junctions is intimately linked to the formation of new microfilaments. As adherens junctions develop between epithelial cells, a large number of proteins are recruited to the inside of the plasma membrane. These include α- and β-catenin, which bind to the cytoplasmic domains of cell-cell adhesion molecules such as E-cadherin. Formin1 binds directly to α-catenin[19] and localizes to the inner face of adherens junctions by means of this association. Since Formin1 nucleates the formation of unbranched actin filaments *in vitro*,[19] it seems likely that its localization in the inner face of cell-cell junctions is an important mechanism for ensuring that the pointed end of a microfilament terminates at the junction. This idea is supported by the effects of transfecting cells with a chimaeric protein that consists of the α-catenin-binding domain of Formin1 conjugated to green fluorescent protein. This chimaera competes for α-catenin with Formin1 itself and displaces Formin1 from the junction (see Figure 2.2.5). Its presence in the cell destabilizes junctions and inhibits the formation of actin cables. Formins such as mDia1 are physically associated with integrin-containing cell-matrix junctions, although there seems not to be any direct evidence for this yet.

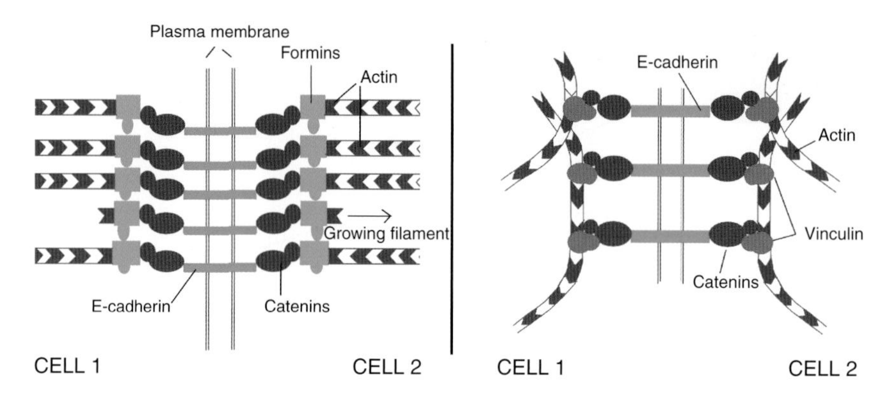

FIGURE 2.2.5 Linkage between cadherins, catenins, Formins (left diagram), vinculin (right diagram), and microfilaments at adherens junctions. The pointed/barbed polarity of actin is depicted by the chevrons in the actin filament.

As well as containing Formins to nucleate unbranched microfilaments, the inner faces of cell-cell adhesions and cell-matrix adhesions contain actin-binding proteins such as vinculin, which is cross-linked to cadherins via catenins,[20] and to integrins via talin.[21] The presence of actin-binding proteins such as vinculin allows junctions to "grab" any microfilaments that happen to reach them. The filaments are bound side-on, rather than barbed end-on, so that they can continue past a junction that they have bound and perhaps bind again to the same junction or to another.

Stress fibers do not exist as single actin filaments but instead consist of bundles of many filaments, typically hundreds, held together by cross-links. Since cross-links can develop between actin filaments aligned in opposite directions, with respect to their pointed-barbed polarity, the bundling of actin allows barbed-end-outwards filaments from different junctions to become connected to form mechanically continuous fibres. Many proteins are able to cross-link microfilaments, but myosin II is particularly important in this respect because the combination of actin and myosin can generate tension. Myosin II self-assembles into short, bipolar filaments, provided that the light chains of the myosin II molecules have been phosphorylated to activate them.[22,23,24,25,26,27,28] The heads of myosin II, which project from the myosin oligomers, contain actin-binding sites and can therefore cross-link actin filaments (see Figure 2.2.6). As well as being an actin-binding protein, myosin is an ATPase, and both its actin-binding activity and its conformation depend on whether it is bound to ATP, ADP, or neither.[29] When myosin binds neither ATP nor ADP, it binds to actin with high affinity and with the head of the myosin molecule turned back somewhat toward the tail of the protein. Once the head of myosin binds ATP, its affinity for actin is reduced so that it tends to let go of the microfilament. The ATPase activity of myosin hydrolyzes its bound ATP to ADP and P_i, and, as it does so, the molecule changes conformation and straightens out slightly so that the head moves approximately 5 nm forwards. Once the myosin head binds to the microfilament again, it releases its P_i, the loss of which is accompanied by an increased affinity of the myosin for actin. It is also accompanied by loss of the ADP and, crucially, a return to the original conformation with the net effect that the microfilament will have been pulled along 5 nm with respect to the myosin. The cycle can repeat as long as there is a supply of ATP. The fact that myosin forms oligomers means

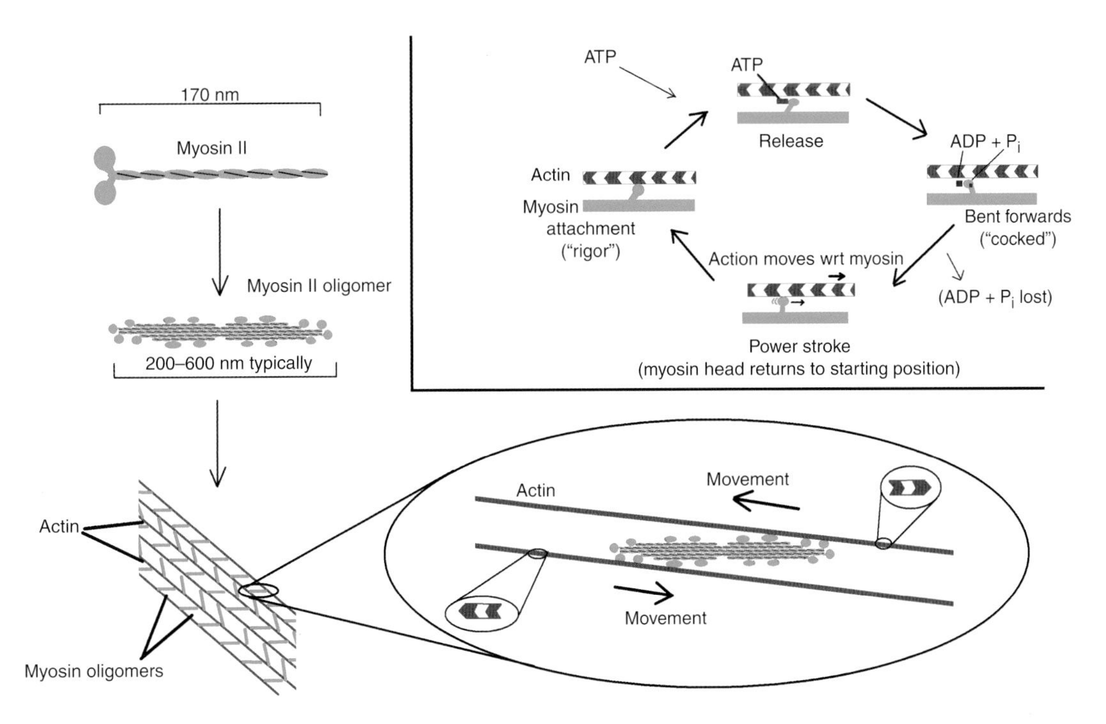

FIGURE 2.2.6 The role of myosin II in generating tension in microfilament bundles. *Main diagram*: Myosin II dimers (top) assemble into oligomers (middle) and these cross-link actin filaments (bottom). The force-generating activity of myosin causes oppositely orientated actin microfilaments to slide past each other if they are free to move, or to experience tension if they are not. *Inset:* The cycle by which a single myosin head exerts force on an actin filament using the energy obtained from hydrolysis of ATP. A single myosin head operating on its own would not be good at exerting tension because the cycle involves a phase when the actin is released. The oligomerization of myosin II into a filament means, however, that some heads will grip the filament while others release so that tension can be maintained.

that different myosin heads are gripping, letting go, and exerting traction at different moments; the net effect is a relatively smooth movement or, if movement is not possible, buildup of strain stored in molecules not quite able to reach their state of minimum energy.

As well as the tension-producing myosin oligomer, cells have many other proteins that can cross-link actin. These include α-actinin, which makes a dumbbell-shaped dimer that has actin-binding domains at each end,[30] fimbrin, which is a small protein with two actin-binding domains that can therefore bundle microfilaments tightly,[31] and filamin. Filamin is unusual in that it forms a V-shaped dimer, which can cross-link microfilaments either in a crisscross array or a parallel array, depending on filamin concentration (high concentrations lead to parallel bundles).[32] The crisscross type of linkage makes actin behave as a gel, which increases the viscosity of the cytoplasm; this type of linkage is common in the cell cortex. Some actin-binding proteins can cross-link microfilaments to other types of cytoskeleton rather than to each other. Plectin, for example, links microfilaments to intermediate filaments.[33]

ADAPTIVE SELF-ORGANIZATION OF THE MICROFILAMENT TENSION SYSTEM

With just the structural elements that have been described above, a cell would be able to build filaments that happen to run between adhesions (see Figure 2.2.7) but would have the problem that most filaments that emerge from junctions would not have the good fortune to run to another junction or to meet cables that do. They would therefore be "wasted" from the point of view of being able to exert useful tension. This problem is solved by an elegant negative feedback system that regulates the formation and stability of microfilaments according to how much mechanical load they are bearing. There are still major gaps in our understanding of how this feedback system works at a molecular level, but recent work seems to have identified a few of the key components.

Evidence that the stability of microfilaments depends on the mechanical forces they carry has come from a number of systems. In epithelial cells, for example, microfilaments run across the cell to link adherens-type cell-cell adhesion complexes on different sides of the cell (the structure of epithelia is described in more detail in Chapter 4.1). These microfilaments are normally under tension, generated internally by myosin II and externally by the pulling of other cells in a tissue under tension, for example in the stretched wall of a full bladder. If the tension is released, by disrupting cell-cell adhesion

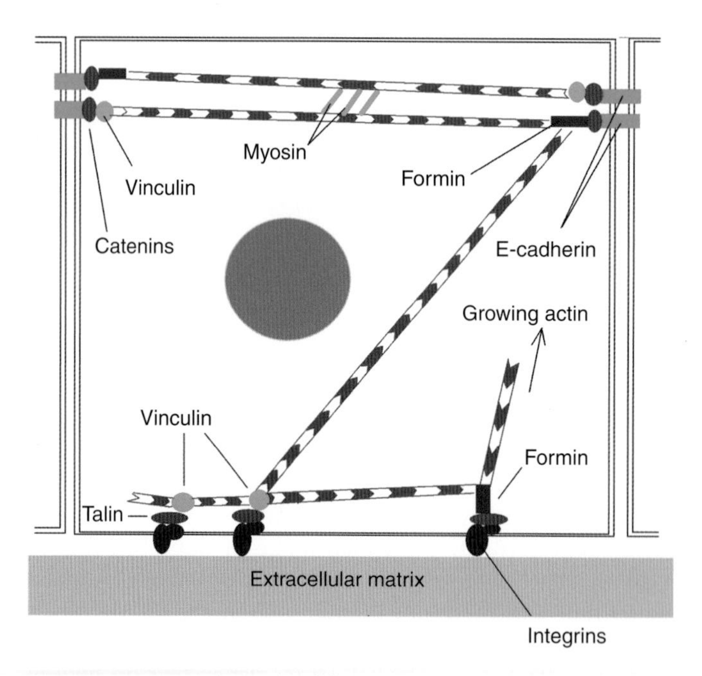

FIGURE 2.2.7 Schematic of the manner in which the microfilament-nucleating activity of Formins and the microfilament-binding activity of vinculin allows microfilaments to span between junctions and across the cell. The diagram depicts an epithelial cell that has both adherens junctions (top) and cell-matrix junctions (bottom). In reality, many microfilaments would reach only partway across the cell and would be cross-linked to each other by myosin to form a continuous, tension-generating cable.

using antibodies directed against the extracellular domains of adhesion molecules, the actin microfilaments become unstable and disappear.[34] If, on the other hand, an experimenter applies a stretching force to an epithelium, more and thicker filaments develop along the direction of the force where they are best placed to resist it.[35,36] Microfilaments therefore depend on mechanical force at cell-cell junctions. Cell-cell junctions show a reciprocal dependence on the actin cytoskeleton and, if filamentous actin is depolymerized using cytochalasin D, cell junctions fall apart.[37]

A similar story emerges from studies of integrin-mediated adhesions between cells and their substrates; disrupting integrin-mediated adhesion disrupts the microfilament system that would otherwise connect with integrin-rich adhesion sites.[38] Reciprocally, disrupting microfilaments causes loss of cell-matrix adhesions.[39] The mere presence of microfilaments is not enough to stabilize junctions; if the ability of microfilaments to generate tension is removed either by inhibiting the activity of myosin or by blocking the activation of myosin by ROCK (see below), junctions disassemble.[40,41,42,43] Junctions therefore need to be under mechanical tension to develop and be stable.

The mechanism by which mechanical loads affect junctional stability has been investigated using integrin-mediated focal adhesions to the extracellular matrix as a model.[44] Mammalian fibroblasts grown on a two-dimensional substrate and *starved of serum* do not form the mature focal adhesions associated with cells grown in serum, but instead make small focal complexes around their edges. These focal complexes are characteristic of the leading edge of motile cells (see Chapter 3.2). If such a cell is poked gently with a sticky pipette, the parts of the cell distal to the pipette experience increased mechanical force (see Figure 2.2.8). In the experiment being described here,[44] this force is approximately 10 nN per focal contact, which is approximately equal to the 1- to 10-nN range of forces that have been measured at mature focal adhesions of cells cultured in serum.[45,46] Within one minute of the application of the pipette, the small focal complexes, in that part of the cell only, mature into large focal adhesions that are attached to thick actin cables. Assembly of these focal adhesions depends on an intact actin cytoskeleton (it is blocked by actin depolymerizing drugs), but provided the source of external tension is maintained, it does not depend on myosin activity. This suggests that junctions respond to tension whether generated internally or externally and supports the idea that the requirement for myosin activity is mechanical rather than biochemical. The development of mature junctions does, however, depend on the activity of the Formin mDia1, which is located at the junction. Unfortunately, it is not

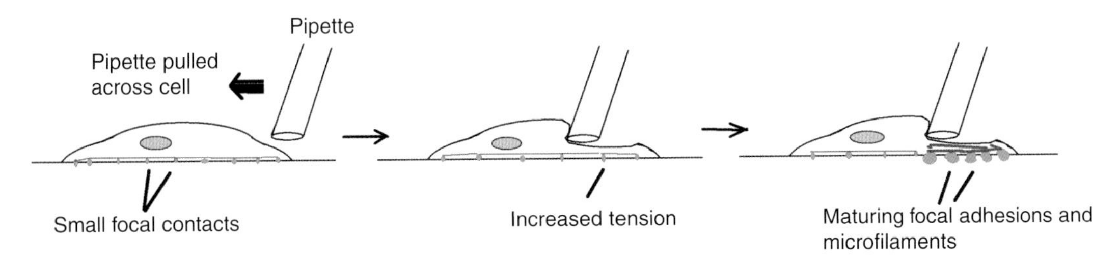

FIGURE 2.2.8 Experiment[44] demonstrating that increased tension alone is sufficient to promote the enlargement and stability of focal adhesions and stress fibres.

yet clear how mechanical force produces a biochemical change in Formin activity, nor how it alters the biochemical stability of microfilaments.

The reciprocal dependence of the stability of microfilaments and cell-cell junctions creates a highly responsive system for the self-organization of the microfilament-cell adhesion system. Microfilaments that are in the right place to connect active adhesion sites and therefore have something against which their tension can pull are comparatively stable, whereas those that are not, either because they fail to connect to adhesion complexes or because adhesion at that complex has failed, are unstable and disappear. Their actin monomers will be recycled for use in the development of new microfilaments, which will themselves survive or perish according to whether they achieve stable connections. The repeated "speculative" nucleation of new microfilaments allows the cytoskeletal system to explore the cell continually and to optimize its arrangement.

It is important that even the "stable" microfilaments and adhesion are not absolutely permanent; they merely turn over much more slowly than ones that fail to be stabilized. The destruction of perfectly good microfilaments may be expensive to the cell, but it is required for the cell to be able to find its optimal arrangement. The fitness of all possible arrangements of a cell can be expressed by adding an extra axis to the phase space graph described in Chapter 1.3, fitness being represented by the height along that axis. For any complex system, the landscape up the fitness axis will be complicated and, as well as including one large peak of fitness, it will include a large number of other peaks of much lower fitness than the main peak but of significantly higher fitness than the states in phase space that immediately surround them. These small peaks, called "local maxima" in the jargon of phase space, are a danger to any self-organizing system because a system that seeks to maximize fitness can be trapped by a local maximum, unable to cross a zone of lower fitness to a much greater peak elsewhere. Systems whose structures are unstable even when in a fitness peak are forced to leave their local maximum from time to time anyway, and, as they seek again a maximum, they have the chance to find the highest peak of all. Because the optimum fitness peak tends to have a broad base (in most biological situations), systems that have found it tend to find it again after small perturbations caused by their inherent instabilities. Therefore, however counterintuitive it may seem, instability of even optimal cellular structures is important to the ability of a cell to find its optimum arrangement. The instability is also very important to the ability of a cell to adapt to changing circumstances.

Naturally, the frequency with which microfilaments are nucleated from particular sites is modulated by biochemical signalling. That is why stress fibres of fibroblasts are more numerous in the presence of serum than in its absence,[44] as mentioned above. Of the various regulators of the stress fibre formation, the small GTPase, Rho, is of outstanding importance to morphogenesis and it will feature in many of the following chapters. Rho can be regulated by a variety of extracellular signals, and it regulates the microfilament system by two main routes (see Figure 2.2.9). Rho encourages the formation of new microfilaments by activating Formins such as mDia1.[47,48] Acting via ROCK (=Rho-dependent kinase), Rho also controls the tension developed by microfilaments by two synergistic processes. ROCK activates myosin light chain kinase,[49] which phosphorylates myosin light chain and activates the tension-producing activity of myosin. ROCK also phosphorylates the myosin-binding subunit of myosin phosphatase and consequently inactivates it; this prevents myosin phosphatase from dephosphorylating the myosin light chain and, by preventing this, it maintains myosin in an active state.[50]

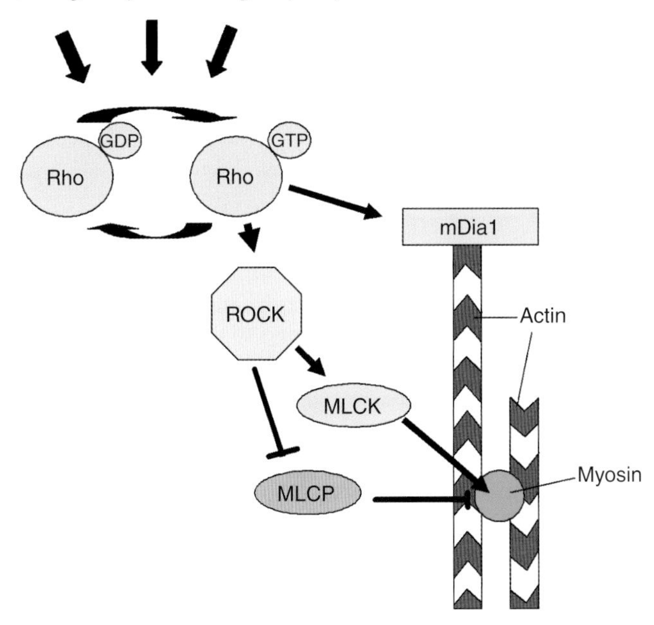

FIGURE 2.2.9 The small GTPase, Rho, controls microfilaments by regulating both their assembly and their ability to generate tension. Rho is itself controlled by Guanidine nucleotide exchange factors, which move Rho into its activated, GTP-bound state and GTPase-activating proteins that promote the reverse transition. These proteins are themselves modulated by extracellular signals, as illustrated in later chapters. In this diagram, as in the rest of the book, arrows ($\downarrow$) imply activation and blocks ($\perp$) imply inhibition.

ASSEMBLY OF THE MICROTUBULE SYSTEM

Microtubules are polymers of tubulin molecules, which themselves consist of a heterodimer of α- and β-subunits. Tubulin molecules are asymmetric, having ends conventionally designated "+" and "−", and their polymers retain this +/− polarity. Tubulin monomers have an affinity for guanosine phosphates and normally bind either two molecules of GTP, one on each protein subunit, or one molecule of GTP and one of GDP. In the latter case, the GTP is bound to the α-tubulin subunit, where it is buried and stable, and the GDP is bound to the β-subunit. The β-subunit has an inherent weak GTPase activity, so that GTP that is bound to it is hydrolyzed to GDP and P_i.[51] The GDP so formed can be exchanged with GTP in the cytoplasm surrounding the tubulin, and the cycle repeats (at the cost of the energy required to synthesize GTP). The GTP bound to the α-subunit is buried and cannot be exchanged; it will therefore be ignored for the rest of this discussion and tubulin will be described as "GTP-tubulin" or "GDP-tubulin" according to the status of the β-subunit only.

Microtubules form when GTP-tubulin molecules come together by self-assembly. The reaction will take place *in vitro*, as long as the concentration of GTP-tubulin is high enough, the + end growing faster than the − end. Linear protofilaments formed in this way associate into hollow cylinders of, usually, 13 protofilaments with a diameter of

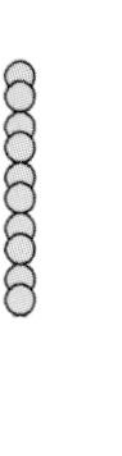

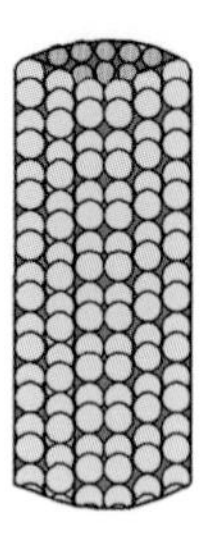

Protofilament Microtubule

FIGURE 2.2.10 The assembly of tubulin into hollow cylindrical associations of protofilaments; microtubules. *In vivo*, microtubules grow by addition of tubulin dimers to their ends so that their constituent protofilaments elongate, not by coming together of protofilaments.

about 25 mn. This cylinder is called the microtubule (see Figure 2.2.10). The association of protofilaments is usually present from the inception of microtubule development *in vivo*. The arrangement of tubulin protofilaments as a hollow tube produces a structure of great stiffness.♣ The flexural rigidity of a microtubule has been measured at about 2×10^{-23} Nm^2 compared with about 7×10^{-26} Nm^2 for actin microfilaments.[52] Microtubules are therefore about 300 times more rigid than microfilaments. This makes them much more suitable for bearing compressive loads without buckling (buckling is never a problem for tensile loads, of course). The Young's moduli (resistances to stretching) of microtubules and microfilaments are 1.2 GPa and 2.6 GPa, respectively,[52] emphasizing that it is the arrangement of tubulin as a broad cylinder, and not inherent properties of the protofilaments, that gives microtubules their rigidity. Having a higher Young's modulus, microfilaments are better suited to withstanding tensile loads without stretching far than are microtubules. It is interesting to note, in passing, that the Young's modulus of microtubules is similar to that of ordinary engineering steel, whereas the Young's modulus of microfilaments is similar to that of piano wire, which has been developed to withstand extremely high tension.[53]

In vivo, the formation of new microtubules is nucleated by microtubule-organizing centres (MTOCs), which contain short cylinders of γ-tubulin that are stabilized by their associated proteins. These γ-tubulin cylinders act as nucleation sites on which α-β tubulin dimers can construct mature microtubules. In animal cells, MTOCs are found typically in the immediate environment of the centrosomes near the nucleus; they may be found in basal bodies of cilia as well. The main microtubule network that is relevant to tensegrity originates from the pericentriolar zone, and it therefore radiates through the cell from one point (see Figure 2.2.3).

The binding of a new GTP-tubulin unit to the growing microtubule is stable. Once further growth has buried the unit into a microtubule so that it is no longer at the end, hydrolysis of its GTP to GDP has no effect, and it remains stable (the energy of hydrolysis is stored as strain in the microtubule). If, however, hydrolysis of GTP takes place on a tubulin monomer while it is still in a terminal position, its binding to the microtubule becomes unstable and it falls away. Statistically, it is likely that, by the time

♣Classroom demonstration: Cut a 5-cm (2-inch) square out of a piece of paper. Attempt to use it to lift an empty coffee cup and you will fail—it bends. Then bend it into a cylinder, taping the edges together; used this way and poked through the cup handle, it can easily lift the weight of the cup without collapsing.

a still-terminal tubulin has lost its GTP, the tubulin units behind it, which have been present in the microtubule for longer, will also have hydrolyzed their GTP to GDP (they cannot exchange it for a fresh GTP while they are in the microtubule). Once the terminal tubulin falls away, the next tubulin behind it, now terminal, will fall away, too, and the whole microtubule will unravel until it either reaches a zone that still has GTP or until new GTP-tubulins happen to cap an unravelling end. Left to themselves, microtubules are therefore always either growing or collapsing catastrophically; standing still is not an option. This state of affairs is called "dynamic instability."[54,55]

The dynamic instability of microtubules can be modulated by microtubule-associated proteins, and it is this modulation that enables the microtubule system to use adaptive self-organization to construct itself into an optimum arrangement. In particular, "+" end capping proteins can stabilize "+" ends and prevent their catastrophic collapse. Provided that these capping proteins are located at sites that require connection to the microtubule system and not elsewhere, the MTOC can nucleate microtubules at random, and only those that happen to reach appropriate targets will be stabilized. The correct microtubule anatomy will therefore be created automatically, without the MTOC having to have any specific information about the shape of the cell in which it finds itself.

For the tensegrity model to work, it is important that microtubules connect to the same membrane structures to which microfilaments connect. If microtubules were not to reach the membrane region at all, then tension in the microfilaments would cause the cell to collapse inwards until the membrane reached the microtubules. If microtubules reached different parts of the membrane, then the membrane would be expected to buckle, and perhaps to tear, between the points of attachment of tension and compression elements. Fortunately, there is ample evidence for microtubules being associated with cell-cell and cell-matrix junctions.[56]

The protein ACF7 (a mammalian homologue of *Drosophila melanogaster*'s Kakapo protein) is a member of the plakin family of junction-associated proteins.[57] It is associated with integrin-containing focal contacts and contains both actin-binding and microtubule-binding domains and can be seen associated with the "+" ends of microtubules at these junctions. It is likely that ACF7 is at least partly responsible for binding microtubules to the microfilament system at these junctions and stabilizing them. In adherens junctions, the cytoplasmic domain of the cell-cell adhesion molecule, E-cadherin, binds to β-catenin, which in turn binds to the microtubule-associated protein, dynein.[58] The location of dynein at the junctions does not depend on the presence of microtubules but does depend on the presence of the microfilament system; it can therefore be there before microtubules have reached the junction. Dynein is known to stabilize microtubules *in vitro*,[59] and if it has this activity *in vivo*, it may play a part in stabilizing microtubules that happen to have found their way to an adherens junction, so that they survive longer than those that make no contact with stabilizing proteins. The microfilament-nucleating Formin, mDia1, associates with microtubules also and can cause their alignment along pre-existing microfilament tracks.[60] The details of this alignment have yet to be worked out, but it may be a way of assisting the correct placement of microtubules.

The microtubule system of cells, like the microfilament system, also seems to be responsive to a changing mechanical load, although it is still not clear precisely what is being sensed. In response to a number of experimental procedures designed to increase tension in a specific part of a cell, microtubules become more numerous in that part of

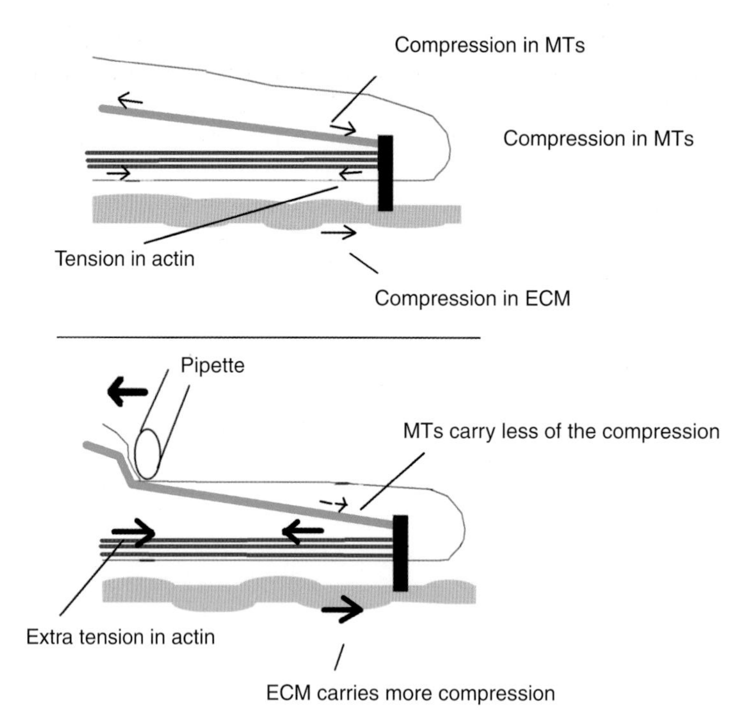

FIGURE 2.2.11 The methods for increasing local microfilament tension, used in the study of microtubule response to mechanical stress described in the main text, will have the effect of reducing compression in microtubules if the tensegrity model is correct. The diagram depicts the method of pushing with a pipette, but the other methods described in the text would have the same net result.

the cell.[61] These procedures include poking the cell with a pipette, stretching the matrix underneath the cell, and holding back the cell body of cells that are migrating forwards. One thing that all of these have in common is that, by applying an extra external force against which microfilament tension can act, they would be expected to *reduce* the compression forces experienced by the microtubules. This is because the matrix and the microtubules share the task of resisting microfilament tension. In the same way that reducing the ability of microtubules to resist compression by depolymerizing them with drugs increases the compressive forces borne by the extracellular matrix,[8] increasing the compressive forces borne by the matrix would be expected to decrease the compression in the microtubules (see Figure 2.1.11). It may, therefore, be that the increase in the number of tubules is a response to reduced compression in the microtubule system rather than increased tension in the microfilaments. It is not, however, clear why this should be required, although it is obvious that the opposite relationship—increase in compression members being encouraged by increased compression—would be bad for the cell because this positive feedback and that operating on the microfilament system would couple to create runaway development of the cytoskeletal system in response to minor fluctuations in force.

The increase in microfilaments observed in stressed parts of cells may have less to do with mechanics and more to do with other functions of microtubules, particularly transport of components within the cell. Given that junctional components such as cadherins and catenins can hitch a ride to the membrane aboard microtubules by being attached to the motor protein kinesin,[62] it may be that the main purpose of an increase in microtubules in tensed areas of cells is to assist with transport of material needed to bolster the tension-bearing network.

The main tension and compression systems of the cytoskeleton, then, organize themselves automatically according to the forces that act on them and the placement of the junctions that transmit those forces. The feedback loops that ensure automatic placement of microfilaments and microtubules are the main way in which embryos solve the problem of having to place internal elements with an accuracy exceeding that of the tissue-level gradients that drive pattern formation. The evidence for this feedback is strong, and, although it has been presented here in the context of the tensegrity model for global cytoskeletal function, it does not depend on that model standing the test of time and of improved biophysical measurements. "Special" morphogenesis by cells proceeds mainly by specific regulation of the generic systems that have been described in this chapter: they will be described in the next chapters.

Reference List

1. Thompson, DW (1961) On growth and form. Abridged: in Bonner JT (Cambridge University Press)
2. Colucci-Guyon, E, Gimenez, Y, Ribotta, M, Maurice, T, Babinet, C, and Privat, A (1999) Cerebellar defect and impaired motor coordination in mice lacking vimentin. *Glia.* **25**: 33–43
3. Colucci-Guyon, E, Portier, MM, Dunia, I, Paulin, D, Pournin, S, and Babinet, C (18-11-1994) Mice lacking vimentin develop and reproduce without an obvious phenotype. *Cell.* **79**: 679–694
4. Ingber, DE (1993) Cellular tensegrity: Defining new rules of biological design that govern the cytoskeleton. *J. Cell Sci.* **104 (Pt 3)**: 613–627
5. Ingber, DE (1-4-2003) Tensegrity I. Cell structure and hierarchical systems biology. *J. Cell Sci.* **116**: 1157–1173
6. Fuller, B (2004) Tensegrity. *Portfolio Artnews Annual.* **4**: 112–127
7. Anon. (2004) Kenneth Snelson Website. *http://www.kennethsnelson.net/.*
8. Kolodney, MS and Wysolmerski, RB (1992) Isometric contraction by fibroblasts and endothelial cells in tissue culture: A quantitative study. *J. Cell Biol.* **117**: 73–82
9. Elgsaeter, A, Stokke, BT, Mikkelsen, A, and Branton, D (5-12-1986) The molecular basis of erythrocyte shape. *Science.* **234**: 1217–1223
10. Jay, DG (20-9-1996) Role of band 3 in homeostasis and cell shape. *Cell.* **86**: 853–854
11. Brown, KD, Zinkowski, RP, Hays, SE, and Binder, LI (1993) Actin-binding protein is a component of bovine erythrocytes. *Cell Motil. Cytoskeleton.* **24**: 100–108
12. Colin, FC and Schrier, SL (1-12-1991) Myosin content and distribution in human neonatal erythrocytes are different from adult erythrocytes. *Blood.* **78**: 3052–3055
13. Lange, Y, Hadesman, RA, and Steck, TL (1982) Role of the reticulum in the stability and shape of the isolated human erythrocyte membrane. *J. Cell Biol.* **92**: 714–721
14. Evangelista, M, Pruyne, D, Amberg, DC, Boone, C, and Bretscher, A (2002) Formins direct Arp2/3-independent actin filament assembly to polarize cell growth in yeast. *Nat. Cell Biol.* **4**: 260–269
15. Tolliday, N, VerPlank, L, and Li, R (29-10-2002) Rho1 directs Formin-mediated actin ring assembly during budding yeast cytokinesis. *Curr. Biol.* **12**: 1864–1870
16. Winter, D, Podtelejnikov, AV, Mann, M, and Li, R (1-7-1997) The complex containing actin-related proteins Arp2 and Arp3 is required for the motility and integrity of yeast actin patches. *Curr. Biol.* **7**: 519–529

17. Pruyne, D, Evangelista, M, Yang, C, Bi, E, Zigmond, S, Bretscher, A., and Boone, C (26-7-2002) Role of formins in actin assembly: Nucleation and barbed-end association. *Science.* **297**: 612–615
18. Sagot, I, Klee, SK, and Pellman, D (2002) Yeast formins regulate cell polarity by controlling the assembly of actin cables. *Nat. Cell Biol.* **4**: 42–50
19. Kobielak, A, Pasolli, HA, and Fuchs, E (2004) Mammalian Formin-1 participates in adherens junctions and polymerization of linear actin cables. *Nat. Cell Biol.* **6**: 21–30
20. Angst, BD, Marcozzi, C, and Magee, AI (2001) The cadherin superfamily. *J. Cell Sci.* **114**: 625–626
21. Izard, T, Evans, G, Borgon, RA, Rush, CL, Bricogne, G, and Bois, PR (8-1-2004) Vinculin activation by talin through helical bundle conversion. *Nature.* **427**: 171–175
22. Miyahara, M and Noda, H (1977) Self-assembly of myosin in vitro caused by rapid dilution. Effects of hydrogen ion, potassium chloride, and protein concentrations. *J. Biochem. (Tokyo).* **81**: 285–295
23. Condeelis, JS (1977) The self-assembly of synthetic filaments of myosin isolated from Chaos carolinensis and Amoeba proteus. *J. Cell Sci.* **25**: 387–402
24. Kuczmarski, ER and Spudich, JA (1980) Regulation of myosin self-assembly: Phosphorylation of Dictyostelium heavy chain inhibits formation of thick filaments. *Proc. Natl. Acad. Sci. U.S.A.* **77**: 7292–7296
25. Niederman, R. and Peters, LK (15-11-1982) Native bare zone assemblage nucleates myosin filament assembly. *J. Mol. Biol.* **161**: 505–517
26. Cross, RA, Citi, S, and Kendrick-Jones, J (1988) How phosphorylation controls the self-assembly of vertebrate smooth and non-muscle myosins. *Biochem. Soc. Trans.* **16**: 501–503
27. Cross, RA, Hodge, TP, and Kendrick-Jones, J (1991) Self-assembly pathway of nonsarcomeric myosin II. *J. Cell Sci. Suppl.* **14**: 17–21
28. de la Roche, MA, Smith, JL, Betapudi, V, Egelhoff, TT, and Cote, GP (2002) Signaling pathways regulating Dictyostelium myosin II. *J. Muscle Res. Cell Motil.* **23**: 703–718
29. Gulick, AM and Rayment, I (1997) Structural studies on myosin II: Communication between distant protein domains. *Bioessays.* **19**: 561–569
30. Djinovic-Carugo, K, Young, P, Gautel, M, and Saraste, M (20-8-1999) Structure of the alpha-actinin rod: Molecular basis for cross-linking of actin filaments. *Cell.* **98**: 537–546
31. Klein, MG, Shi, W, Ramagopal, U, Tseng, Y, Wirtz, D, Kovar, DR, Staiger, CJ, and Almo, SC (2004) Structure of the actin crosslinking core of fimbrin. *Structure (Camb.).* **12**: 999–1013
32. Tseng, Y, An, KM, Esue, O, and Wirtz, D (16-1-2004) The bimodal role of filamin in controlling the architecture and mechanics of F-actin networks. *J. Biol. Chem.* **279**: 1819–1826
33. Sevcik, J, Urbanikova, L, Kost'an, J, Janda, L, and Wiche, G (2004) Actin-binding domain of mouse plectin. Crystal structure and binding to vimentin. *Eur. J. Biochem.* **271**: 1873–1884
34. Danjo, Y and Gipson, IK (1998) Actin 'purse string' filaments are anchored by E-cadherin-mediated adherens junctions at the leading edge of the epithelial wound, providing coordinated cell movement. *J. Cell Sci.* **111 (Pt 22)**: 3323–3332
35. Nagai, H and Kalanins, VI (25-2-1996) An apical tension-sensitive microfilament system in retinal pigment epithelial cells. *Exp. Cell Res.* **223**: 63–71
36. Kalnins, VI, Sandig, M, Hergott, GJ, and Nagai, H (1995) Microfilament organization and wound repair in retinal pigment epithelium. *Biochem. Cell Biol.* **73**: 709–722
37. Quinlan, MP and Hyatt, JL (1999) Establishment of the circumferential actin filament network is a prerequisite for localization of the cadherin-catenin complex in epithelial cells. *Cell Growth Differ.* **10**: 839–854
38. Castel, S, Pagan, R, Garcia, R, Casaroli-Marano, RP, Reina, M, Mitjans, F, Piulats, J, and Vilaro, S (2000) Alpha v integrin antagonists induce the disassembly of focal contacts in melanoma cells. *Eur. J. Cell Biol.* **79**: 502–512
39. Folsom, TD and Sakaguchi, DS (1997) Characterization of focal adhesion assembly in XR1 glial cells. *Glia.* **20**: 348–364

40. Chrzanowska-Wodnicka, M and Burridge, K (1996) Rho-stimulated contractility drives the formation of stress fibers and focal adhesions. *J. Cell Biol.* **133**: 1403–1415
41. Helfman, DM, Levy, ET, Berthier, C, Shtutman, M, Riveline, D, Grosheva, I, Lachish-Zalait, A, Elbaum, M, and Bershadsky, AD (1999) Caldesmon inhibits nonmuscle cell contractility and interferes with the formation of focal adhesions. *Mol. Biol. Cell.* **10**: 3097–3112
42. Volberg, T, Geiger, B, Citi, S, and Bershadsky, AD (1994) Effect of protein kinase inhibitor H-7 on the contractility, integrity, and membrane anchorage of the microfilament system. *Cell Motil. Cytoskeleton.* **29**: 321–338
43. Citi, S, Volberg, T, Bershadsky, AD, Denisenko, N, and Geiger, B (1994) Cytoskeletal involvement in the modulation of cell-cell junctions by the protein kinase inhibitor H-7. *J. Cell Sci.* **107 (Pt 3)**: 683–692
44. Riveline, D, Zamir, E, Balaban, NQ, Schwarz, US, Ishizaki, T, Narumiya, S, Kam, Z, Geiger, B, and Bershadsky, AD (11-6-2001) Focal contacts as mechanosensors: Externally applied local mechanical force induces growth of focal contacts by an mDia1-dependent and ROCK-independent mechanism. *J. Cell Biol.* **153**: 1175–1186
45. Galbraith, CG and Sheetz, MP (19-8-1997) A micromachined device provides a new bend on fibroblast traction forces. *Proc. Natl. Acad. Sci. U.S.A.* **94**: 9114–9118
46. Dembo, M and Wang, YL (1999) Stresses at the cell-to-substrate interface during locomotion of fibroblasts. *Biophys. J.* **76**: 2307–2316
47. Watanabe, N, Kato, T, Fujita, A, Ishizaki, T, and Narumiya, S (1999) Cooperation between mDia1 and ROCK in Rho-induced actin reorganization. *Nat. Cell Biol.* **1**: 136–143
48. Watanabe, N, Madaule, P, Reid, T, Ishizaki, T, Watanabe, G, Kakizuka, A, Saito, Y, Nakao, K, Jockusch, BM, and Narumiya, S (2-6-1997) p140mDia, a mammalian homolog of *Drosophila diaphanous*, is a target protein for Rho small GTPase and is a ligand for profilin. *EMBO J.* **16**: 3044–3056
49. Totsukawa, G, Yamakita, Y, Yamashiro, S, Hartshorne, DJ, Sasaki, Y, and Matsumura, F (21-8-2000) Distinct roles of ROCK (Rho-kinase) and MLCK in spatial regulation of MLC phosphorylation for assembly of stress fibers and focal adhesions in 3T3 fibroblasts. *J. Cell Biol.* **150**: 797–806
50. Kimura, K, Ito, M, Amano, M, Chihara, K, Fukata, Y, Nakafuku, M, Yamamori, B, Feng, J, Nakano, T, Okawa, K, Iwamatsu, A, and Kaibuchi, K (12-7-1996) Regulation of myosin phosphatase by Rho and Rho-associated kinase (Rho-kinase). *Science.* **273**: 245–248
51. O'Brien, ET, Voter, WA, and Erickson, HP (30-6-1987) GTP hydrolysis during microtubule assembly. *Biochemistry.* **26**: 4148–4156
52. Gittes, F, Mickey, B, Nettleton, J, and Howard, J (1993) Flexural rigidity of microtubules and actin filaments measured from thermal fluctuations in shape. *J. Cell Biol.* **120**: 923–934
53. Gordon, JE (1976) The new science of strong materials. **2**: (Penguin)
54. Mitchison, T and Kirschner, M (15-11-1984) Dynamic instability of microtubule growth. *Nature.* **312**: 237–242
55. Cassimeris, LU, Walker, RA, Pryer, NK, and Salmon, ED (1987) Dynamic instability of microtubules. *Bioessays.* **7**: 149–154
56. Kaverina, I, Rottner, K, and Small, JV (13-7-1998) Targeting, capture, and stabilization of microtubules at early focal adhesions. *J. Cell Biol.* **142**: 181–190
57. Karakesisoglou, I, Yang, Y, and Fuchs, E (3-4-2000) An epidermal plakin that integrates actin and microtubule networks at cellular junctions. *J. Cell Biol.* **149**: 195–208
58. Ligon, LA, Karki, S, Tokito, M, and Holzbaur, EL (2001) Dynein binds to beta-catenin and may tether microtubules at adherens junctions. *Nat. Cell Biol.* **3**: 913–917
59. Ohba, S, Kamata, K, and Miki-Noumura, T (28-11-1993) Stabilization of microtubules by dynein-binding in vitro. Stability of microtubule-dynein complex. *Biochim. Biophys. Acta.* **1158**: 323–332
60. Ishizaki, T, Morishima, Y, Okamoto, M, Furuyashiki, T, Kato, T, and Narumiya, S (2001) Coordination of microtubules and the actin cytoskeleton by the Rho effector mDia1. *Nat. Cell Biol.* **3**: 8–14

61. Kaverina, I, Krylyshkina, O, Beningo, K, Anderson, K, Wang, YL, and Small, JV (1-6-2002) Tensile stress stimulates microtubule outgrowth in living cells. *J. Cell Sci.* **115**: 2283–2291
62. Chen, X, Kojima, S, Borisy, GG, and Green, KJ (10-11-2003) p120 catenin associates with kinesin and facilitates the transport of cadherin-catenin complexes to intercellular junctions. *J. Cell Biol.* **163**: 547–557

CHAPTER 2.3

CELLULAR MORPHOGENESIS IN ANIMALS

The last chapter described the mechanisms that determine the shape of "ordinary" cells. This chapter will consider how those mechanisms may be adapted to make specific departures from the ordinary. In most cases, changes of cell shape are led by the development of compression elements that can push the plasma membrane outwards, although cell wedging, described in Chapter 4.4, may be an exception. The compression elements involved may be formed by microtubules, which normally bear compression according to the tensegrity model, or they may be formed by microfilaments that are deployed in a manner specially adapted to bear compressive loads.

MICROFILAMENT-BASED STRUCTURES FORMED BY INSIDE→OUT POLYMERIZATION

Cells form processes and out-pushings for a variety of purposes ranging from locomotion (Chapters 3.1 and 3.2) to sensation to fertilization, and many of these are based on microfilament skeletons that are orientated with their pointed ends toward the main body of the cell and their barbed ends toward the tip of the process. Within this broad category of processes there are two different classes: those in which filaments are initiated at their inner (cytoplasmic) ends and extend outwards, and those in which filaments are initiated at their membrane ends and attempt to penetrate inwards (see Figure 2.3.1). The latter are probably more common, but this chapter will begin by considering a system that grows from the inside outwards because it has been well-studied and it allows key principles to be introduced without too many complications. It is also highly relevant to development because it concerns the meeting between sperm and egg in echinoderms.

Thyone briareus (syn: *Holothuria briareus*) is a type of marine echinoderm, sometimes called a sea-cucumber, that lives in marine mud and is abundant at the famous Woods Hole laboratory, where it has been studied extensively. Echinoderm eggs are covered in thick, protective jelly, which creates a problem for sperm attempting to fertilize them. The problem is solved in two main ways. First, the head of the sperm bears an acrosomal vesicle that is full of enzymes capable of digesting the jelly coat, and these enzymes are released when receptors on the sperm detect that they have bound the coat

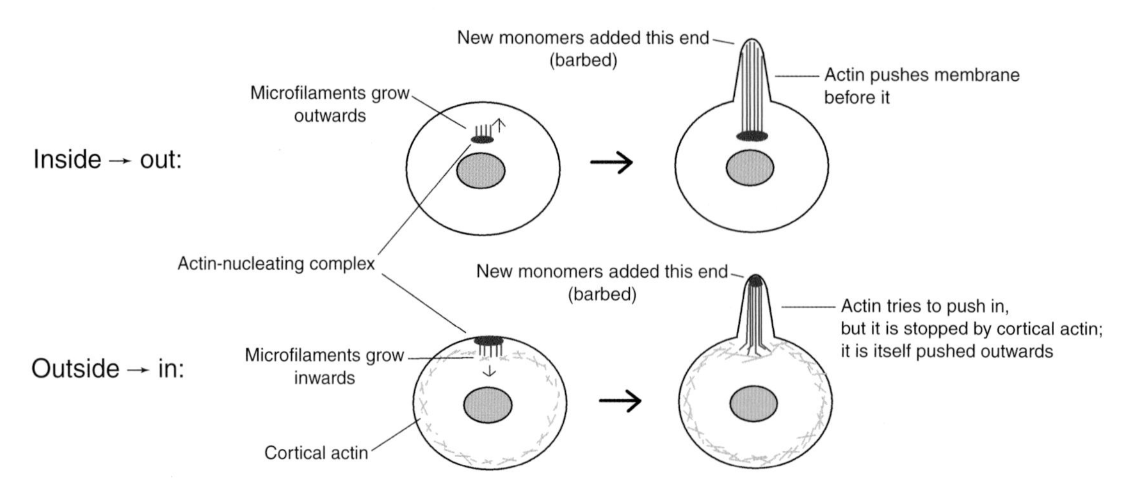

FIGURE 2.3.1 Inside→out and outside→in elongation of microfilaments. In both cases, the barbed end is at the top of the diagrams.

of an egg. Second, the sperm produces a long, rigid process, the acrosomal process, which forces its way through the digesting jelly coat to meet and fuse with the surface of the egg. Formation of this acrosomal process is rapid and dramatic; it extends from nothing to become 90 μm long within a mere 10 seconds of sperm meeting egg coat.[1,2] The acrosomal process is constructed mainly from microfilaments.

Before binding, actin is present at high concentrations near the acrosome, but it is complexed with profilin and other proteins. A small amount of the actin is already polymerized into a bundle of about 25 microfilaments, arranged with their barbed ends facing the membrane. The microfilaments are complexed with other proteins. Not all of these have been identified, but their presence shows up in transmission electron microscopy. This actin bundle is called the "actomere," and it acts as an initiating site for actin polymerization,[3] the polarity of its filaments determining the barbed-end-to-membrane polarity of the filaments of the acrosomal process itself. Once sperm-egg binding has taken place, actin polymerization proceeds extremely quickly (see Figure 2.3.2). Profilin skews the efficiency of the binding of actin monomers to barbed and pointed ends so that only the barbed ends grow despite the high concentration of actin present. This can be shown by adding either actin or actin-profilin complexes to an acrosomal bundle *in vitro*: actin alone will extend it from both ends (demonstrating that neither end is blocked by capping proteins), whereas actin-profilin extends it almost exclusively from the barbed end.[4] In order to understand how the system is set up in the first place, it is obviously necessary to understand how the actomere is constructed and how it is positioned so accurately beneath the acrosomal vesicle; unfortunately, almost nothing is known about this yet.

The direction of actin polymerization in the acrosomal processes of *T. briareus* creates two potential problems for the cell: restricted transport of unpolymerized actin to the growing tip, and restricted access to filaments even when actin has got there.

Actin monomers are added to the distal end of the growing process, which gets further and further from the main body of the cell as polymerization proceeds. This creates an increasingly serious problem in transporting the actin-profilin complexes

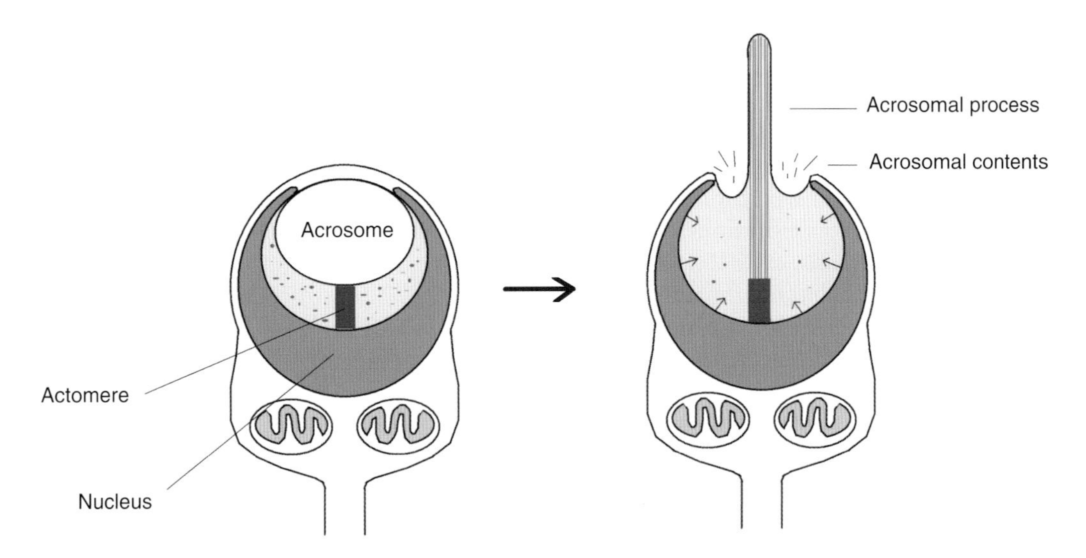

FIGURE 2.3.2 Growth of the acrosomal process in sperm of the sea cucumber *Thyone briareus*.

from the main body of the cell to the tip of the acrosomal process, where they are needed. There seems to be no specific transport system in the acrosomal process, so transport has to rely on diffusion. The flow of unpolymerized actin by diffusion can be approximated by the one-dimensional form of Fick's law of diffusion, since the acrosomal process is essentially one-dimensional in form once it has grown out some way:

$$J = -D \cdot dC/dx$$

where J is the flow of monomer, D is the diffusion constant, C is the local concentration of monomer, and x is the distance along the process. If the system were at steady state (which it will not be—see below) the flow of actin when the acrosomal process had reached a length L would be:

$$J_{\text{steady state}} = -D \cdot (C_{\text{cell}} - C_{\text{tip}})/L$$

where C_{tip} is the concentration of monomer at the tip of the process and C_{cell} is the concentration in the reservoirs in the cell. The key point made by this equation is that, as the acrosomal process grows longer, the rate of monomer delivery falls. In fact, the problem is worse than this because although C_{tip} will remain approximately constant, being set by the association constant of actin-profilin with the barbed end of the filament, C_{cell} will decrease as the cell's reserves of actin are used up (more precise models of the diffusion, which take this into account, have been worked out[5]). *T. briareus* seems to have evolved an adaptation to fight back against the diminishing returns of the diffusion equation by using water flows to maintain the effective concentration of actin monomers in the cellular stores. At the very beginning of the acrosomal reaction, ion channels open and allow saltwater into the periacrosomal region, which doubles in volume (this halves the concentration of actin, which may be important in preventing addition to the pointed end even in the presence of profilin). As acrosome process formation proceeds, the ion channels close and the pumps that are normally active in the cell remove the salts, which also draws out the water by osmosis. The dwindling pool of unpolymerized actin at the base of the growing acrosomal process is therefore confined in a diminishing volume of cytoplasm so that its concentration is kept higher than would otherwise be expected.[5] This solves the problem of diminishing effective concentrations, though the limiting effect of increasing length on actin transport remains.

The fact that microfilaments are used as compression structures creates a second potential problem for actin-profilin complexes even when they have reached the tip of the acrosomal process: if the barbed ends of the filament are pushing against the plasma membrane to force it forwards, how can new monomers fit onto those ends? The probable solution makes elegant use of the random thermal movements inherent in all matter at microscales and gains from movements that happen to be directed in the forward direction while blocking those in the reverse direction. This action is analogous to that of an electronic rectifying diode, which makes pulses of direct current that flows only one way by accepting phases of an alternating current source that are polarized one way while rejecting those polarized the other way. For this reason, the action mechanism of filament elongation is sometimes called "rectification," although it is more usually known by the term used in the rest of this book: the "Brownian ratchet."[6] Brownian motion is the random movement of a small object in suspension, and it results from the averaged impacts of small molecules of water and so on that collide constantly with the object. The membrane is itself a "small object" and, being flexible, its shape will be continuously changing in response to the battering it receives from the molecules that surround it. From time to time, therefore, the membrane will be moved forwards and the space between it and the barbed end of a filament end and the membrane will open up; at that moment, a new actin monomer can join the filament (see Figure 2.3.3). When the membrane "tries" to move back, its way is now blocked by the elongated filament, and its mean position has therefore effectively been moved forward by the length of one actin unit. The process then repeats. This basic mechanism remains valid whether

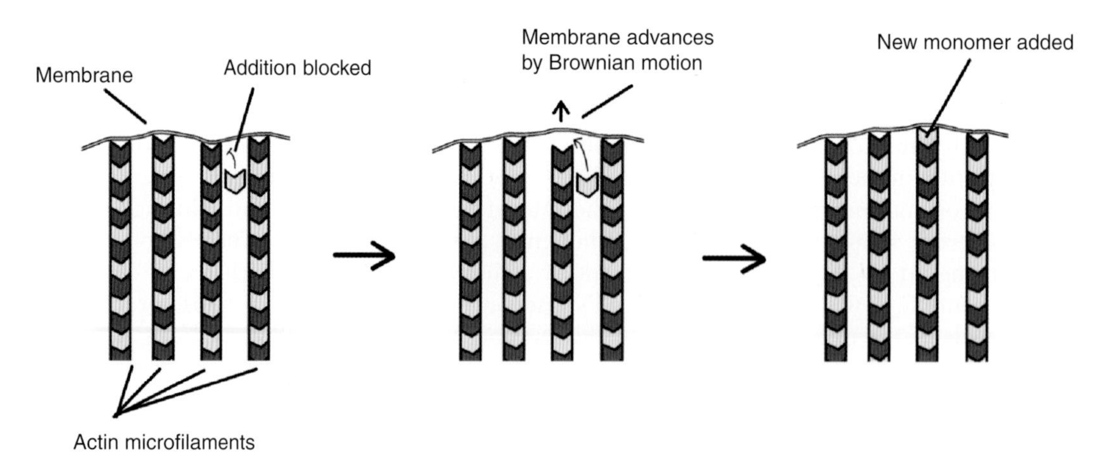

FIGURE 2.3.3 The Brownian ratchet at the tip of the growing acrosomal process.

the barbed end of the filament actually contacts the membrane itself or contacts some cortical protein that in turn pushes on the membrane.

The explanation above focused on the membrane alone; in reality, the terminal parts of the filaments, too, will show random changes in length as they are compressed and stretched by the molecules that hit them, and the ratchet mechanism is the result of both sets of random motions.

Not all acrosomal processes develop by rapid polymerization of actin. In species such as the horseshoe crab, *Limulus polyphemus*, the acrosomal actin is pre-assembled but is coiled up in a tight spiral between the nucleus and the acrosome. In this cell, the actin is cross-linked by two proteins, calmodulin and scruin, which form a 1:1 complex. The complex changes conformation depending on whether Ca^{2+} is present in the cytoplasm. Receptors for the outer surface of the egg trigger release of Ca^{2+} in the sperm cytoplasm and a change in conformation of the calmodulin/scruin complex. This relaxes the coiled actin bundle and it projects forward to penetrate the egg's jelly coat.[7]

PROCESSES FORMED BY OUTSIDE→IN POLYMERIZATION OF MICROFILAMENT BUNDLES: Small Microvilli and Their Derivatives

One of the simplest types of protrusion is the small microvillus. Small microvilli, which are shorter than 500 nm and are distinct from larger microvilli, are present in vast numbers at the surface of most animal cells.[8] They are highly dynamic and short-lived structures; each lasts only about 12 minutes in both amphibian and mammalian cell lines but, because formation and destruction of small microvilli are not synchronized across a cell, the overall population on the plasma membrane remains about the same.

The dominant structure of a small microvillus is a bundle of cross-linked microfilaments that are packed in a hexagonal array. The microfilaments are orientated with their barbed ends toward the membrane and they grow from their barbed ends so that the already-polymerized microfilaments are pushed toward the centre of the cell. Normally, these filaments make connections with the cortical actin system and therefore end up pushing against it rather than through it. The result is that the pointed ends fail to move inwards and the barbed ends therefore have to move outwards, thus creating a cell process (see Figure 2.3.4).

Small microvilli begin to form when microfilaments are nucleated by a patch of proteins just under plasma membrane. This patch is of unknown composition but is detectable as an electron-dense zone in transmission electron microscopy. The polymerization of actin at the tip of the microvillus features the same problems that were discussed for the *T. briareus* sperm above. The problem of actin transport is present without the water export mechanism to ameliorate it, and direct observations on the rate of extension of small microvilli of different lengths confirm that the rate of elongation varies approximately inversely with microvillus length,[8] as would be expected from the transport equation quoted above. The problem of access may also be worse—presumably, whatever membrane-associated nucleating factors initiate microtubule polymerization must let go and get out of the way of the barbed ends to allow fresh monomers to be added.

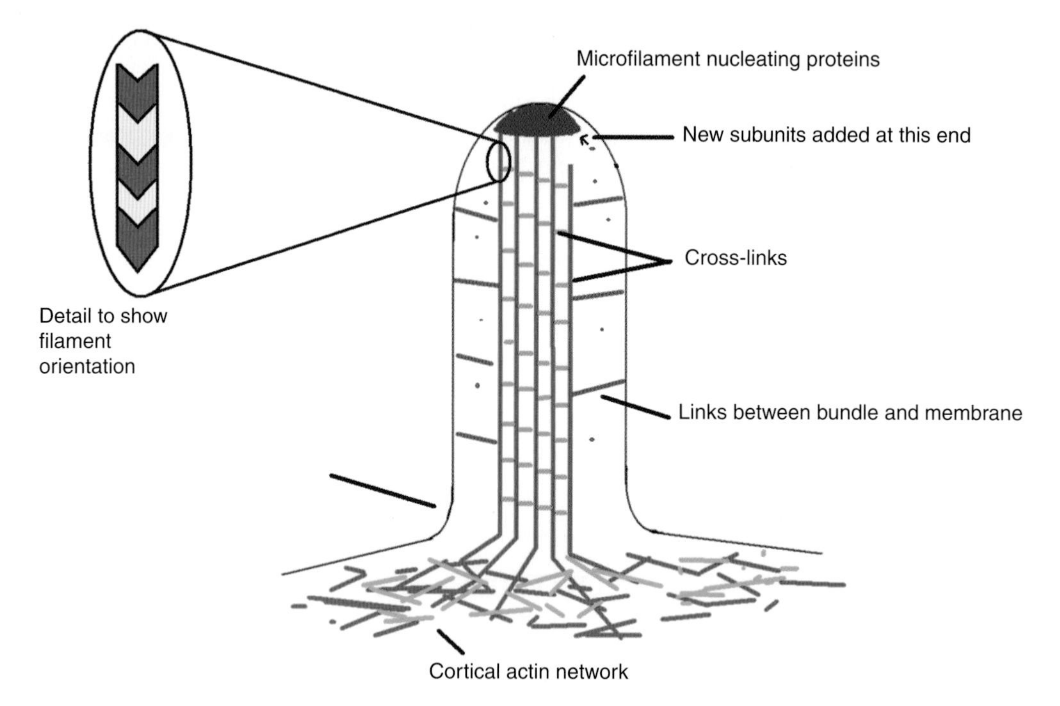

FIGURE 2.3.4 The basic structure of a small microvillus. What cannot easily be captured in the diagram is the fact that the filament system is not static, but is being built at the barbed end and depolymerized at the pointed end. Even the rates of these processes are inconstant, and most small microvilli last only minutes.

When the bundles of actin in microvilli are examined closely, they are seen to have fewer microfilaments at the tip end than at the base. Since it is unlikely that new filaments would be nucleated part of the way along a filament, it seems probable that barbed ends face a certain probability of being capped and therefore blocked from polymerization. If this capping were to take place randomly, it would result in the observed thinning of filament bundles toward the microvillus' tip (see Figure 2.3.5).

Small microvilli usually last a short time and then regress. This regression takes place at the steady state rate of pointed end depolymerization, which suggests that the destruction of a microvillus takes place simply by preventing further polymerization at the barbed end of the microfilaments. The control of polymerization at the barbed end must be highly localized to one part of the cell membrane because some small microvilli regress at the same time that others are growing. The general pattern will be under global controls too, though—an obvious example of a global control would be the availability of free actin.

Small microvilli are a cell's main source of bundled microfilaments that are capable of bearing compression.[9] Sometimes these elements are detached from the villus that bore them, perhaps by the mechanism illustrated in Figure 2.3.5, and are used as building blocks for larger structures, as happens in the construction of insect sensory bristles (see below). Sometimes the small microvilli survive and are expanded to become large microvilli and related cellular processes.

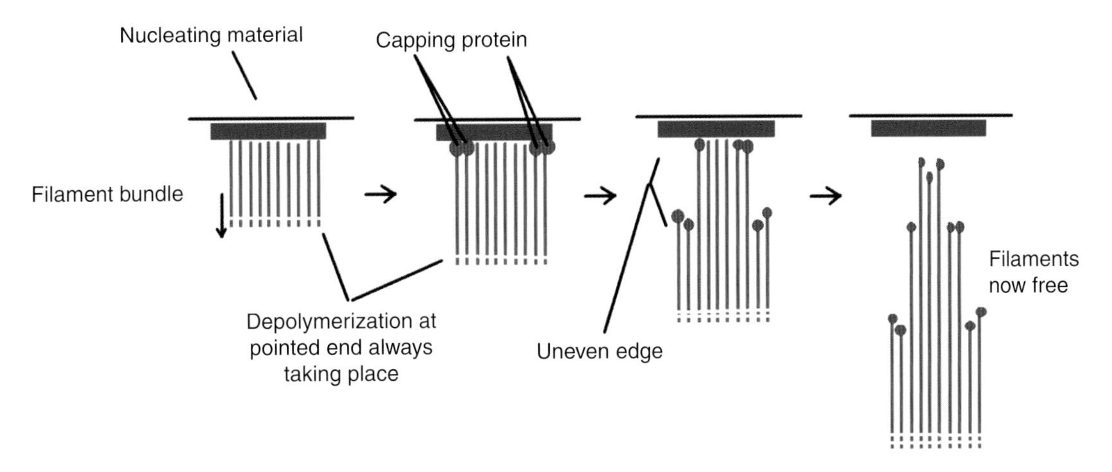

FIGURE 2.3.5 A possible explanation for the uneven appearance of the barbed ends of microfilament bundles in small microvilli and their derivatives. The actin filaments (red) are united by cross-links (not shown) that keep capped filaments moving downwards with the bundle.

LARGE MICROVILLI

Some epithelial cells, such as those of the intestine and of the proximal convoluted tubule of the kidney, have a "brush border" that consists of very large numbers of microvilli. Each of these is about 1 μm long and contains about 20 microfilaments that are cross-linked into a bundle with hexagonal packing. Some cells, such as those lining the human caput epididymis (part of the testis), have microvilli over 80 μm long. The microfilament bundle also contains the actin cross-linking proteins fimbrin and villin. The actin array at the core of the villus is attached to the membrane by arrays of cross-links every 33 nm; these cross-links consist of a complex between the Ca^{2+}-sensitive protein, calmodulin, and brush border myosin I (BBMI, also called MYO1A and M1A).[10] The basic structure of the microvillus core can self-assemble *in vitro*. If actin is mixed with fimbrin and villin, it forms cross-linked microfilament bundles by spontaneous self-assembly and, on addition of the calmodulin/BBMI complex, it acquires the 33 nm repeated side bridges that would connect with the plasma membrane.[11] Expression of high levels of villin in fibroblasts causes them to produce large numbers of microvilli, suggesting that villin has the ability to trigger self-organization of these structures.[12] When examined *in vitro*, villin shows a large number of activities associated with its ability to bind actin, including cross-linking, barbed-end capping, nucleation of new filaments, and severing of existing filaments.[13,14,15] Site-directed mutagenesis of villin demonstrates that its cross-linking activities are important to the formation of microvilli in cultured cells, but the nucleation activity is not.[16] Surprisingly, though, microvilli form normally in villin-knockout mice,[17] suggesting that some other protein, perhaps fimbrin, must be able to substitute for villin *in vivo*.

The microfilament core of a microvillus is a highly labile structure that is being remodeled constantly; in this respect it is similar to that small microvillus from which it developed. This can be demonstrated by transfecting cells with GFP-labelled♣ BBMI

♣GFP = green fluorescent protein.

or with GFP-actin so that their microvilli fluoresce green. If a patch of one microvillus is photobleached with bright light (which destroys the GFP), recovery of fluorescence is rapid. Since the only likely mechanism for recovery of fluorescence would be incorporation of fresh GFP-BBMI or GFP-actin, it is clear that the filaments and the proteins bound to them are being made and replaced very quickly[18] (this analysis makes the assumption that photobleaching does not itself destabilize the bleached structures). The microfilaments of the microvillus core are therefore not stable units but undergo "treadmilling," being extended at their barbed ends while being depolymerized at their pointed ends.

STEREOCILIA

Despite their misleading name (they are *not* cilia!), stereocilia are microfilament-based extensions to the sensory cells in the cochlea of the inner ear. They are tuned to sounds of a particular frequency,[19] and the length of each stereocilium is controlled precisely and varies along the length of the ear; in hamsters, the shortest stereocilia are about 1.2 μm long at the base of the cochlea and 5 μm at the apex.[20] Stereocilia have a core similar to that of microvilli, but their microfilaments are cross-linked by a combination of fimbrin[21,22] and espin.[23] Espin is clearly very important to this system because mutations in the gene cause deafness.[23] The protein seems to be able to organize the basic small microvillus to form stereocilia instead. Transfection of kidney proximal tubule cells, which normally form brush borders of microvilli about 1 μm long, with the espin gene causes them to produce very elongated protrusions that are over 7 μm long and that look similar to stereocilia.[24] Mutation of the promoter used to drive the espin gene in the transfection experiments results in lower levels of espin expression and allows the effects of different levels of espin expression to be compared. Lower amounts of espin result in shorter "stereocilia," implying that the amount of espin controls stereocilium length. Significantly, the level of espin expression changes along the length of the cochlea, and it correlates well with the length of the stereocilia.

How might espin control length? Like the cores of microvilli, the cores of stereocilia are in a state of continuing flux despite the constant lengths of each stereocilium; if GFP-actin is transfected into even cells bearing mature stereocilia, fluorescence is incorporated into the stereocilium rapidly and travels from the tip toward the base.[25] The length of such a structure might, in principle, be controlled by varying the rates of polymerization or of depolymerization. For a constant "half-life" of a particular actin unit to be in the microfilament, increasing the rate of addition at the barbed end would increase the distance that unit will have been pushed toward the cell before it is removed from the filament (see Figure 2.3.6). Conversely, for a fixed rate of addition at the barbed end, the shorter the half-life of a unit, the shorter the distance it will be pushed toward the cell. Measurements of rates of microfilament polymerization and degradation in the presence of espin have shown that it produces a small ($<$ twofold) reduction in the rate of microfilament depolymerization *in vitro*. This effect may be modest, but at the measured rates of treadmilling in the cells being examined, a twofold reduction of the rate of depolymerization would cause an elongation of 1 μm filament bundles to 8 μm in about 1 hour.[24] It is therefore possible that length is determined simply by altering rates of depolymerization in this way. The sensitivity of filament bundle length to very small changes in biochemical rates makes the control systems

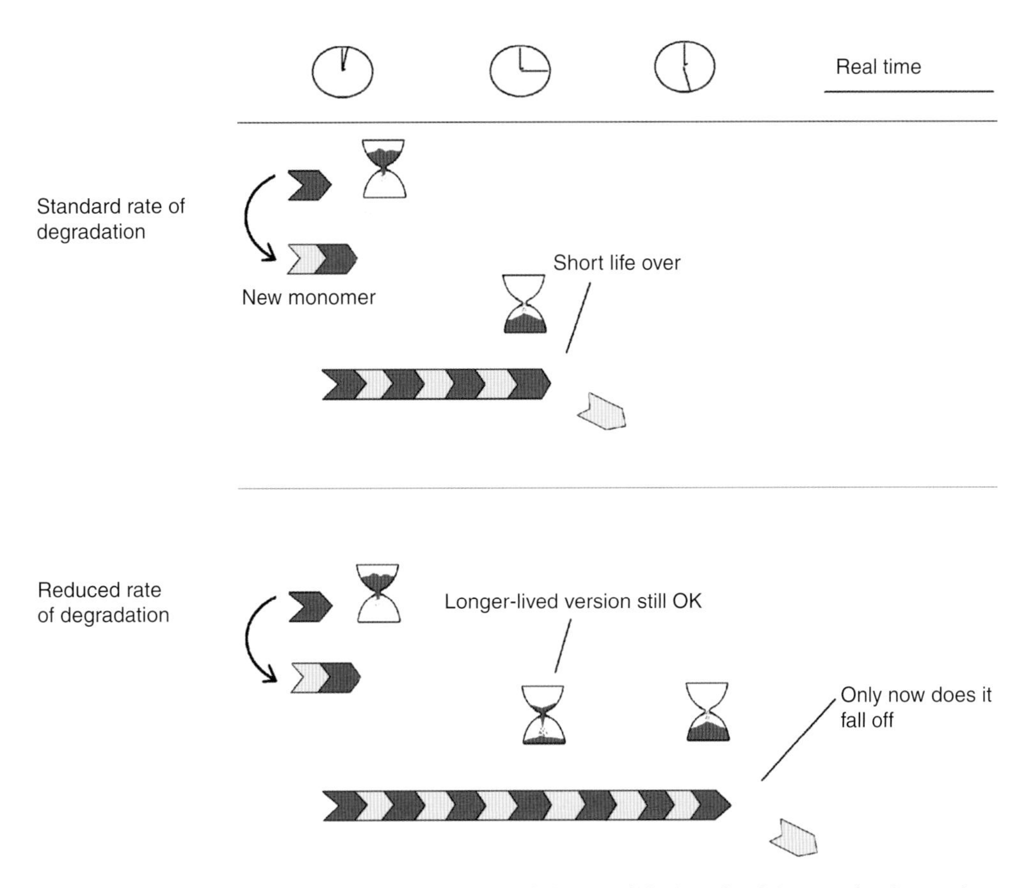

FIGURE 2.3.6 Control of filament length by modulation of the length of time a subunit remains in the filament.

very difficult for cell biologists to measure key parameters accurately. It must also be difficult for cells themselves to achieve the accuracy necessary for acute hearing, and the concentrations of espin are probably regulated by extremely tight feedback loops.

SENSORY BRISTLES IN *DROSOPHILA MELANOGASTER*

The epidermis of flies such as *Drosophila melanogaster* is covered in sensory bristles that respond to sound waves, to air turbulence, and to contact. A small set of cells underneath the bristle, arranged in a stereotypical manner that is discussed in detail in Chapter 5.2, forms a sensory organ that relays information about bristle movement to the central nervous system. The bristle itself is a process from a single cell. Like the microvillus-derived structures of vertebrates, the bristle process is supported by cores of actin bundles as it is growing; in the case of bristles, there are 8 to 12 bundles of filaments,[26,27] each of which contains about 250[ref28] microfilaments in a hexagonal array.[29] The filament bundles are arranged around the circumference of the growing bristle and are associated with longitudinal grooves in the plasma membrane

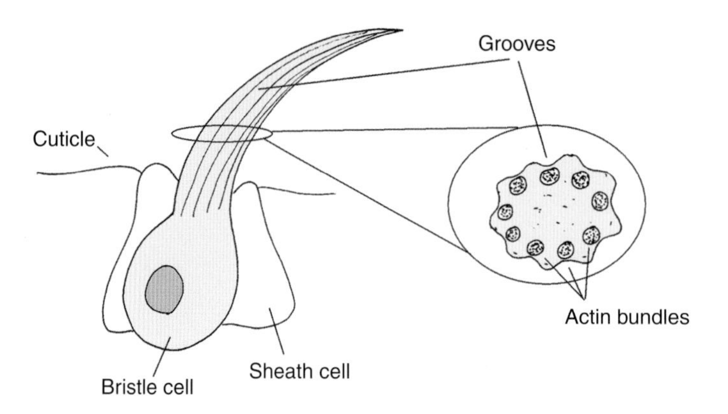

FIGURE 2.3.7 Overview of sensory bristle anatomy in *D. melanogaster*.

(see Figure 2.3.7). The growth of these filaments is not as simple as it is in microvilli, though, because the filament bundles of bristles are made by the concatenation of sub-assemblies rather than by the polymerization of continuous microfilaments.

The microfilament bundles that form the long cores of bristles originate in the cores of small microvilli. These emerge in high density on the surface of the bristle-forming cell shortly before bristle formation begins, and remain on the bristle tip throughout its growth ("remain" is used here in the sense of the population and does not imply that individual microvilli are stable). The microvilli form small bundles of 10 or so microfilaments that are cross-linked by proteins that have not yet been identified but that are known *not* to be the Forked and Fascin proteins that play later roles in bristle assembly.[30] There are about 50 of these small bundles, distributed fairly evenly around the circumference of the bristle tip.

Bundles that form in the microvilli, which are about 3 μm long, seem to be cast off from them so that they are free to move in the cytoplasm (the details of this process remain unclear but may proceed as in Figure 2.3.5). As is normally the case for microvilli, the microfilaments are arranged barbed-end-outwards. These small bundles of filaments are then cross-linked by the actin-binding protein Forked, which is expressed in the bristle-forming cells but not their neighbours.[30,31] The action of the Forked protein produces 8 to 12 thicker bundles, each of which contains about 50 microfilaments. Once this has taken place, another cross-linking protein, Fascin, is recruited to each of the bundles, and it causes their microfilaments to be packed in a regular hexagonal array[28] (see Figure 2.3.8). These hexagonally packed microfilament bundles are the sub-assemblies that are concatenated to form 60- to 70-μm bundles that run the length of the bristle. Newly formed sub-assemblies overlap the already-assembled filament and become cross-linked to it, again using the Forked protein. When sub-assemblies have only recently joined on to a filament, it is still possible to detect "knuckles" of overlap where the join is. Additional actin polymerization from the ends of the overlapping new sub-assembly toward the base of the bristle then smoothes out the "knuckles" and also has the effect of increasing the number of filaments in the bundle, so that each has about 500 filaments by the time the base of the bristle is reached (see Figure 2.3.8).

There are many questions still to be answered about the dynamics of actin during the formation of bristles. For example, the lengths of sub-assemblies vary but always

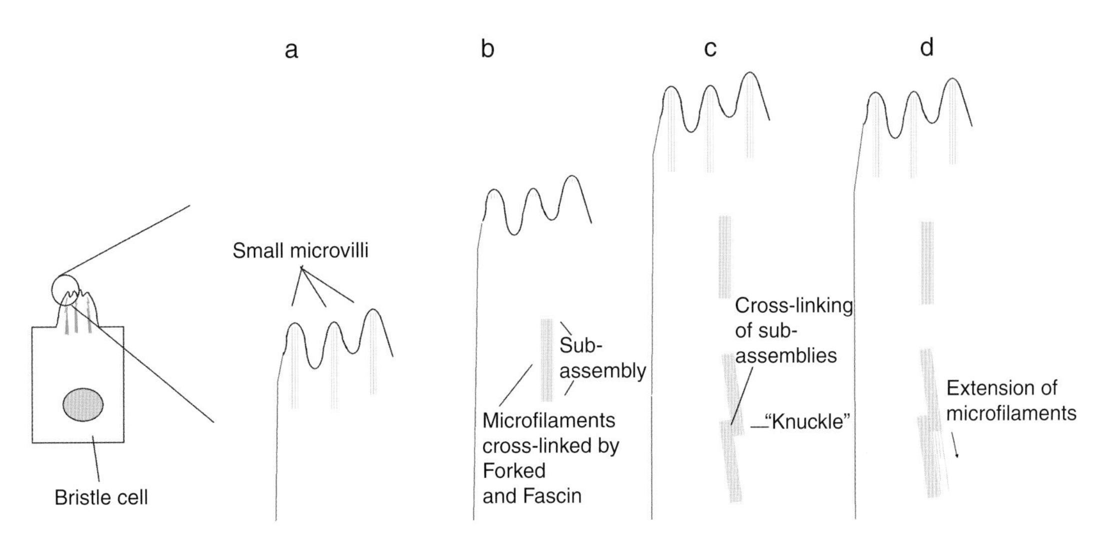

FIGURE 2.3.8 Bristle growth by concatenation of sub-assemblies. The low-magnification diagram on the left shows the area depicted by the high-magnification diagrams a–d. (a) Many small bundles of microfilaments are nucleated within small microvilli. (b) These are then cross-linked in two stages to make sub-assemblies of about 50 microfilaments in a hexagonal array. (c) New sub-assemblies are then cross-linked to those that have already formed. (d) Further growth of the microfilaments smoothes the "knuckles" left by joining of subassemblies and thickens the forming filament.

seem to be the same at a given level of bristle so that the joins in different filament bundles are in perfect register. How? And, for that matter, why? At the moment, it appears that the modular construction of microfilament cores is a feature of insect bristle development that is not seen in actin-based cell processes in vertebrates. It may be, however, that a concerted attempt to find it in vertebrates will detect concatenation there too and that the development of vertebrate microvillus derivatives described above will need to be revised.

MICROTUBULE-BASED STRUCTURES

As well as forming the compression structures of the cell's tensegrity system and providing transport routes for microtubule-associated vesicles, microtubules are also responsible for the construction of specific structures in the cell. Some, such as cilia and flagella, are very obvious but are of relatively modest importance to embryonic morphogenesis (although ciliary beat can be important in the transport of molecules involved in pattern formation, for example in the right/left axis system of mammals[32]). One structure that is very important to embryogenesis is the mitotic spindle, which separates chromosomes and allows cell division to take place and sets the spatial direction of division (see Chapter 5.1). This is one reason that it is worth considering the construction of the spindle in this chapter; the other is that the spindle illustrates very well the power and costs of morphogenesis based on adaptive self-organization.

The main purpose of the mitotic spindle is to provide a means of separating the sister chromatids of a chromosome. Before mitotic anaphase, when separation takes place, these sister chromatids are attached to each other with cohesin[33] so that they move together as a single unit, the mitotic chromosome. Each chromatid contains a centromere DNA sequence, and on this sequence is assembled a complex of proteins called the kinetochore. Between them, these kinetochore proteins are responsible for the mechanical attachment of the chromatid to microtubules and also for regulating the progression through the phases of mitosis.

The microtubules that construct the spindle are nucleated by the microtubule organizing centers (MTOCs) that are associated with the centrioles, which duplicate before mitosis and go to the opposite ends of the cell. Before mitosis begins, the mother and daughter centrioles are held together by the protein C-Map1, which binds and possibly cross-links the ends of centrioles. C-Map1 is phosphorylated by the kinase Nek2A, which is itself activated at the beginning of mitosis in response to cell cycle regulators, and this phosphorylation of C-Map1 causes it to stop holding the centrioles together so that they are free to separate.[34] The forces responsible for separating the centrioles probably arise from the microtubules that each centriole nucleates,[35] and there is evidence that some of these microtubules interact with, and are pulled by, the actin-based cytoskeleton in the cortex of the cell.[36] A simple model for centriole separation (probably too simple) combines traction toward the cell cortex, driven by microfilaments, with the mutual repulsion of the microtubules that emanate from each centriole and push on each other if they collide.

The tubules produced by the MTOCs associated with each centriole have to fulfil one of two possible functions: some need to locate and bind to chromosomes near the centre of the cell, whereas others need to project toward the cell cortex to make contact with the spindle-alignment machinery that is important in setting the direction of cell division (see Chapter 5.2). The microtubules that interact with the cortex are called "astral microtubules" (because, as a population, they look star-like as they radiate from the MTOC) and the others are called "spindle microtubules." They are probably drawn from the same initial population.

The task of constructing a spindle presents the cell with several problems (see Figure 2.3.9). First, the spindle microtubules have to make contact with the kinetochores of the chromosomes, each of which occupies less than 0.001 percent of the cell volume and does not occupy a fixed position. Second, the spindle microtubules need to gain a configuration in which sister chromatids are attached to microtubules that emanate from opposite poles of the spindle: their both being attached to the same pole would be useless. Third, the cell has to wait until all chromatids are held correctly, but not wait any longer. The solutions to all of these problems depend on the dynamic instability inherent in microtubule biochemistry (see Chapter 2.2).

The solution to the problem of finding kinetochores is perhaps the simplest. The MTOCs nucleate MTOCs in random directions, and these microtubules will extend and then collapse catastrophically when their terminal GTP-tubulin becomes GDP-tubulin. Only if they bind to microtubule-stabilizing proteins will they be saved from this collapse. Kinetochores contain multiple microtubule-stabilizing protein complexes that can help to stabilize the microtubules.[37] Therefore, the simple act of locating a kinetochore will select a microtubule, which set off randomly from the kinetochores, for survival, while its less fortunate colleagues will fail catastrophically (unless they happen to find another kinetochore or a cortical target that stabilizes them as astral tubules).

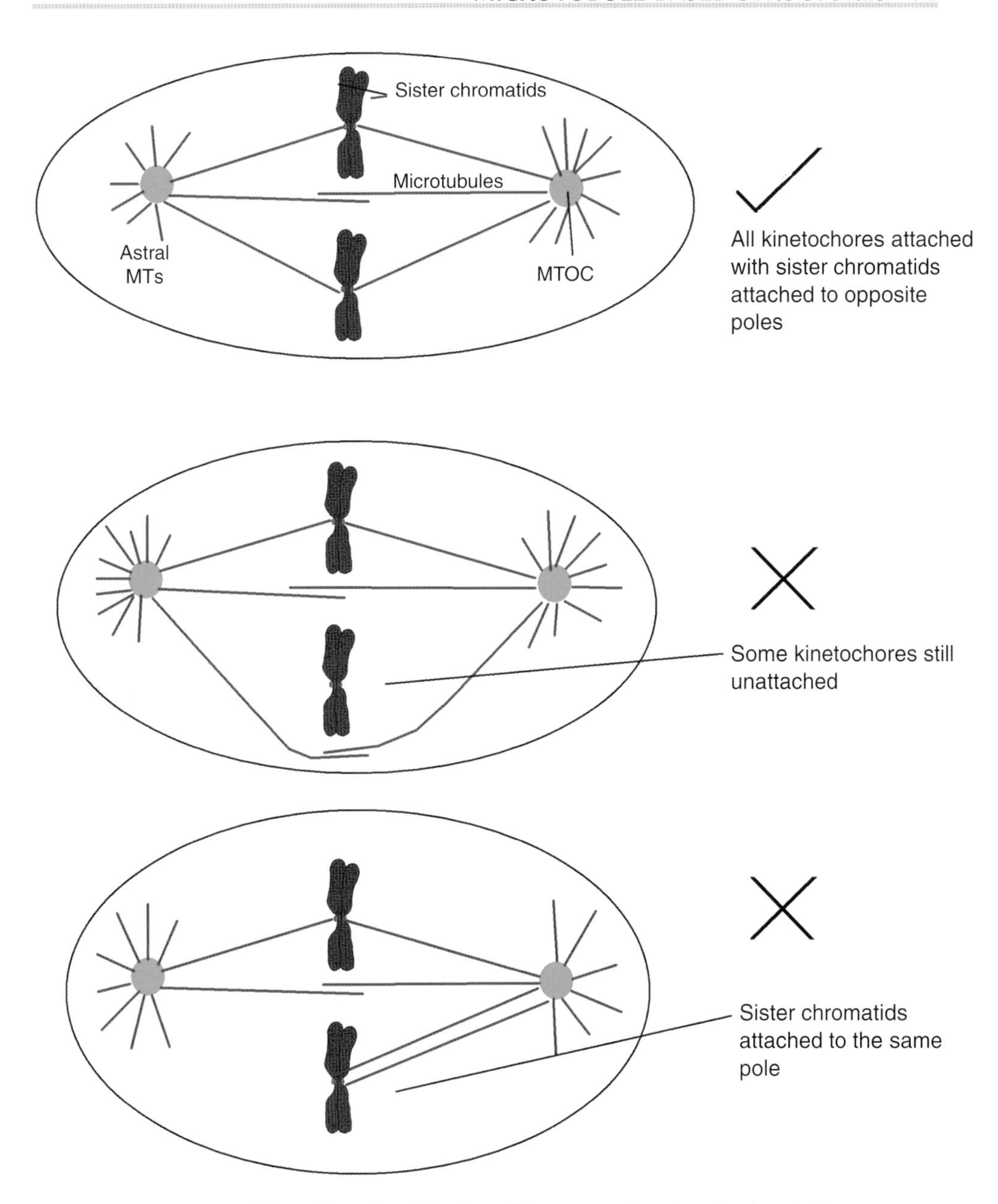

FIGURE 2.3.9 Potential problems faced by the cell in constructing the mitotic spindle.

Simply binding to a kinetochore is not itself enough to fully stabilize microtubules: if it were, there would be a serious risk of both sister chromatids being attached to the same pole of the cell. The tension that is generated by attachment of one chromatid to one pole and the other to the other pole is needed for microtubules to be properly stabilized (tension is generated by microtubule motors). This can be demonstrated using the meiotic spindles of the male mantis *Mantis religiosa*. In this species, males have the chromosome constitution XXY and, during meiosis, a trivalent complex of these chromosomes forms on the spindle equator. Normally, the two X chromosomes proceed together to one daughter cell and the Y to the other, but the process is error-prone and sometimes the cell forms an XY bivalent and leaves the second X chromosome unpaired. Attachment of this chromosome to the microtubules coming from one pole is not sufficient to lift the block on progress, but if an experimenter pulls on the chromosome with a micropipette to simulate the tension that would have been generated if it were in a trivalent, the block on progress is lifted and cell division continues.[38] Tension between sister chromatids seems to be sensed by Aurora B, a protein kinase located at kinetochores that can destabilize incorrectly located microtubules, probably by activating centromere-associated kinesin that has microtubule-destabilizing activity.[39]

The solution to detecting when all chromosomes have been bound is solved not by a counting process (which, apart from being difficult to implement, would run into problems as chromosome number changed over evolutionary time) but by the silencing of "not bound yet" signals that emanate from kinetochores that have yet to be captured. Yeast kinetochores contain a complex of the proteins BubR1, Bub1p, and Mad1. This complex activates the protein Mad2, which inhibits a cell cycle protein called cdc20, activity of which is vital for progress to anaphase and completion of mitosis. As long as Mad2 is activated, anaphase cannot be entered. The BubR1/Bub1p/Mad1 signalling complex is silenced only when each microtubule-binding site on the kinetochore is bound by microtubules that are under tension.

The problems of making a spindle are therefore solved by the MTOCs generating random microtubule architectures, and the targets (kinetochores) selecting only those microtubules that happen to connect with them correctly for stabilization. This illustrates the power of adaptive self-organization very well, and also illustrates its cost: every time tubulin from collapsed microtubules is made available for making new tubules, the cell has to pay the price of making a GTP from GDP. The system is highly flexible, but very expensive in terms of energy.

MICROTUBULES AND ELONGATION OF CELLS

Elongation of cells is a common feature of animal development. It is seen, for example, in ectoderm epithelium that is about to undergo neural differentiation; the epithelium typically switches morphology from having approximately cubical cells to having tall, thin ones. This is sometimes called "palisading."[40]♣

The lens of the vertebrate eye forms from the surface ectoderm epithelium that overlies the eye itself. In response to inductive signals emanating from the optic vesicle,[41] which were detected indirectly more than a century ago and were the first inductive signals ever to be described,[42,43] that part of the ectoderm begins a program of differentiation and morphogenesis to form a lens. The earliest morphological event is cell elongation (see Figure 2.3.10), which is associated with the construction of a large number of microtubule bundles parallel to what will be the long axis of the cell.[44] These microtubules all have the same polarity, the "−" end being toward the front of the lens and the "+" end being toward the back.[45] These microtubules are therefore probably nucleated at one end of the cell rather than being part of the system that centres on

♣This word is often spelled, wrongly, with a double "l"; it is worth bearing this in mind when performing literature searches.

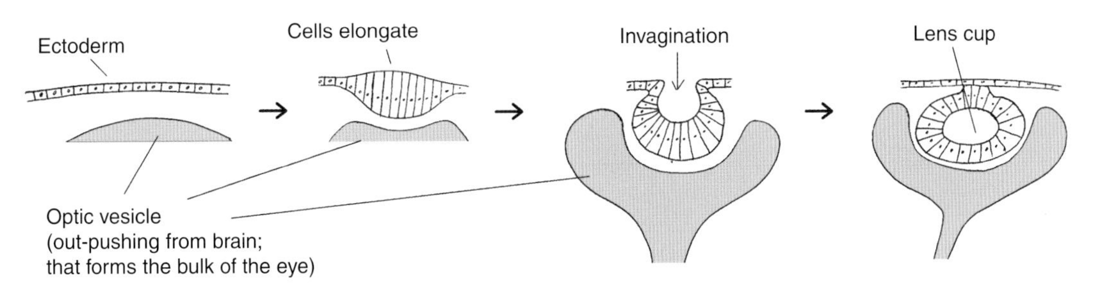

FIGURE 2.3.10 The main morphogenetic movements during early development of the vertebrate lens. The elongation of cells, depicted in the second stage, is the most relevant stage to this discussion.

the centrioles. Lens epithelia will elongate normally in organ culture, but if cultures are treated with the microtubule-depolymerizing drug colchicine, cell elongation fails.[46] These observations strongly suggest that elongation of microtubules may be the driving force for elongation of the entire cell. This could either be by direct mechanical means, their growth literally pushing the cell to be longer, or it could be by targeting microtubule-associated transport vesicles to the ends of the cell rather than to the lateral domains. In support of this latter possibility, transport vesicles have been observed in close association with microtubule bundles in elongating lens cells.[45]

More recent work has, however, cast doubt on the hypothesis that microtubules are the cause of cell elongation. Treating lens cells with nocadazole, which also depolymerizes microtubules, does not cause the failure of elongation that is observed when cells are treated with colchicine: it is therefore possible that, in this system, colchicine acts on a completely different cellular target.[47] One possible target is ion transport, which is likely to be important in the control of cell volume, and which has been shown to be affected by colchicine in this system.[48]

Cell elongation is also an early event in the development of the neural tube, which, like the lens, arises from ectoderm and undergoes cell elongation followed by invagination (see Chapter 4.4). Bundled longitudinal microtubules are also a feature of elongating ectoderm as it begins to form a neural tube: depolymerizing these causes a reduction in elongation but does not return cells to isotropy, again suggesting that the physical presence of the microtubules cannot be the only thing maintaining the shapes of the cells.[49]

There are, of course, many other examples of morphogenesis within animal cells. The ones discussed in this chapter have been chosen for their ability to illustrate basic principles and for their relevance to the embryo-scale morphogenetic events described in the following chapters.

Reference List

1. Inoue, S and Tilney, LG (1982) Acrosomal reaction of Thyone sperm. I. Changes in the sperm head visualized by high resolution video microscopy. *J. Cell Biol.* **93**: 812–819
2. Tilney, LG and Inoue, S (1982) Acrosomal reaction of Thyone sperm. II. The kinetics and possible mechanism of acrosomal process elongation. *J. Cell Biol.* **93**: 820–827
3. Tilney, LG (1978) Polymerization of actin. V. A new organelle, the actomere, that initates the assembly of actin filaments in Thyone sperm. *J. Cell Biol.* **77**: 851–864

4. Tilney, LG, Bonder, EM, Coluccio, LM, and Mooseker, MS (1983) Actin from Thyone sperm assembles on only one end of an actin filament: A behavior regulated by profilin. *J. Cell Biol.* **97**: 112–124
5. Olbris, DJ and Herzfeld, J (1999) An analysis of actin delivery in the acrosomal process of thyone. *Biophys. J.* **77**: 3407–3423
6. Peskin, CS, Odell, GM, and Oster, GF (1993) Cellular motions and thermal fluctuations: The Brownian ratchet. *Biophys. J.* **65**: 316–324
7. Sherman, MB, Jakana, J, Sun, S, Matsudaira, P, Chiu, W, and Schmid, MF (19-11-1999) The three-dimensional structure of the Limulus acrosomal process: A dynamic actin bundle. *J. Mol. Biol.* **294**: 139–149
8. Gorelik, J, Shevchuk, AI, Frolenkov, GI, Diakonov, IA, Lab, MJ, Kros, CJ, Richardson, GP, Vodyanoy, I, Edwards, CR, Klenerman, D, and Korchev, YE (13-5-2003) Dynamic assembly of surface structures in living cells. *Proc. Natl. Acad. Sci. U.S.A.* **100**: 5819–5822
9. DeRosier, DJ and Tilney, LG (10-1-2000) F-actin bundles are derivatives of microvilli: What does this tell us about how bundles might form? *J. Cell Biol.* **148**: 1–6
10. Coluccio, LM and Bretscher, A (18-12-1990) Mapping of the microvillar 110K-calmodulin complex (brush border myosin I). Identification of fragments containing the catalytic and F-actin-binding sites and demonstration of a calcium ion dependent conformational change. *Biochemistry.* **29**: 11089–11094
11. Coluccio, LM and Bretscher, A (1989) Reassociation of microvillar core proteins: Making a microvillar core in vitro. *J. Cell Biol.* **108**: 495–502
12. Friederich, E, Kreis, TE, and Louvard, D (1993) Villin-induced growth of microvilli is reversibly inhibited by cytochalasin D. *J. Cell Sci.* **105 (Pt 3)**: 765–775
13. Bretscher, A and Weber, K (1980) Villin is a major protein of the microvillus cytoskeleton which binds both G and F actin in a calcium-dependent manner. *Cell.* **20**: 839–847
14. Craig, SW and Powell, LD (1980) Regulation of actin polymerization by villin, a 95,000 dalton cytoskeletal component of intestinal brush borders. *Cell.* **22**: 739–746
15. Glenney, JR, Jr, Kaulfus, P, and Weber, K (1981) F actin assembly modulated by villin: Ca^{++}-dependent nucleation and capping of the barbed end. *Cell.* **24**: 471–480
16. Friederich, E, Vancompernolle, K, Louvard, D, and Vandekerckhove, J (17-9-1999) Villin function in the organization of the actin cytoskeleton. Correlation of in vivo effects to its biochemical activities in vitro. *J. Biol. Chem.* **274**: 26751–26760
17. Pinson, KI, Dunbar, L, Samuelson, L, and Gumucio, DL (1998) Targeted disruption of the mouse villin gene does not impair the morphogenesis of microvilli. *Dev. Dyn.* **211**: 109–121
18. Tyska, MJ and Mooseker, MS (2002) MYO1A (brush border myosin I) dynamics in the brush border of LLC-PK1-CL4 cells. *Biophys. J.* **82**: 1869–1883
19. Pickles, JO and Corey, DP (1992) Mechanoelectrical transduction by hair cells. *Trends Neurosci.* **15**: 254–259
20. Kaltenbach, JA, Falzarano, PR, and Simpson, TH (8-12-1994) Postnatal development of the hamster cochlea. II. Growth and differentiation of stereocilia bundles. *J. Comp. Neurol.* **350**: 187–198
21. Daudet, N and Lebart, MC (2002) Transient expression of the t-isoform of plastins/fimbrin in the stereocilia of developing auditory hair cells. *Cell Motil. Cytoskeleton.* **53**: 326–336
22. Tilney, MS, Tilney, LG, Stephens, RE, Merte, C, Drenckhahn, D, Cotanche, DA, and Bretscher, A (1989) Preliminary biochemical characterization of the stereocilia and cuticular plate of hair cells of the chick cochlea. *J. Cell Biol.* **109**: 1711–1723
23. Zheng, L, Sekerkova, G, Vranich, K, Tilney, LG, Mugnaini, E, and Bartles, JR (4-8-2000) The deaf jerker mouse has a mutation in the gene encoding the espin actin-bundling proteins of hair cell stereocilia and lacks espins. *Cell.* **102**: 377–385
24. Loomis, PA, Zheng, L, Sekerkova, G, Changyaleket, B, Mugnaini, E, and Bartles, JR (8-12-2003) Espin cross-links cause the elongation of microvillus-type parallel actin bundles in vivo. *J. Cell Biol.* **163**: 1045–1055
25. Schneider, ME, Belyantseva, IA, Azevedo, RB, and Kachar, B (22-8-2002) Rapid renewal of auditory hair bundles. *Nature.* **418**: 837–838

26. Overton, J (1967) The fine structure of developing bristles in wild type and mutant *Drosophila melanogaster*. *J. Morphol.* **122**: 367–379
27. Appel, LF, Prout, M, Abu-Shumays, R, Hammonds, A, Garbe, JC, Fristrom, D, and Fristrom, J (1-6-1993) The *Drosophila* stubble-stubbloid gene encodes an apparent transmembrane serine protease required for epithelial morphogenesis. *Proc. Natl. Acad. Sci. U.S.A.* **90**: 4937–4941
28. Tilney, LG, Connelly, P, Smith, S, and Guild, GM (1996) F-actin bundles in *Drosophila* bristles are assembled from modules composed of short filaments. *J. Cell Biol.* **135**: 1291–1308
29. Tilney, LG, Tilney, MS, and Guild, GM (1995) F actin bundles in *Drosophila* bristles. I. Two filament cross-links are involved in bundling. *J. Cell Biol.* **130**: 629–638
30. Tilney, LG, Connelly, PS, and Guild, GM (15-7-2004) Microvilli appear to represent the first step in actin bundle formation in *Drosophila* bristles. *J. Cell Sci.* **117**: 3531–3538
31. Tilney, LG, Connelly, PS, Vranich, KA, Shaw, MK, and Guild, GM (5-10-1998) Why are two different cross-linkers necessary for actin bundle formation in vivo and what does each cross-link contribute? *J. Cell Biol.* **143**: 121–133
32. Levin, M (2004) The embryonic origins of left-right asymmetry. *Crit. Rev. Oral. Biol. Med.* **15**: 197–206
33. Uhlmann, F (15-5-2004) The mechanism of sister chromatid cohesion. *Exp. Cell Res.* **296**: 80–85
34. Faragher, AJ and Fry, AM (2003) Nek2A kinase stimulates centrosome disjunction and is required for formation of bipolar mitotic spindles. *Mol. Biol. Cell.* **14**: 2876–2889
35. Waters, JC, Cole, RW, and Rieder, CL (1993) The force-producing mechanism for centrosome separation during spindle formation in vertebrates is intrinsic to each aster. *J. Cell Biol.* **122**: 361–372
36. Buendia, B, Bre, MH, Griffiths, G, and Karsenti, E (1990) Cytoskeletal control of centrioles movement during the establishment of polarity in Madin-Darby canine kidney cells. *J. Cell Biol.* **110**: 1123–1135
37. Yasuda, S, Oceguera-Yanez, F, Kato, T, Okamoto, M, Yonemura, S, Terada, Y, Ishizaki, T, and Narumiya, S (15-4-2004) Cdc42 and mDia3 regulate microtubule attachment to kinetochores. *Nature.* **428**: 767–771
38. Li, X and Nicklas, RB (16-2-1995) Mitotic forces control a cell-cycle checkpoint. *Nature.* **373**: 630–632
39. Lampson, MA, Renduchitala, K, Khodjakov, A, and Kapoor, TM (2004) Correcting improper chromosome-spindle attachments during cell division. *Nat. Cell Biol.* **6**: 232–237
40. Schoenwolf, GC (1991) Cell movements driving neurulation in avian embryos. *Development.* **Suppl 2**: 157–168
41. Baker, CV and Bronner-Fraser, M (1-4-2001) Vertebrate cranial placodes I. Embryonic induction. *Dev. Biol.* **232**: 1–61
42. Spemann, H (1901) Uber Correlationen in der Entwickelung des Auges. *Verh. Anat. Ges.* **15**: 61–79
43. Mencl, E (1903) Ein fall von beiderseitiger augenlinsenausbildung wahrend der abwesenheit von augenblasen. *Arch. Entwicklungsmechanik* **16**: 328–339
44. Piatigorsky, J, Webster, HD, and Craig, SP (1972) Protein synthesis and ultrastructure during the formation of embryonic chick lens fibers in vivo and in vitro. *Dev. Biol.* **27**: 176–189
45. Lo, WK, Wen, XJ, and Zhou, CJ (2003) Microtubule configuration and membranous vesicle transport in elongating fiber cells of the rat lens. *Exp. Eye Res.* **77**: 615–626
46. Piatigorsky, J, Webster, Hde F, and Wollberg, M (1972) Cell elongation in the cultured embryonic chick lens epithelium with and without protein synthesis. Involvement of microtubules. *J. Cell Biol.* **55**: 82–92
47. Beebe, DC, Feagans, DE, Blanchette-Mackie, EJ, and Nau, ME (16-11-1979) Lens epithelial cell elongation in the absence of microtubules: Evidence for a new effect of colchicine. *Science.* **206**: 836–838

48. Beebe, DC and Cerrelli, S (1989) Cytochalasin prevents cell elongation and increases potassium efflux from embryonic lens epithelial cells: Implications for the mechanism of lens fiber cell elongation. *Lens Eye Toxic. Res.* **6**: 589–601
49. Schoenwolf, GC and Powers, ML (1987) Shaping of the chick neuroepithelium during primary and secondary neurulation: Role of cell elongation. *Anat. Rec.* **218**: 182–195

CHAPTER 2.4

CELLULAR MORPHOGENESIS IN PLANTS

The development of multicellular animals makes much use of animal cells' ability to move and to exchange neighbours. Plant cells cannot move, so a change in the shape of a plant tissue bears a very direct relationship to changes in the shapes of its constituent cells (see Chapter 2.1). Serious technical problems inherent in plant cell biology, chiefly problems of access, mean that the study of plant cell morphogenesis has lagged behind that of animal cells, although it is now catching up very quickly. It is therefore not yet possible to describe morphogenetic systems in the same detail that is available for animal cells, and because of this it is more difficult to step back from the details to recognize emerging principles. Nevertheless, the importance of cell morphogenesis to plant development makes it worth devoting a chapter of this book to some of the systems that have been studied most closely. They reveal some surprising similarities between plant and animal morphogenesis but also stress some important differences.

Plant cells differ from those of animals in many respects, but the most important difference from the point of view of morphogenesis is that a plant cell is surrounded by a rigid cell wall (see Figure 2.4.1). This wall is made mainly from cellulose, with hemicellulose and pectin also being present in significant quantities. Cellulose, a polymer of glucose containing 500 to 15,000 glucose monomers, forms quasi-crystalline bundles of about 70 cellulose chains to produce cellulose microfibrils. These tend to be laid down in flat, parallel arrays, in which microfibrils are cross-linked by branched molecules of hemicellulose. The flat arrays of cellulose microfibrils are laid on top of each other at different angles, and hemicellulose and pectins are used to form cross-links between as well as within the layers. The angles between different layers give the whole wall an ability to resist forces from all directions. The wall is so good at resisting forces that it can contain cells that are under positive pressure due to osmotic swelling of the cell; this pressure can reach up to 10 bars.[1] In plant cells, therefore, the main tension-bearing elements are external.

The shapes of plant cells are determined by the shapes and mechanical properties of their cell walls. During the period of plant cell morphogenesis, the cell is surrounded only by its primary cell wall; later on, a thick secondary cell wall may be added on the inside for strength, but the secondary cell wall is largely beyond the scope of this chapter.

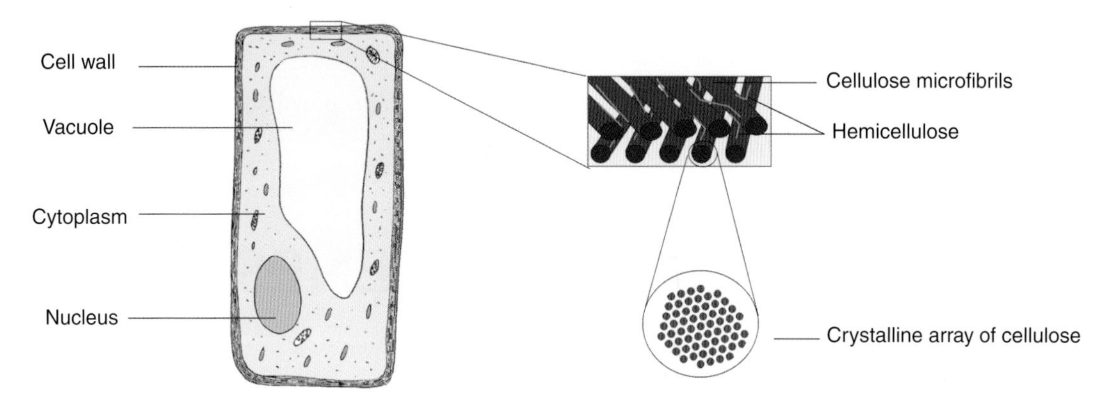

FIGURE 2.4.1 The structure of a generalized plant cell and the main components of its cell wall.

Therefore, wherever this discussion mentions a cell wall, the primary cell wall should be assumed.

Plant cells synthesize cellulose directly at the surface of their plasma membranes, rather than in the endoplasmic reticulum/golgi system where protein-based extracellular matrix molecules are made. New cellulose is made from cytoplasmic UDP-glucose by rosette-like complexes of cellulose synthase.[2] As soon as new cellulose polymers are synthesized and leave the rosettes, they associate with one another to form microfibrils and take their place in the microfibril array. Because the already-synthesized section of a cellulose polymer rapidly associates with the other molecules of the cell wall, parallel to them, the continuing polymerization of the cellulose tends to push the cellulose synthase complex along the membrane, parallel to the microfibrils in the layer of the wall nearest the plasma membrane unless any other influences intervene.

DIFFUSE CELL ELONGATION IN PLANTS

Much of the elongation of plant tissues is driven by diffuse elongation of their constituent cells, a process that takes place after mitotic proliferation has ended. By definition, diffuse elongation takes place throughout the cell and is achieved by allowing cell growth more in one axis than in the others. This distributed pattern of growth contrasts with focused elongation, described later in this chapter, in which elongation takes place only at one part of a cell.

The force that drives any expansion of plant cells, whether diffuse or focused, is mainly osmotic turgor pressure. Because the inside of the cell is liquid, turgor pressure pushes equally in all directions so, for plant cells to expand more in one axis than others, cell walls have to yield to the internal pressure more easily in some directions than in others. A sheet composed of parallel arrays of filaments, like the microfilaments of the cell wall, will usually show most resistance to stretching along the axis of the filaments and least resistance orthogonal to that axis.[3] Therefore, if the direction in which cellulose microfilaments are laid down is biased during the growth of a cell wall, the direction in which the internal pressure of the cell can force the wall to expand will

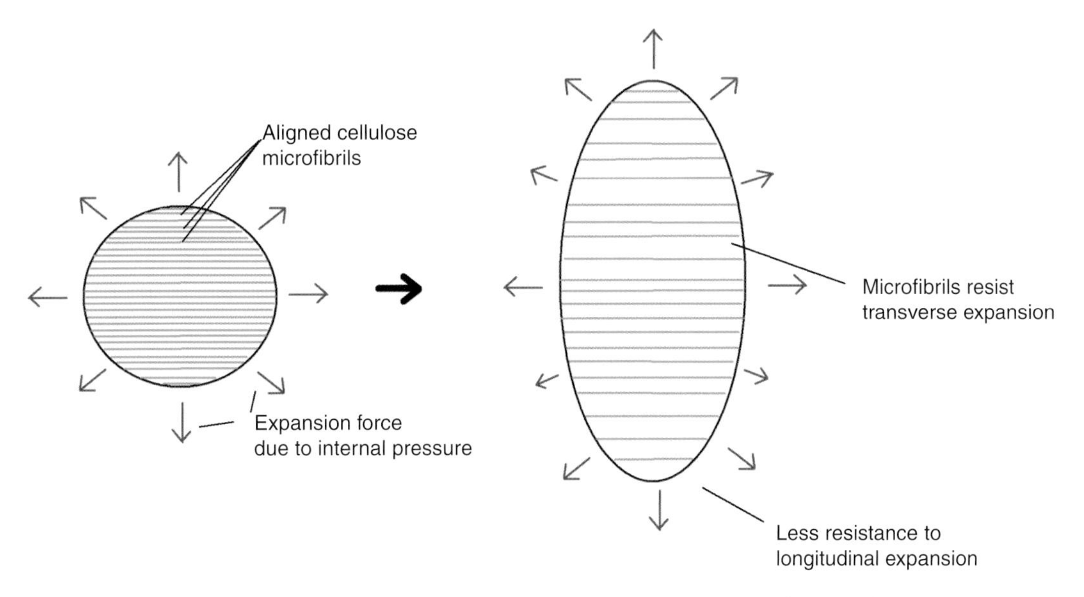

FIGURE 2.4.2 A bias in the direction of tension-bearing microfibrils causes directional expansion in response to internal pressure. In the interests of simplification, this diagram shows all fibrils orientated one way (in practice, the bias will not be absolute, lest the cell burst) and depicts the cell as round (most plant cells are cuboidal, as depicted in the rest of the diagrams in this chapter).

be set orthogonally to the microfilaments (see Figure 2.4.2).♣ Detailed examination of the alignment of cellulose microfibrils in actively elongating plant cells has confirmed that they are indeed aligned predominantly at 90 degrees to the axis of expansion.[4,5] If cellulose walls are taken away, expansion is isotropic.[6] According to one popular model, the problem of setting the direction of plant cell elongation therefore reduces to that of setting the direction of cellulose microfibrils, which in turn reduces to that of setting the direction in which the cellulose synthase complex moves through the plasma membrane.

In cells that are undergoing elongation, there is a clear anatomical correlation between the direction in which cellulose microfibrils are laid down outside the plasma membrane and the direction in which microtubules run just inside the plasma membrane.[7] The correlation can also be demonstrated functionally, by mutations in the angiosperm *Arabodopsis thaliana*. The *fass* mutation results in severe disruption to the organization of the cortical microtubule cytoskeleton[8] and is associated with defective in cell elongation. Similarly, the *spiral1* mutation causes microtubules to line up obliquely rather than transversely across the cell and creates a matching defect in elongation.[9] Unfortunately, the molecular basis of neither of these mutations is understood, so that the correlation between orientation of microtubules and cellulose fibrils remains just a correlation, with no indication of the direction of cause and effect. Oblique alignment of microfibrils may, however, be a "default" pattern when guidance fails: it is seen when

♣A quick classroom demonstration of this can be made by partially inflating a balloon, wrapping parallel lengths of sticky tape around it like lines of latitude on a globe to represent filaments, then inflating the balloon more. Its expansion will be anisotropic, at least until the extreme stresses at the tape edges cause it to burst (this end-point will, at least, wake the sleepy souls at the back of the room).

elongation ceases in normal development[7] as well as when mutations or drugs damage the microtubule system,[9] and once normal cells produce oblique microfibrils, which happens when elongation has ceased and cell walls are thickening, the relationship between the orientation of microfibrils and that of microtubules seems to be lost.[9,10]

There is both morphological and biochemical evidence that microtubules determine the orientation of cellulose microfibrils. The morphological evidence is that, in at least some plant cells, microtubules are aligned more precisely with each other than are microfilbrils.[10] This would be difficult to explain if microfibrils were the source of alignment information. The biochemical evidence comes mainly from experiments in which microtubule-depolymerizing drugs are used to perturb the microtubule systems of plant cells, and their effect on cell elongation and cellulose alignment is studied. In some plant cells, treatment with colchicines causes the alignment of microfibrils to become random, as would be predicted from the microtubules-align-microfibrils model.[11] In other systems, local alignment remains across areas of cells but the global alignment across cells and tissues is lost.[12] If *A. thaliana* embryos are grown in agar that contains the microtubule-depolymerizing drug propyzamide, the elongation of their cells becomes disregulated so that they grow in a spiral rather than straight manner.[9] This supports the idea that the default orientation of microfibrils is spiral, and that microtubules are normally used to steer the system away from its default. The microtubule-depolymerizing drug olyzalin has a similar effect on microfibril orientation in growing roots.[13]

If the orientation of microtubules does determine the orientation of cellulose microfibrils, the connection between the two may be explained by a simple model that uses the fact that cellulose synthase molecules have to move as they do their work. Parallel arrays of microtubules just under the membrane will impede the passage of a cellulose synthase complex across them but will not impede its passage parallel to them (see Figure 2.4.3). With all of the cellulose synthase enzymes constrained to travel in the same direction, the cellulose microfibrils will have to be laid down in this direction, and the required anisotropy will be built into the cell wall.[14]

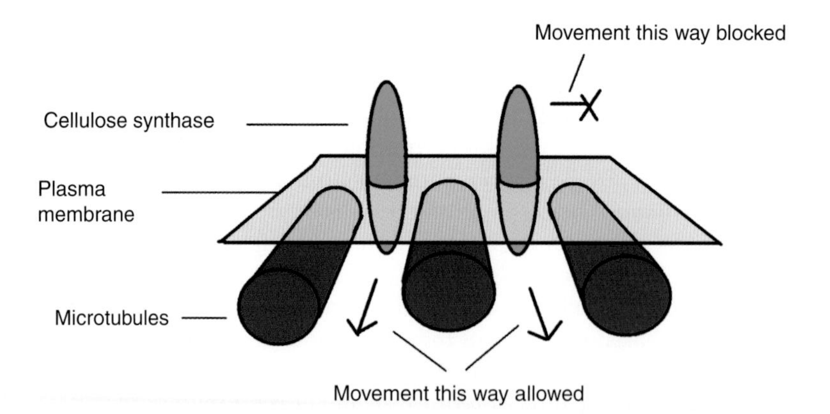

FIGURE 2.4.3 A mechanical model for the guidance of cellulose synthase complexes by cortical microtubules. The movement of the synthase complexes will determine the orientation of the cellulose polymers they leave behind. This model depicts a passive mechanical constraint: it is possible that the synthase complexes are linked to microtubules via kinesin-type motors and are therefore guided along them.

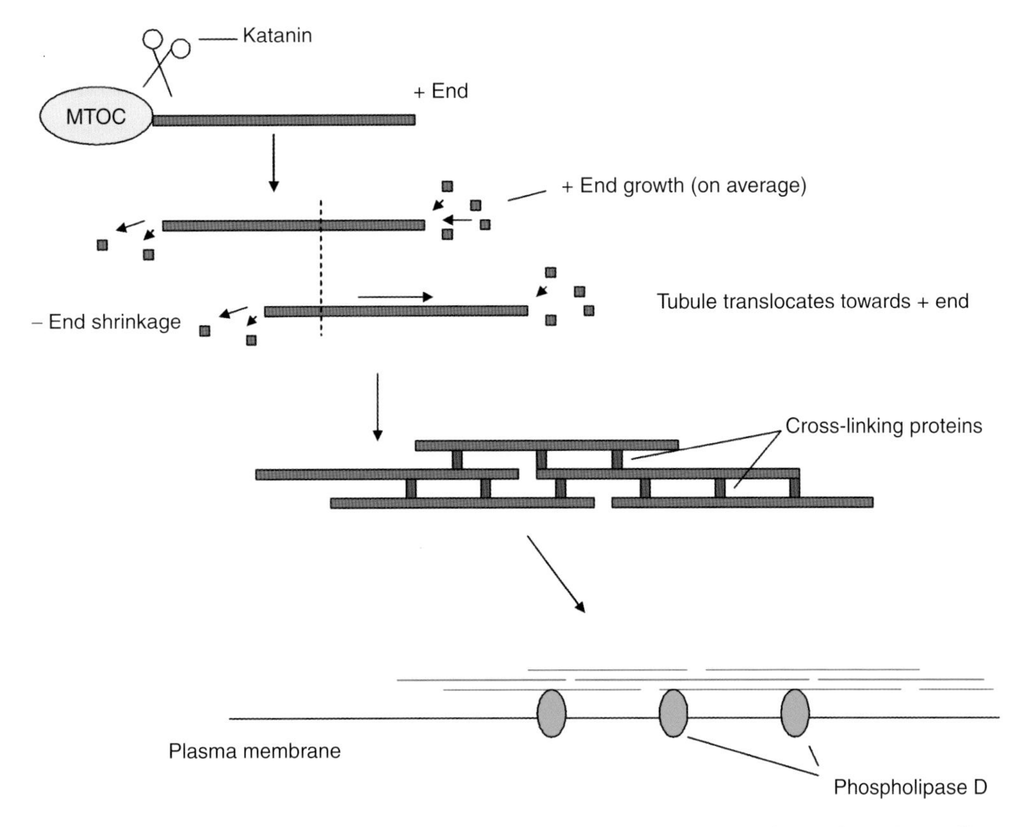

FIGURE 2.4.4 The processes involved in production of cortical microtubules (red) in plant cells. Nascent microtubules are cut off from their organizing Centro and undergo net depolymerization at their "–" ends and net polymerization (with occasional catastrophic collapse) at their "+" ends, so that they move away from the place of their birth. They become cross-linked to each other by microtubule-associated proteins and cross-linked to the membrane by phospholipase D.

Little is known about precisely how microtubules are patterned according to the general growth axis of the tissue. Microtubules seem to be nucleated in the cell cortex itself[15] and are later cut free from their nucleation sites by the protein Katanin. They then show the "+" end dynamic instability typical of microtubules in animal cells (see Chapter 2.2), and they also show a slow shortening from their "−" ends. Their centre of mass therefore translocates, on average, toward the "+" direction, and they can therefore disperse throughout the cortex[16] (see Figure 2.4.4). When they meet, they can be cross-linked into bundles by a variety of microtubule-binding proteins such as MAP65 and MOR1.[15] Microtubules are also linked to the membrane itself, and the most likely candidate for the linker is, surprisingly, the "signalling protein" phospholipase D.[17] Activating this enzyme pharmacologically, by adding *n*-butanol to cells, causes cortical microtubules to be released from the plasma membrane over the course of minutes.[18] Inhibiting the enzyme with 1-butanol, on the other hand, disrupts the organization of cortical microtubules so that they have oblique and longitudinal orientation as well as transverse, in zones of the plant where elongation is meant to take place.[19] This disorganizes the morphology of the resultant plant.

Linkage via a signalling protein makes functional sense because the alignment of cellulose microfibrils is known to be responsive to hormonal cues. For example, the growth-associated hormone giberellic acid, which causes elongation of stems,[20] induces microtubules to become transverse and so to cause cell elongation within the stem.[21] The wound-associated Ethylene, on the other hand, causes microtubules to shift from a transverse to a longitudinal orientation.[22,23] This presumably allows cells to expand sideways to re-fill the wound space. Cessation of cell elongation caused by reduction of turgor induced experimentally by a raise in environmental concentration of KCl or sorbitol also causes a reorientation of microtubules from transverse to longitudinal, although in this case reorientation follows cessation of growth and not the other way around.[24] During these re-orientations, cells are never without microtubules, suggesting that the reorientation takes place gradually, rather than by scrapping the existing cytoskeleton and building a new one *de novo*. Detailed confocal observations have been made of plant cells that are undergoing a transition from transverse to longitudinal orientation and have been injected with fluorescent tubulin. These show that, while some destruction of microtubules does take place, many existing tubules survive to be re-orientated to line up with new microtubules that have formed in the new orientation.[25]

Reorientation of existing microtubules would require their release from the plasma membrane, and this presumably requires the activation of phospholipase D.[18] Failure to release microtubules, as when plants are treated with the phospholipase D inhibitor 1-butanol, would prevent inappropriately aligned microtubules from correcting themselves, and would therefore account for the disorganized cytoskeleton in such plants.[19] If phospholipase D is controlled by plant hormones, as it is by growth factor pathways in animals,[26] then one of the actions of the hormones may be to provide a controlled release of microtubules from the cortex to allow their repositioning. What is missing from this account is the mechanism that actually performs reorientation. If microtubules with the "new" orientation were somehow fixed firmly in the cell, then association with them, via cross-linking proteins, may be enough to re-orientate released microtubules that had the old orientation. This mechanism cannot work if all microtubules are equal, though, because the more common existing ones would be expected to pull the new microtubules to the old orientation, the opposite effect from the one required. An alternative possibility is that a motor system is aligned with respect to the axis of the cell and pulls any free microtubules, whether new or old, into the new desired orientation. The problem, in that case, is to find that motor system and to understand how it comes to be aligned.

Microtubules are often found, especially when they can be imaged clearly, to be arranged obliquely in resting plant cells so that they form a helix.[27] It may be that cortical microtubules (and the microfibrils above them) are in fact always arranged helically in interphase plant cells, but that the pitch of the helix can vary from very small, making the microtubules seem transverse, to very large, making them seem longitudinal. In this case, the mechanism for changing between "transverse" to "oblique" or "longitudinal" orientation would be to alter the natural pitch of the spiral, presumably through altering microtubule-associated proteins. Such a mechanism would have the aesthetic advantage that it is relatively simple, global, and would scale with any size of cell, but it is so far unsupported by biochemical evidence.

The microtubule-led model for control of cell elongation has gained widespread acceptance in textbooks, but there is a growing body of evidence to suggest that the real mechanism may not be so simple. The construction of cell walls where there were none before can be studied using protoplasts, which are cultured plant cells from which existing cell walls have been removed enzymatically. Protoplasts of *Nicotiana tabacum* will regenerate their walls and, having done so, their cells elongate normally. If construction of the cortical microtubule system lies upstream of the alignment of cellulose microfibrils in a hierarchy of spatial patterning processes, the microtubule system in regenerating protoplasts ought to be normal regardless of whether cellulose biosynthesis is possible. Cellulose biosynthesis can be inhibited using the herbicide Isoxaben. As expected, treatment of cells with Isoxaben causes cells that ought to elongate in one direction to expand in all directions instead.[28] Protoplasts regenerating in the presence of Isoxaben do construct a cortical microtubule system, but it is ordered randomly and never acquires the ordered orientation it acquires in control cells.[29] This effect does not depend on starting from protoplasts, for even if cells that still have their cell walls are treated with Isoxaben, microtubule order is lost over the course of a few days. Unless Isoxaben has an additional, uncharacterized effect beyond that on cellulose synthesis—and attempts to find such an effect have so far failed—we are forced to conclude that cellulose microfibrils determine the direction of the microtubules. If the original idea that microtubules determine the direction of cellulose microfibrils is to be retained, then there must at the very least be a powerful feedback system in which each component shapes the other.

Detailed observations on the rates of elongation and radial expansion of cells and on the orientations of microtubules and microfibrils as cells move through the growth zone of the maize root suggest another problem: although the alignment of the fibres is compatible with the model, changes in the alignment lag behind changes in the rate of elongation.[10] This observation casts doubt on the whole idea that changes in the direction of microfibril deposition regulate elongation. It may be that other changes in the cell wall, connected with the degree and strength of cross-linking, are the real determinants of the direction in which the wall will yield to turgor pressure. Some secreted enzymes, such as expansins, promote the creep between components of the cell wall and therefore allow the wall to stretch in response to applied force.[30] It may be that tightly regulated expression or secretion of these enzymes plays a major role in controlling the rates of cell expansion.

FOCUSED CELL GROWTH: Root Hairs, Pollen Tubes, and Trichomes

In focused cell growth, expansion takes place from only one surface of the cell, and the other walls remain as they are. This mode of elongation seems not to be used to drive the expansion of whole tissues, at least not in angiosperms and gymnosperms.♣ It is used instead to produce long, thin cell processes such as root hairs that increase the surface area available for absorption of nutrients from the soil. These processes grow from their tips, in a manner morphologically analogous to the growth of neuronal processes in animals (see Chapter 3.2). In an expanding root hair, the microtubule cytoskeleton is aligned approximately longitudinally with respect to the axis of elongation,[31] an

♣Angiosperms are flowering plants and gymnosperms are coniferous plants and their relatives: together they constitute the "higher" plants.

orientation that emphasizes that focused growth is different from diffuse elongation. The microtubule cytoskeleton seems to play a comparatively minor role in the growth of root hair tips: depolymerizing microtubules of root hair cells in *A. thaliana* using olyzalin does not block root hair growth.[31] It does, however, perturb the directionality of growth so that normally straight root hairs become wavy and some root hairs develop multiple growing tips.

Actin microfilaments are found in a similar orientation as microtubules. Actin also forms a diffuse network at the growing tip.[32] There are some differences in the organization of microfilaments in growing root hairs of different species. Root hairs of angiosperms tend to have arrays that run approximately longitudinally, whereas rhizoids of bryophytes,♦ which are essentially similar tissues, have arrays that are obviously helical.[33] Again, it is possible that helices are present in both cases but that their pitch is very different. Whatever their precise arrangement, microfilaments are vital to the process of tip growth. If they are depolymerized with cytochalasins or latrunculins, root hair extension fails.[34,31]

The longitudinal microfilaments are associated with secretory vesicles that contain material for the building of new cell wall, and it is possible that their main role in tip extension is to ensure delivery of these vesicles to the right place. In root hairs growing normally, cytoplasmic vesicles and organelles are transported in an "inverse fountain" pattern, moving toward the tip in the cortical zone of the cytoplasm and back toward the cell body along the middle of the cell (some material remains at the tip to be incorporated into new wall).[32] This movement of material depends on the presence of an intact microfilament system, and it ceases if root hairs of *Hydrocharis* are treated with cytochalasin B.[35] This observation suggests that an important mechanism of focal growth is the accurate targeting of vesicles to the growing region, and nowhere else. As well as microfilament bundles being needed for transport of vesicles, it is possible that the cortical microfilament mesh helps to prevent inappropriate fusion of vesicles to other parts of the membrane by standing physically in the way.

Another example of focal growth in plant cells is provided by the pollen tubes of angiosperms. The male haploid stage of the life cycle of angiosperms is represented by the pollen grain, which consists in most species of just two cells. Its development begins when it lands on the stigma of a flower, the stigma being a "landing pad" that is attached, via a column called the "style," to the ovule that contains the female gametes (see Figure 2.4.5). The largest of these cells, the tube cell, undergoes a dramatic focal growth to produce a long tube that invades the stigma and grows along the style toward the ovary. It carries with it the smaller cell, the sperm cell, which divides into two at this time to produce two sperm cells, each of which is used to fertilize a cell of the female gametophyte. (Flowering plants have a double fertilization: one sperm fuses with the egg cell to produce the zygote proper, the other fuses with "polar nuclei" in the female gametophyte to produce a triploid nutritive tissue, the endosperm.)

The growing pollen tube is structurally similar to a growing root hair (see Figure 2.4.6). Microtubules and microfilaments are aligned longitudinally, again sometimes with a hint of helical arrangement.[36,32] As in root hairs, transport—and even the formation—of secretory vesicles is inhibited in the presence of cytochalasin D, and growth of the pollen tube fails.[37]

♦Bryophytes are mosses and liverworts.

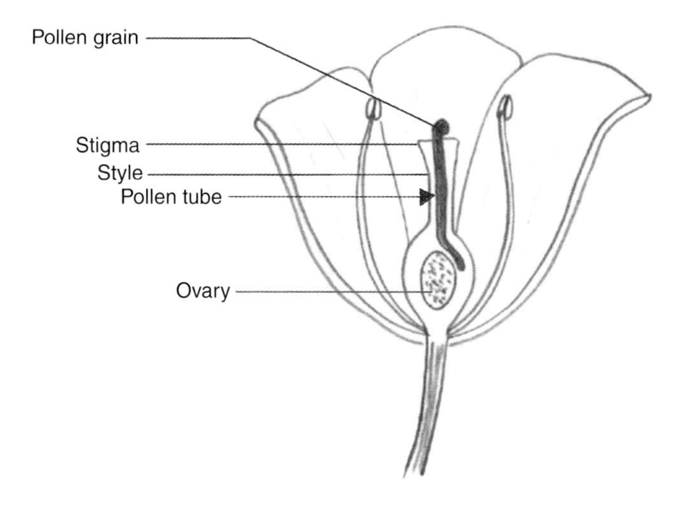

FIGURE 2.4.5 Growth of the pollen tube from the pollen grain, on the stigma, through the style toward the ovary of a flowering plant.

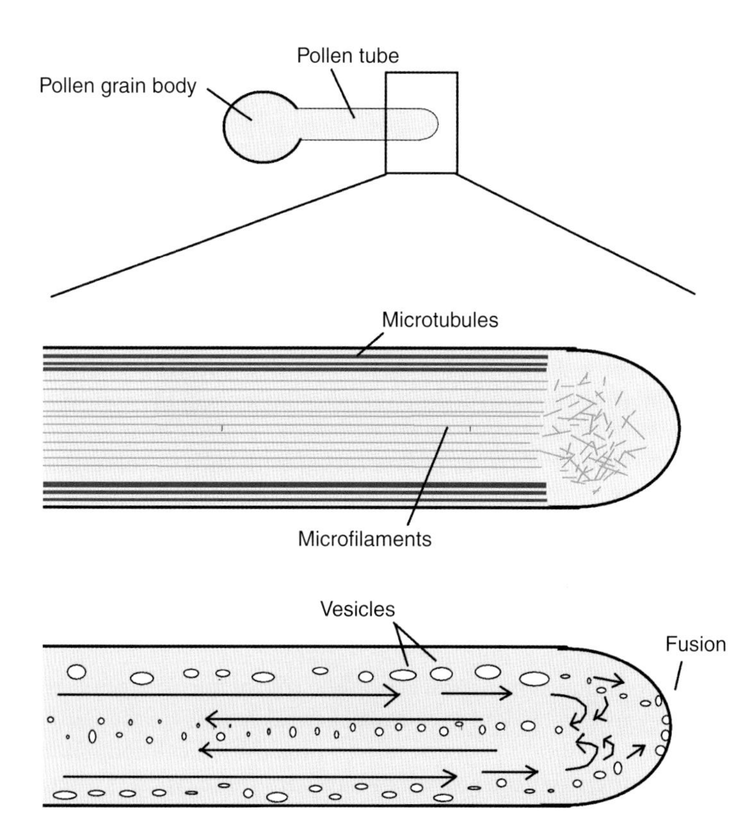

FIGURE 2.4.6 The cytoskeleton and cytoplasmic streaming near the tip of a pollen tube. Cortical microtubules and cytoplasmic microfilaments run longitudinally, or nearly so. Vesicles move in an "inverse fountain," coming forward in the cortical regions of the cell and returning near the axis.

The control of microfilament polymerization is not understood nearly as well in plants as it is in animals, but a combination of genetic and cell biological studies have succeeded in identifying some of the key components that are involved in both root hairs and pollen tubes. *A. thaliana* has homologues of the Rho family of small GTPases that are important in organizing stress fibres in animal cells (see Chapter 3.2); they are called Rop proteins (from "Rho of plants"). Rop1 and Rop5 are enriched in the tips of growing pollen tubes and Rop2 in the tips of growing root hairs.[38,39,40] Dominant negative mutants of Rop1 block the growth of pollen tubes.[39] Constitutive activation of Rop1, on the other hand, results in isotropic growth so that the pollen tube swells up to become a bulbous cell rather than a fine cylinder. Even overexpression of normal Rop1 causes growth to become delocalized, apparently because too much Rop1 is associated with the plasma membrane in regions other than the growth tip.[39] Together, these observations suggest that Rop1 encourages cell expansion and that localization of Rop1 activity is important if growth is to be localized.

How might Rop1 act on microfilaments? One possibility might be to encourage actin-myosin binding as Rho does in animal cells. Another possibility is that it acts by repressing actin-depolymerizing factors (ADFs). ADFs are cytoplasmic proteins that can both nucleate and sever microfilaments, and they can be inactivated by Rop activity.[41] ADFs are associated with the zone immediately distal to the tip of a pollen tube and may be responsible for cutting filaments there into pieces and preventing them from penetrating all the way forward.[41] It is known that the pH of the cytoplasm in the tip zone is lower than that elsewhere. It is also known that ADFs work best in alkaline conditions. It may be, therefore, that the pH gradient ensures that the ADFs can associate with actin and promote depolymerization and severing of filaments most in this zone, and that they leave the microfilaments in the shaft of the tube alone. Overexpression of ADF1 in the pollen tube of *Nicotiana tabacum* results in ADF/microfilament interaction everywhere, rather than just in the zone at the tip, and this causes a reduction in the number of longitudinal microfilaments in the main body of the pollen tube. Significantly, this overexpression of ADF can counteract the cell-ballooning effect induced by overexpression of Rop1[41] in pollen tubes of *N. tabacum*. This suggests that the main target for this Rop is ADF, and that the Rop controls microfilament dynamics by regulating the destruction of microfilaments in the zone just behind the tip.

There is a gradient of cytoplasmic Ca^{2+} in pollen tubes, with the concentration highest at the tip and decreasing toward the main cell body of the pollen grain. The establishment and/or maintenance of this gradient requires the action of Rop1: if dominant negative Rop1 is expressed in the pollen tube, both Ca^{2+} influx at the tip and subsequent tube growth are markedly inhibited.[39] Growing pollen tubes in high-Ca^{2+} media rescues growth, suggesting that the critical activity of Rop1 is to promote Ca^{2+} influx. How it does so will remain obscure until the Ca^{2+} channel used at the tube tip has been identified. The link may or may not involve actin. The role of the Ca^{2+}gradient is not fully understood, but it seems to be linked to the cytoskeletal system via the Ca^{2+}-sensitive protein calmodulin; experimentally induced release of activated calmodulin in other parts of the pollen tube re-orientates the growth axis toward those parts.[42] There is probably a feedback system, therefore, in which Rop proteins set up the gradient of Ca^{2+} and control the cytoskeleton, and the gradient of Ca^{2+}, via calmodulin, controls the activation of Rop.

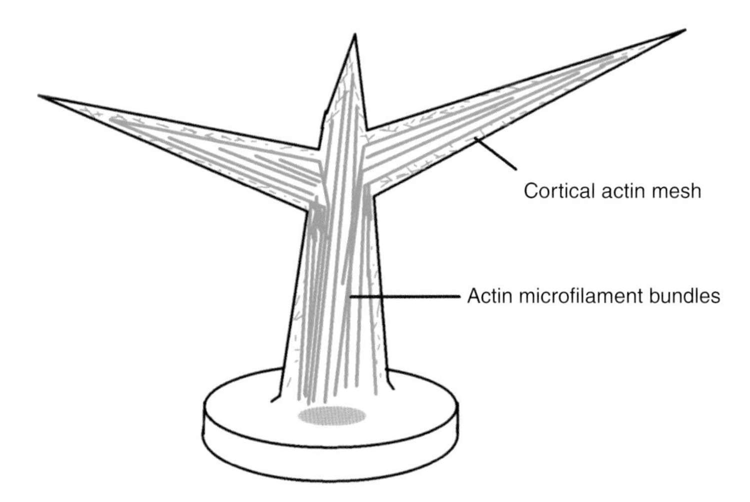

FIGURE 2.4.7 The arrangement of microfilaments in a developing trichome of *A. thaliana*. The diagram is a schematic representation of data presented in J. Mathur et al.[45]

Root hairs and pollen tubes normally grow as single tubes without branching. Some plant cells, such as the trichomes♣ of *A. thaliana*, can produce branched processes. Cells that have made the decision to form trichomes stop dividing but continue to replicate their DNA by endoreduplication until they reach a DNA content of 32C.[43] As this takes place, the cell expands away from the surface of the leaf and emits two new processes, each being a branch from the existing one. Then each process elongates rapidly. The number of branches is highly controlled and can be increased or decreased by specific mutations.[44] Both the microfilament and microtubule cytoskeletons are required for normal branching morphogenesis to take place.

Microfilaments show a relatively diffuse organization until the trichome process emerges; they then form a more organized longitudinal array leading from the nucleus toward the growing tip. Intense foci of microfilaments are seen at the two future branch points, and as soon as elongation begins, there is a clear arrangement of longitudinal microfilaments leading along the processes. There is also a mesh-like network of cortical actin (see Figure 2.4.7).

Depolymerization of microfilaments using cytochalasin D, latrunculin, phalloidin, or jasplakinolide inhibits the elongation of trichomes and distorts their shapes, but the number of branches is still correct.[45] Actin is therefore important to trichome elongation but is not essential to branching. It is interesting that the cortical actin cytoskeleton, rather than the central filaments, may be particularly important. Actin meshworks tend to be nucleated by the Arp2/3 protein complex, which is discussed in detail in Chapter 3.2. Mutations that block the activation or function of *A. thaliana*'s Arp2/3 complex cause trichomes to be distorted in a manner very similar to that seen with cytochalasins and similar drugs.[46,47]

Treating trichomes with the microtubule-disrupting drugs colchicine, olyzalin, propyzamide, or taxol at an early stage of their development causes them to bulge

♣Trichomes are "leaf hairs" that stick out from the surface of the leaf and modulate airflows across it.

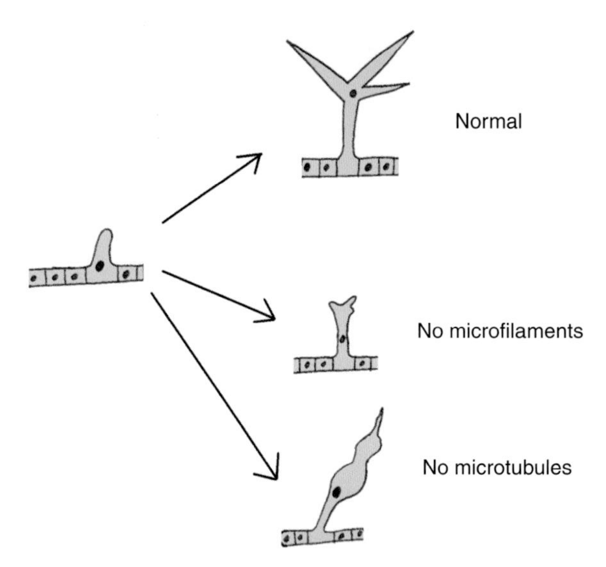

FIGURE 2.4.8 The differing effects of disrupting microfilaments and microtubules during development of trichomes in *A. thaliana*. This diagram is based on the discussion in J. Mathur et al.[45]

and to expand isotropically. Treating them later, once the process has already begun to head away from the plane of the leaf but before branching has taken place, causes them to fail to branch, although elongation happens normally.[45] It therefore seems that microtubules are particularly important in setting up the initial process and also the "new" processes that become the branches, while microfilaments are most important in elongation, as they are in root hairs and pollen tubes (see Figure 2.4.8). This role for microtubules is supported by the phenotypes of mutants in the Katanin-P60 gene, and weak alleles of the TFCA and TFCC proteins that assist the formation of tubulin dimers that are the precursors of microtubules; both types of mutant are characterized by too few trichome branch points.[45]

Although the roles of the two types of cytoskeleton have been separated in the discussion above, they interact strongly, and mutations that disrupt the microfilaments, particularly cortical microfilaments, cause a distortion of microtubules too.[47] Further work on this interaction may reveal much about the control of plant cell morphogenesis.

As I stated at the beginning of this chapter, our state of knowledge about morphogenesis in plant cells is not yet at the stage that it is clear where control lies, and whether and how processes such as adaptive self-organization are used to provide fine adjustments. The rapid growth of plant molecular cell biology, particularly in *A. thaliana*, promises to rectify that situation very soon; indeed, by the time it is printed, this chapter will probably be out of date.

Reference List

1. Refregier, G, Pelletier, S, Jaillard, D, and Hofte, H (2004) Interaction between wall deposition and cell elongation in dark-grown hypocotyl cells in Arabidopsis. *Plant Physiol.* **135**: 959–968
2. Brett, CT (2000) Cellulose microfibrils in plants: Biosynthesis, deposition, and integration into the cell wall. *Int. Rev. Cytol.* **199**: 161–199

3. Gordon, JE (1976) The new science of strong materials. **2**: (Penguin)
4. Delmer, DP and Amor, Y (1995) Cellulose biosynthesis. *Plant Cell.* **7**: 987–1000
5. Green, PB (1980) Organogenesis—a biophysical view. *Ann. Rev. Plant Physiol.* **31**: 51–82
6. Shedletzky, E, Shmuel, M, Delmer, DP, and Lamport, DTA (1990) Adaptation and growth of tomato cells on the herbicide 2, 6-dichlorobenzonitrile leads to production of unique cell walls virtually lacking a cellulose-xyloglucan network. *Plant Physiol.* **94**: 980–987
7. Sugimoto, K, Williamson, RE, and Wasteneys, GO (2000) New techniques enable comparative analysis of microtubule orientation, wall texture, and growth rate in intact roots of Arabidopsis. *Plant Physiol.* **124**: 1493–1506
8. McClinton, RS and Sung, ZR (1997) Organization of cortical microtubules at the plasma membrane in Arabidopsis. *Planta.* **201**: 252–260
9. Furutani, I, Watanabe, Y, Prieto, R, Masukawa, M, Suzuki, K, Naoi, K, Thitamadee, S, Shikanai, T, and Hashimoto, T (2000) The SPIRAL genes are required for directional control of cell elongation in Arabidopsis thaliana. *Development.* 127: 4443–4453
10. Baskin, TI, Meekes, HT, Liang, BM, and Sharp, RE (1999) Regulation of growth anisotropy in well-watered and water-stressed maize roots. II. Role of cortical microtubules and cellulose microfibrils. *Plant Physiol.* **119**: 681–692
11. Hogetsu, T and Shibaoka, H (1978) Effects of colchicine on cell shape and on microfibril arrangement in the cell wall of Closterium acerosum. *Planta.* **140**: 15–18
12. Itoh, T (1976) Microfibrillar orientation of radially enlarged cells of coumarin- and colchicine-treated pine seedlings. *Plant Cell Physiol.* **16**: 385–398
13. Baskin, TI, Beemster, GT, Judy-March, JE, and Marga, F (2004) Disorganization of cortical microtubules stimulates tangential expansion and reduces the uniformity of cellulose microfibril alignment among cells in the root of Arabidopsis. *Plant Physiol.* **135**: 2279–2290
14. Giddings, TH and Staehelin, LA (1991) Microtubule-mediated control of microfibril deposition: A re-examination of the hypothesis. In Lloyd, CW, The cytoskeletal basis of plant growth and form (Academic Press)
15. Lloyd, C and Chan, J (2004) Microtubules and the shape of plants to come. *Nat. Rev. Mol. Cell Biol.* **5**: 13–22
16. Shaw, SL, Kamyar, R, and Ehrhardt, DW (13-6-2003) Sustained microtubule treadmilling in Arabidopsis cortical arrays. *Science.* **300**: 1715–1718
17. Gardiner, JC, Harper, JD, Weerakoon, ND, Collings, DA, Ritchie, S, Gilroy, S, Cyr, RJ, and Marc, J (2001) A 90-kD phospholipase D from tobacco binds to microtubules and the plasma membrane. *Plant Cell.* **13**: 2143–2158
18. Dhonukshe, P, Laxalt, AM, Goedhart, J, Gadella, TW, and Munnik, T (2003) Phospholipase D activation correlates with microtubule reorganization in living plant cells. *Plant Cell.* **15**: 2666–2679
19. Gardiner, J, Collings, DA, Harper, JD, and Marc, J (2003) The effects of the phospholipase D-antagonist 1-butanol on seedling development and microtubule organisation in Arabidopsis. *Plant Cell Physiol.* **44**: 687–696
20. Thomas, SG and Sun, TP (2004) Update on gibberellin signaling. A tale of the tall and the short. *Plant Physiol.* **135**: 668–676
21. Akashi, T and Shibaoka, H (2004) Effects of gibberellin on the arrangement and the cold stability of cortical microtubules in epidermal-cells of PEA internodes. *Plant and Cell Physiology.* **28**: 339–348
22. Lang, JM, Eisinger, WR, and Green, PB (1982) Effects of ethylene on the orientation of microtubules and cellulose microfibrils of PEA epicotyl cells with polylamellate cell-walls. *Protoplasma.* **110**: 5–14
23. Alberts, B, Bray, D, Lewis, J, Raff, M, Roberts, K, and Watson, JD (1994) Molecular biology of the cell. **3rd**: (Garland)
24. Blancaflor, EB and Hasenstein, KH (1995) Growth and microtubule orientation of Zea mays roots subjected to osmotic stress. *Int. J. Plant Sci.* **156**: 774–783

25. Yuan, M, Shaw, PJ, Warn, RM, and Lloyd, CW (21-6-1994) Dynamic reorientation of cortical microtubules, from transverse to longitudinal, in living plant cells. *Proc. Natl. Acad. Sci. U.S.A.* **91**: 6050–6053
26. Zhou, BH, Chen, JS, Chai, MQ, Zhao, S, Liang, J, Chen, HH, and Song, JG (2000) Activation of phospholipase D activity in transforming growth factor-beta-induced cell growth inhibition. *Cell Res.* **10**: 139–149
27. Lloyd, C (1999) How I learned to love carrots: The role of the cytoskeleton in shaping plant cells. *Bioessays.* **21**: 1061–1068
28. Lefebvre, A, Maizonnier, D, Gaudry, JC, Clair, D, and Scalla, R (1987) Some effects of the herbicide EL-107 on cellular growth and metabolism. *Weed Res.* **27**: 125–134
29. Fisher, DD and Cyr, RJ (1998) Extending the microtubule/microfibril paradigm. Cellulose synthesis is required for normal cortical microtubule alignment in elongating cells. *Plant Physiol.* **116**: 1043–1051
30. Cosgrove, DJ, Li, LC, Cho, HT, Hoffmann-Benning, S, Moore, RC, and Blecker, D (2002) The growing world of expansins. *Plant Cell Physiol.* **43**: 1436–1444
31. Bibikova, TN, Blancaflor, EB, and Gilroy, S (1999) Microtubules regulate tip growth and orientation in root hairs of Arabidopsis thaliana. *Plant J.* **17**: 657–665
32. Geitmann, A and Emons, AM (2000) The cytoskeleton in plant and fungal cell tip growth. *J. Microsc.* **198 (Pt 3)**: 218–245
33. Alfano, F, Russell, A, Gambardella, R, and Duckett, JG (2004) The actin cytoskeleton of the liverwort riccia-fluitans—effects of cytochalasin-b and aluminum ions on rhizoid tip growth. *J. Plant Physiol.* **142**: 569–594
34. Miller, DD, de Ruijter, NCA, Bisseling, T, and Emons, AMC (1999) The role of actin in root hair morphogenesis: Studies with lipochito-oligosaccharide as a growth stimulator and cytochalasin as an actin perturbing drug. *AMC.* **17**: 141–154
35. Shimmen, T, Hamatani, M, Saito, S, Yokota, E, Mimura, T, Fusetani, N, and Karaki, H (1995) Roles of actin-filaments in cytoplasmic streaming and organization of transvacuolar strands in root hair-cells of hydrocharis. *Protoplasma.* **185**: 188–193
36. Cai, G, Moscatelli, A, and Cresti, M (1997) Cytoskeletal organization and pollen tube growth. *Trends Plant Sci.* **2**: 86–91
37. Shannon, TM, Picton, JM, and Steer, MW (1984) The inhibition of dictyosome vesicle formation in higher plant cells by cytochalasin D. *Eur. J. Cell Biol.* **33**: 144–147
38. Molendijk, AJ, Bischoff, F, Rajendrakumar, CS, Friml, J, Braun, M, Gilroy, S, and Palme, K (1-6-2001) Arabidopsis thaliana Rop GTPases are localized to tips of root hairs and control polar growth. *Embo. J.* **20**: 2779–2788
39. Li, H, Lin, Y, Heath, RM, Zhu, MX, and Yang, Z (1999) Control of pollen tube tip growth by a Rop GTPase-dependent pathway that leads to tip-localized calcium influx. *Plant Cell.* **11**: 1731–1742
40. Lin, Y, Wang, Y, Zhu, K, and Yang, Z (1996) Localization of a Rho GTPase implies a role in tip growth and movement of the generative cell in Pollen tubes. *Plant Cell.* **8**: 293–303
41. Chen, CY, Wong, EI, Vidali, L, Estavillo, A, Hepler, PK, Wu, HM, and Cheung, AY (2002) The regulation of actin organization by actin-depolymerizing factor in elongating pollen tubes. *Plant Cell.* **14**: 2175–2190
42. Rato, C, Monteiro, D, Hepler, PK, and Malho, R (2004) Calmodulin activity and cAMP signalling modulate growth and apical secretion in pollen tubes. *Plant J.* **38**: 887–897
43. Hulskamp, M, Misra, S, and Jurgens, G (11-2-1994) Genetic dissection of trichome cell development in Arabidopsis. *Cell.* **76**: 555–566
44. Bouyer, D and Hulskamp, M (2005) Branching of single cells in Arabidopsis. In Davies, JA, Branching morphogenesis (Landes Bioscience)
45. Mathur, J, Spielhofer, P, Kost, B, and Chua, N (1999) The actin cytoskeleton is required to elaborate and maintain spatial patterning during trichome cell morphogenesis in Arabidopsis thaliana. *Development.* **126**: 5559–5568

46. El Assal, Sel, Le, J, Basu, D, Mallery, EL, and Szymanski, DB (10-8-2004) Arabidopsis GNARLED encodes a NAP125 homolog that positively regulates ARP2/3. *Curr. Biol.* **14**: 1405–1409
47. Saedler, R, Mathur, N, Srinivas, BP, Kernebeck, B, Hulskamp, M, and Mathur, J (2004) Actin control over microtubules suggested by DISTORTED2 encoding the Arabidopsis ARPC2 subunit homolog. *Plant Cell Physiol.* **45**: 813–822

3

Cell Migration

CHAPTER 3.1

CELL MIGRATION IN DEVELOPMENT: A Brief Overview

Cell migration takes place at least at some point in the life cycle of almost all animals. For some, such as coelenterates, it is an important but rather rare event, whereas for others, such as vertebrates, cell migration is involved in the morphogenesis of most parts of the body. This chapter will present an overview of the broad types of cell migration that are common in development; the chapters that follow will examine in detail the cellular mechanisms involved in their guidance and regulation.

Cell migration of one sort or another is an almost universal attribute of sexual reproduction, for the obvious reason that two gametes produced in different places, and usually in different individuals, have to unite. Some organisms, usually ones that are anatomically simple, produce gametes that are morphologically identical (isogamous) even if they happen to be divided into two biochemically distinct "mating types." In these cases, the gametes tend to share the task of finding each other. This does not always involve cell migration; in the filamentous alga *Spyrogyra*, for example, gametes remain in the organism that produces them and find each other simply by extending a cell process toward the gametes of an adjacent individual[1] (see Figure 3.1.1a). In most cases, though, adults are not present at a high enough density for this to be feasible, and gametes have to migrate by crawling or swimming. In both plants and animals, migration has favoured the loss of isogamy and the specialization of gametes either to be nutrient-rich and relatively immobile (ova), or to be numerous, small, and specialized for migration (sperm).

Sperm use a variety of methods for migration (see Figure 3.1.1). Those of nematodes are approximately amoeboid in shape and move by amoeboid motion (see Figure 3.1.1b). Those of decapod crustaceans (crabs, lobsters, and so on) have a large number of actin-containing processes[2] (see Figure 3.1.1c). Mammalian sperm, and those of many other animals and algae, move by the whiplash action of flagella (see Figure 3.1.1d). In many species, sperm is attracted toward eggs of the correct species by chemotaxis (see Chapter 3.3). Male gametes of primitive plants, such as liverworts (*hepaticae*), also swim using flagella, but in flowering plants, the meeting of sperm and egg nuclei is achieved by the extension of a tube from the pollen, which invades

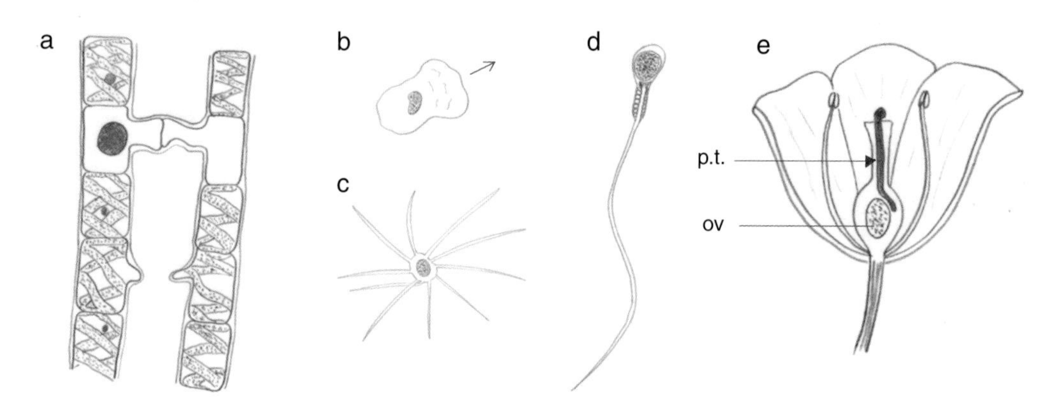

FIGURE 3.1.1 Mechanisms for bringing gametes together. (a) Union of gametes by extension of cell processes in the isogametic alga *Spyrogyra*. The upper conjugation is seen at a late stage, after union of gametes, whereas the lower one is just beginning. (b) Amoeboid sperm of the nematode *Caenorhabditis elegans*, drawn from a phase contrast micrograph.[3] (c) Sperm of the decapod *Inacnus*, showing actin-rich extensions. (d) Human sperm, showing a single long flagellum. (e) The pollen tube (p.t.) of a flowering plant, growing toward the ovule (ov); the size of the pollen tube is exaggerated for clarity.

the female parts of the flower and brings the male gamete nucleus to the ovum (see Figure 3.1.1e).

Important as it is to development, fertilization is not morphogenesis. During post-fertilization development, cell migration and locomotion contribute directly to morphogenesis in four main ways:

1. Bringing dispersed cells together to aggregate in one place
2. Moving a group of cells from one place in the body to another
3. Dispersing cells created in one place to a variety of sites
4. Connecting cells and cell processes together in a specific network, such as the nervous system

There are also many cell rearrangements that operate on too short a range to be considered true migrations and are therefore discussed in Chapter 4.2 rather than here.

MORPHOGENESIS BY COALESCENCE OF DISPERSED CELLS

One of the simplest examples of cell migration effecting truly morphogenetic change is seen in the social amoeba *Dictyostelium discoideum*. Where food sources are plentiful, *D. discoideum* exists as a community of scattered myxamoebae, which are amoeba-like cells that roam decaying wood on forest floors and prey on bacteria. When food becomes scarce, the myxamoebae begin to emit a chemotactic signal and, through sensing and information-processing systems that will be discussed in detail in Chapter 3.3, they move toward sources of this signal and therefore toward each other. Their migration results in the formation of converging streams of moving cells that converge to produce

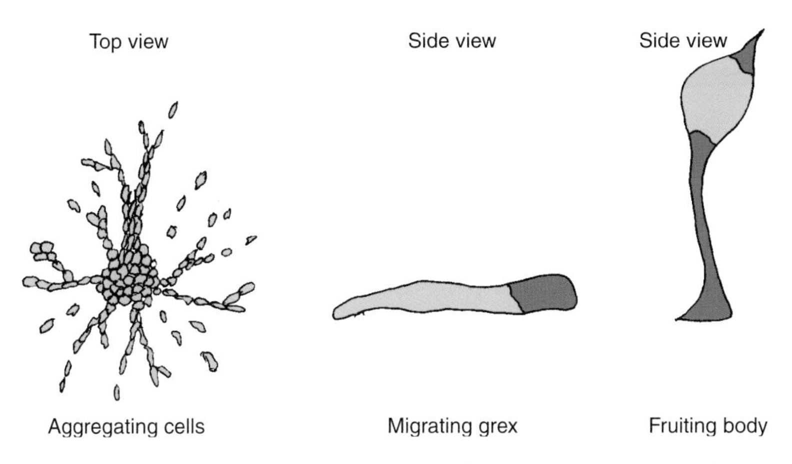

FIGURE 3.1.2 Formation of the grex of *D. discoideum* by chemotaxis-driven aggregation of myxamoebal cells.

a tight, conical aggregate (see Figure 3.1.2). This aggregate then behaves as a single organism, and later processes of differentiation within it turn it into a migratory grex, sometimes called a "slug" because of its superficial resemblance to these animals. The grex moves away, and, as it does so, some cells differentiate to form spores. It then stops migrating, raises these spores into the air atop a long stalk, and releases them so that they disperse. Where they land, they form further myxamoebae, and the cycle, which depends absolutely on cell migration for grex formation, repeats.

Morphogenesis by coalescence of migrating cells is seen many times in the development of higher animals, too. The great blood vessels of vertebrates, for example, form in this way. In embryos of the frog *Xenopus laevis*, precursor cells of the vascular system arise in the lateral mesoderm, 200 to 300 μm[ref4] from where they will eventually be needed (see Figure 3.1.3). They migrate from these sites toward the axis of the embryo, specifically toward the hypochord area, which lies just ventral to the notochord. Once they have converged on the midline, the cells differentiate into endothelium and adhere to each other to form the dorsal aorta, the vessel that carries high-pressure blood from the heart to the trunk of the body. The hypochord expresses a diffusible growth factor (VEGF), and the precursor cells express a receptor for it (flk-1), suggesting that the cells navigate up the concentration gradient of VEGF. This hypothesis is strengthened by the observations that if an ectopic source of VEGF is applied to an *X. laevis* embryo, the vessel precursor cells migrate toward this as well,[4] and that dorsal aorta development fails in $vegf^{-/-}$ mice.[5] The same general pool of vascular precursors in the lateral mesoderm also gives rise to other parts of the vasculature, and the process by which the precursor pool is divided is not yet understood. Morphogenesis of blood vessels by coalescence of migratory cells, usually called vasculogenesis, is not the only way that blood vessels form; smaller vessels often form by branching from existing vessels in a process called "angiogenesis," and that will be described in Chapter 4.6.

Coalescence of cells is not confined to foetal development; the coming together of migrating cells is an important mechanism in the morphogenesis of lymphatic tissue, which continues to develop and remodel through adult life. In the spleen, arterioles that derive from the splenic artery ramify through the organ and become sheathed by

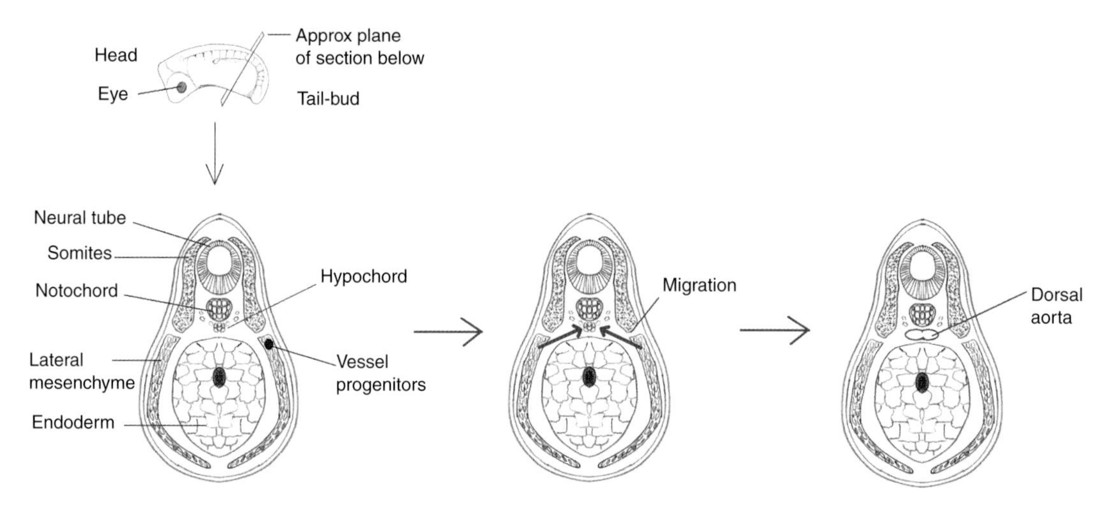

FIGURE 3.1.3 Morphogenesis of the dorsal aorta of *Xenopus laevis* by coalescence of migrating angioblasts, the precursors of blood vessels, which arise in the lateral mesoderm.

lymphoid cells, mainly in the form of lymphoid follicles. The follicles form when stromal cells at the sites of presumptive follicles secrete a chemokine, CXCL13.[6] B lymphocytes bear a receptor for CXCL13, CXCR5,[6] and it is assumed CXCL13, perhaps with other chemokines, is responsible for attracting migrating B cells to the presumptive follicle. Certainly, B cells will migrate toward sources of CXCL13 in culture,[7] and mice lacking CXCR5 fail to form follicles.[8] Relatively short-range movements of mesenchymal cells are also involved in the formation of cell "condensations," a process so important to development that it will be discussed separately in Chapter 3.7.

TRANSLOCATION OF GROUPS OF CELLS FROM ONE PLACE TO ANOTHER

As well as driving morphogenesis through cell aggregation, cell migration can be used as a means of cell dispersal or for the translocation of cells from the place of their birth to the place of their final use. An illustration is provided by the primordial germ cells of vertebrate embryos, which give rise to the gametes discussed at the beginning of this chapter. Primordial germ cells arise, early in development, far outside the gonads that will eventually be their home (see Figure 3.1.4). In frogs, for instance, they develop in the posterior region of the gut and then migrate anteriorly along its dorsal surface, out along the dorsal mesentery (a sheet of tissue that connects the gut to the rest of the abdomen), up the abdominal walls into the genital ridges, and along the ridges to the developing gonads.[9,10,11] In mammals, primordial germ cells arise near the junction between the embryo-proper and the allantois, an extraembryonic structure, and invade the endoderm of the embryo, which will form the gut.[12] Once there, they migrate anteriorly along the gut and enter the genital ridges[13] (see Figure 3.1.4). It used to be believed that mammalian primordial germ cells migrated via the allantois, but it now seems that those cells taking this route are in fact "lost" and do not contribute

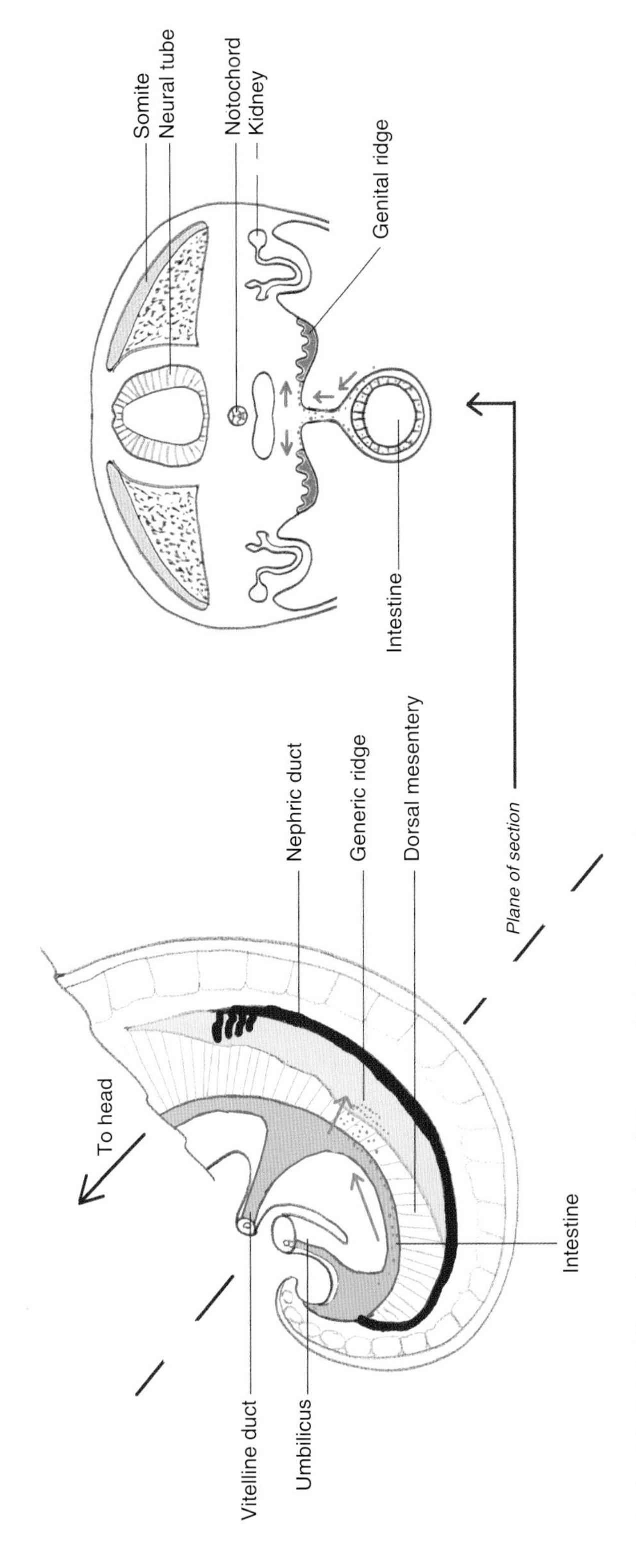

FIGURE 3.1.4 Migration of primordial germ cells in the mouse embryo.

to the germ line; cells migrating into the endoderm are the true source of mammalian gametes.[12,14]

DISPERSAL OF CELLS FROM ONE SITE TO THE REST OF THE BODY

One of the most spectacular demonstrations of cell migration in development is provided by the vertebrate neural crest. The neural crest arises from the dorsal-most zone of the neural tube (presumptive brain and spinal cord) (see Figure 3.1.5), and it gives rise to a multitude of cell types in locations often far removed from the neural tube itself. The face, for example, is produced primarily from neural crest cells that migrate around from dorsal zone of the cranial neural tube and produce craniofacial mesenchyme; this mesenchyme subsequently differentiates into the cartilage, bones, teeth, glial cells, neurons, and connective tissues.[15] Crest from the neck region migrates to colonize the gut to produce the enteric nervous system.[16] In the trunk region of birds and mammals, the neural crest follows two main migration routes, one a dorsolateral route that runs outside the somites and under the epidermis, the other a ventral route that takes it through the sclerotomes of the anterior halves of each somite[17] (see Figure 3.1.5). Some cells remain in the sclerotomes and aggregate together to form the dorsal root ganglia, which process sensory information from the periphery and relay it to the spinal cord. Others settle more ventrally to produce the sympathetic nervous system and the inner parts of the adrenal gland. The cells that migrate between the outside of the somites and the skin give rise mainly to melanocytes, responsible for the pigmentation of the skin. The vast range of tissues produced by the migratory crest cells is so important to vertebrate development that one commentator has suggested that the crest be regarded as a fourth germ layer (to be added to the traditional three of endoderm, mesoderm, and ectoderm).[18]

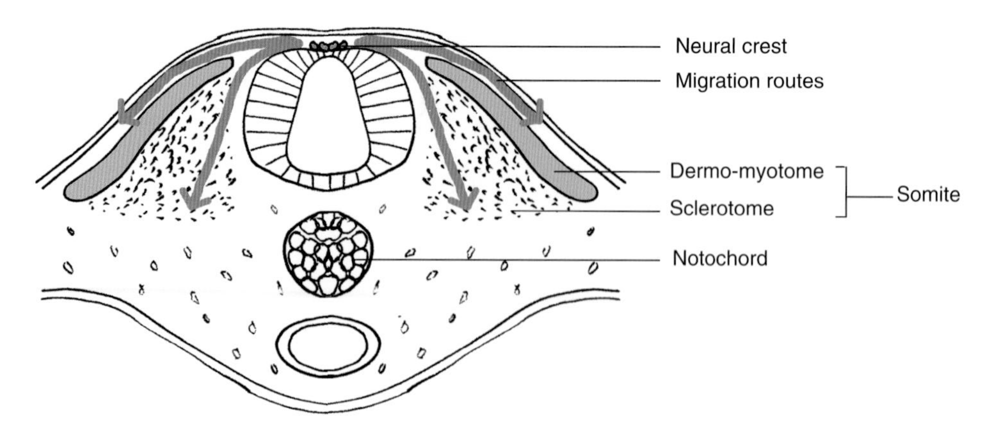

FIGURE 3.1.5 Principal migration routes of the neural crest within the trunk of the embryo.

MIGRATION BY CELL PROCESSES

Development of the nervous system requires migration both of whole cells and of cell processes. Many neurons of the peripheral nervous system arise from migratory cells of the neural crest, but even when these cells have found their ultimate location and settled down to differentiate, directed locomotion continues as cell processes emerge and navigate the embryo to set up the "wiring pattern" of neural connections. The principle processes of neuronal cells, axons, are laid down behind a quasi-autonomous migratory system called a growth cone. Growth cones, which will be described in more detail in Chapters 3.2 and 3.5, move in a manner similar to complete cells, and they have been described as "fibroblasts on a leash." They use both local, contact-mediated cues and diffusible molecules to navigate to targets that may be at a considerable distance from the cell body (for example, the muscles at the end of the limb). What is more, navigation of axons has to be fairly precise in order for the correct neural connections to be made, although mechanisms do exist to deal with occasional wrong connections (see Chapter 5.3). Different neurons can also navigate accurately in quite opposite directions within the same tissue. For example, within the spinal cord, axons from commissural neurons, the cell bodies of which are located in the dorsal spinal cord, grow ventrally toward the floor plate of the cord, being lured there by the locally produced chemoattractant, netrin 1, and netrin 2 produced in the areas adjacent to the floorplate.[19,20,21] Some motor axons, for example those of trochlear motor neurons, are repelled by netrin 1 and migrate in the opposite direction,[22] although their apparently normal navigation in *netrin1-/-* mice suggests that other factors must also be involved[21] (see Figure 3.1.6).

The foregoing examples of migration were chosen to illustrate typical developing events that make use of directed cell movement, and this short chapter is in no way intended to be a comprehensive list. Many more examples, with more details on pathways and anatomy, can be found in general textbooks of embryology and in specialist texts on the development of specific systems such as the brain. Having established that cell migration is an important morphogenetic mechanism, this section of the book will go on to focus on the processes by which cells can move, and then on the systems that allow them to navigate accurately through the complex environment of a growing embryo.

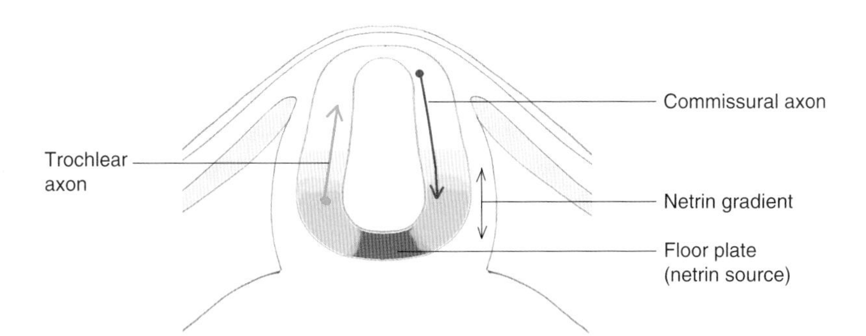

FIGURE 3.1.6 Netrin-1 attracts commissural neurons but repels trochlear motor axons during development of the spinal cord/hindbrain. The source of netrin is known, but the existence of a gradient is assumed rather than proven.

Reference List

1. Bell, P and Woodcock, C (1983) The diversity of green plants. (Arnold)
2. Wilson, EB (1896) The cell in development and inheritance. (Macmillan)
3. Zannoni, S, L'Hernault, SW, and Singson, AW (3-12-2003) Dynamic localization of SPE-9 in sperm: A protein required for sperm-oocyte interactions in Caenorhabditis elegans. *BMC Dev. Biol.* **3**: 10
4. Cleaver, O and Krieg, PA (1998) VEGF mediates angioblast migration during development of the dorsal aorta in Xenopus. *Development.* **125**: 3905–3914
5. Carmeliet, P, Ferreira, V, Breier, G, Pollefeyt, S, Kieckens, L, Gertsenstein, M, Fahrig, M, Vandenhoeck, A, Harpal, K, Eberhardt, C, Declercq, C, Pawling, J, Moons, L, Collen, D, Risau, W, and Nagy, A (4-4-1996) Abnormal blood vessel development and lethality in embryos lacking a single VEGF allele. *Nature.* **380**: 435–439
6. Gunn, MD, Ngo, VN, Ansel, KM, Ekland, EH, Cyster, JG, and Williams, LT (19-2-1998) A B-cell-homing chemokine made in lymphoid follicles activates Burkitt's lymphoma receptor-1. *Nature.* **391**: 799–803
7. Legler, DF, Loetscher, M, Roos, RS, Clark-Lewis, I, Baggiolini, M, and Moser, B (16-2-1998) B cell-attracting chemokine 1, a human CXC chemokine expressed in lymphoid tissues, selectively attracts B lymphocytes via BLR1/CXCR5. *J. Exp. Med.* **187**: 655–660
8. Forster, R, Mattis, AE, Kremmer, E, Wolf, E, Brem, G, and Lipp, M (13-12-1996) A putative chemokine receptor, BLR1, directs B cell migration to defined lymphoid organs and specific anatomic compartments of the spleen. *Cell.* **87**: 1037–1047
9. Wylie, CC and Heasman, J (1976) The formation of the gonadal ridge in Xenopus laevis. I. A light and transmission electron microscope study. *J. Embryol. Exp. Morphol.* **35**: 125–138
10. Wylie, CC, Bancroft, M, and Heasman, J (1976) The formation of the gonadal ridge in Xenopus laevis. II. A scanning electron microscope study. *J. Embryol. Exp. Morphol.* **35**: 139–148
11. Heasman, J, Hynes, RO, Swan, AP, Thomas, V, and Wylie, CC (1981) Primordial germ cells of Xenopus embryos: The role of fibronectin in their adhesion during migration. *Cell.* **27**: 437–447
12. Anderson, R, Copeland, TK, Scholer, H, Heasman, J, and Wylie, C (1-3-2000) The onset of germ cell migration in the mouse embryo. *Mech. Dev.* **91**: 61–68
13. Clark, JM and Eddy, EM (1975) Fine structural observations on the origin and associations of primordial germ cells of the mouse. *Dev. Biol.* **47**: 136–155
14. Downs, KM and Harmann, C (1997) Developmental potency of the murine allantois. *Development.* **124**: 2769–2780
15. Le Lievre, CS and Le Douarin, NM (1975) Mesenchymal derivatives of the neural crest: Analysis of chimaeric quail and chick embryos. *J. Embryol. Exp. Morphol.* **34**: 125–154
16. Le Douarin, NM and Teillet, MA (1973) The migration of neural crest cells to the wall of the digestive tract in avian embryo. *J. Embryol. Exp. Morphol.* **30**: 31–48
17. Weston, JA (1963) A radioautographic analysis of the migration and localization of trunk neural crest cells in the chick. *Dev. Biol.* **6**: 279–310
18. Hall, BK (2000) The neural crest as a fourth germ layer and vertebrates as quadroblastic not triploblastic. *Evol. Dev.* **2**: 3–5
19. Kennedy, TE, Serafini, T, de la Torre, Jr, and Tessier-Lavigne, M (12-8-1994) Netrins are diffusible chemotropic factors for commissural axons in the embryonic spinal cord. *Cell.* **78**: 425–435
20. Serafini, T, Kennedy, TE, Galko, MJ, Mirzayan, C, Jessell, TM, and Tessier-Lavigne, M (12-8-1994) The netrins define a family of axon outgrowth-promoting proteins homologous to C. elegans UNC-6. *Cell.* **78**: 409–424
21. Serafini, T, Colamarino, SA, Leonardo, ED, Wang, H, Beddington, R, Skarnes, WC, and Tessier-Lavigne, M (13-12-1996) Netrin-1 is required for commissural axon guidance in the developing vertebrate nervous system. *Cell.* **87**: 1001–1014
22. Colamarino, SA and Tessier-Lavigne, M (19-5-1995) The axonal chemoattractant netrin-1 is also a chemorepellent for trochlear motor axons. *Cell.* **81**: 621–629

CHAPTER 3.2

CELL MIGRATION: The Nano-Machinery of Locomotion

The mechanisms that control the extent and direction of cell migration in a developing embryo do so by selective and local modification of the processes of cell locomotion. It is therefore sensible, to avoid an overload of information when guidance mechanisms are being discussed, to describe the basic machinery of cell migration first, without reference to navigation, and to refer back to this foundation in the subsequent chapters on guidance and navigation.

Of the various types of cell locomotion seen across the phylogenetic tree, such as flagellar swimming, ciliary gliding, and amoeboid crawling, by far the most significant to developmental biology is crawling. The other mechanisms may be faster, but they tend to be used only extracorporeally, for the dispersal of gametes. There are several components to crawling motion. The first is protrusion of a specialized area of the cell periphery, the lamellipodium, forward on to the substrate ("forward" being defined by the orientation of the cell's polarity, which will be discussed in more detail later). This is followed by advance of the main parts of the cell (nucleus, organelles, and so on) into the lamellipodium, and the drawing up of the trailing edge of the cell. These three phases may occur cyclically, but they are often overlapping or simultaneous, and their separation in this chapter is mostly for convenience of discussion.

PROTRUSION: The Actin-Based Nano-Machinery of the Leading Edge

Lamellipodia

The lamellipodium of a cell migrating on a flat substrate is a flat plate of cytoplasm, only 100 to 200 nm thick,[1] the internal structure of which is dominated by protein filaments. In most cells, these are actin microfilaments, although some cells use other proteins for a broadly similar function; an example is the crawling sperm of the parasitic nematode *Ascaris suum*, which uses major sperm protein (MSP) instead.[2]

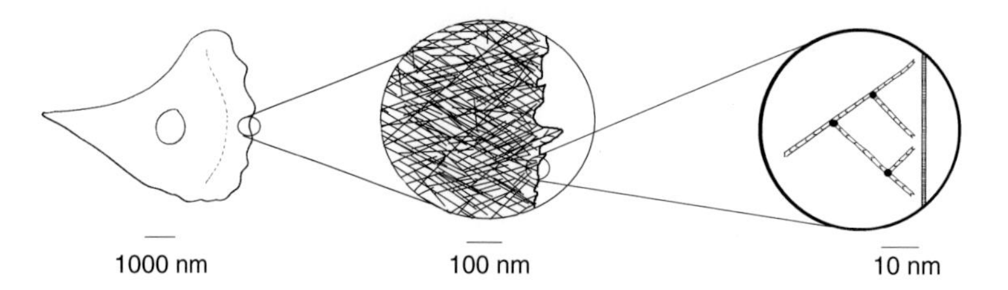

FIGURE 3.2.1 The internal structure of the lamellipodium: the thin (100- to 200-nm) sheet at the leading edge of crawling cells contains a dense network of branched actin filaments, arranged with their barbed ends facing the edge of the membrane.

The microfilaments at the leading edge meet the membrane at an angle of about 40 to 60 degrees.[3] It is the presence of these filaments that prevents the membrane from collapsing inwards under its own surface tension (see Figure 3.2.1).

Neither the filaments nor the membrane are frozen, but they exist in the liquid environment of a living cell and are therefore subject to constant thermodynamic battering by molecules around them. Actin filaments have a Young's modulus ("springiness")* of 2.6 GPa,[ref4] which is approximately the same as that of the plastic used to make the cheap rulers and protractors used by schoolchildren.[5] Anyone who has, as a schoolchild, "twanged" such a ruler over the edge of a desk or has used it to catapult pieces of paper at Teacher knows the ease with which thin pieces of material with this Young's modulus can be bent. Actin filaments, being only 8 nm in diameter,[6] can be bent significantly even by the motion of the solvent around them, and this random bending and snaking has been observed by direct time-lapse micrography for filaments labelled fluorescently.[4] Such vibration of actin rods toward and away from the membrane would take place at the leading edge of a cell too, but the density of filaments near the membrane is very high—in the region of one filament every 4 nm[1]—and the population of filaments would therefore push the membrane outwards relatively stably, even though each individual filament makes a varying contribution to the push.

The random bending and lashing of the actin fibres that make oblique contact with the membrane has a further important consequence: at moments when the filament is temporarily bent away from the membrane, the barbed end is exposed and can grow by the addition of further actin monomers. When the filament springs back again, it is now longer and therefore pushes the membrane out a little further. This process, called the "elastic Brownian ratchet,"[7] is the basis for membrane advance (see Figure 3.2.2).

The functioning of the elastic Brownian ratchet depends on two key features of the actin filaments: that they grow when their barbed ends are exposed, and that they are of the right size and orientation to push the membrane. Growth of the filament depends on delivery of actin to the barbed end; it is usually delivered as a complex with profilin, which targets the actin to barbed ends while preventing it from polymerizing spontaneously in free cytoplasm. The angle at which filaments meet the plasma membrane (see Figure 3.2.3) is important to the rate of advance, and is a compromise between the rate at which monomers can be added and how much the membrane moves forward

*Young's modulus, E = stress/strain, where stress is the force applied to a substance per unit area, and strain is the resultant change in length per unit length.

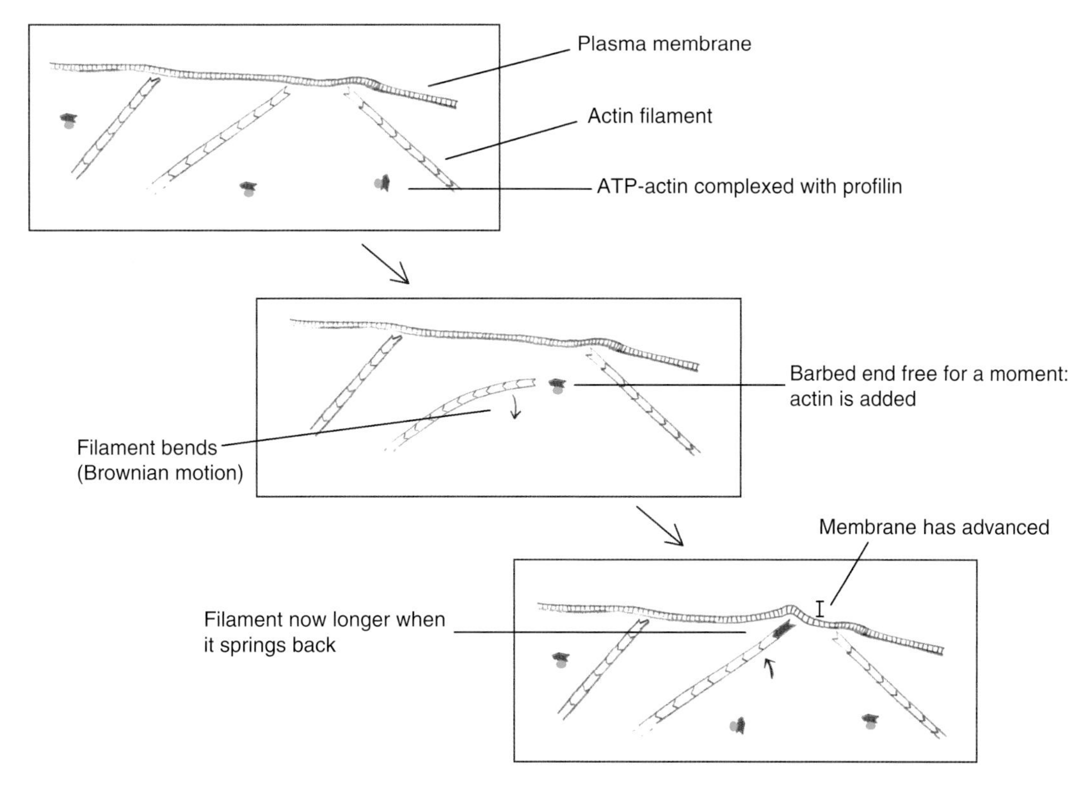

FIGURE 3.2.2 The elastic Brownian ratchet at the leading edge. Random bending motions of peripheral actin fibres, driven by Brownian motion, fleetingly expose their ends; this allows them to be extended so that they are longer by the time they spring back and push against the membrane again. This diagram is a simplified representation; all components are in continuous random motion and the average of the activities of all of the Brownian ratchets results in cell advance.

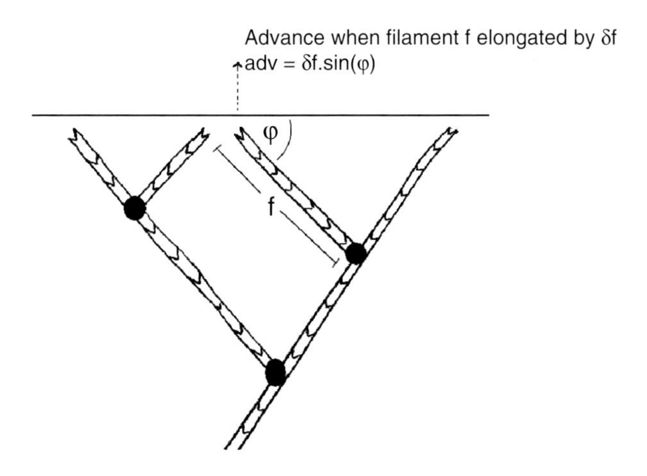

FIGURE 3.2.3 The geometry of actin's pressure on the plasma membrane.

per monomer added. Considering the action of one single filament, to start with, elementary trigonometry dictates that the amount by which the membrane must advance forward when a filament meeting it at an angle φ extends is proportional to $\sin(\varphi)$ (see Figure 3.2.3), and would therefore be greatest when the filament meets the membrane at 90 degrees. In such an orientation, though, there would be little opportunity for new monomers to find space to attach. All else being equal, the probability of the end being free and amenable to a new monomer entering would be expected to be proportional to $\cos(\varphi)$, although some steric effects very close to the membrane may modify this slightly. For that one filament, the rate of advance would therefore be proportional to $\sin(\varphi) \cdot \cos(\varphi)$, and would peak at $\varphi = 45$ degrees.[†] This back-of-the-envelope calculation may provide a teleological explanation for the range of angles actually observed. The flexibility of actin filaments, so necessary for the working of the Brownian ratchet, also acts as a limitation for the length of actin filament that can push on the membrane. This is because very long filaments will simply bend over and start to extend uselessly along from the membrane or just buckle. The maximum practical length that a filament can protrude beyond its last anchor point is about 150 nm,[7] and this means that new filaments must be produced constantly from the actin network and that the network itself must advance close behind the leading edge of the cell.

New actin filaments are nucleated on the sides of existing actin filaments by the activity of the Arp2/3 protein complex.[8,9,10] This complex caps the pointed end of new actin filaments and allows their barbed ends to be free for new growth, and it directs the new filament at an angle of 70 degrees to the axis of the mother filament.[11,12] Once nucleated, new filaments grow by addition of actin to their barbed ends and continue to do so until their barbed ends are capped by barbed end-capping proteins. The length of a new actin filament is therefore set by the relative probabilities of new monomer addition and barbed end capping. To complicate matters, the rates of capping may not be spatially uniform since the activity of barbed end capping proteins is antagonized by phosphatidyl inositides such as PI-4-P and PI-4,5-P_2, which are located at the inner face of the plasma membrane (this will be discussed further in Chapter 3.3).[13] Also, of course, barbed end capping of a filament directly abutting the membrane can, like monomer addition, take place only when the filament is transitorily sprung away from the membrane by the motion of the Brownian ratchet. This spatial modulation of capping rates is useful because it allows filaments pushing usefully against the membrane to be somewhat protected from capping, while those that happen to be directed uselessly away from the membrane are capped quickly and do not therefore grow and waste reserves of actin (see Figure 3.2.4).

Arp2/3, which drives the production of new branches, is not constitutively active but requires activation by other proteins such as WASP (Wiscott-Aldrich syndrome protein[14]) and SCAR ("suppressor of cAR"). These proteins tend to be complexed to activating signal transduction molecules at or very near the plasma membrane, and this means that Arp2/3 tends to be active only near the plasma membrane so that new branches are only nucleated where they may be of some use (see Figure 3.2.5). Together, these features of Arp2/3-driven actin branching produce a powerful self-organizing system; filaments near enough to the leading edge to be useful can give rise to daughter filaments, and those that project to the leading edge at the optimum angle

---

[†]Because $\sin(\varphi)\cos(\varphi) = 0.5\sin(2\varphi)$—"the sine rule"; if $f(2\varphi) = 0.5\sin(2\varphi)$, then the first derivative $f'(2\varphi) = 0.5\cos(2\varphi)$. At the peak of the function, $f'(2\varphi) = 0$, so $\cos(2\varphi) = 0$, so $2\varphi = 90$ degrees making $\varphi = 45$ degrees.

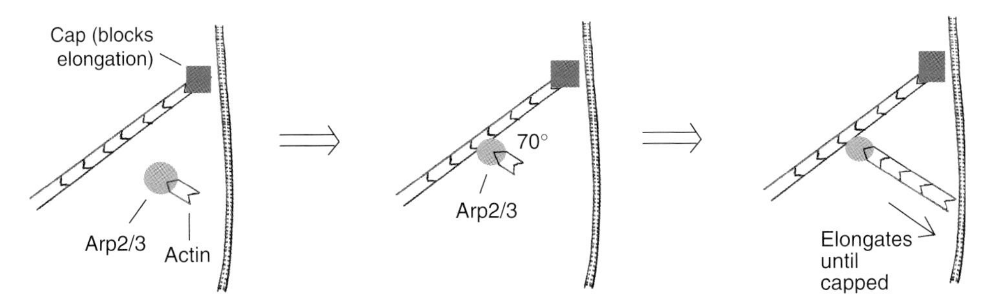

FIGURE 3.2.4 Arp2/3-mediated nucleation of new filaments as side branches on existing filaments. The mother filament is shown as being capped by capping protein, although it need not be; the daughter filament will grow until it is itself capped, and can give rise to further daughters at any time.

grow most and last longest and can in turn give rise to daughter filaments. Badly placed or badly directed filaments are capped and fail to grow.

The association between branches and their mother filament is not stable forever; *in vitro*, the half-life of a branch before it dissociates from the mother filament is about 500 seconds.[15] This is in remarkably close agreement with the half-life of the γ-phosphate remaining on the actin's bound ATP, and it is altered by experimental treatments that alter the γ-phosphate half-life, in exactly the way one would expect if γ-phosphate dissociation were to be the trigger for branch dissocation.[15] In living cells, the half-life of the branched filaments is much shorter (factors such as ADF/cofilins, which are located just behind the leading edge, accelerate γ-phosphate dissociation). The freed filaments are potentially able to join up by associating pointed end to barbed end, to make long, unbranched filaments; this works *in vitro*, although it is not yet clear how capping would not make the barbed ends inaccessible to being rejoined *in vivo*. Either way, it is clear that a little way back from the leading edge of a lamellipodium, the highly branched network of short actin filaments is remodelled into a zone of long and unbranched filaments. Those filaments that do not survive to become long ones are broken up by ADF/cofilin. Profilin binds the actin from dissociated filaments to enable it to bind fresh ATP so that it can be used again.

With respect to the reference frame of the cell, specific pieces of actin therefore move backwards, becoming polymerized at the leading edge of the cell, being pushed back away from the edge by their daughters, and eventually becoming unpolymerized (or perhaps rearranged) at the trailing edge of the lamellipodium. This "treadmilling," which can be demonstrated by photobleaching and by birefringence imaging,[16,17] usually takes place with respect to the reference frame of the substrate (Petri-dish, and so on) as well as that of the cell.[17]

FILOPODIA IN CELL CRAWLING

The lamellipodium is not the only structure that may be found at the leading edge of a migrating cell, and some leading edges, particularly those of growth cones, also bear many filopodia (see Figure 3.2.6). Filopodia are long spikes supported internally by

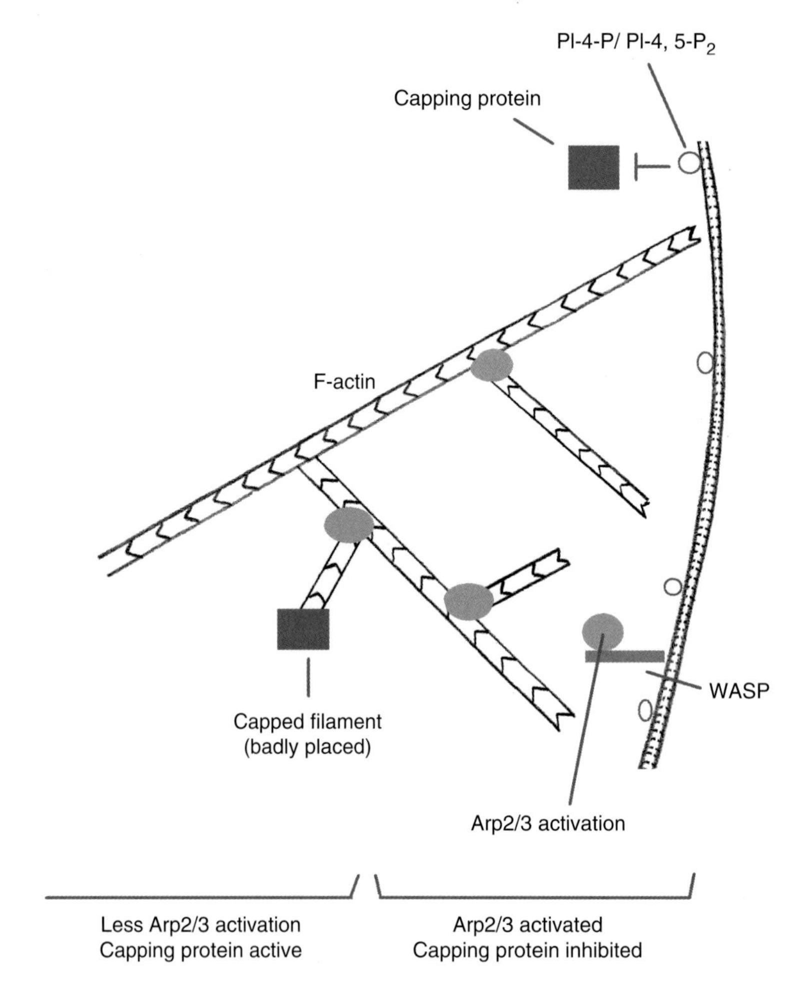

FIGURE 3.2.5 The effects of membrane-located molecules on Arp2/3 and capping proteins mean that Arp2/3 is active mainly near the membrane and capping protein is active mainly further away. This allows filaments growing usefully toward the membrane to grow, and caps those growing pointlessly further back.

parallel filaments of actin, arranged in the barbed-end-outwards orientation typical of cellular protrusions (see Chapter 2.3). Filopodia and lamellipodia are both based on actin, but filopodia have no Arp2/3 and almost none of the capping proteins that are so common in lamellipodia. They are, instead, relatively rich in fascin,[18] which is a protein that cross-links filaments in a bundle. Although a leading edge or growth cone may possess filopodia for a long time, the life of an individual filopodium is short, and each generally follows a relatively rapid course of protrusion and contraction.[19] The timing and rates of these phases are determined locally and can vary between different filopodia, even between different filopodia that exist simultaneously on the same growth cone of a single neuron.[20]

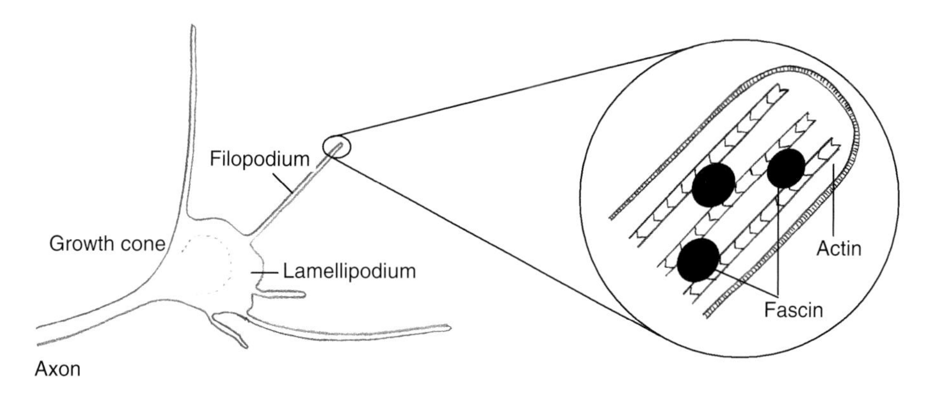

FIGURE 3.2.6 A typical growth cone, showing both lamellipodia and filopodia that contain long, parallel bundles of actin cross-linked by fascin.

Despite their morphological differences, lamellipodia and filopodia have much in common. Both are based on actin (albeit arranged differently), and the actin in both shows treadmilling. In filopodia, as in lamellipodia, actin moves backwards toward the main body of the cell as new filaments are added to the outer, barbed ends.[17,20] The speed of retrograde flow is more or less independent of the rate of advance of the cell as a whole. The two structures also have an intimate relationship with each other in the cell: filopodia are rooted in lamellipodia and seem to arise by modification of the lamellipodial network. The first sign of filopodial growth is the emergence of small Λ-shaped groups of actin filaments, distinct at their apices but melting into the general pool of lamellipodial filaments at their bases. The apices of the Λ-precursors, as they are called,[21] are enriched in proteins associated with filopodia, such as Ena-VASP and fascin, but are not associated with the branch-initiator, Arp2/3. The Λ-precursor therefore seems to represent a gradual transition between lamellipodium-like character at its base and filopodium-like character at its apex. Actin filaments then elongate considerably, outwards from the apical ends of Λ-precursors, and, as this happens, other filaments cluster with them. This takes place in an initially fascin-independent manner, but the filaments later become cross-linked with fascin.[21] Cross-linking of filaments gives the assembly more strength and resistance to bending than is available from a single filament, and the filaments can therefore support a long extension of the plasma membrane. As treadmilling takes place, the original associations between the filopodium and the underlying lamellipodial network are lost, and the relationship becomes less obvious.

The formation of both lamellipodia-like and filopodia-like arrays of actin can be demonstrated from the same brain extract *in vitro*. When beads coated in either of the Arp2/3 activators, WASP or SCAR, are placed in rat brain extracts, two distinct actin morphologies result around beads. One, common in undiluted extracts, is a lamellipodium-like network of short filaments, whereas the other, more common in diluted extracts, is a star-shaped arrangement of "arms" composed of long, parallel actin filaments arranged and orientated as they would be in filopodia.[22] Both depend on the presence of Arp2/3. High-resolution electron microscopy of the system reveals that a lamellipodium-like network forms initially, and filopodial bundles form from this by addition of new monomers to the uncapped ends of filaments and by the "zippering" of adjacent filaments. The key determinant of whether the short filaments of the lamellipodium-like network are able to go on to produce filopodia seems to be the availability of capping protein; if more capping protein is added to the system, lamellipodial organization predominates, whereas if it is depleted, filopodial organization is seen.

The choice between lamellipodial and filopodial organization therefore probably depends critically on the concentration of capping protein. The formation of proper filopodia depends on extension of filaments taking place at a few discrete sites rather than being spaced out all around the periphery of the lamellipodium. This may arise if the edge of the cell is anyway inhomogeneous, perhaps having locally high concentrations of molecules such as Ena-VASP and Formins, both of which can bind barbed ends and shield them from being capped while still allowing them to elongate.[23,24] Such inhomogeneities may be the result of extracellular signals (see Chapters 3.3 and 3.5). There may be some degree of self-organization, too, though. Consider a cell with a homogenous distribution of Ena-VASP: once filaments have associated with it and have started to form Λ-precursors and to extend, they will push the local reserves of Ena-VASP out of the bulk cytoplasm, producing a temporary local lowering of the concentration of these molecules in the vicinity of their "roots" and thus inhibiting the formation of any more filopodia close by. Thus, some degree of spacing may be inherent in the system, even in the absence of external cues.

CONTROL OF FORMATION OF LAMELLIPODIA/FILOPODIA

Formation of both lamellipodia and filopodia requires a supply of the basic raw materials, such as actin, and also requires the activity of branch-initiating complexes such as Arp2/3. Arp2/3 is activated mainly by proteins of the WASP/WAVE/SCAR family of proteins,[25] and WAVE/SCAR are in turn activated by proteins such as Abi, Nap1 (=Kette), and Sra1.[26,27] These proteins are induced to interact with WAVE/SCAR by the activated, GTP-bound form of the small GTPase, Rac. Thus, Rac activation acts as a "master regulator" of lamellipodium construction, and at least some cells fail to migrate in the absence of Rac activity.[28,29] Not surprisingly, murine development fails spectacularly if Rac is deleted from the genome.[30]

Rac is a member of a family of GTPases, other members of which, such as cdc42, can also lead to activation of Arp2/3. Although both Rac and cdc42 lead to activation of Arp2/3, Rac activation tends to be associated with the formation of lamellipodia and cdc42 with filopodia. Until recently, it was assumed that Rac and cdc42 acted in completely separate parallel pathways to drive formation of lamellipodia and filopodia, respectively, but a number of experiments published in the last two years have suggested that both act on Arp2/3 in the same way to achieve the same nucleation of actin branches.[31] The difference between the favoured actin morphologies seems to arise mainly from other activities of Rac and cdc42 that have diverged between the two proteins (see Figure 3.2.7). Filopodia are favoured when growing actin filaments are protected from capping, and cdc42 can provide this protection by signalling via IRSp53 to cause recruitment of Mena to the filament tips at the ends of filopodia: Mena protects

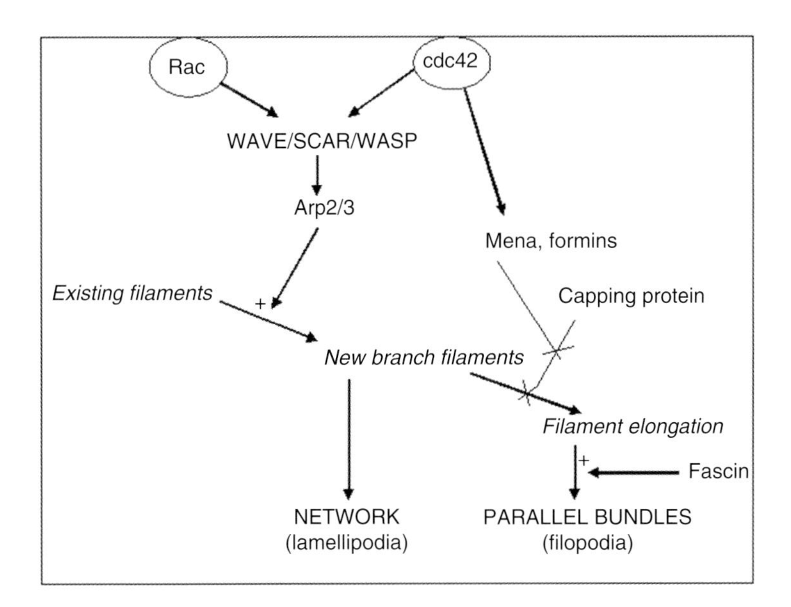

FIGURE 3.2.7 A simplified overview of the networks that control actin assembly at the leading edge. Activation of Arp2/3 and formation of new branch filaments is essential to both lamellipodia and filopodia, but modulation of filament capping tips the balance between these two morphologies.

the filament ends from capping.[23,32,33] Some Formins are also effectors of cdc42 and encourage elongation rather than capping of actin filaments.[34]

ADVANCE OF THE CELL BODY

Locomotion of crawling cells is possible only if they can exert force on their substratum; by elementary Newtonian physics, they can move forward only by pushing their environment backward. The forces that cells exert on their environment can be visualized by growing cells on very thin, deformable films. The idea goes back to 1980, when K.G. Harris and colleagues grew cells on thin silicone films that wrinkled in response to tension.[35] The system worked well enough to indicate when tension was being generated, but the large size of the wrinkles limited spatial resolution, and the highly non-linear physics of wrinkling made absolute measurements of force difficult to obtain. Silicone has therefore been replaced by pre-tensed polyacrylamide films, with a Young's modulus as low as 1 to 10 kN/m^2, in which tiny fluorescent marker beads are embedded. Up to the elastic limit of the film, movement of the beads is proportional to the force exerted so that forces can be calculated from displacement of the beads in the substrate.[36] The leading edge of advancing 3T3 fibroblast cells produces strong traction directed inward toward the centre of the cell body, but inside the lamellipodium proper is a zone with precisely the opposite orientation (so that the substrate is being tugged toward the front of the cell; see Figure 3.2.8).

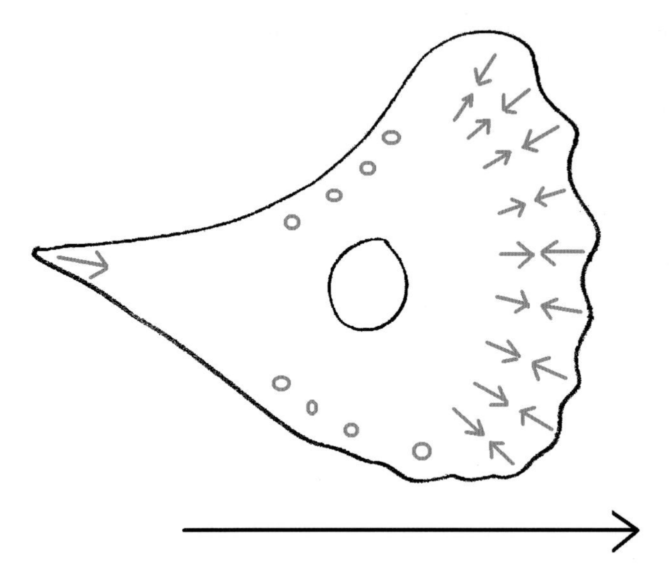

FIGURE 3.2.8 Tensions exerted on the substratum by a migrating cell. Arrows show the direction of tension, while circles are placed on areas in which there seems to be no reproducible direction of tension. The cell is advancing in the direction of the large arrow on the substrate.

Several things may be deduced from this observation. The transmission of these forces to the substrate at all indicates that the cell must have strong mechanical connections with the substrate at each side of the lamellipodium/cell body boundary. The convergence of force on this boundary has been taken to indicate that much of the mechanical work done by the cell takes place there.[36] From the point of view of tension-generating mechanisms that pull the substrate toward this boundary, this must be so. It should be remembered, though, that the compressive forces generated by extension of the lamellipodium against the surface tension of the membrane system will tend to push the substrate backward toward the boundary and will also result in an inward force on the substrate under the lamellipodium itself.

Whereas actin filaments in the lamellipodium are all arranged with their barbed ends pointing outward, filaments in the cell body are orientated both ways around.[37] This allows them to line up in an anti-parallel fashion cross-linked with myosin. The myosin will "try" to slide the filaments past each other if the filaments are free, or, if the filaments are attached, it will place the filaments under tension. Unlike the filaments in the lamellipodium, which show a net rearward motion, the filaments in the cell body near the lamellipodium are stationary with respect to the substrate and even move forward in the heart of the cell body and its trailing edge.[37] There is also little evidence of constant and directed traction in the cell body except at the very tip of the tail, which pulls on the substrate as if it is reluctant to let go as it follows the cell body forward.[36,38] Together, these observations suggest that tension generated against the substrate adhesions just behind the lamellipodium pulls a rather passive cell body along.

Mechanical attachment of cells' actin cytoskeletons to the substrate is usually via integrin-containing junctions called "focal adhesions." As the lamellipodium advances, new focal adhesions are formed. Their development requires Rac or cdc42 activity,[29,39,40] but it is not clear whether this is a direct requirement or an indirect one in which Rac is needed for formation of new lamellipodium, and new lamellipodium is needed for formation of focal adhesions.[41] The new focal adhesions formed at lamellipodia are smaller than those in the cell body, but they exert much stronger forces on the substrate than do the larger and mature adhesions in the cell body,[38] providing further evidence that the cell is pulled from its leading edge (or the region just behind it) and that the cell body itself produces little propulsive force. Although the nascent focal adhesions include many of the same proteins found in all focal adhesions (such as talin, paxillin, and low levels of vinculin and FAK[42]), they are devoid of some "normal" components such as zyxin. They also form connections with microtubules that would be expected, from the tensegrity model, to be in compression.[43]

The organization of actin into tension-generating structures is under the control of another small GTPase of the Rac family, Rho. Rho encourages formation of stress fibres by a variety of means, including regulation of filament length and regulation of tension itself. Activated Rho inhibits cofilin (via LIMK), which would otherwise depolymerize actin; Rho therefore encourages the accumulation of long actin fibres.[44,45] Rho also acts via ROCKs (Rho-kinases) to phosphorylate, and therefore activate, myosin light chains by both stimulating myosin light chain kinase and by inhibiting myosin light chain phosphatase.[46,47,48] Rho therefore increases the tension in the stress fibre system and, because of the positive feedback involved in the stress-fibre/adhesion system described in Chapter 2.2, it also increases cell-substrate adhesion. This is not the only way in which the stress-fiber-inducing activity of Rho can be antagonistic to the functions of Rac and cdc42; by locking up actin in stress fibres, Rho decreases the amount available for the protrusion of lamellipodia and filopodia. Locomotion therefore depends on an appropriate balance in the activities of different small GTPases. Mechanisms exist for shutting Rho down in the micro-environment of the leading edge; one example is the protein Smurf1, which is activated by cdc42 and causes degradation of RhoA.[49]

RETRACTION OF THE REAR OF THE CELL

The dominant model for retraction of the trailing edge of a migrating cell is that it is pulled along by the advancing front of the cell, and it has to break its adhesions with the substratum in order for the cell not to be anchored forever to one spot. The direct measurements of tension described above confirm that, for at least some cell types, the rear of the cell does indeed pull on the substrate.[38] Myosin-generated tension is not the only way in which tension on the substrate might be generated, however; disassembly of the cytoskeleton would reduce the compressive elements that are able to withstand surface tension from the plasma membrane and would therefore cause the cell to pull more on the substrate, which is the only compressive element left. The crawling sperm of the nematode *Ascaris suum* uses a specialized cytoskeleton based on Major Sperm Protein rather than actin.[50] Protrusion of its lamellipodium can be de-coupled from retraction of the trailing edge with a simple change of pH; under mildly acidic conditions, cytoskeletal polymerization and therefore lamellipodial protrusion stops, but retraction of the trailing edge continues, powered by cytoskeletal disassembly.[51] An *in vitro* system based on the sperm's cytoskeletal system now suggests that retraction can be driven by cytoskeletal depolymerization alone.[52]

In vertebrate cells, inhibition of myosin activity, using drugs such as KT5926, dramatically reduces the tension exerted on the substrate (measured by the film deformation method described above) but fails to eliminate cell locomotion, although large organelles such as the nucleus lag behind the general advance of the cell. This suggests that the role of tension may be to keep these structures correctly positioned but that a cell is quite capable of moving without myosin activity.[53] Myxamoebal cells of *D. discoideum* that lack myosin heavy chain are still capable of locomotion, although they are incompetent navigators.[54,55,56] Careful analysis of the motions of these cells and the forces generated by them suggests that, in cells equipped with myosin II, the retraction of the rear of the cell is vigorous and can even help push the anterior parts of the cell forward, whereas in the myosin II knockouts, retraction is more gentle and seems not to assist the protrusion of the front of the cell.[57] On very sticky substrates, the migration of myxamoebae lacking myosin II slows down markedly, suggesting that myosin-generated tension may be important in releasing cells from adhesion so that their trailing edges can retract properly.[58]

The forces driving retraction of the rear of the cell are therefore probably a combination of those driven by myosin activity and by actin disassembly (against the background of constant surface tension from the membrane).

The adhesion sites that are so important to cell advance at the leading edge would become a nuisance if they survived intact even when the cell had moved on because they would anchor the cell and prevent any more forward motion. They therefore have to be destroyed. The destruction of focal adhesions seems to require the protease calpain, without which cells show dramatically reduced trailing edge detachment and strongly inhibited migration.[59] Calpain can cleave many constituents of focal adhesions, such as integrins (on the cytoplasmic side),[60] ezrin,[61] talin,[62] and FAK,[63] so it is an obvious potential mediator of focal adhesion release. What is not yet clear is how its activity is targeted to the adhesions at the rear of the cell, which need to be released, rather than those at the front that are still needed for cell advance.

KEY POINTS TO TAKE FORWARD INTO THE NEXT CHAPTERS

The above brief account on cell locomotion was not meant to be a comprehensive review of cell locomotion, but rather provides an overview of those parts of the cell migration nano-machinery whose regulation has been shown to be key to morphogenetic events. The most important points to take forward into the next chapter are:

- Protrusion of the leading edge of the cell depends on Arp2/3-driven formation of a branched actin network, and the balance of elongation and capping activity.
- Arp2/3, and other aspects of the lamellipodium/filopodia, are controlled by the small GTPases Rac and cdc42.
- Integrin-mediated cell adhesion is key to crawling locomotion, and tensions generated against the substrate through focal adhesions are used to move the bulk of the cell forward toward the lamellipodium.
- Release of adhesion via the calpain and other systems is critical to continuous cell movement.

Reference List

1. Abraham, VC, Krishnamurthi, V, Taylor, DL, and Lanni, F (1999) The actin-based nanomachine at the leading edge of migrating cells. *Biophys. J.* **77**: 1721–1732
2. Roberts, TM and Stewart, M (3-4-2000) Acting like actin. The dynamics of the nematode major sperm protein (msp) cytoskeleton indicate a push-pull mechanism for amoeboid cell motility. *J. Cell Biol.* **149**: 7–12
3. Maly, IV and Borisy, GG (25-9-2001) Self-organization of a propulsive actin network as an evolutionary process. *Proc. Natl. Acad. Sci. U.S.A.* **98**: 11324–11329
4. Gittes, F, Mickey, B, Nettleton, J, and Howard, J (1993) Flexural rigidity of microtubules and actin filaments measured from thermal fluctuations in shape. *J. Cell Biol.* **120**: 923–934
5. Gordon. JE (2004) The new science of strong materials. 2nd: (Penguin)
6. Alberts, B, Bray, D, Lewis, J, Raff, M, Roberts, K, and Watson, JD (1994) Molecular biology of the cell. **3rd**: (Garland)
7. Mogilner, A and Oster, G (1996) Cell motility driven by actin polymerization. *Biophys. J.* **71**: 3030–3045
8. Kelleher, JF, Atkinson, SJ, and Pollard, TD (1995) Sequences, structural models, and cellular localization of the actin-related proteins Arp2 and Arp3 from Acanthamoeba. *J. Cell Biol.* **131**: 385–397
9. Machesky, LM, Atkinson, SJ, Ampe, C, Vandekerckhove, J, and Pollard, TD (1994) Purification of a cortical complex containing two unconventional actins from Acanthamoeba by affinity chromatography on profilin-agarose. *J. Cell Biol.* **127**: 107–115
10. Mullins, RD, Stafford, WF, and Pollard, TD (27-1-1997) Structure, subunit topology, and actin-binding activity of the Arp2/3 complex from Acanthamoeba. *J. Cell Biol.* **136**: 331–343
11. Mullins, RD, Heuser, JA, and Pollard, TD (26-5-1998) The interaction of Arp2/3 complex with actin: Nucleation, high affinity pointed end capping, and formation of branching networks of filaments. *Proc. Natl. Acad. Sci. U.S.A.* **95**: 6181–6186
12. Blanchoin, L, Amann, KJ, Higgs, HN, Marchand, JB, Kaiser, DA, and Pollard, TD (27-4-2000) Direct observation of dendritic actin filament networks nucleated by Arp2/3 complex and WASP/Scar proteins. *Nature.* **404**: 1007–1011
13. Schafer, DA, Jennings, PB, and Cooper, JA (1996) Dynamics of capping protein and actin assembly in vitro: Uncapping barbed ends by polyphosphoinositides. *J. Cell Biol.* **135**: 169–179
14. Rengan, R and Ochs, HD (2000) Molecular biology of the Wiskott-Aldrich syndrome. *Rev. Immunogenet.* **2**: 243–255
15. Blanchoin, L, Pollard, TD, and Mullins, RD (19-10-2000) Interactions of ADF/cofilin, Arp2/3 complex, capping protein and profilin in remodeling of branched actin filament networks. *Curr. Biol.* **10**: 1273–1282
16. Welch, MD, Mallavarapu, A, Rosenblatt, J, and Mitchison, TJ (1997) Actin dynamics in vivo. *Curr. Opin. Cell Biol.* **9**: 54–61
17. Katoh, K, Hammar, K, Smith, PJ, and Oldenbourg, R (1999) Birefringence imaging directly reveals architectural dynamics of filamentous actin in living growth cones. *Mol. Biol. Cell.* **10**: 197–210
18. Kureishy, N, Sapountzi, V, Prag, S, Anilkumar, N, and Adams, JC (2002) Fascins, and their roles in cell structure and function. *Bioessays.* **24**: 350–361
19. Bray, D and Chapman, K (1985) Analysis of microspike movements on the neuronal growth cone. *J. Neurosci.* **5**: 3204–3213
20. Mallavarapu, A and Mitchison, T (6-9-1999) Regulated actin cytoskeleton assembly at filopodium tips controls their extension and retraction. *J. Cell Biol.* **146**: 1097–1106
21. Svitkina, TM, Bulanova, EA, Chaga, OY, Vignjevic, DM, Kojima, S, Vasiliev, JM, and Borisy, GG (3-2-2003) Mechanism of filopodia initiation by reorganization of a dendritic network. *J. Cell Biol.* **160**: 409–421

22. Vignjevic, D, Yarar, D, Welch, MD, Peloquin, J, Svitkina, T, and Borisy, GG (17-3-2003) Formation of filopodia-like bundles in vitro from a dendritic network. *J. Cell Biol.* **160**: 951–962
23. Bear, JE, Svitkina, TM, Krause, M, Schafer, DA, Loureiro, JJ, Strasser, GA, Maly, IV, Chaga, OY, Cooper, JA, Borisy, GG, and Gertler, FB (17-5-2002) Antagonism between Ena/VASP proteins and actin filament capping regulates fibroblast motility. *Cell.* **109**: 509–521
24. Pring, M, Evangelista, M, Boone, C, Yang, C, and Zigmond, SH (21-1-2003) Mechanism of Formin-induced nucleation of actin filaments. *Biochemistry.* **42**: 486–496
25. Miki, H and Takenawa, T (2003) Regulation of actin dynamics by WASP family proteins. *J. Biochem. (Tokyo).* **134**: 309–313
26. Kunda, P, Craig, G, Dominguez, V, and Baum, B (28-10-2003) Abi, Sra1, and Kette control the stability and localization of SCAR/WAVE to regulate the formation of actin-based protrusions. *Curr. Biol.* **13**: 1867–1875
27. Steffen, A, Rottner, K, Ehinger, J, Innocenti, M, Scita, G, Wehland, J, and Stradal, TE (5-2-2004) Sra-1 and Nap1 link Rac to actin assembly driving lamellipodia formation. *Embo. J.* **25**: 749–759
28. Nobes, CD and Hall, A (22-3-1999) Rho GTPases control polarity, protrusion, and adhesion during cell movement. *J. Cell Biol.* **25**: 749–759
29. Allen, WE, Jones, GE, Pollard, JW, and Ridley, AJ (1997) Rho, Rac and Cdc42 regulate actin organization and cell adhesion in macrophages. *J. Cell Sci.* **110 (Pt 6)**: 707–720
30. Sugihara, K, Nakatsuji, N, Nakamura, K, Nakao, K, Hashimoto, R, Otani, H, Sakagami, H, Kondo, H, Nozawa, S, Aiba, A, and Katsuki, M (31-12-1998) Rac1 is required for the formation of three germ layers during gastrulation. *Oncogene.* **17**: 3427–3433
31. Biyasheva, A, Svitkina, T, Kunda, P, Baum, B, and Borisy, G (15-3-2004) Cascade pathway of filopodia formation downstream of SCAR. *J. Cell Sci.* **117**: 837–848
32. Krugmann, S, Jordens, I, Gevaert, K, Driessens, M, Vandekerckhove, J, and Hall, A (30-10-2001) Cdc42 induces filopodia by promoting the formation of an IRSp53:Mena complex. *Curr. Biol.* **11**: 1645–1655
33. Lanier, LM, Gates, MA, Witke, W, Menzies, AS, Wehman, AM, Macklis, JD, Kwiatkowski, D, Soriano, P, and Gertler, FB (1999) Mena is required for neurulation and commissure formation. *Neuron.* **22**: 313–325
34. Peng, J, Wallar, BJ, Flanders, A, Swiatek, PJ, and Alberts, AS (1-4-2003) Disruption of the Diaphanous-related Formin Drf1 gene encoding mDia1 reveals a role for Drf3 as an effector for Cdc42. *Curr. Biol.* **13**: 534–545
35. Barlev, NA, Liu, L, Chehab, NH, Mansfield, K, Harris, KG, Halazonetis, TD, and Berger, SL (2001) Acetylation of p53 activates transcription through recruitment of coactivators/histone acetyltransferases. *Mol. Cell.* **8**: 1243–54
36. Dembo, M and Wang, YL (1999) Stresses at the cell-to-substrate interface during locomotion of fibroblasts. *Biophys. J.* **76**: 2307–2316
37. Cramer, LP, Siebert, M, and Mitchison, TJ (24-3-1997) Identification of novel graded polarity actin filament bundles in locomoting heart fibroblasts: implications for the generation of motile force. *J. Cell Biol.* **136**: 1287–1305
38. Beningo, KA, Dembo, M, Kaverina, I, Small, JV, and Wang, YL (14-5-2001) Nascent focal adhesions are responsible for the generation of strong propulsive forces in migrating fibroblasts. *J. Cell Biol.* **153**: 881–888
39. Nobes, CD and Hall, A (7-4-1995) Rho, rac, and cdc42 GTPases regulate the assembly of multimolecular focal complexes associated with actin stress fibers, lamellipodia, and filopodia. *Cell.* **81**: 53–62
40. Rottner, K, Hall, A, and Small, JV (17-6-1999) Interplay between Rac and Rho in the control of substrate contact dynamics. *Curr. Biol.* **9**: 640–648
41. Ridley, AJ (2001) Rho GTPases and cell migration. *J. Cell Sci.* **114**: 2713–2722
42. Zaidel-Bar, R, Ballestrem, C, Kam, Z, and Geiger, B (15-11-2003) Early molecular events in the assembly of matrix adhesions at the leading edge of migrating cells. *J. Cell Sci.* **116**: 4605–4613

43. Kaverina, I, Rottner, K, and Small, JV (13-7-1998) Targeting, capture, and stabilization of microtubules at early focal adhesions. *J. Cell Biol.* **142**: 181–190
44. Maekawa, M, Ishizaki, T, Boku, S, Watanabe, N, Fujita, A, Iwamatsu, A, Obinata, T, Ohashi, K, Mizuno, K, and Narumiya, S (6-8-1999) Signaling from Rho to the actin cytoskeleton through protein kinases ROCK and LIM-kinase. *Science.* **285**: 895–898
45. Sumi, T, Matsumoto, K, Takai, Y, and Nakamura, T (27-12-1999) Cofilin phosphorylation and actin cytoskeletal dynamics regulated by rho- and Cdc42-activated LIM-kinase 2. *J. Cell Biol.* **147**: 1519–1532
46. Katoh, K, Kano, Y, Amano, M, Onishi, H, Kaibuchi, K, and Fujiwara, K (30-4-2001) Rho-kinase–mediated contraction of isolated stress fibers. *J. Cell Biol.* **153**: 569–584
47. Kaibuchi, K, Kuroda, S, and Amano, M (1999) Regulation of the cytoskeleton and cell adhesion by the Rho family GTPases in mammalian cells. *Annu. Rev. Biochem.* **68**: 459–486
48. Amano, M, Fukata, Y, and Kaibuchi, K (25-11-2000) Regulation and functions of Rho-associated kinase. *Exp. Cell Res.* **261**: 44–51
49. Wang, HR, Zhang, Y, Ozdamar, B, Ogunjimi, AA, Alexandrova, E, Thomsen, GH, and Wrana, JL (5-12-2003) Regulation of cell polarity and protrusion formation by targeting RhoA for degradation. *Science.* **302**: 1775–1779
50. Roberts, TM and Stewart, M (3-4-2000) Acting like actin. The dynamics of the nematode major sperm protein (MSP) cytoskeleton indicate a push-pull mechanism for amoeboid cell motility. *J. Cell Biol.* **149**: 7–12
51. Italiano, JE, Jr, Stewart, M, and Roberts, TM (6-9-1999) Localized depolymerization of the major sperm protein cytoskeleton correlates with the forward movement of the cell body in the amoeboid movement of nematode sperm. *J. Cell Biol.* **146**: 1087–1096
52. Miao, L, Vanderlinde, O, Stewart, M, and Roberts, TM (21-11-2003) Retraction in amoeboid cell motility powered by cytoskeletal dynamics. *Science.* **302**: 1405–1407
53. Pelham, RJ, Jr and Wang, Y (1999) High resolution detection of mechanical forces exerted by locomoting fibroblasts on the substrate. *Mol. Biol. Cell.* **10**: 935–945
54. Knecht, DA and Loomis, WF (29-5-1987) Antisense RNA inactivation of myosin heavy chain gene expression in *Dictyostelium discoideum*. *Science.* **236**: 1081–1086
55. Wessels, D, Soll, DR, Knecht, D, Loomis, WF, De Lozanne, A, and Spudich, J (1988) Cell motility and chemotaxis in *Dictyostelium* amebae lacking myosin heavy chain. *Dev. Biol.* **128**: 164–177
56. De Lozanne, A and Spudich, JA (29-5-1987) Disruption of the *Dictyostelium* myosin heavy chain gene by homologous recombination. *Science.* **236**: 1086–1091
57. Uchida, KS, Kitanishi-Yumura, T, and Yumura, S (1-1-2003) Myosin II contributes to the posterior contraction and the anterior extension during the retraction phase in migrating *Dictyostelium* cells. *J. Cell Sci.* **116**: 51–60
58. Jay, PY, Pham, PA, Wong, SA, and Elson, EL (1995) A mechanical function of myosin II in cell motility. *J. Cell Sci.* **108 (Pt 1)**: 387–393
59. Huttenlocher, A, Palecek, SP, Lu, Q, Zhang, W, Mellgren, RL, Lauffenburger, DA, Ginsberg, MH, and Horwitz, AF (26-12-1997) Regulation of cell migration by the calcium-dependent protease calpain. *J. Biol. Chem.* **272**: 32719–32722
60. Du, X, Saido, TC, Tsubuki, S, Indig, FE, Williams, MJ, and Ginsberg, MH (3-11-1995) Calpain cleavage of the cytoplasmic domain of the integrin beta 3 subunit. *J. Biol. Chem.* **270**: 26146–26151
61. Potter, DA, Tirnauer, JS, Janssen, R, Croall, DE, Hughes, CN, Fiacco, KA, Mier, JW, Maki, M, and Herman, IM (4-5-1998) Calpain regulates actin remodeling during cell spreading. *J. Cell Biol.* **141**: 647–662
62. Beckerle, MC, Burridge, K, DeMartino, GN, and Croall, DE (20-11-1987) Colocalization of calcium-dependent protease II and one of its substrates at sites of cell adhesion. *Cell.* **51**: 569–577
63. Cooray, P, Yuan, Y, Schoenwaelder, SM, Mitchell, CA, Salem, HH, and Jackson, SP (15-8-1996) Focal adhesion kinase (pp125FAK) cleavage and regulation by calpain. *Biochem. J.* **318 (Pt 1)**: 41–47

CHAPTER 3.3

CELL MIGRATION: Navigation by Chemotaxis

Chemotaxis is the directional migration of cells, or parts of cells, in response to a gradient of concentration of a diffusible molecule. It is the diffusibility of the guidance molecule(s) that distinguishes chemotaxis from haptotaxis, in which cells are guided by molecules anchored on a substrate (see Chapter 3.5). Chemotaxis is an important mechanism for cell guidance during development and continues to be used in some aspects of adult life, such as the recruitment of neutrophils in the vertebrate inflammatory response.

For chemotaxis to work, a developing system must possess the following features:

- A mechanism for creating the gradient of the chemoattractant (chemorepellant)
- A mechanism that allows parts of migrating cells to sense local chemoattractant concentrations
- A mechanism that translates typically shallow external gradients in external chemoattractant into typically steep gradients in protrusive activity within a cell

This chapter will consider each of these elements in turn and will go on to address how chemotaxis is identified experimentally. Finally, some examples of chemotaxis in development will be reviewed.

THE CHEMOTACTIC GRADIENT

There are many possible ways of producing a chemotactic gradient, but most biological examples studied to date seem to rely on producing a chemoattractant from a localized source and then allowing it to diffuse away. The rate of diffusion is determined by the chemoattractant's diffusion constant and by the local concentration gradient. The change in concentration over time at any point in the field is given by Fick's law, which states for one-dimensional systems that

$$dm/dt = -D(d^2m/dx^2)$$

where m is the local concentration of the molecule and D its diffusion constant. The equation scales to three dimensions in the obvious way:*

$$dm/dt = -D\nabla^2 m$$

Real chemotactic fields usually involve a source that produces chemoattractant continuously (or at least over a time that is long compared to the response time of migrating cells) and a sink that destroys it. The sink may be localized, but it is commonly "distributed," for example as an enzyme activity scattered homogeneously throughout the field. The rate of production (P) and the degradation constant (k_{deg}) are important constants in solutions of Fick's equation in biological situations. Gradients in real tissues are rarely measured quantitatively, but they are likely to be very complex. It is useful, however, to consider the parameters that shape a gradient in a very simple model space because the insights from this exercise will be important to the discussion of gradient sensing later in this chapter.

Consider a cylinder of tissue of length L, one end of which is a disc of cells that acts as a uniform source of chemoattractant (see Figure 3.3.1). Assume that degradative enzymes are distributed evenly throughout the tissue and that its sides are impermeable so that overall diffusion is simply along the cylinder's axis. For such a simple system, the solution to the diffusion equation when the system has reached a steady state is[1]:

$$m(r) = \frac{P}{k_{deg}} \frac{L}{\lambda} \frac{\cosh((L - r)/\lambda)}{\sinh(L/\lambda)}$$

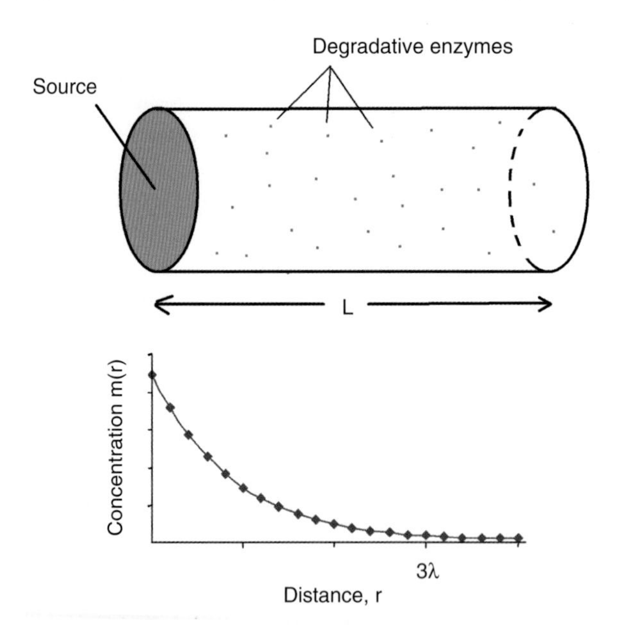

FIGURE 3.3.1 Diffusion from a local source in a simple system.

*The ∇^2 operator, often called the Laplacian, represents in a single symbol the rates of change of concentration gradients in each axis, $(\delta^2/\delta x^2, \delta^2/\delta y^2, \delta^2/\delta z^2)$.

where m(r) is the concentration m, r micrometers from the source and λ is defined by

$$\lambda = \sqrt{(D/k_{deg})}$$

In this equation, λ dominates the shape of the concentration curve with respect to r and is called the "space constant"; its dimensions ("units") are just (micro)meters. About 95 percent of the molecules are localized within a distance of 3λ from the source (see Figure 3.3.1). For a given rate of degradation, k_{deg}, slow rates of diffusion lead to a high value for λ and thence to a steep concentration gradient (see Figure 3.3.2).

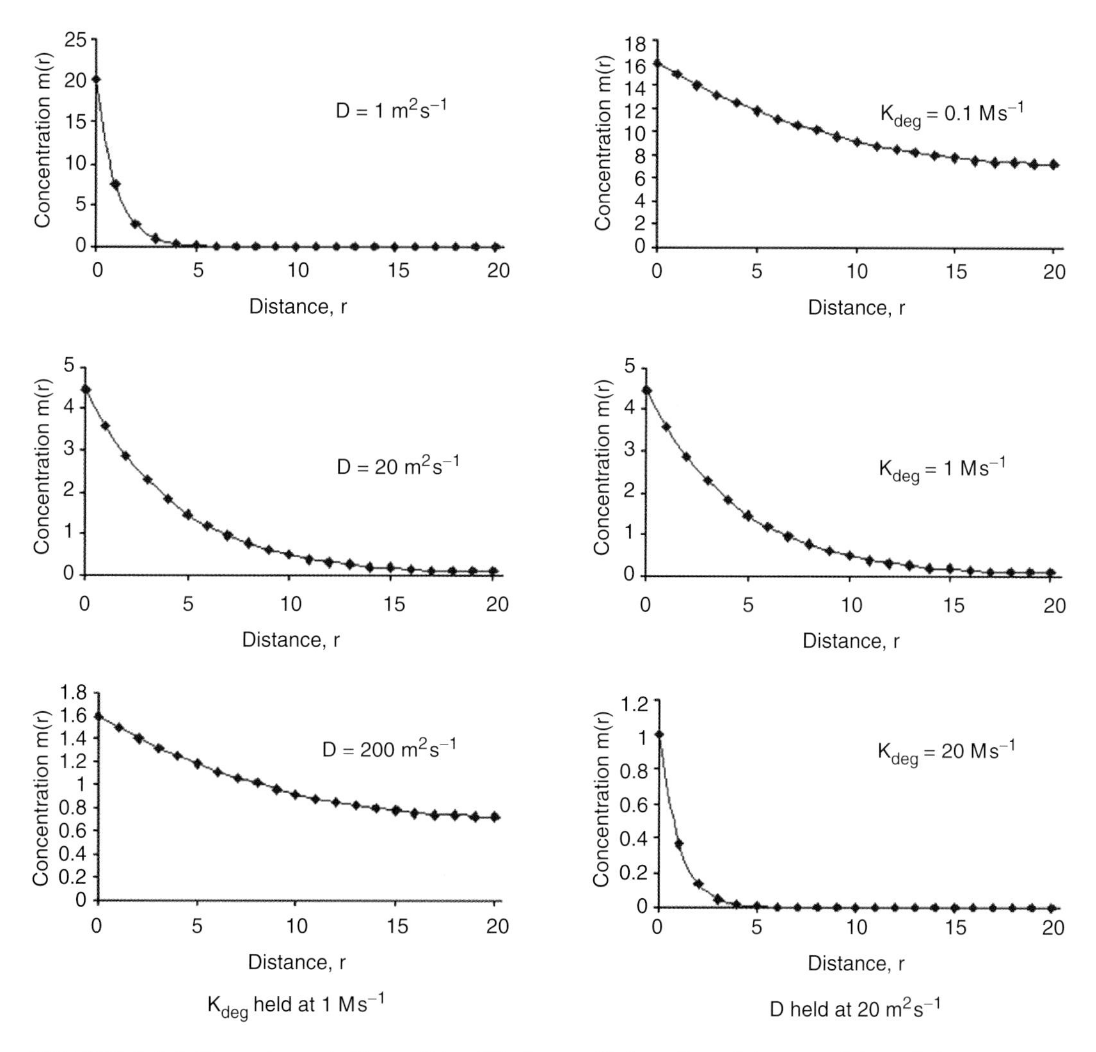

FIGURE 3.3.2 The reciprocal effects of the diffusion coefficient, D, and the rate of degradation, K_{deg}, on the steady-state concentrations of a molecule in the simple diffusion system illustrated in Figure 3.2.1. For this large-scale example, units of concentration are in M, units of distance in m, and the "fixed" variables were as follows: L = 20 m, P = 1 Ms^{-1}.

Slow rates of diffusion (large D) have the opposite effect. The effect of the degradation constant, k_{deg}, is complementary to that of the diffusion constant—high rates of degradation diminish λ and therefore sharpen the gradient (see Figure 3.2.2).

To sum up, slow diffusion and fast destruction create steep gradients, whereas fast diffusion and slow destruction create shallow ones.

The diffusion constants of some real signalling molecules have been determined experimentally, and they differ by orders of magnitude. Large proteins that associate with extracellular matrices have D values of less than 1 $\mu m^2/s$, soluble proteins have D values around 10 $\mu m^2/s$,[ref2] whereas small molecules such as cyclic adenosine monophosphate (cAMP) have D values $>100\ \mu m^2/s$.[ref3,4] For a given rate of destruction, larger molecules produce steeper gradients. Rates of destruction can alter according to context, however, so that a small molecule such as cAMP may act as a shallow-gradient, long-range signalling molecule extracellularly where it is relatively long-lived, or as a steep gradient signalling molecule in the cytoplasm where it is quickly destroyed.

These are general principles of chemotactic gradients, but the ultra-simple model system used to demonstrate them mathematically is not typical of real biological fields. In life, the arrangement of the source cells can be complicated, the diffusion "constant" may in fact vary with direction (as when tissues provide channels to facilitate diffusion in one direction and barriers to impede it in another), and the gradient may be altered by the very cells that are migrating along it; an example of this will be discussed below.

READING THE CHEMOTACTIC GRADIENT

Chemotactic signals normally act over a range that is large compared to the size of an individual cell. That means that their space constant, λ, is large and the gradient correspondingly shallow. For typical chemotactic gradients, the difference between concentrations at the front and at the back of a cell is less than 2 percent.[5] This presents the cell with a formidable problem in navigation. In the systems that have been studied at a deep level—mainly the aggregating myxamoebae of *Dictyostelium discoideum* but also a few vertebrate cell types such as neutrophils—the problem seems to be solved by re-encoding the shallow external gradient as a steep internal one and using feedback to control accurately the direction of that internal gradient.

Aggregating myxamoebae of *D. discoideum* migrate chemotactically toward a source of cyclic adenosine monophosphate (cAMP), which they detect by means of a serpentine transmembrane receptor, cAR1. cAR1 functions as a conventional G-protein-coupled receptor; its binding by extracellular cAMP causes it to activate and release its bound heterotrimeric G protein $G\alpha_2\beta\gamma$ (see Figure 3.3.3). The $G\alpha_2\beta\gamma$ complex initiates several signal transduction cascades, one of the most significant of which (from the point of view of chemotaxis) results in the activation of PI-3-kinase. PI-3-kinase phosphorylates the phospholipid $PI(4,5)P_2$, a minor component of the inner face of the plasma membrane, to produce $PI(3,4,5)P_3$. $PI(3,4,5)P_3$is a ligand for the plekstrin homology (PH) domains that are a feature of many signal transduction proteins, and these PH domain-containing proteins therefore become located and activated at regions of membrane where $PI(3,4,5)P_3$ is present. The parts of the cell that are high in $PI(3,4,5)P_3$ produce lamellipodia by a mechanism that will be discussed later. This regulatory effect of $PI(3,4,5)P_3$ has been demonstrated dramatically in other cell types,

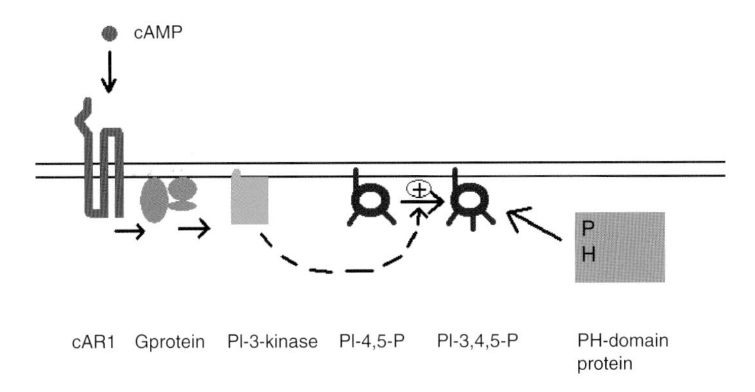

FIGURE 3.3.3 The signalling cascade activated by cAMP in *Dictyostelium discoideum.*

such as vertebrate fibroblasts and neutrophils, which can be induced to activate lamellipodia and move in response to a membrane-permeable analogue of $PI(3,4,5)P_3$ in the absence of any other directional cues.[6,7]

The signal transduction cascade therefore links the external cAMP chemoattractant to the mechanism of cell movement. This might be enough to make cAMP a motogen (an agent that activates cell movement), but it would not be sufficient to cause chemotaxis. For chemotaxis, there must be a mechanism to restrict motile activity to only that part of the cell facing the highest concentration of cAMP. There are two elements to this: the shallow external cAMP gradient must be represented as a steep internal gradient inside the cell (lest the entire cell try to become a lamellipodium), and this steep internal gradient must be aligned accurately with the shallow external one.

The creation of a steep internal gradient is achieved partly by a clever biophysical sleight of hand inherent in the signal transduction cascade itself. The diffusion coefficient, D, of cAMP is 270 $\mu m^2 s^{-1}$[ref3] but that of the $PI(3,4,5)P_3$ used to represent it internally is only about 1 $\mu m^2 s^{-1}$[ref8], mainly because it is attached to the membrane. If for the moment we just *assume* that the signal transduction pathway of Figure 3.3.3 is active only at the point of the cell circumference facing the highest concentration of cAMP, then with all other things equal, the space constant λ for the internal gradient would be less than a tenth of that of the external gradient (see Figure 3.3.4). Gradient sharpening is therefore an inherent property of the signal transduction cascade, provided that the assumption that the cascade is confined to just one part of the cell circumference is satisfied. Without this confinement, however, each part of the cell circumference would act as its own source of diffusing $PI(3,4,5)P_3$ in proportion to the amount of external cAMP at that site; the gradient would be steep enough in a centripetal direction to confine lamellipodial activity to regions just behind the membrane, but in a tangential direction, the relative amounts of $PI(3,4,5)P_3$ would resemble the external gradient of cAMP and the cell would try to move in all directions at once (see Figure 3.3.5). Understanding the mechanism that confines signal transduction to one area of the plasma membrane is therefore critical to understanding chemotactic behaviour in this system.

Direct evidence that the signal transduction cascade triggered by cAMP is localized to the up-gradient end of the cell has been provided by experiments in which specific

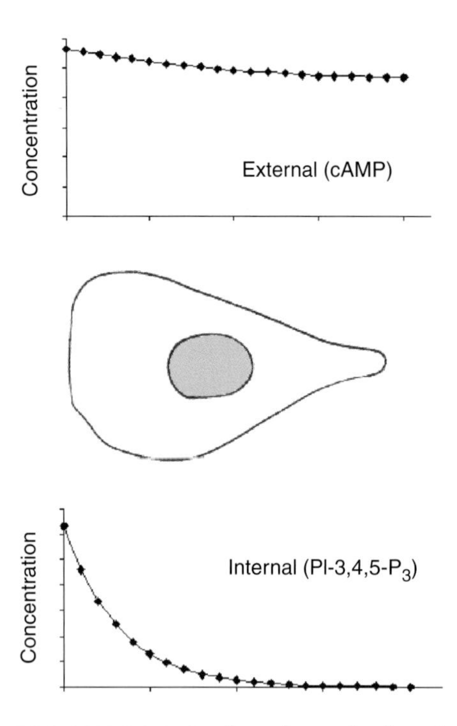

FIGURE 3.3.4 Gradient sharpening by using a lower diffusion constant for the internal representation of the gradient. This diagram considers only a single dimension along the axis of the cell. The reason why there are not competing gradients coming from the sides of the cell is considered in the main text.

components of the cascade were tagged with green fluorescent protein (GFP). In myxamoebae that are migrating up a cAMP gradient, the cAMP receptor cAR1 and its G protein are distributed evenly around the membrane.[9] PI-3-kinase, however, has a markedly inhomogeneous distribution and is concentrated at the up-gradient edge.[10] This is true only when external cAMP shows a concentration gradient; if cells are placed in a homogenous bath of cAMP, PI-3-kinase still locates to the membrane but does so around the circumference of the cell.[10] It has not yet been possible to image the location of PI(3,4,5)P_3 directly, but GFP-tagged proteins with PH domains are targeted to the leading edge of the cell where PI-3-kinase is located, strongly suggesting that this is where PI(3,4,5)P_3 is mainly found.[11] It has been suggested on theoretical grounds that a positive feedback loop, in which PI-3-kinase activity and generation of PI(3,4,5)P_3 encourages location of further active PI-3-kinase molecules to the area, might be an important mechanism in sharpening the gradient.[1] Recent experimental evidence has shown this not to be the case, however; cells treated with the PI-3-kinase inhibitor LY294002 still show normal location of PI-3-kinase protein.[12]

The activity of PI-3-kinase is opposed by the phosphatase PTEN, which dephosphorylates at the 3 position of PI(3,4,5)P_3. PTEN shows a distribution complementary to PI-3-kinase; in cells not treated with cAMP, it is located at the membrane all around the cell and is active. In cells treated with homogenous cAMP, it leaves the membrane and loses its activity. In those treated with a gradient of cAMP, it is located and active

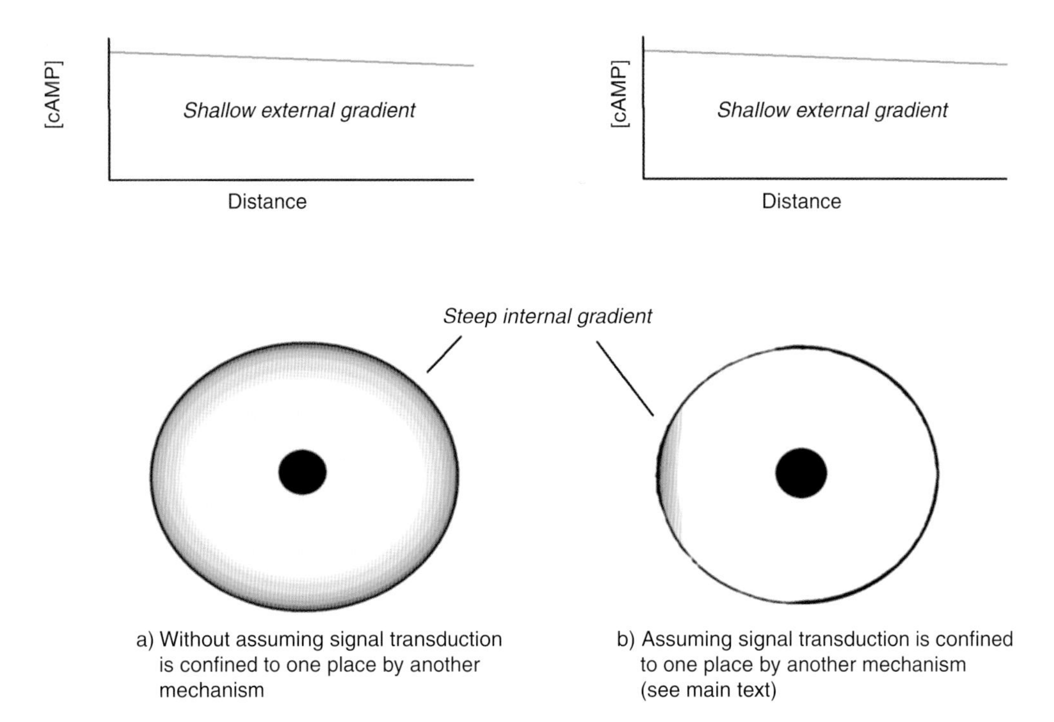

FIGURE 3.3.5 The gradient-sharpening effect of transducing an external signal to an internal one carried by a molecule with a lower coefficient of diffusion, D.

in all regions of the membrane *except* the leading edge. Both the membrane location and activity of PTEN are regulated by a PI(4,5)P_2-binding domain.[13] The fact that PI-3-kinase removes PI(4,5)P_2 from its locale, by phosphorylating it to make PI(3,4,5)P_3, suggests one mechanism by which PTEN may be excluded from PI-3-kinase–rich areas. In the absence of a strong activation of PI-3-kinase, PTEN-rich areas may maintain themselves by undoing the work of any PI-3-kinase that does happen to be around, again helping to separate areas of strong PI(3,4,5)P_3 production, such as the leading edge of the cell, from areas in which it is virtually absent. However useful these feedback loops are, they cannot be the whole reason for the complementary distribution of PI-3-kinase and PTEN, because blocking the activity of either enzyme, by mutation in the case of PTEN or with the drug LY294002 in the case of PI-3-kinase, fails to randomize the distribution of the proteins, although their domains may change a little in size.

At the time of writing, it is still not clear exactly how PI-3-kinase becomes so highly localized to the up-gradient end of the cell.

LINKING THE INTERNAL REPRESENTATION OF THE EXTERNAL GRADIENT TO MOTILITY

Work in *D. discoideum* and in other cells suggests a relatively simple link between the signal transduction system of Figure 3.3.3 and the production of a lamellipodium.

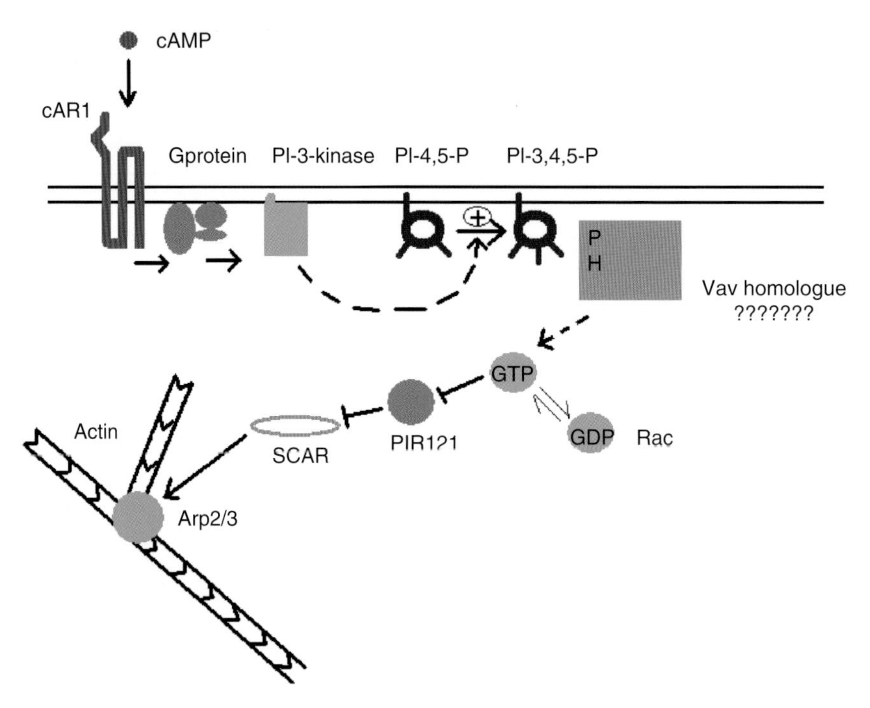

FIGURE 3.3.6 Hypothetical link between the cAMP sensing pathway, known as far as PI(3,4,5)P_3, and the control of lamellipodial actin, known from Rac-GTP onwards. The suggested link is based on data from vertebrate cells.[14,16]

There is good evidence, from a large number of mutant and other studies, that *D. discoideum* assembles its lamellipodium using Rac-mediated activation of actin filament branching, via SCAR and Arp2/3, that is described in Chapter 3.2. The existence of a simple linkage between PI(3,4,5)P_3 and Rac has not yet been demonstrated in this organism, but mammalian cells contain proteins such as vav that bind PI(3,4,5)P_3 by means of PH domains and activate Rac.[14] If homologues of these proteins exist in *D. discoideum*, then the complete pathway may be relatively direct, as depicted in Figure 3.3.6. As this diagram would predict, mutations that block Rac activity block cell migration, whereas mutations that result in hyperactive Rac de-couple motility from sensing and cause the appearance of lamellipodium-like membrane ruffles all round the cell, not just in the direction of a cAMP gradient.[15]

Cells migrating by chemotaxis normally have only one protrusive leading edge, at the place on the cell's edge that faces directly up the external gradient of cAMP. This implies that other parts of the cell must be prevented from making a leading edge of their own. The cell is unlikely to use PI(3,4,5)P_3 to repress protrusive activity elsewhere, because the very short range of the PI(3,4,5)P_3 gradient that is so useful shaping activity at the leading edge is useless for signalling across the entire cell. Without a cell-wide signal, though, how is a point on the side of the cell receiving some cAMP to "know" that another part of the cell is receiving more and is already a leading edge?

The available evidence suggests that long-range inhibition of leading edge production is mediated by cyclic GMP, which, like cyclic AMP, has a high coefficient of diffusion and will therefore travel quickly throughout the cytoplasm. The cytoplasmic concentration of cGMP is controlled by the opposed activities of guanyl cyclase, which synthesizes it, and phosphodiesterases that destroy it. Guanyl cyclase is activated by the cAMP sensing system, although the details of this link are not yet known. The dynamics of activation are unusual, in that presenting a cell with a constant concentration of external cAMP causes a transitory rise in internal cGMP, which then falls away; only if the cell is presented with a steadily *rising* concentration of external cAMP does the concentration of internal cGMP stabilize.[17] Effectively, the cGMP system computes the differential of external cAMP with respect to time:

$$[\mathrm{cGMP}] \propto \frac{\mathrm{d}}{\mathrm{dt}}[\mathrm{cAMP}]$$

This connection with the time domain is probably very important to navigation and will be discussed further below.

One of the effects of cGMP is activation of cGMP-dependent myosin light chain kinases that activate myosin and allow it to form filaments that associate with actin.[18] The formation of these filaments, which are seen in the lateral and trailing edges of migrating cells (i.e., everywhere except for the protrusive leading edge), is important for the restriction of protrusive activity. As long as actin is organized as actin/myosin filaments, formation of a branched protrusive network is inhibited. However, once this organization is lost, the cortex is able to accommodate protrusive actin. Mutant *D. discoideum* that have no functional myosin II cannot produce myosin filaments; they form lateral pseudopodia that can change the direction of cell movement and confuse chemotaxis[19] (see Figure 3.3.7). Conversely, experimentally driven increase in the cytoplasmic levels of cGMP increases the speed of chemotaxis through reduced "wasted" protrusion in the lateral parts of the cell.[18] The efficiency of cGMP-driven myosin activation is useful for preventing lateral extension, but would be disastrous if it were allowed to convert even the leading edge cytoskeleton to actin/myosin bundles. This is prevented by another kinase, myosin heavy chain kinase A (MHCK-A), which locates

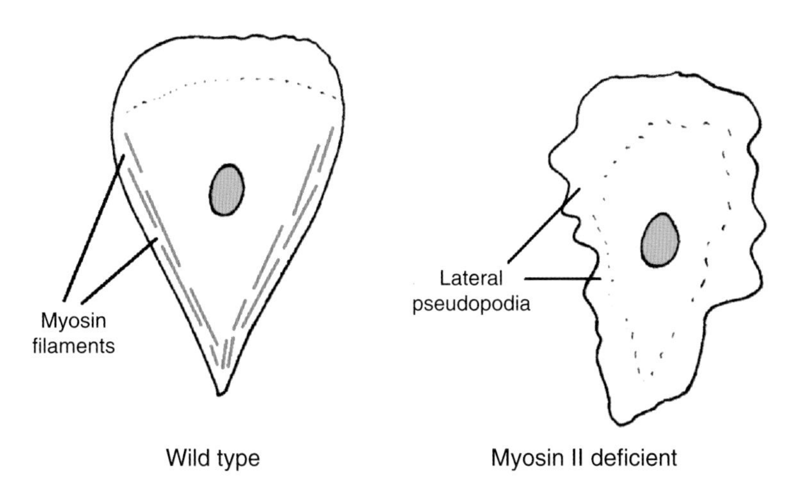

FIGURE 3.3.7 The role of myosin II filaments in suppressing formation of lateral pseudopodia.

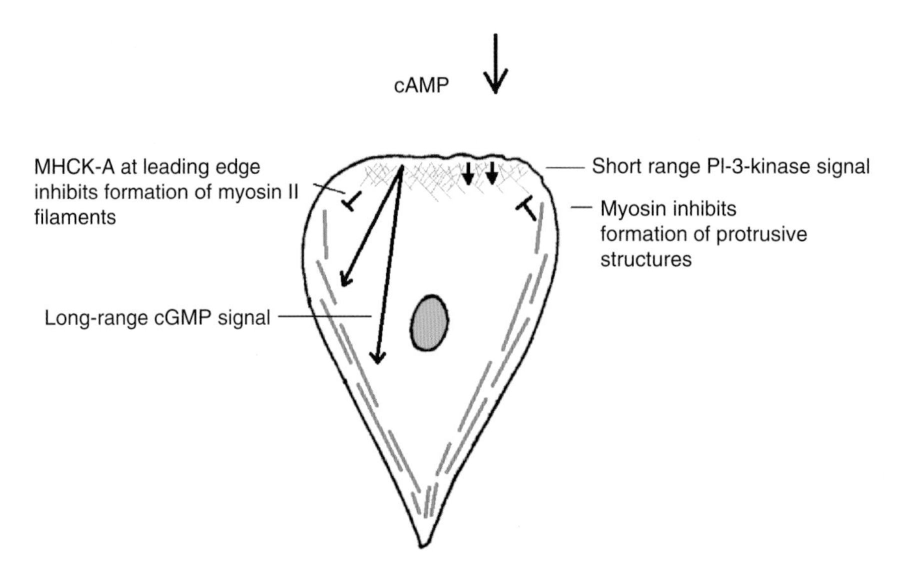

FIGURE 3.3.8 Signalling pathways responsible for separating zones of protrusion from zones of myosin II contraction.

to the leading edge because it can bind to the fine filaments of F-actin that dominate that area.[20,21,22] MHCK-A phosphorylates myosin II heavy chains in a manner that prevents them from forming filaments, even when the cGMP pathway is active. The leading edge is therefore "safe."

The cGMP signalling system is at the heart of a feedback system that operates in both the space and time domains. The spatial function ensures that, in a cell migrating up a chemotactic gradient, protrusive activity is confined to the leading edge. The fine actin filaments already present there recruit MHCK-A to protect the area from formation of myosin II filaments, but all other parts of the cell are encouraged by cGMP to form myosin II filaments and are therefore prevented from producing new protrusions of their own. Assuming that the cGMP is produced mainly at the leading edge, this system would allow an established leading edge to hold back any potential rivals, and is an example of lateral inhibition (see Figure 3.3.8).

The importance of mutual antagonism between the domains of the cell full of protrusive actin and those full of myosin filaments has been demonstrated dramatically in small fragments of fish keratinocytes, highly mobile cells that are unusually large and accessible. When the cells are divided into fragments, some are inherently polarized and some are not because their cytoskeletons are more or less homogenous. If a transient mechanical stimulus is applied to one part of an un-polarized cell fragment, the cytoskeleton responds by organizing myosin locally to resist deformation of the cytoskeleton (see Chapter 2.2). Remarkably, this induced anisotropy then polarizes the cell so that a leading edge forms, and the cell begins to move; it then retains its polarity even in the absence of the pressure that caused it.[23] Once their symmetry is broken, cells can therefore set up a polarized locomotion system by adaptive self-organization.

The dynamics of cGMP in the time domain suggest the possibility of error-checking negative feedback. As was mentioned above, cGMP concentrations reflect the rate of change of external cAMP and are stable only when the absolute concentration of cAMP is increasing. As long as a cell is migrating up a cAMP gradient, this concentration will be increasing, cGMP will be stable, and the leading edge will have no rivals trying to pull the cell another way. If the concentration of external cAMP ceases to increase, though—either because the cell has made an error and is migrating across or even down the gradient, or because the source has moved—internal levels of cGMP will fall, and lateral domains of the membrane will be allowed to begin protrusive exploration. Those facing up the new gradient will be most likely to organize themselves into a new leading edge, and the cell will set off in another direction. Cell turning by decay of the original leading edge and protrusion of a new one has been directly observed in *D. discoideum*,[24] although less abrupt turning of the whole cell, so that the existing leading edge continues to lead, can be seen, too.

HOW GOOD A MODEL IS *D. DISCOIDEUM* FOR OTHER SPECIES?

D. discoideum has been chosen as a model to illustrate the principles of chemotaxis simply because it is the system that is understood in most detail. The complexity of metazoan development makes detailed analysis of the cellular mechanisms of chemotaxis difficult in the context of an embryo. Some chemotactic cells of adults, such as neutrophil polymorphonuclear leukocytes of the mammalian immune system, can be obtained in large enough quantities for their chemotactic mechanisms to be studied in detail so that they can be compared with those of *D. discoideum*.

Neutrophils migrate chemotactically toward N-formyl peptides such as fMet-Leu-Phe, which are produced by bacteria at sites of infection, and will show robust responses to such chemoattractants in simple culture systems. Migrating neutrophils have a general form similar to myxamoebae, having a leading edge full of a fine network of actin where protrusion is concentrated and a myosin-rich trailing edge. When presented with a shallow external gradient of chemoattractant, neutrophils generate a steep gradient of internal PI(3,4,5)P_3, which can be demonstrated by tagging the PI(3,4,5)P_3-seeking PH domain of Akt.[25] Furthermore, experiments with inhibitors confirm that production of PI(3,4,5)P_3 is necessary for chemotaxis.[26] As in myxamoebae, PI(3,4,5)P_3 production leads to activation of Rac[27] and thence production of a protrusive network of actin. Direct local application of a membrane-permeable complex of PI(3,4,5)P_3 to fibroblasts, even in the absence of an external gradient of chemoattractant, triggers polarization of the cell and chemotactic movement.[28] All of this suggests that the chemotaxis of neutrophils is remarkably similar to that of myxamoebae. There are some differences, however. Neutrophils seem not to segregate PTEN to their sides and trailing edges and rely completely on myosin to prevent lateral protrusion. The pathway that controls myosin II via Rho is quite separate from that controlling the leading edge, and the pathways diverge right up at the level of the G proteins. The PI(3,4,5)P_3/Rac path proceeds via Gi, while two other G proteins, G12 and G13, act via Rho to control myosin. Again, the "front" and "back" signals antagonize one another as they do in myxamoebae.[29] One other interesting difference between neutrophils and myxamoebae is that the leading edge of neutrophils is much more sensitive to chemoattractants than the trailing edge. This creates a marked reluctance of the cells to re-polarize, and they tend to respond to sudden changes in external gradient by performing a U-turn rather than by reversing.[30] Overall, though, the similarities between neutrophils and myxamoebae are striking, and while the vertebrate cells seem to be complicated by several parallel pathways triggered by different signals,[31] the basic ideas remain the same.

Not all chemotactic signals proceed via G proteins. The mammalian wound-healing response includes chemotactic migration of fibroblasts up a gradient of platelet derived growth factor (PDGF), which is released by the wound itself. PDGF signals using receptor tyrosine kinases rather than G-proteins. Nevertheless, the chemotactic signalling mechanism still relies on PI-3-kinases, a requirement that can be demonstrated by pharmacological inhibition of PI-3-kinase[32] and by mutants that cannot activate PI-3-kinase in response to PDGF signalling.[33,34,32] It also involves the translocation of Akt to the leading edge of the cell, where the (shallow) external gradient of PDGF is at its highest, a fact that has been demonstrated using Akt tagged with green fluorescent protein.[35] This translocation fails when PI-3-kinase is inhibited.[35] Again, the internal gradient of Akt is far steeper than the external gradient of PDGF.

There has been too little work done on metazoan chemotaxis for any certain conclusions to be drawn. However, the balance of evidence so far suggests that there is essentially one broad mechanism for chemotactic crawling migration that is conserved between cell types within an organism and that has been conserved for the hundreds of millions of years that separate *D. discoideum* and mammals.

CHEMOREPULSION

Chemotaxis does not always involve chemoattraction, and some morphogenetic events in metazoan development depend on the ability of tissues to repel specific migratory cells. Sometimes repulsion is mediated by fixed molecules (see Chapter 3.5), but in other cases, diffusible chemorepellants are used.

One example of a diffusible chemorepellant system is provided by the Slit family of secreted proteins. In mammals, these proteins repel many cell types, including neutrophils,[36] dendritic cells,[37] endothelial cells,[38] glia,[39] and neurons.[40,41] As well as being widespread within one organism, Slit proteins are used as repellents in a broad range of other metazoa, including insects (data from *Drosophila melanogaster*)[42] and nematodes (data from *Caenorhabditis elegans*).[43] Slit proteins signal via transmembrane receptors of the Robo (*Roundabout*) family.[44] Binding of Slit to Robo activates a specific intracellular domain of Robo, CC3, that allows it to interact with the SH3 domain of a GTPase activating protein (GAP) that increases the intrinsic GTPase activity of cdc42.[45] This alters the cdc42-GTP ↔ cdc42-GDP equilibrium to the right and reduces the ability of cdc42 to activate WASP and thence Arp2/3-mediated polymerization of protrusive actin filaments. Slit therefore directly antagonizes the main pathway that encourages leading edge advance, which presumably explains why it antagonizes migration. What is not yet clear, though, is how shallow external gradients of Slit can steer migratory cells as well as just slowing them down. Perhaps there are further internal feedback systems to turn small differences in external concentration into strong internal gradients of second messengers, or perhaps the activity of Robo is itself enough to bias and re-orient the feedback systems (PI(3,4,5)P_3/cGMP, and so on) described earlier in this chapter in the section on chemoattraction.

Semaphorins constitute another important family of repulsive molecules, some of which are secreted and act as chemorepellants, whereas others remain membrane-bound and act by contact inhibition. Semaphorin III (synonyms: collapsin I, semaphorin D) is a

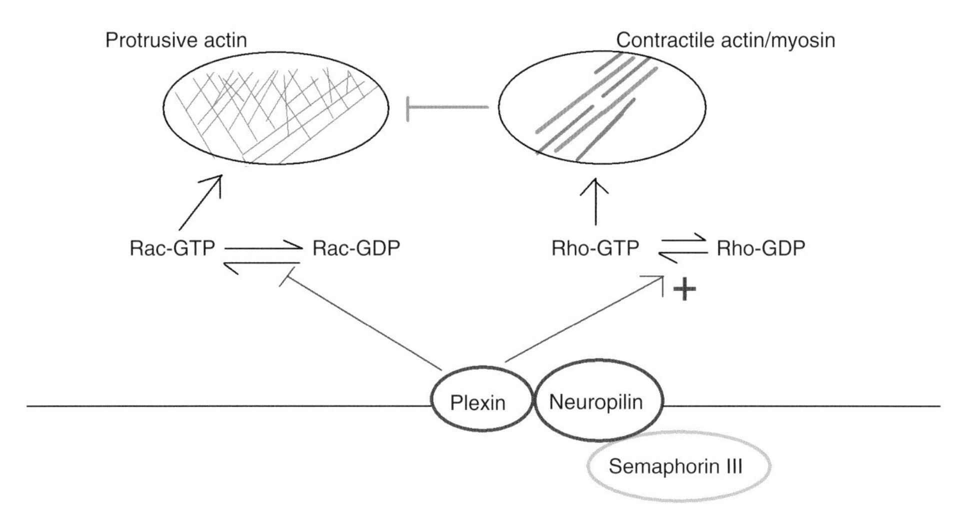

FIGURE 3.3.9 Plexins inhibit protrusive activity both directly via Rho and indirectly by encouraging the formation of contractile myosin filaments that themselves inhibit formation of a protrusive actin network.

secreted semaphorin that will repel the growth cones of sensory and sympathetic neurons in culture.[46,47] It signals via a complex of a transmembrane protein, neuropilin,[48,49] which is expressed on subsets of neurons and also on other cells, such as regulatory T cells of the immune system,[50] and which binds semaphorin III, and a plexin protein that generates intracellular signals in response to semaphorin III/neuropilin binding. Plexins show GAP activity[51] and interact directly with Rac, presumably activating its GTPase activity[52] and also perhaps competing with GNFs for Rac binding.[53] Experiments using dominant negative Rac confirm that Rac is needed for plexins to inhibit growth cone advance.[54] Plexins therefore act in a manner similar to Robo in that they antagonize the ability of Rac to organize a protrusive leading edge. In addition to having a Rac binding domain, plexins have an additional domain that interacts with Rho (at least, *D. melanogaster* plexin does, and conservation of amino acid structures suggests that those of other animals do as well). Genetic analyses, using Rho mutants, suggest that Plexin activates Rho and, therefore, assembly of contractile actin-myosin systems.[53] The receptor system for semaphorins therefore mounts a two-pronged attack on the protrusive activity of a leading edge, reducing the cell's ability to make Arp2/3 nucleated actin branches and also increasing the activity of Rho-mediated polymerization of contractile actin-myosin filaments that give trailing rather than leading edge character to the cell cortex (see Figure 3.3.9).

MULTIPLE SOURCES OF CHEMOREPELLANT CAN DEFINE A PATHWAY IN A WAY THAT MULTIPLE SOURCES OF CHEMOATTRACTANT CANNOT

Cells and growth cones steer away from sources of chemorepellants. If they are in an area of the embryo in which multiple sources of chemorepellants exist, then (in the absence of any additional cues) migrating cells will avoid them all by keeping to a pathway defined by the local minimum in the concentration field. Two extended sources can define a path in two-dimensional space (cells constrained to migrate on a flat sheet of matrix are, effectively, in two-dimensional space), while three extended sources can define a unique path in three-dimensional space. Because the "force" of repulsion rises if a cell blunders nearer the source, the system provides rapid feedback to correct the course of the cell, and the guidance system is stable.

It might be supposed at first glance that multiple sources of chemoattractant could also define a pathway on the principle of Buriden's ass,♣ a cell being unwilling to go toward one source of attractant because such an action would take it further from the others. Such a system would only be stable if the attraction of a place increased the further a cell strayed from it (as if, for example, the cell were attached to that place with a mechanical spring). In chemotaxis, the "force" of attraction is a function of the concentration gradient, which reduces as the cell moves further away. The equilibrium of a cell between two sources is therefore inherently unstable, and if the cell strays even slightly toward one source, the attraction of that source increases while that of the rival source decreases; the cell has been captured. Attractive and repulsive systems are not, therefore, equivalent in their abilities; surround-repulsion can define pathways in a way that attraction cannot.

THE USEFULNESS OF NOISE TO DECISION MAKING BY MIGRATING CELLS

The activity at the leading edge of the cell, while regulated to some extent by external signals, also reflects moment-to-moment fluctuations in local concentrations of key components and of the chance nucleation of actin branching, actin severance, and so on. The probabilities of activations, nucleations, and cuttings are biased by signalling pathways but are still subject to random fluctuations of thermodynamic origin. In other words, the system is "noisy." As noted, noise is often thought of as an enemy of order and a potential nuisance to be eliminated as much as possible from any good control system. It is a very useful attribute of cell navigation, however, because it protects a migrating cell from the fatal indecision of Buriden's ass. Even if a cell is, for a moment, placed exactly between two equally strong sources of chemoattractant, thermal noise will create a temporary random bias in one direction or another. As soon as the cell moves in that direction, the balance of attractions will be broken and a clear environmental cue will be present.

INTEGRATION OF CHEMOTAXIS AND CONTACT GUIDANCE

In real embryos, chemotaxis is seldom the sole source of navigational information, and cells migrating in response to chemotaxis are generally crossing substrates that are also providing navigational information of some sort, even if only of a low-grade "you may

♣The ass in Buriden's fable died of starvation because it was precisely between two bales of hay so was unable to decide which to eat first.

crawl across this area/you may not crawl across that area" sort. Generally, chemotaxis is better adapted to provide long-range navigational information and contact-guidance, short-range information, and the cell has to integrate both types of cue. Generally, since both cues tend to converge on the Rac/Rho/cdc42 signalling systems, integration of directional information is automatic. Other branches of intracellular pathways can also determine the ways in which signals are integrated; for example, the binding of a particular component of the extracellular matrix might result in a change in gene transcription to cause expression of a receptor for an already-present chemoattractant for the first time. In this way, cells can use local information about their current location to "decide" which long-range cues to follow. The ability of a cell to focus on different navigational cues depending on where it finds itself is the basis of the "waypoint navigation" discussed in more detail in Chapter 3.6.

This chapter has presented an account of chemotaxis using just a few examples, chosen to be those that are known in most detail. There are many other cell types that are presumed to respond by chemotaxis to a range of other signalling molecules. The striking similarity of the systems studied so far and their convergence on the same intracellular mediators of cytoskeletal activity suggests that all other examples will work by the same general mechanisms. Nevertheless, it is worth keeping an open mind on that issue, at least until more examples have been studied in detail.

Reference List

1. Postma, M and Van Haastert, PJ (2001) A diffusion-translocation model for gradient sensing by chemotactic cells. *Biophys. J.* **81**: 1314–1323
2. Arrio-Dupont, M, Foucault, G, Vacher, M, Devaux, PF, and Cribier, S (2000) Translational diffusion of globular proteins in the cytoplasm of cultured muscle cells. *Biophys. J.* **78**: 901–907
3. Chen, C, Nakamura, T, and Koutalos, Y (1999) Cyclic AMP diffusion coefficient in frog olfactory cilia. *Biophys. J.* **76**: 2861–2867
4. Allbritton, NL, Meyer, T, and Stryer, L (11-12-1992) Range of messenger action of calcium ion and inositol 1,4,5-trisphosphate. *Science.* **258**: 1812–1815
5. Chung, CY, Funamoto, S, and Firtel, RA (2001) Signaling pathways controlling cell polarity and chemotaxis. *Trends Biochem. Sci.* **26**: 557–566
6. Niggli, V (12-5-2000) A membrane-permeant ester of phosphatidylinositol 3,4,5-trisphosphate (PIP(3)) is an activator of human neutrophil migration. *FEBS Lett.* **473**: 217–221
7. Derman, MP, Toker, A, Hartwig, JH, Spokes, K, Falck, JR, Chen, CS, Cantley, LC, and Cantley, LG (7-3-1997) The lipid products of phosphoinositide 3-kinase increase cell motility through protein kinase C. *J. Biol. Chem.* **272**: 6465–6470
8. Almeida, PFF and Vaz, WLC (1995) Lateral diffusion in membranes. (Elsevier)
9. Comer, FI and Parent, CA (31-5-2002) PI 3-kinases and PTEN: How opposites chemoattract. *Cell.* **109**: 541–544
10. Funamoto, S, Meili, R, Lee, S, Parry, L, and Firtel, RA (31-5-2002) Spatial and temporal regulation of 3-phosphoinositides by PI 3-kinase and PTEN mediates chemotaxis. *Cell.* **109**: 611–623
11. Parent, CA, Blacklock, BJ, Froehlich, WM, Murphy, DB, and Devreotes, PN (2-10-1998) G protein signalling events are activated at the leading edge of chemotactic cells. *Cell.* **95**: 81–91
12. Iijima, M, Huang, YE, Luo, HR, Vazquez, F, and Devreotes, PN (5-2-2004) Novel mechanism of PTEN regulation by its PIP2 binding motif is critical for chemotaxis. *J. Biol. Chem.* **279**: 16606–16613.

13. Iijima, M and Devreotes, P (31-5-2002) Tumor suppressor PTEN mediates sensing of chemoattractant gradients. *Cell.* **109**: 599–610
14. Hornstein, I, Alcover, A, and Katzav, S (2004) Vav proteins, masters of the world of cytoskeleton organization. *Cell Signal.* **16**: 1–11
15. Chung, CY, Lee, S, Briscoe, C, Ellsworth, C, and Firtel, RA (9-5-2000) Role of Rac in controlling the actin cytoskeleton and chemotaxis in motile cells. *Proc. Natl. Acad. Sci. U.S.A.* **97**: 5225–5230
16. Kim, C, Marchal, CC, Penninger, J, and Dinauer, MC (15-10-2003) The hemopoietic Rho/Rac guanine nucleotide exchange factor Vav1 regulates N-formyl-methionyl-leucyl-phenylalanine-activated neutrophil functions. *J. Immunol.* **171**: 4425–4430
17. Van Haastert, PJ and Van der Heijden, PR (1983) Excitation, adaptation, and deadaptation of the cAMP-mediated cGMP response in *Dictyostelium discoideum*. *J. Cell Biol.* **96**: 347–353
18. Bosgraaf, L, Russcher, H, Smith, JL, Wessels, D, Soll, DR, and Van Haastert, PJ (2-9-2002) A novel cGMP signalling pathway mediating myosin phosphorylation and chemotaxis in *Dictyostelium*. *Embo J.* **21**: 4560–4570
19. Stites, J, Wessels, D, Uhl, A, Egelhoff, T, Shutt, D, and Soll, DR (1998) Phosphorylation of the *Dictyostelium* myosin II heavy chain is necessary for maintaining cellular polarity and suppressing turning during chemotaxis. *Cell Motil. Cytoskeleton.* **39**: 31–51
20. Steimle, PA, Licate, L, Cote, GP, and Egelhoff, TT (10-4-2002) Lamellipodial localization of *Dictyostelium* myosin heavy chain kinase A is mediated via F-actin binding by the coiled-coil domain. *FEBS Lett.* **516**: 58–62
21. de la Roche, MA and Cote, GP (15-3-2001) Regulation of *Dictyostelium* myosin I and II. *Biochim. Biophys. Acta.* **1525**: 245–261
22. Liang, W, Licate, L, Warrick, H, Spudich, J, and Egelhoff, T (24-7-2002) Differential localization in cells of myosin II heavy chain kinases during cytokinesis and polarized migration. *BMC Cell Biol.* **3**: 19
23. Verkhovsky, AB, Svitkina, TM, and Borisy, GG (14-1-1999) Self-polarization and directional motility of cytoplasm. *Curr. Biol.* **9**: 11–20
24. Wessels, D, Vawter-Hugart, H, Murray, J, and Soll, DR (1994) Three-dimensional dynamics of pseudopod formation and the regulation of turning during the motility cycle of Dictyostelium. *Cell Motil. Cytoskeleton.* **27**: 1–12
25. Servant, G, Weiner, OD, Herzmark, P, Balla, T, Sedat, JW, and Bourne, HR (11-2-2000) Polarization of chemoattractant receptor signalling during neutrophil chemotaxis. *Science.* **287**: 1037–1040
26. Wang, F, Herzmark, P, Weiner, OD, Srinivasan, S, Servant, G, and Bourne, HR (2002) Lipid products of PI(3)Ks maintain persistent cell polarity and directed motility in neutrophils. *Nat. Cell Biol.* **4**: 513–518
27. Benard, V, Bohl, BP, and Bokoch, GM (7-5-1999) Characterization of rac and cdc42 activation in chemoattractant-stimulated human neutrophils using a novel assay for active GTPases. *J. Biol. Chem.* **274**: 13198–13204
28. Weiner, OD, Neilsen, PO, Prestwich, GD, Kirschner, MW, Cantley, LC, and Bourne, HR (2002) A PtdInsP(3)- and Rho GTPase-mediated positive feedback loop regulates neutrophil polarity. *Nat. Cell Biol.* **4**: 509–513
29. Xu, J, Wang, F, Van Keymeulen, A, Herzmark, P, Straight, A, Kelly, K, Takuwa, Y, Sugimoto, N, Mitchison, T, and Bourne, HR (25-7-2003) Divergent signals and cytoskeletal assemblies regulate self-organizing polarity in neutrophils. *Cell.* **114**: 201–214
30. Zigmond, SH, Levitsky, HI, and Kreel, BJ (1981) Cell polarity: An examination of its behavioral expression and its consequences for polymorphonuclear leukocyte chemotaxis. *J. Cell Biol.* **89**: 585–592
31. Chodniewicz, D and Zhelev, DV (15-9-2003) Novel pathways of F-actin polymerization in the human neutrophil. *Blood.* **102**: 2251–2258

32. Wennstrom, S, Hawkins, P, Cooke, F, Hara, K, Yonezawa, K, Kasuga, M, Jackson, T, Claesson-Welsh, L, and Stephens, L (1-5-1994) Activation of phosphoinositide 3-kinase is required for PDGF-stimulated membrane ruffling. *Curr. Biol.* **4**: 385–393
33. Kundra, V, Escobedo, JA, Kazlauskas, A, Kim, HK, Rhee, SG, Williams, LT, and Zetter, BR (3-2-1994) Regulation of chemotaxis by the platelet-derived growth factor receptor-beta. *Nature.* **367**: 474–476
34. Wennstrom, S, Siegbahn, A, Yokote, K, Arvidsson, AK, Heldin, CH, Mori, S, and Claesson-Welsh, L (1994) Membrane ruffling and chemotaxis transduced by the PDGF beta-receptor require the binding site for phosphatidylinositol 3' kinase. *Oncogene.* **9**: 651–660
35. Haugh, JM, Codazzi, F, Teruel, M, and Meyer, T (11-12-2000) Spatial sensing in fibroblasts mediated by 3' phosphoinositides. *J. Cell Biol.* **151**: 1269–1280
36. Ford, D, Easton, DF, Stratton, M, Narod, S, Goldgar, D, Devilee, P, Bishop, DT, Weber, B, Lenoir, G, Changclaude, J, Sobol, H, Teare, MD, Struewing, J, Arason, A, Scherneck, S, Peto, J, Rebbeck, TR, Tonin, P, Neuhausen, S, Barkardottir, R, Eyfjord, J, Lynch, H, Ponder, BAJ, Gayther, SA, Birch, JM, Lindblom, A, Stoppalyonnet, D, Bignon, Y, Borg, A., Hamann, U, Haites, N, Scott, RJ, Maugard, CM, and Vasen, H (1998) Genetic heterogeneity and penetrance analysis of the BRCA1 and BRCA2 genes in breast cancer families. *Am. J. Hum. Genet.* **62**: 676–689
37. Guan, H, Zu, G, Xie, Y, Tang, H, Johnson, M, Xu, X, Kevil, C, Xiong, WC, Elmets, C, Rao, Y, Wu, JY, and Xu, H (15-12-2003) Neuronal repellent Slit2 inhibits dendritic cell migration and the development of immune responses. *J. Immunol.* **171**: 6519–6526
38. Park, KW, Morrison, CM, Sorensen, LK, Jones, CA, Rao, Y, Chien, CB, Wu, JY, Urness, LD, and Li, DY (1-9-2003) Robo4 is a vascular-specific receptor that inhibits endothelial migration. *Dev. Biol.* **261**: 251–267
39. Auld, VJ (2001) Why didn't the glia cross the road? *Trends Neurosci.* **24**: 309–311
40. Shu, T, Sundaresan, V, McCarthy, MM, and Richards, LJ (3-9-2003) Slit2 guides both precrossing and postcrossing callosal axons at the midline in vivo. *J. Neurosci.* **23**: 8176–8184
41. Bagri, A, Marin, O, Plump, AS, Mak, J, Pleasure, SJ, Rubenstein, JL, and Tessier-Lavigne, M (17-1-2002) Slit proteins prevent midline crossing and determine the dorsoventral position of major axonal pathways in the mammalian forebrain. *Neuron.* **33**: 233–248
42. Rothberg, JM, Hartley, DA, Walther, Z, and Artavanis-Tsakonas, S (23-12-1988) Slit: An EGF-homologous locus of *D. melanogaster* involved in the development of the embryonic central nervous system. *Cell.* **55**: 1047–1059
43. Hao, JC, Yu, TW, Fujisawa, K, Culotti, JG, Gengyo-Ando, K, Mitani, S, Moulder, G, Barstead, R, Tessier-Lavigne, M, and Bargmann, CI (11-10-2001) C. elegans slit acts in midline, dorsal-ventral, and anterior-posterior guidance via the SAX-3/Robo receptor. *Neuron.* **32**: 25–38
44. Kidd, T, Brose, K, Mitchell, KJ, Fetter, RD, Tessier-Lavigne, M, Goodman, CS, and Tear, G (23-1-1998) Roundabout controls axon crossing of the CNS midline and defines a novel subfamily of evolutionarily conserved guidance receptors. *Cell.* **92**: 205–215
45. Wong, K, Ren, XR, Huang, YZ, Xie, Y, Liu, G, Saito, H, Tang, H, Wen, L, Brady-Kalnay, SM, Mei, L, Wu, JY, Xiong, WC, and Rao, Y (19-10-2001) Signal transduction in neuronal migration: Roles of GTPase activating proteins and the small GTPase Cdc42 in the Slit-Robo pathway. *Cell.* **107**: 209–221
46. Messersmith, EK, Leonardo, ED, Shatz, CJ, Tessier-Lavigne, M, Goodman, CS, and Kolodkin, AL (1995) Semaphorin III can function as a selective chemorepellent to pattern sensory projections in the spinal cord. *Neuron.* **14**: 949–959
47. Shepherd, IT, Luo, Y, Lefcort, F, Reichardt, LF, and Raper, JA (1997) A sensory axon repellent secreted from ventral spinal cord explants is neutralized by antibodies raised against collapsin-1. *Development.* **124**: 1377–1385
48. Kolodkin, AL, Levengood, DV, Rowe, EG, Tai, YT, Giger, RJ, and Ginty, DD (22-8-1997) Neuropilin is a semaphorin III receptor. *Cell.* **90**: 753–762

49. He, Z and Tessier-Lavigne, M (22-8-1997) Neuropilin is a receptor for the axonal chemorepellent Semaphorin III. *Cell.* **90**: 739–751
50. Bruder, D, Probst-Kepper, M, Westendorf, AM, Geffers, R, Beissert, S, Loser, K, von Boehmer, H, Buer, J, and Hansen, W (2004) Neuropilin-1: A surface marker of regulatory T cells. *Eur. J. Immunol.* **34**: 623–630
51. Rohm, B, Rahim, B, Kleiber, B, Hovatta, I, and Puschel, AW (1-12-2000) The semaphorin 3A receptor may directly regulate the activity of small GTPases. *FEBS Lett.* **486**: 68–72
52. Vikis, HG, Li, W, He, Z, and Guan, KL (7-11-2000) The semaphorin receptor plexin-B1 specifically interacts with active Rac in a ligand-dependent manner. *Proc. Natl. Acad. Sci. U.S.A.* **97**: 12457–12462
53. Hu, H, Marton, TF, and Goodman, CS (11-10-2001) Plexin B mediates axon guidance in *Drosophila* by simultaneously inhibiting active Rac and enhancing RhoA signalling. *Neuron.* **32**: 39–51
54. Jin, Z and Strittmatter, SM (15-8-1997) Rac1 mediates collapsin-1-induced growth cone collapse. *J. Neurosci.* **17**: 6256–6263

CHAPTER 3.4

CELL MIGRATION: Galvanotaxis

Galvanotaxis, the guidance of cells by electric fields, has long been a Cinderella subject in developmental biology. The ability of electric fields to guide cells has been shown convincingly in culture, and there is good evidence for the existence of electric fields *in vivo*; for these reasons, galvanotaxis is championed as a developmental mechanism by a small band of enthusiasts. Nevertheless, it is generally ignored in texts and reviews of development, largely because of an enduring suspicion that although electric fields have the potential to steer cells, they are not used for this purpose by real embryos. There is not enough evidence to settle the question yet, but there is, in my opinion, enough to justify this brief review of what may soon be seen to be an important mechanism in development, or at least an important tool in tissue engineering.[1]

CELL MOVEMENT IN RESPONSE TO ELECTRIC FIELDS

Many migratory cell types will respond galvanotactically to electric fields in culture (see Figure 3.4.1). When a small DC field of 600 V/m is applied to a culture of embryonic quail fibroblasts, for example, they migrate toward the cathode.[2] This field strength is very low compared to the fields across a typical plasma membrane (60 mV across 7 nm $\approx$ 8,600,000 V/m), and, to give a human-scaled example, the field applied to the fibroblasts is approximately equal to that between the conductor rails in the tracks of the London Underground. Other cells, such as avian neural crest,[3] mammalian keratinocytes,[4] corneal epithelium cells,[5] and neuronal growth cones,[6] exhibit similar behaviour, albeit with minor differences in timing and minimum effective field strength (often down to 10 V/m).[7] The response of cells to electric fields depends on the substratum on which they are growing; some substrates cause cells to migrate toward the cathode, whereas others cause the same cells to migrate toward the anode.[8]

Not all cells show galvanotaxis; human skin-derived skin melanocytes[9] and fibroblasts[5] fail to respond to fields that are capable of steering keratinocytes from the same kind of skin. This variation in galvanotactic ability supports the assertion that galvanotaxis is a "deliberate" facility possessed by just some cell types rather than an "accidental" property that emerges from the general features of cell biology. It therefore suggests, but does not prove, that galvanotaxis is really used *in vivo*.

Cells do not appear to have any structures that are obviously dedicated to detection of large-scale DC electric fields, and the work on chemotaxis discussed in Chapter 3.3 has shown that cells' sense of direction is encoded by internal chemical gradients.

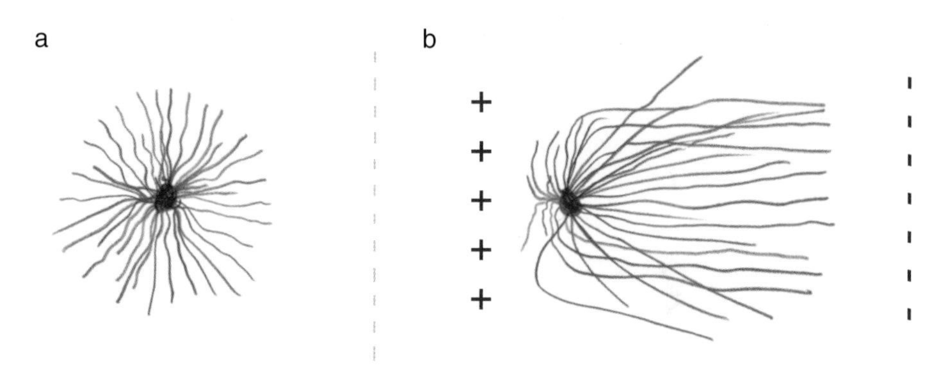

FIGURE 3.4.1 Galvanotactic response of neurites to a DC electric field. (a) In the absence of an external field, neurites grow equally well in all directions from an embryonic ganglion placed in culture. (b) In the presence of an electric field, their direction of growth is biased, typically toward the cathode. That neurites growing toward the cathode are longer than controls with no field is also typical, at least for most tissue culture substrates.

Together, these facts imply that the most likely mechanism for galvanotaxis is one by which electric fields can influence the orientation of chemical gradients inside the cell. Several experimental findings support this idea. It might be expected that, if electric fields steer growth cones by biasing the direction of chemical signalling, they will be more influential when there is a strong chemical signal on which to work. The turning of neuronal growth cones by electric fields will take place at lower field strengths if neurotrophins are also supplied homogeneously rather than in a gradient.[10] Corneal epithelial cells are unresponsive to electric fields in serum-free media but become responsive when serum or EGF is added.[11] Chemical activity caused by application of neurotransmitters has a similar effect. Blocking acetylcholine signalling, arising from small amounts of acetylcholine released by the neurons themselves, using turbocurarine inhibits cathode-directed turning by growth cones,[12] as does blocking the rise in cytoplasmic calcium elicited by acetylcholine signalling.[13] Similarly, blocking the activity of EGF receptors on human keratinocytes with the specific pharmacological inhibitor PD158780 renders the cells insensitive to electric fields, although they still migrate.[14]

The most obvious way in which DC electric fields might influence a chemical system is by electrophoresis, the attraction of negatively charged molecules toward the anode and positively charged ones toward the cathode. This could operate either at the level of biasing the distribution of extracellular chemoattractants or at the level of directly biasing components of intracellular signalling. Direct injection of fluorescently labelled charged proteins into early amphibian limb buds, where high electric fields naturally exist (see below), results in a diffusion pattern that is indeed biased along the electric field, rather like the patterns seen on rocket electrophoresis (see Figure 3.4.2). In principle, electric fields could therefore act indirectly by giving rise to chemotactic gradients through concentrating a chemoattractant to one end of the tissue.

There is also some evidence for electrophoresis of cellular components; cellular glycoproteins capable of binding the lectin concanavalin A (Con A) accumulate on the cathodal side of neurons placed in an electric field.[15] What is more, blocking this accumulation abolishes the galvanotactic response. The principle Con A-binding receptors of these growth cones are the acetylcholine receptors,[13] suggesting that, in this case at

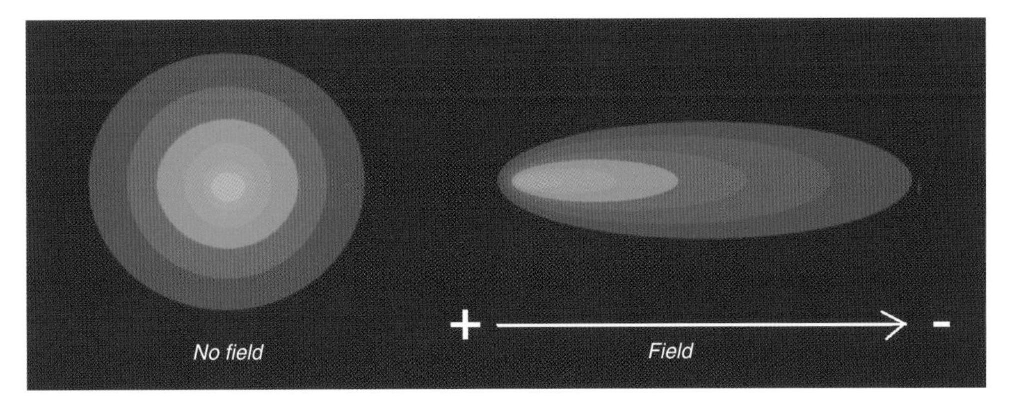

FIGURE 3.4.2 A DC electric field can cause asymmetric diffusion of a charged molecule. The figure assumes a net positive charge, which is typical of growth factors.

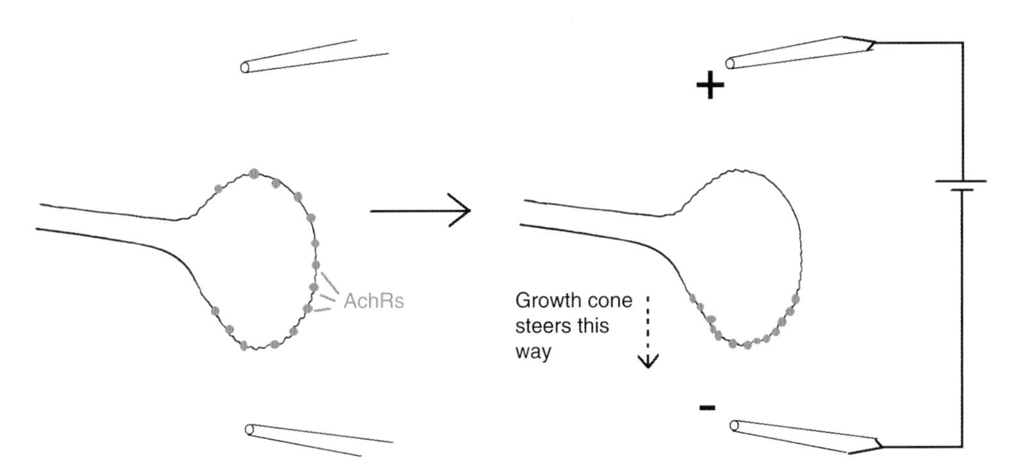

FIGURE 3.4.3 DC electric fields bias the distribution of acetylcholine receptors (AchR) on growth cones. They have a similar effect on EGF receptors on keratinocytes.

least, the electric field controls the direction of migration by causing an unequal distribution of receptors for chemical attractants. Similarly, the EGF receptors of human keratinocytes are drawn to the cathodal side of the cell.[11,14] As far as the internal machinery of a cell is concerned, an induced accumulation of receptors for a potential chemoattractant would have the same result as an external gradient of that chemoattractant; symmetry would be broken, and internal gradient would be established and the cell would move in that direction (see Figure 3.4.3).

The redistribution of receptors as a means of coupling electric fields to directed migration is an intriguing model but needs to be interpreted with caution because simple correlation of receptor accumulation with cell polarization does not discriminate cause from consequence. In the keratinocyte experiments described above, cathodal accumulation of EGF receptors does not take place if the activity of those receptors is blocked by PD158780. This small molecule would not be predicted to have a dramatic effect on the charge or electrophoretic mobility of EGF receptors, so its ability to block their cathodal clustering suggests that EGF receptor activity is essential for clustering to take place. This may be because activation of EGF receptors causes them to move from bulk membrane to lipid rafts,[16] which may alter their mobility, or it may be because the movement of the receptors is a *response* to the setting up of a cathodally directed internal migration orientation system rather than a cause of it. It does not depend on polarization of the actin cytoskeleton, though, because active EGF receptors and the Erk MAP kinase through which they signal polarize in electric fields even when actin polymerization is inhibited.[11]

Myxamoebae of *Dictyostelium discoideum* will also undergo galvanotaxis, but attempts to analyze this genetically argue strongly that galvanotaxis does not operate simply by modulating the chemotactic systems of the cell. Null mutants in the cAMP receptor, the Gα protein and the Gβ protein, all of which are essential for chemotaxis up gradients of cAMP, still undergo galvanotaxis, although its efficiency is reduced in the Gβ mutants. Furthermore, cells polarized by galvanotaxis do not show the strong accumulation of PI(3,4,5)P_3at their leading edges that is such a hallmark of chemotaxis (see Chapter 3.3). The basic machinery of locomotion, such as protrusive actin polymerization at the leading edge, is apparently identical in myxamoebae migrating by chemotaxis and galvanotaxis. Overall, these data suggest that electric fields are sensed by a mechanism quite distinct from that used for chemotaxis, and that the pathways converge on the locomotory machinery of the cell at a deep level, downstream of G proteins and PI-3-kinase.

The search for a mechanism of galvanotaxis has therefore generated several models and, at the moment, some confusion. Simple *in vivo* experiments suggest that action by electrophoresis of a chemoattractant may be possible. Detailed studies of receptors suggest that biasing of receptor distribution may be a mechanism, although it is not clear whether this bias is a cause or a consequence of cell polarization in the electric field. Finally, genetic evidence suggests that cells can orientate themselves in an electric field independently of their chemotactic sensor systems. These views may be reconciled by showing that receptor redistribution is secondary to cell polarization, or it may be that all of these mechanisms are used at different times and places.

ELECTRIC FIELDS IN LIVING SYSTEMS

Biological membranes are equipped with selective channels and pumps that are designed to separate ions and create chemiosmotic gradients. The separation of charges involved in the production of these gradients gives rise to strong electric fields; the 60-mV potential difference present across a typical 7-nm thick plasma membrane produces a field-strength of about 8,600,000 V/m, more than a thousand times the peak field strengths that can be found underneath overhead power lines. Most biological electric fields are local and operate over tiny distances, but where ion fluxes are maintained across a complete tissue, as is the case for many epithelia, small electric fields can be detected at the tissue scale. Mammalian skin and cornea, for example, have a potential of difference of up to 100 mV between the outside (−) and the inside (+),[17] generating a field strength of up to 1,000 V/m.♣

♣This can be demonstrated by placing a high-impedance (battery-operated) voltmeter between a saline solution in which a finger is dipped and a fine hypodermic needle that is stuck through the skin of the hand,

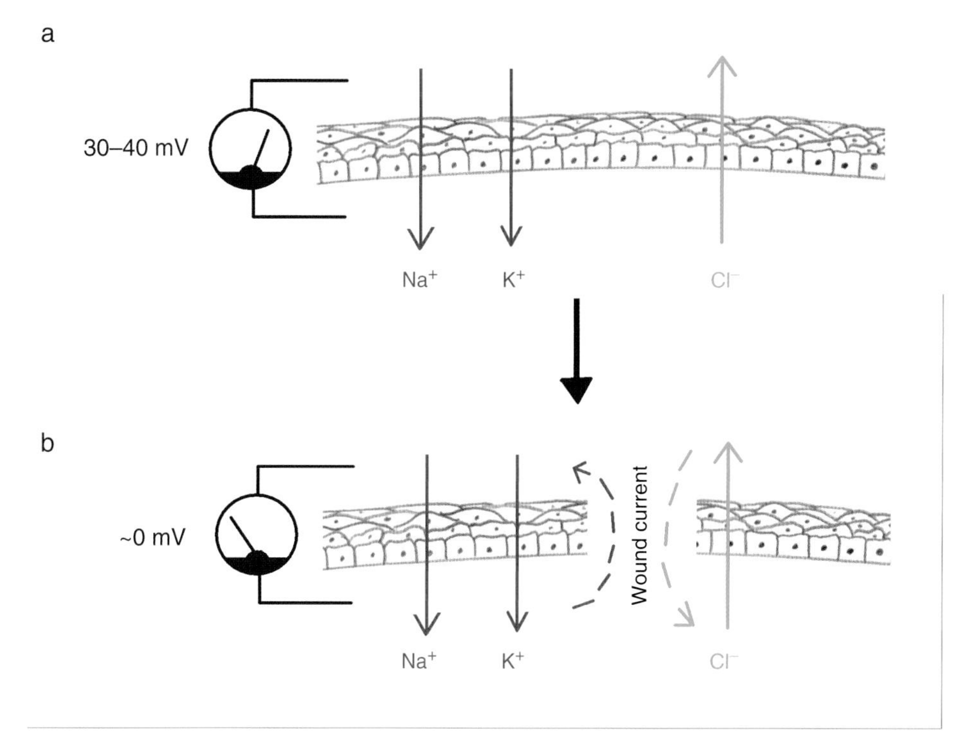

FIGURE 3.4.4 (a) The activity of ion pumps creates a transepithelial potential difference of 30 to 40 mV across the intact cornea (and up to 100 mV across skin). (b) Wounding the tissue allows current to flow, shorting out the potential difference across the epithelium in the immediate vicinity of the wound.

The maintenance of transepithelial potential differences depends absolutely on the resistance of intercellular tight junctions to leakage of ions in the opposite direction to which they are being pumped. Wounding such an epithelium allows ions to equilibrate again as they flow in a "wound current" through the breach in the tissue (see Figure 3.4.4). This collapses the potential difference across the epithelium in the vicinity of the wound, creating a change in fields dramatic enough to be monitored even with simple equipment; for this reason, electrical changes associated with wounding were described as early as 1843.[18]

Because the body fluids under- and overlying the wounded epithelium are not perfect conductors, the electrical effect of the breach diminishes with distance, and by about 200 μm away from a wound, the transepithelial potential is within about 5 percent of normal. If the inside of an epithelium such as the cornea remains at +40 mV with respect to the outside (far from the wound site) and at $\approx$0 mV at the wound site itself, it follows that wounding creates a new electric field between the wound site and "distant" unaffected cornea, tangential to the corneal surface (see Figure 3.4.5). This new field is complex in that it is non-linear and converging on

preferably though an insulating layer such as candle wax. (The author and publisher stress that you, and not they, are responsible for seeing that you meet all local ethical and safety policies if you do this.)

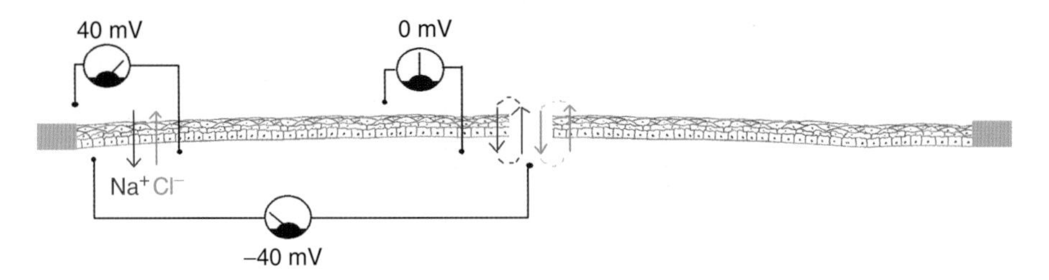

FIGURE 3.4.5 Wounding swaps the transepithelial electric field for a tangential one.

the wound. Critically, from the point of view of a cell at the inner surface of the epithelium, the new field is orientated so that the wound is effectively a cathode and the unaffected epithelium an anode. Any cells that undergo cathodally directed galvanotaxis would therefore be attracted to the wound, and they would receive the signal to move almost instantly, without having to wait for any slow changes in gene expression caused by chemical signalling from the damaged cells (electric fields travel at the speed of light).

Healing of damaged rat cornea can be monitored by marking cells fluorescently. One of the responses seen in normal healing is a vigorous sprouting of axons toward the wound. Inhibiting the generation of transepithelial electric fields (and, consequently, inhibiting the generation of the tangential wound field that depends on them) with the channel blocker ouabain reduces the amount of sprouting and greatly reduces the accuracy of navigation of the sprouts that do still form.[7] It also reduces the rate of wound healing. Conversely, increasing the transepithelial electric field using aminophylline, which enhances efflux of Cl^-, increases the rate of healing.[7] These findings provide circumstantial evidence that the tangential field really is used for guidance.

Regardless of whether electric fields are an important guidance mechanism in normal wound healing, they have been used to promote healing in experimental systems. The spinal cord of the larva of the sea lamprey *Petromyzon marinus* is an unusually accessible experimental system because it can be maintained for many days in culture and it contains axons large enough to distinguish in whole-mount observation. If a cultured *P. marinus* spinal cord is severed, a strong current flows from the medium into the cut end of the cord; the current falls to about 1/20th of its initial value after a few hours and is then maintained at that level. It leaves the sides of the cord, starting some distance away from the cut site.[19] This flow of current into the transected spinal cord implies an electric field orientated the wrong way for re-growth of the cut axon stumps into the wound site, and indeed the axons die back. Experimental application of a counter-current so that the local field reverses and the wound is cathodic relative to the axon strongly promotes regeneration by growth of the axon stumps into the wound.[20] The growing axons have large active growth cones, absent in controls, suggesting that galvanotactic encouragement of migration may be one mechanism by which healing is promoted, but it remains possible that healing is promoted by entirely different mechanisms instead. For example, the imposed current strongly reduces how far the axons die back following injury, probably by controlling Ca^{2+} buildup, rather than any galvanotactic response.[21] There is growing evidence that imposed electric currents of various sorts may also promote healing from spinal injuries in mammals,[22,23,24] although the research is still at a fairly early stage.

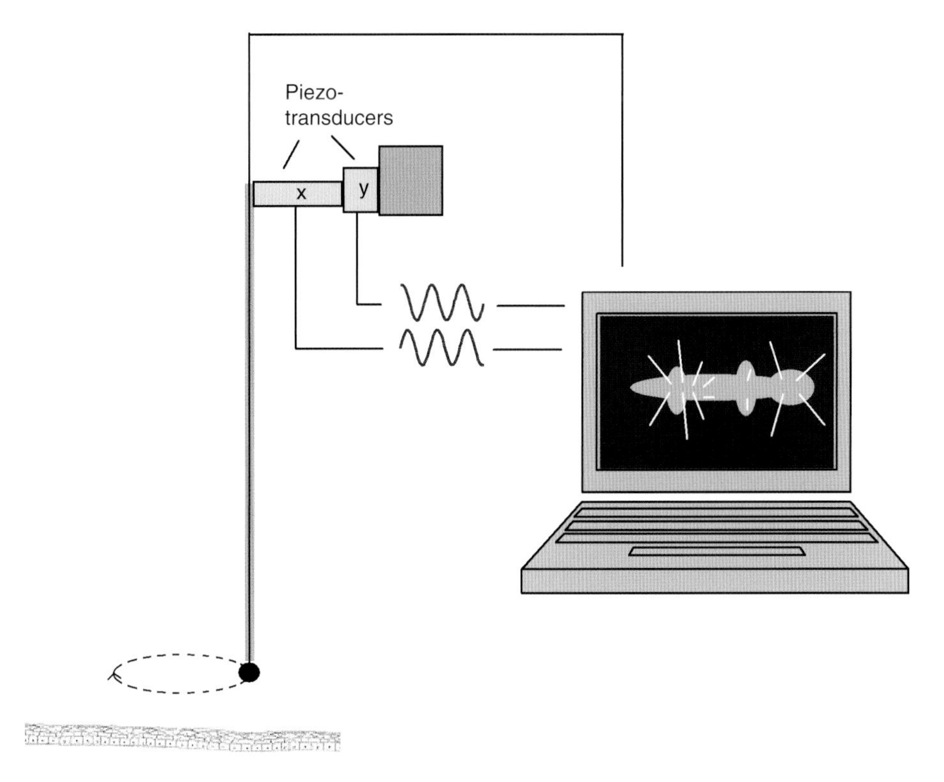

FIGURE 3.4.6 Vibrating probe apparatus used to measure embryo-scale electric fields.

The electric fields of embryos can be studied by using a vibrating probe to compare the potential difference between two closely spaced points just above a tissue of interest. A modern vibrating probe apparatus[25] typically consists of a conventional reference electrode immersed in the bulk medium in which the embryo is placed, and a very fine insulated probe electrode that terminates in a small un-insulated platinum sphere (see Figure 3.4.6). The probe electrode is supported by two piezoelectric crystals arranged at 90 degrees to each other, one of which can impart a tiny movement in the x axis and the other in the y axis. The crystals are supplied with sine wave currents 90 degrees out of phase with each other at the resonant frequency of the probe; the result is that the probe tip describes a tiny circular movement. The voltage sensed by the probe tip as it orbits, and the sine waves that drive it, are monitored by a computer; from the places at which the tip senses the maximum and minimum voltages, and from the differences between these voltages, the computer can calculate the direction and magnitude of the local electric field vector and plot it on a video image of the embryo. The currents that must be flowing locally can be inferred from the fields and the conductivity of the medium.

An example of an embryonic electric field that has much in common with wound fields is seen at the blastopore of embryos of the frog *Xenopus laevis*. The blastopore is a pit in the side of the embryo through which cells fated to be endodermal flow so that they leave the outer surface of the embryo and can create a new inner surface; it persists from the beginning of gastrulation to late neurulation (see Figure 3.4.7).

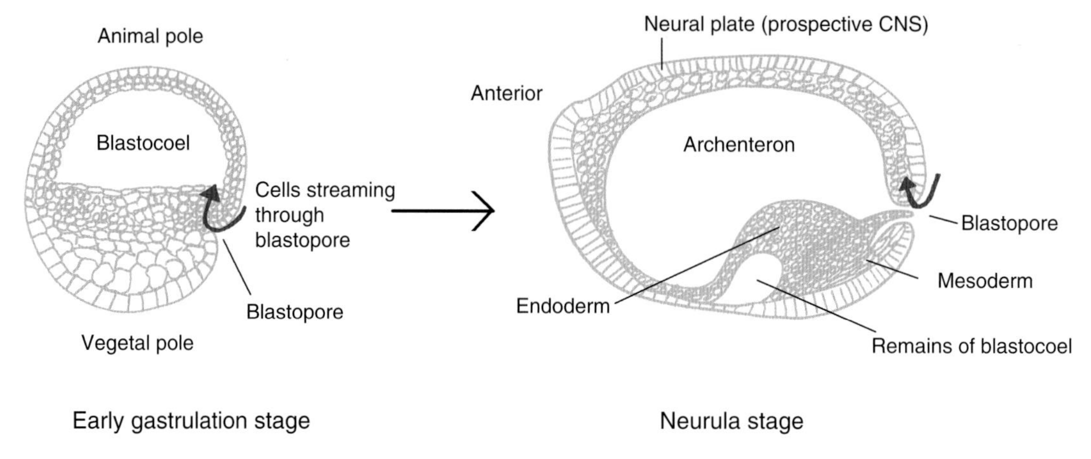

FIGURE 3.4.7 The blastopore in embryos of *X. laevis*.

There is a strong outward flow of conventional current through the blastopore throughout its existence, which peaks at over 1 A/m^2 and at least 50 percent of which is carried by sodium ions.[26] Currents therefore flow in the same direction in the blastopore as they do in a corneal wound, and any tangential field will be in the same direction, the pore appearing as a cathode (sink for positive ions) from the point of view of cells on the inside and appearing as an anode (source of positive ions) from the outside. This highlights an important difference between the systems: in corneal wound healing, cells migrate toward the cathode, whereas in *X. laevis*, they are travelling toward an effective anode when moving toward the pore on the outside, when they move through it and also when moving away on the inside. Either these cells prefer to migrate toward the anode or the field, if at all relevant to gastrulation movements, is actually used to limit the rate of their migration. There is evidence that the electric field is used for some developmental purpose in these embryos; if it is opposed by an artificially applied field, about 75 percent of embryos develop abnormally.[26] Unfortunately, the abnormalities are variable, possibly because precise placing of electrodes is difficult, which precludes the identification of a single developmental event that requires the electric field.

Endogenous electric fields and currents have also been detected in chick embryos, particularly in the region of the posterior intestinal portal, where the developing alimentary canal is open to the ventral surface of the embryo (see Figure 3.4.8). Again, conventional current is directed primarily out of the pole and back in through other regions of the embryo, and again at least half is carried by sodium ions.[27] If these currents are shorted out by bridging the embryonic epithelium with fine glass needles filled with conductive agarose gel, over 80 percent of embryos develop abnormally, especially in the region of the tail. Control glass needles that are non-conducting cause only 11 percent of embryos to become abnormal, confirming that abnormalities are caused by the shorting of the electric fields rather than simple surgical damage.[28] Again, the spectrum of abnormalities observed is too broad to pin down one single mechanism, and the effect may have nothing to do with control of cell movement.

There is evidence for an electric field in one of developmental biology's most studied systems, the vertebrate limb bud. The embryos of both mice and chicks show

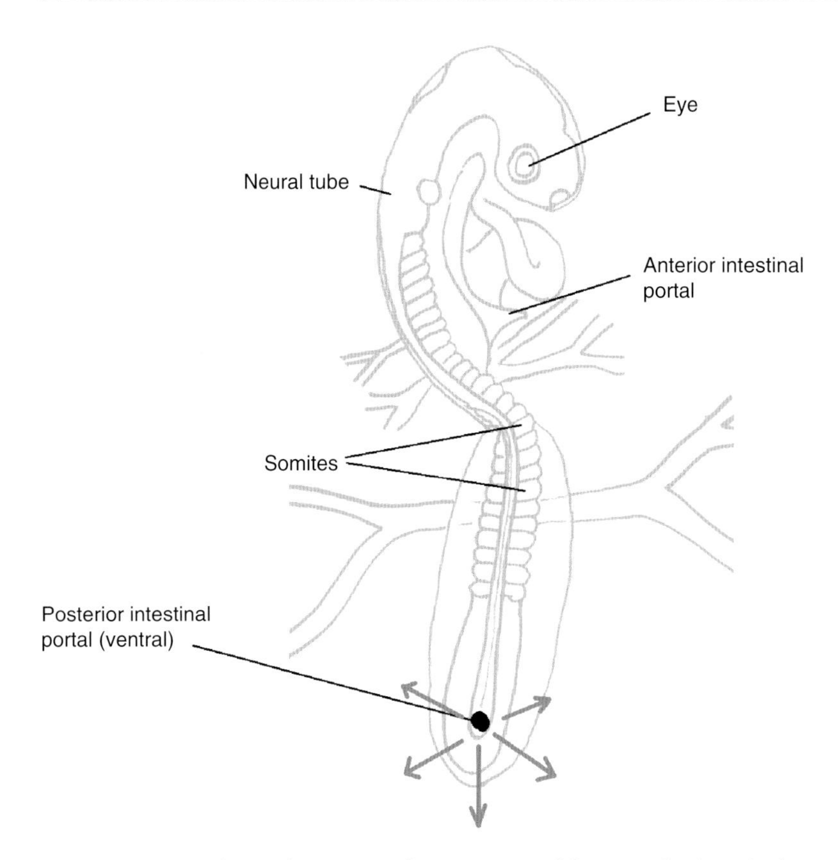

FIGURE 3.4.8 Flows of current (red arrows) out of the posterior intestinal portal of a chick embryo.

an inwardly directed conventional current along the flank of the embryo, far from the limb buds, and an outward current along the limb buds.[29] The cause of the current is not clear, but it may be due to the "normal" inward flux of Na^+ ions and their leakage from imperfectly sealed epithelia in the limb bud; alternatively, the bud may possess active mechanisms to drive these currents. Strikingly, the outward current associated with limb buds begins in the presumptive limb bud regions of the flank several hours before overt limb bud emergence. Reversing this current, using fine electrodes implanted in chick embryos, causes limbs to develop abnormally, although controls that suffer the same surgery but without current reversal show no effects. Again, it is not yet clear precisely which developmental events are being affected, but the electric fields and currents must be critical for some process in the limb.

The whole issue of galvanotaxis in development therefore remains uncertain. The ability of cells to navigate along an electric field vector has been proven beyond reasonable doubt, as has the existence of such fields in real embryos. There is strong evidence that these fields are important to development, but in no system studied to date has it been shown that endogenous fields form a necessary component of a cell navigation system. This may change very soon.

Reference List

1. Borgens, RB (1999) Electrically mediated regeneration and guidance of adult mammalian spinal axons into polymeric channels. *Neuroscience.* **91**: 251–264
2. Nuccitelli, R and Erickson, CA (1983) Embryonic cell motility can be guided by physiological electric fields. *Exp. Cell Res.* **147**: 195–201
3. Gruler, H and Nuccitelli, R (1991) Neural crest cell galvanotaxis: New data and a novel approach to the analysis of both galvanotaxis and chemotaxis. *Cell Motil. Cytoskeleton.* **19**: 121–133
4. Nishimura, KY, Isseroff, RR, and Nuccitelli, R (1996) Human keratinocytes migrate to the negative pole in direct current electric fields comparable to those measured in mammalian wounds. *J. Cell Sci.* **109 (Pt 1)**: 199–207
5. Farboud, B, Nuccitelli, R, Schwab, IR, and Isseroff, RR (2000) DC electric fields induce rapid directional migration in cultured human corneal epithelial cells. *Exp. Eye Res.* **70**: 667–673
6. Patel, NB and Poo, MM (1984) Perturbation of the direction of neurite growth by pulsed and focal electric fields. *J. Neurosci.* **4**: 2939–2947
7. McCaig, CD, Rajnicek, AM, Song, B, and Zhao, M (2002) Has electrical growth cone guidance found its potential? *Trends Neurosci.* **25**: 354–359
8. Rajnicek, AM, Robinson, KR, and McCaig, CD (15-11-1998) The direction of neurite growth in a weak DC electric field depends on the substratum: Contributions of adhesivity and net surface charge. *Dev. Biol.* **203**: 412–423
9. Grahn, JC, Reilly, DA, Nuccitelli, RL, and Isseroff, RR (2003) Melanocytes do not migrate directionally in physiological DC electric fields. *Wound Repair Regen.* **11**: 64–70
10. McCaig, CD, Sangster, L, and Stewart, R (2000) Neurotrophins enhance electric field-directed growth cone guidance and directed nerve branching. *Dev. Dyn.* **217**: 299–308
11. Zhao, M, Pu, J, Forrester, JV, and McCaig, CD (2002) Membrane lipids, EGF receptors, and intracellular signals colocalize and are polarized in epithelial cells moving directionally in a physiological electric field. *Faseb J.* **16**: 857–859
12. Erskine, L and McCaig, CD (1995) Growth cone neurotransmitter receptor activation modulates electric field-guided nerve growth. *Dev. Biol.* **171**: 330–339
13. Stewart, R, Erskine, L, and McCaig, CD (1995) Calcium channel subtypes and intracellular calcium stores modulate electric field-stimulated and -oriented nerve growth. *Dev. Biol.* **171**: 340–351
14. Fang, KS, Ionides, E, Oster, G, Nuccitelli, R, and Isseroff, RR (1999) Epidermal growth factor receptor relocalization and kinase activity are necessary for directional migration of keratinocytes in DC electric fields. *J. Cell Sci.* **112 (Pt 12)**: 1967–1978
15. Patel, N and Poo, MM (1982) Orientation of neurite growth by extracellular electric fields. *J. Neurosci.* **2**: 483–496
16. Waugh, MG, Minogue, S., Anderson, JS, dos Santos, M, and Hsuan, JJ (2001) Signalling and non-caveolar rafts. *Biochem. Soc. Trans.* **29**: 509–511
17. Barker, AT, Jaffe, LF, and Vanable, JW, Jr. (1982) The glabrous epidermis of cavies contains a powerful battery. *Am. J. Physiol.* **242**: R358–R366
18. Du Bois-Raymond, E (1843) *Ann. Phys. und Chem.* **58**: 1–30
19. Borgens, RB, Jaffe, LF, and Cohen, MJ (1980) Large and persistent electrical currents enter the transected lamprey spinal cord. *Proc. Natl. Acad. Sci. U.S.A.* **77**: 1209–1213
20. Borgens, RB, Roederer, E, and Cohen, MJ (7-8-1981) Enhanced spinal cord regeneration in lamprey by applied electric fields. *Science.* **213**: 611–617
21. Roederer, E, Goldberg, NH, and Cohen, MJ (1983) Modification of retrograde degeneration in transected spinal axons of the lamprey by applied DC current. *J. Neurosci.* **3**: 153–160
22. Borgens, RB, Blight, AR, and McGinnis, ME (16-10-1987) Behavioral recovery induced by applied electric fields after spinal cord hemisection in guinea pig. *Science.* **238**: 366–369

23. Borgens, RB, Blight, AR, and McGinnis, ME (22-6-1990) Functional recovery after spinal cord hemisection in guinea pigs: The effects of applied electric fields. *J. Comp. Neurol.* **296**: 634–653
24. Borgens, RB, Toombs, JP, Breur, G, Widmer, WR, Waters, D, Harbath, AM, March, P, and Adams, LG (1999) An imposed oscillating electrical field improves the recovery of function in neurologically complete paraplegic dogs. *J. Neurotrauma.* **16**: 639–657
25. Jaffe, LF and Nuccitelli, R (1974) An ultrasensitive vibrating probe for measuring steady extracellular currents. *J. Cell Biol.* **63**: 614–628
26. Hotary, KB and Robinson, KR (1994) Endogenous electrical currents and voltage gradients in Xenopus embryos and the consequences of their disruption. *Dev. Biol.* **166**: 789–800
27. Hotary, KB and Robinson, KR (1990) Endogenous electrical currents and the resultant voltage gradients in the chick embryo. *Dev. Biol.* **140**: 149–160
28. Hotary, KB and Robinson, KR (1992) Evidence of a role for endogenous electrical fields in chick embryo development. *Development.* **114**: 985–996
29. Altizer, AM, Moriarty, LJ, Bell, SM, Schreiner, CM, Scott, WJ, and Borgens, RB (2001) Endogenous electric current is associated with normal development of the vertebrate limb. *Dev. Dyn.* **221**: 391–401

CHAPTER 3.5

CELL MIGRATION: Navigation by Contact

Most migrating cells migrate across and through solid tissues, and these tissues generally show numerous fine-scale variations, consisting as they do of different types of cells and matrices. These variations can be used as a navigation system by migrating cells, either in conjunction with chemotactic and galvanotactic signals or as a complete guidance system in their own right.

HAPTOTAXIS

Haptotaxis♣ is the guidance by means of a gradient of adhesion. It is a phenomenon that can affect purely physical as well as biological systems and relies on another version of a Brownian Ratchet (see Chapters 2.3 and 3.3). Adhesions between surfaces occur when the free energy of the system (the surfaces plus solvent) is lower when the surfaces are in contact than when they are separated. If this difference in free energy is not much greater than thermal energy in the system, adhesive bonds will not be permanent and will be continuously made and broken. The stronger an adhesive contact (the greater the difference in free energy), the longer, on average, it will resist disruption. Consider an object able to roll, or to deform like a membrane vesicle or cell, on a gradient of an adhesive molecule. The object will be buffeted by random thermal agitation in both up-gradient and down-gradient directions. Buffeting in the up-gradient direction will strain the slightly weaker adhesions, whereas buffeting in the down-gradient direction will strain the slightly stronger ones. The slightly stronger adhesions will resist disruptive forces marginally better and last longer. On average, therefore, the down-gradient bonds will break and allow the object to be moved up-gradient more often than vice versa, and the object will be translocated up the gradient (see Figure 3.5.1). Haptotaxis can be modelled in simple physical vesicles,[1] and it can be demonstrated in a practical class using everyday materials.♥

♣From the Greek: *haptein*, meaning "fasten," and *táxis*, meaning "arrangement."

♥Wrap a piece of paper at least a quarter of the way around a curved surface such as a gas cylinder, and spray it lightly with spray adhesive from a direction that is perpendicular to the surface of one end of the paper. The result will be a piece of paper with an adhesive gradient. Place the paper in a box so that it covers

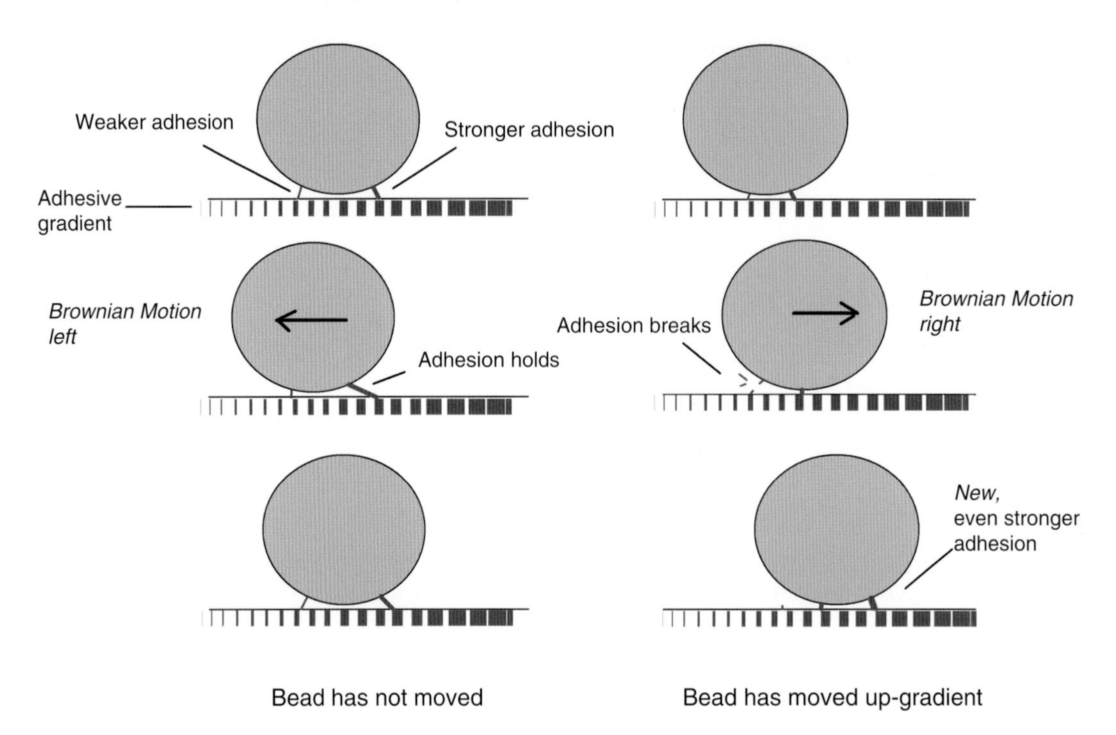

FIGURE 3.5.1 The combination of an adhesive gradient and thermodynamic motion creates a Brownian Ratchet mechanism that can move an object up a gradient. In this diagram, the object is a simple bead that can roll, and kinetic energy from the thermodynamic forces impinging on the bead is shown stored temporarily in stretched adhesive bonds. A deformable object such as a balloon or membrane vesicle can store energy temporarily by deforming its shape into a flatter one.

The very simple form of haptotaxis described above tends not to be used in living embryos because cells that are not actively migratory either form attachments so strong that they are effectively immobilized (as most cells in a mature tissue tend to be) or they tend to form essentially no attachments (e.g., erythrocytes in the blood). Actively migrating cells have labile attachments, not just because of thermal fluctuations but also because of the "deliberate" protrusion-attachment-tension-detachment cycle of locomotion described in Chapter 3.2. These cells, therefore, are well adapted for haptotactic steering.

The leading edge of migratory cells does not just face forwards; the edges of lamellipodia explore the environment a little to each side of the main axis of the cell, and filopodia can extend great distances to the side. In this way, cells can "sample" the environment for navigational cues. An element of competition between different parts of the leading edge follows inevitably from this lateral exploration, since the cell cannot advance simultaneously toward the left-most parts of the leading edge and the right-most parts for long without becoming too spread and stretched (see Figure 3.5.2). Anything that tips the balance of this competition will therefore be able to steer the cell.

---

the bottom of the box, scatter small spherical cake decorations (known as "hundreds and thousands" in the UK) onto the paper, and agitate the box by allowing it to make contact with a vortex machine or some other gentle vibrator. The decorations move up the adhesive gradient.

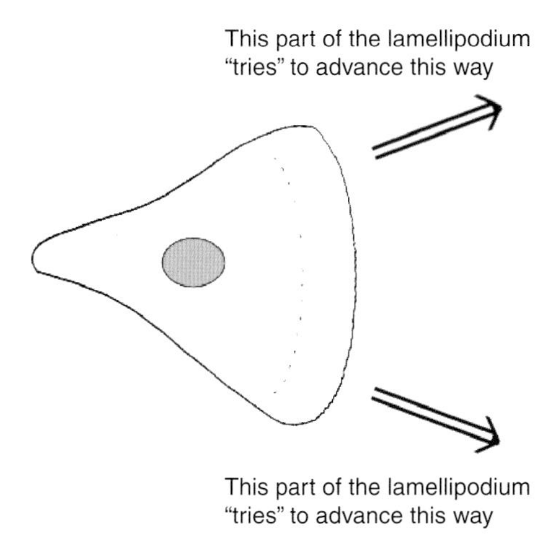

FIGURE 3.5.2 The "tug of war" between rival parts of a lamellipodium.

The back of the leading edge is richly supplied with integrin-containing cell-substrate adhesion complexes, against which the leading edge protrusions push as they advance, and against which the actin/myosin filaments of the rest of the cell pull (see Chapter 3.2). Strong adhesions will support vigorous protrusion and drawing forward of the cell, whereas weaker ones that slip under mechanical load will be less efficient. In the tug of war between rival parts of the leading edge, those parts that have the strongest grip on the substrate would be expected to win (see Figure 3.5.2). This type of guidance has been demonstrated in a variety of culture systems.

One classical method for demonstrating haptotaxis is to produce patterned substrates *in vitro*, on which cells will be confronted by boundaries of different adhesiveness, as in Figure 3.5.3. The relative adhesiveness of different substrates can be assessed semi-quantitatively by plating cells onto them and then by trying to blow them off by directing a jet of fluid at them from a fine pipette (see Figure 3.5.4a); the closer the pipette has to be brought, the more adhesive the substrate. Such simple assays demonstrate that the growth cones of chick embryo dorsal root ganglion cells stick very well to polyornithine, a little less well to collagen, less well still to palladium and tissue culture plastic, and very poorly to bacteriological plastic.[2] Substrates can be coated with palladium using equipment originally designed to produce thin, conductive films of metal on samples for scanning electron microscopy; if a fine metal grid is interposed between the coating gun and the substrate, it will cast palladium-free "shadows" in a grid pattern on an otherwise palladium-coated surface. The growth cones of ganglion cells plated on such a substrate will therefore be presented with many boundaries between palladium and an alternative. If the alternative is tissue culture plastic, which is about equally adhesive as palladium, the growth cones migrate randomly across boundaries. If the alternative is a more adhesive substrate such as polyornithine, they will instead remain on or cross on to the polyornithine and will then follow it, even performing right-angled turns in preference to straying on to the less adhesive substrate[2,3] (see Figure 3.5.4b). At a boundary, filopodia explore the less adhesive as well as the more adhesive substrate, but then slip back. Similar behaviour can be observed in other cells such as fibroblasts,[4]

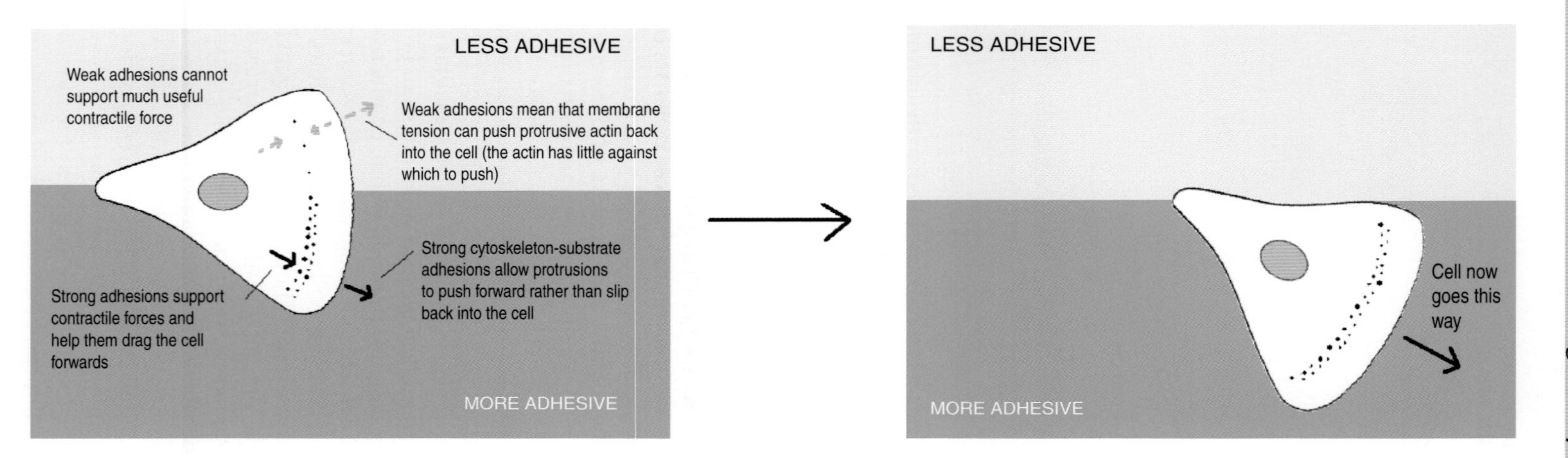

FIGURE 3.5.3 Haptotaxis at a simple boundary. The cell depicted in this diagram is sampling two substrates: one that supports more adhesion and one that supports less adhesion. The parts of the lamellipodium that enjoy more adhesion advance more because the leading edge has something against which to push, while the contractile machinery behind it has something against which to pull. In the "tug of war" between rival parts of the lamellipodium, these parts therefore win, and the cell steers around toward them.

a

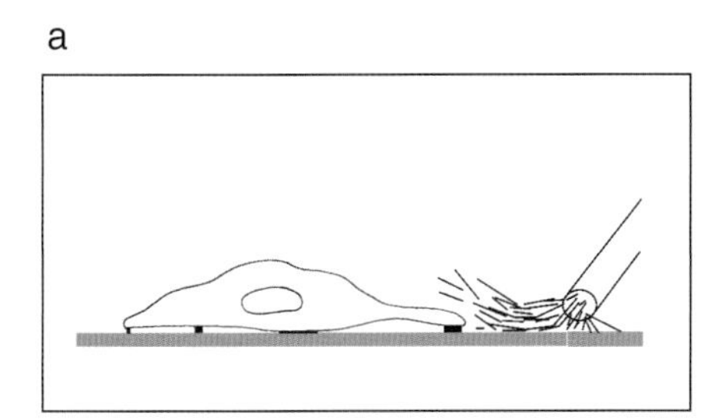

b

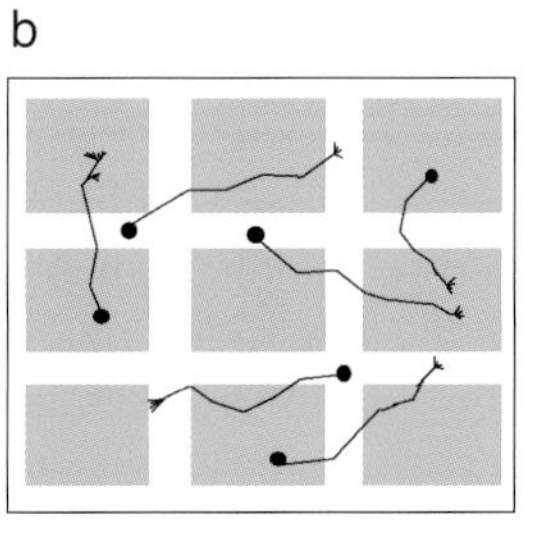

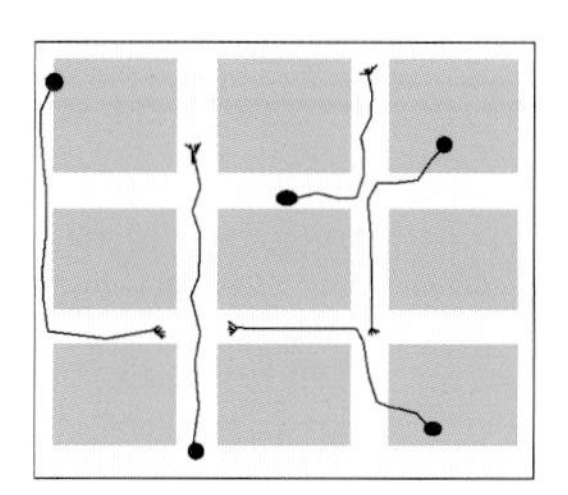

FIGURE 3.5.4 (a) Semi-quantitative method for measuring cell adhesion in which a fluid jet is used to loosen cells. (b) Growth cones cross boundaries between equally adhesive substrates but avoid crossing from a more adhesive to a less adhesive substrate.

although their courses of migration are less obvious than those of growth cones because they leave no axon behind them.

A similar set of boundaries can be made by shining ultraviolet light through a metal grid onto laminin to produce lanes of native laminin that separate UV-denatured laminin. Neurites remain on or move on to the native molecule and stick to it much more strongly (as assessed by the shear force required to pull them away).[5]

Direct and local interference with cell-substrate adhesion can be used to steer growth cones. Cultured embryonic chick parasympathetic neurons produce axons with very wide growth cones. If a fine glass needle is used to lift one side off the substrate, the growth cone responds by extending away from the needle, as would be predicted by haptotaxis. This can be used to steer the growing axon in arbitrary shapes. Similarly, if a needle is used to lift the very middle of a growth cone, each of the sides continues to advance so that the growth cone, and hence the axon, bifurcates[6] (see Figure 3.5.5).

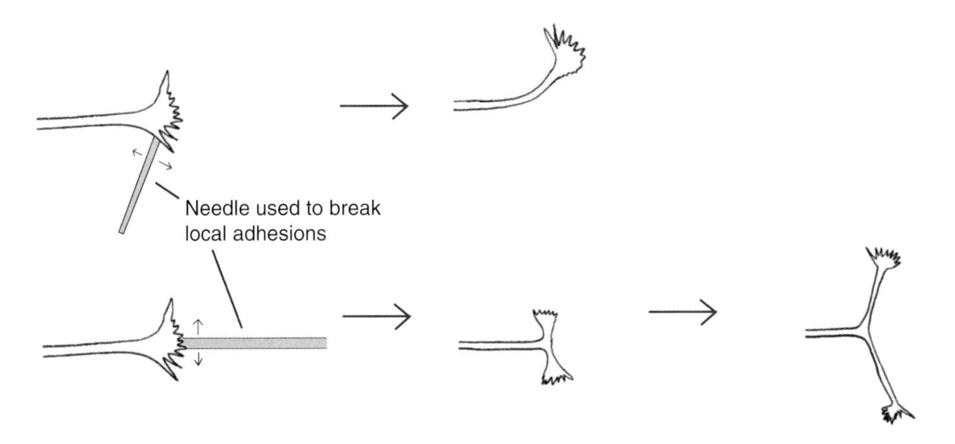

FIGURE 3.5.5 Breaking the adhesion between the substrate and just one part of a growth cone, using a fine needle, causes that part of the growth cone to stop advancing while other parts continue, just as the haptotaxis hypothesis predicts.

It is interesting to note that adhesive surfaces need not be continuous in order to support axon growth. Spaced spots of highly adhesive substrates can allow axons to extend over intervening substrates of very low adhesion as long as the spots are close enough together to be explored by filopodia of growth cones on adjacent spots.[7] This may be highly relevant to the phenomenon of waypoint navigation (see Chapter 3.6).

If the only possible activity of adhesion complexes were to be the transmission of mechanical forces, the case for haptotaxis as a navigational mechanism could be made simply from the observations described above. Integrin-containing adhesion complexes can do much more than transduce force, however; in particular, they can trigger signal transduction pathways via junction-associated molecules such as Focal Adhesion Kinase (FAK), Rho, and Src. Indeed, most types of adhesion complex are associated with some kind of signalling system. This makes guidance directly by haptotaxis difficult to disentangle experimentally from guidance by ligand-receptor signalling.

The key difference between haptotaxis and biochemical substrate-sensing is that, while they both require binding between the cell and extracellular ligands, only the haptotaxis mechanism would require those ligands to be capable of exerting mechanical forces. This fact has been used in an elegant study on the steering of growth cones that demonstrates unequivocally that, at least in one model system, simple ligand-receptor binding is not enough and that physical forces are required. Bag cells of the gastropod sea slug *Aplysia californica* are endowed with very large growth cones that bear a homophilic adhesion molecule of the NCAM superfamily, called apCAM. Small beads coated either with apCAM or with antibodies to it bind the apCAM on the growth cone surface. Once bound, the beads are swept back with the general retrograde flow of actin, showing that they are cross-linked to the cytoskeleton. Under these conditions, the presence of the beads has little effect on the direction of cell movement, showing that the simple act of binding and clustering apCAM is not itself sufficient to guide growth cones. The rearward motion of bound beads can, however, be prevented by physically blocking their progress with a fine glass needle (see Figure 3.5.6). This effectively mimics the attachment of the bead to a substrate. Under these conditions, the growth cone pulls on the bead, bending the needle a little, and then begins to extend and to steer in the direction of the bead.[8] This is precisely what would be expected of true haptotaxis.

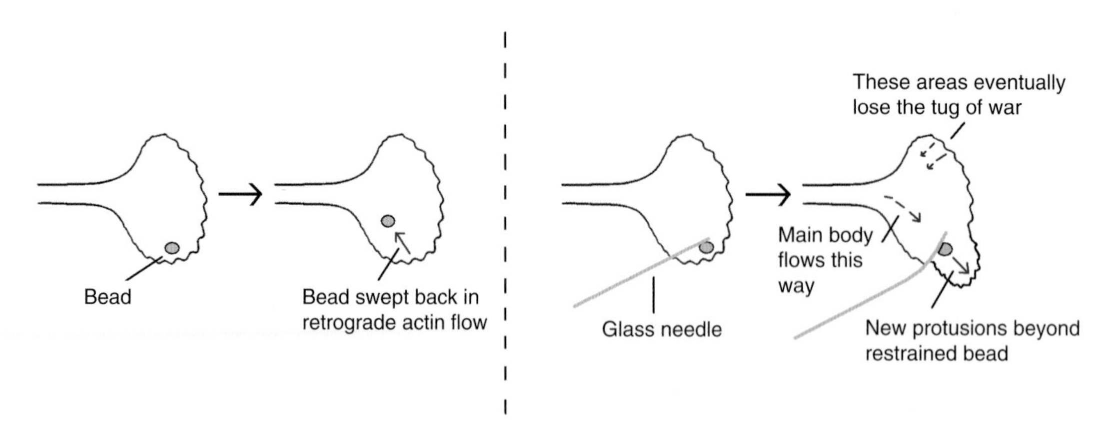

FIGURE 3.5.6 Haptotaxis caused by restraining the movement of beads adhering to the apCAM adhesion complex of *A. californica*'s bag cell growth cones.

DUROTAXIS: Guidance of Cells by Gradients of Mechanical Compliance

Durotaxis,♦ the ability of cells to navigate according to the stiffness of their substrate, is related closely to haptotaxis because substrate flexibility is related to the ability of an adhesion to bear mechanical force. When cells are cultured on substrates that are chemically identical but have different mechanical flexibilities, their rate of migration alters so that they migrate more and faster on stiffer substrates.[10] If a cell were to find itself straddling the boundary between substrates of different flexibilities, one would therefore expect it to migrate on to the stiffer one since the part of its leading edge on the stiffer one would show the stronger migratory response. If this is true of a boundary, then it should also be true of a steady gradient.

Substrates with a gradient of rigidity (Young' modulus) varying from 14 kN/m^2 to 30 kN/m^2 can be made by making a polyacrylamide "gradient gel" with a constant amount of acrylamide but a varying amount of bis-acrylamide cross-linker. When coated evenly with collagen, such gels are suitable substrates for fibroblasts. Cells plated onto such a substrate migrate toward the stiffer substrate as long as they are few enough that deformations of the substrate by the other cells do not confuse the situation.[9]

The mechanism of durotaxis may be entirely mechanical and explicable in terms of how well a section of leading edge can push and pull on the substrate without it "wasting" their force by deforming under them, or by the biochemical transduction of sensed mechanical forces. So far, it is an entirely *in vitro* phenomenon with no known example in development; it has been mentioned here, briefly, in case it does turn out to be important *in vivo* too.

ATTRACTION BY CONTACT-DRIVEN CELL SIGNAL TRANSDUCTION

Integrin-containing adhesive complexes contain, in addition to proteins that link them to the actin cytoskeleton, signalling proteins such as Focal Adhesion Kinase (FAK) that announce to the rest of the cell the binding of integrins to their external ligands.[11] As well as being activated by integrin-matrix binding, FAKs are sensitive to mechanical forces at the site of adhesion. Without them, cells are incapable of durotaxis.[12] Loss of FAK also causes Rho to remain abnormally active when cells are plated on a substrate to which their integrins bind, and active Rho stabilizes their stress fibres and adhesive contacts and tends to inhibit turnover of adhesions.[13] There is growing evidence that FAK signals to Rho by activating the Rho guanidine nucleotide exchange factors, LARG and PDZ-RhoGEF, which return inactive Rho-GDP to its active Rho-GTP form.[14] Integrin-mediated guidance of cells is therefore likely to be more subtle than simple haptotactic models imply.

Other "adhesion molecules" are also involved in cell signalling. The molecules NCAM, N-cadherin, and L1 were first identified as homophilic cell adhesion molecules that are important particularly, but not exclusively, in the nervous system. As well

♦A word of mixed linguistic parentage, from the Latin *durus*, meaning "hard," and the Greek *taxis*, meaning "arrangement." It was coined by the laboratory who made the graded substrates described in this section.[9]

as being adhesive, each of these molecules can signal its homophilic binding to a similar molecule on another cell by activating the FGF receptor (FGFR) tyrosine kinase on its own cell.[15] This action is independent of the presence of FGF. If FGFR function is inhibited, neurons and growing axons lose their ability to respond to NCAM, N-cadherin, or L1 in culture and *in vivo*.[16] At least one pathway linking the FGF receptor to growth cone morphogenesis has been elucidated. FGFR activation signals via phospholipase-C-γ and diacylglycerol lipase to stimulate production of arachidonic acid.[17] In neurites, the presence of arachidonic acid leads to activation of some isoforms of protein kinase C and consequent phosphorylation of the neural growth-associated protein GAP43. What is more, this pathway acts synergistically with free cytoplasmic Ca^{2+}, so that the phosphorylation of GAP43 driven by arachidonic acid is much greater in the presence of free Ca^{2+}.[ref18]

GAP43 is synthesized by neurons as they produce neurites, but not before, and the protein accumulates in the growth cone. Phosphorylated GAP43 binds to and stabilizes actin polymers, protecting them from depolymerization.[19] Local binding of the NCAM on a filopodium, or on part of a lamellipodium, to NCAM on another cell would therefore cause local phosphorylation of GAP43, local stabilization of actin filaments, and an advantage to that filopodium in the competition between filopodia. In support of this idea, antibodies specific for phosphorylated GAP43 show that phosphorylation correlates with areas of growth-cone advance, and unphosphorylated GAP43 correlates with areas of retraction. The adhesion molecule-FGFR-GAP43 signalling path seems to be separate from that driven by integrins. Mutation of GAP43 abolishes the ability of NCAM to support elongation of neurites,[20] but they can still elongate normally over laminin,[21] which uses integrin receptors.

The fact that "cell adhesion molecules" can stimulate neurite outgrowth mainly by signal transduction, rather than mechanically, is emphasized by the fact that application of soluble NCAM-Fc or L1-Fc chimaeras stimulates outgrowth as strongly as do cell-surface NCAM or L1, and direct stimulation of FGFR by FGFs has the same effect.[20] Observations like this challenge the whole concept of what a "cell adhesion molecule" is. The first member of each adhesion molecule family was generally identified in screens for monoclonal antibodies that prevented, for example, fasciculation of neurites in culture. This was naturally assumed to reflect a failure of growth cones to adhere to axons, which in a sense it was, but it is now clear that at least part of this effect must have been due to changes in the abilities of growth cones to navigate by sensing the chemical properties of their surroundings. The balance of mechanical adhesion and signalling in most systems is not clear, and will not be until much more research has been done. Until then, it would be wise to treat the term "adhesion molecule" like the term "growth factor"—a phrase that is appropriate for how the first members of these protein families were identified, but that is not necessarily the best description of their main function *in vivo*.

PATHWAYS OF ATTRACTIVE MOLECULES IN THE EMBRYO

The predilection of cells for substrates that encourage cell-substrate adhesion can be used to guide migration in the embryo (wherever the balance between the mechanisms of haptotaxis and signal transduction lies). Neural crest cells of vertebrates migrate from the dorsal part of the neural tube and, in the trunk, follow two main pathways at first, travelling either ventromedially, close to the neural tube, or dorsolaterally under the skin. Both pathways are rich in components of the extracellular matrix, including fibronectin, laminin, and a few dozen others,[22] suggesting that they form tracks along which migratory cells are meant to travel. Antibodies that interfere with the adhesion of neural crest cells to the matrix reduce crest cell migration markedly,[23] as do antisense oligonucleotides that inhibit the expression of integrin receptors for these matrix components.[24] On the other hand, implanting artificial membranes containing large quantities of fibronectin into embryos of the axolotl *Ambystoma mexicanum* greatly stimulates outgrowth of neural crest cells along the artificial membranes.[25]

One way in which adhesive substrates are used all over the body and across a great range of phyla is in the construction of nerves from individual axons. Nerves are constructed by the process of fasciculation, in which growth cones choose to migrate along existing axons in preference to making their own way across the surrounding substrate. Existing axons are attractive to growth cones because they express cell adhesion molecules such as NCAM and L1, which will bind to the same type of molecule on the growth cone.[26] L1 is expressed in many fibre tracts and is required for their formation; antibodies to L1 reduce fasciculation in cultured embryonic cerebelum,[27] and mouse and human mutants lacking L1 function show severe disruption to central nervous system development.[28] Similarly, NCAM is expressed in the mossy fibre tract, a large tract of fasciculated axons in the mammalian hippocampus, as it forms. In the absence of NCAM function, the fasciculation of these axons fails.[29] In the densely packed neural tissue of the central nervous system, there are many fibre tracts that head in different directions to different destinations. A growth cone is therefore faced with the problem of joining exactly the right tract and ignoring the distractions of others that may lie in its path. In the central nervous system of insects, different fibre tracts express adhesive proteins so that pathways may be labelled uniquely. In the grasshopper *Schistocerca americana* and in the fruit-fly *Drosophila melanogaster*, Fasciclins I, II, III, and IV are expressed on the cell surfaces of different axon bundles.[30,31,32,33] Fasciclin I, for example, is expressed initially in all commissural (midline-crossing) axon bundles and later becomes restricted to a subset of them.[34] Disruption of fasciclin expression or function disrupts fasciculation of axons,[35,36] while forcing abnormally high levels of fasciclin expression prevents defasciculation where the axons ought to leave the fibre bundle to fan out to their targets.[37] Only fibres using the particular fasciclin that is subject to experimental manipulation are affected. (Fasciclins have other functions in later neural development—for example, synaptogenesis—as well.[38])

GUIDANCE OF CELLS BY ALIGNED FIBRES

During amphibian gastrulation, prospective mesodermal cells from the region of the blastopore migrate along the underside of the roof of the blastocoel toward the animal pole (see Figure 3.5.7). Shortly before this takes place, in both *Pleurodeles waltlii* and *Amystoma mexicanum*, the roof of the blastocoel expresses high concentrations of the extracellular matrix molecule fibronectin.[39,40] The fibronectin is important for mesodermal migration, which fails to take place when the embryo is injected with function-blocking anti-fibronectin antibodies.[41] Antibodies that block the function of fibronectin-binding integrins have a similar effect.[42] The blastocoel roof can be dissected from an embryo, cultured inside-down for a while, and then removed from its culture

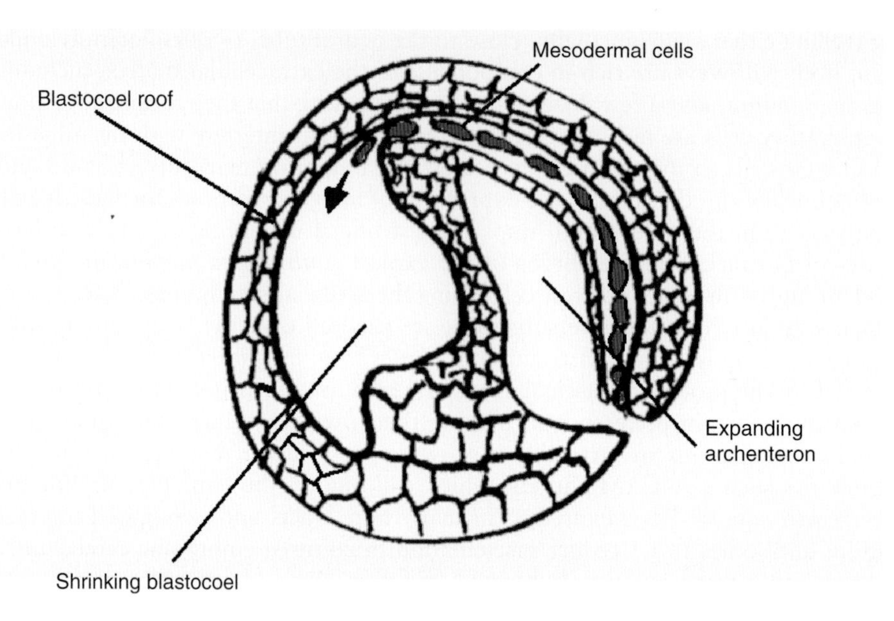

FIGURE 3.5.7 Migration of mesoderm cells under the roof of the blastoderm during amphibian gastrulation. The figure depicts an embryo of *Xenopus laevis* partway through gastrulation. The other embryos mentioned in this section differ in detail but not in the essentials of migration.

dish; what remains on the culture dish is an acellular, fibronectin-rich matrix that was secreted by roof cells and is now "blotted" on to the dish. Remarkably, when mesodermal cells are placed on this matrix, they migrate toward the part that underlay the animal pole, whatever their original orientation.[43] The matrix is therefore a sufficient directional cue. There is, however, no evidence for a large-scale adhesive gradient in this matrix, and cells seem to stick to all of its regions equally.[43]

Electron microscopy reveals a potential directional cue in the microstructure of fibronectin in both living embryos and "blotted" matrices in culture. Its molecules are aligned to produce fibres that lie along the direction of normal cell migration. These fibers are the result of activity in the cells that secrete the fibronectin and seem to rely on mechanical force exerted by the actin/myosin cytoskeleton acting via integrins. Interference either with fibronectin-binding integrins using the competitor peptide sequence GRGDSP or interference with the actin cytoskeleton using cytochalasin B blocks the formation of fibronectin fibres. The matrix laid down by a blastocoel roof that is cultured in the presence of either of these drugs is rich in fibronectin but has no aligned fibers. Significantly, mesodermal cells plated onto it still adhere normally and still move, but their movement is now undirected and random.[43] Clearly, the alignment of fibres is an important directional cue for mesodermal cells. The mechanism by which cells can be guided by aligned fibres has not been investigated in detail. Simple arguments based on adhesion, being more stable on fibres than between them, may explain a preference for following fibres but cannot in itself explain a preference for going one way rather than the other (unless the fibres converge, in which case the overall density of adhesive fibres will be highest where they are coming together—adhesion measurements in the blastocoel roof suggest that adhesion gradients are not present in amphibian gastrulation).

GUIDANCE BY INHIBITION OF LOCOMOTION

As well as being guided by attractive molecules, cells can be repelled by contact with molecules that interfere with the process of locomotion. This is most obvious in parts of the embryo that must not be invaded. These areas are commonly marked by powerful "keep out" signals. One such area is the posterior half-somite of vertebrates. Somites, the (indirect) precursors of vertebrae and segmental musculature, form as a series of blocks on each side of the neural tube. The posterior half of the sclerotome (ventromedial part) of each somite will form the hard parts of the vertebrae, where nerves must not run. The anterior part will form the softer tissue through which spinal nerves should run, and in which the dorsal root ganglia (relay stations that undertake pre-processing of sensory signals) should develop. In the embryo, neural crest cells (progenitors of the dorsal root ganglia, among other things) and axons travel only through the anterior half-somite and avoid entering the posterior half-somite (see Figure 3.5.8).

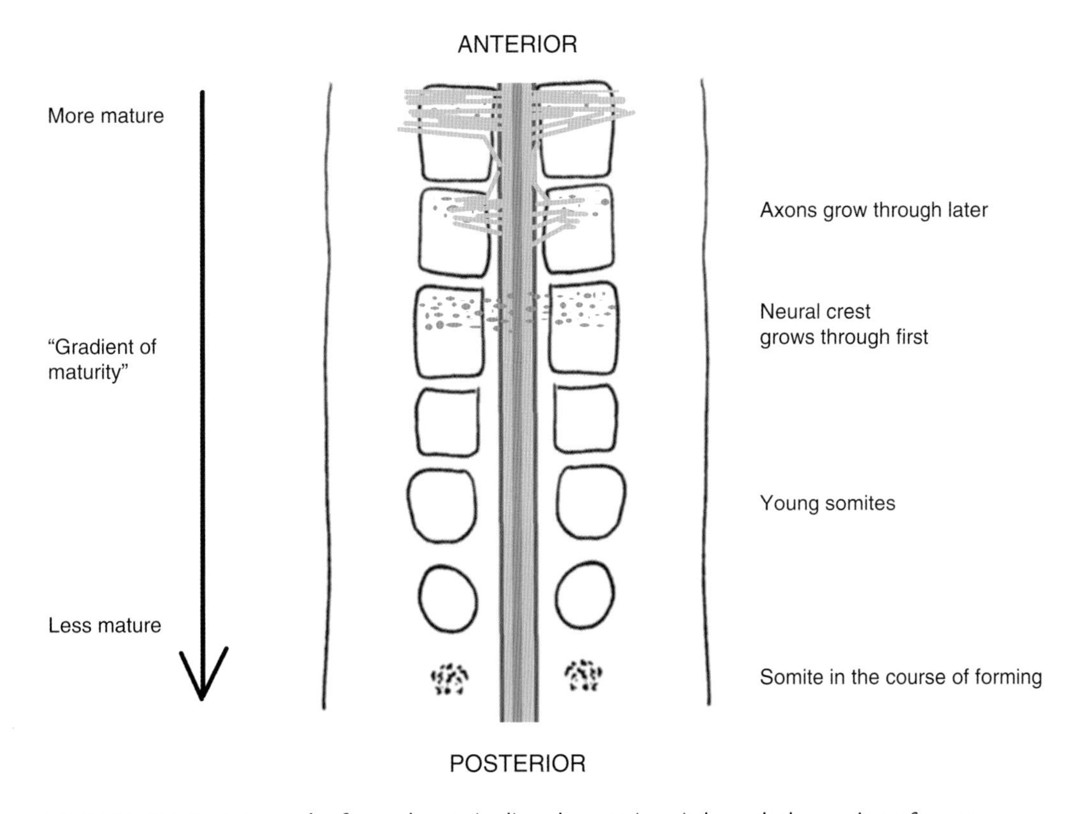

FIGURE 3.5.8 Outgrowth of neural crest (red) and axons (gray) through the somites of a vertebrate embryo. The "gradient of maturity" is not just a device used to represent different stages in one drawing, but a real feature of vertebrate development.

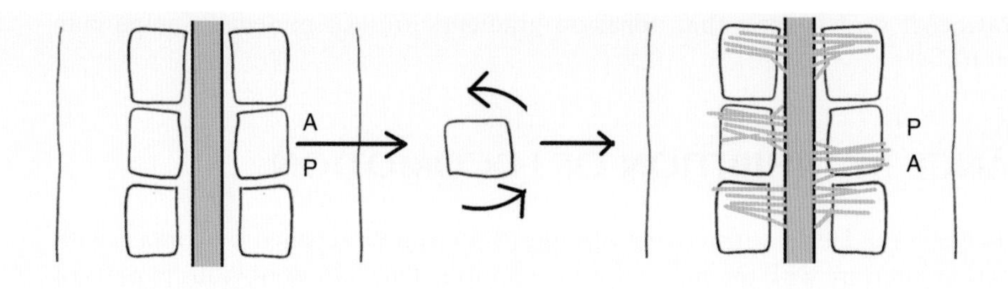

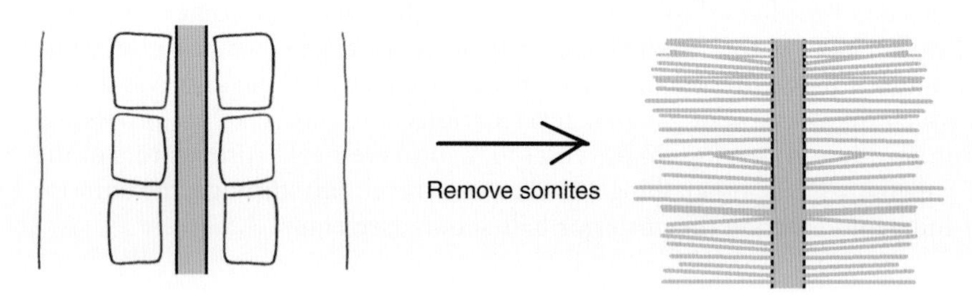

FIGURE 3.5.9 Critical experiments showing that the restriction of axon outgrowth to the anterior half-somite is controlled by a property of the somite itself, not the neural tube.

The restriction of outgrowth to the anterior half-somites is known not to be an inherent property of the neural tube because the pattern reverses if a somite is reversed (see Figure 3.5.9).[44] This suggests that the control lies in the somite itself. Furthermore, if the somites are removed completely, axons emerge from all levels of the neural tube, suggesting that the somites work not by encouraging outgrowth at the anterior half but rather by repressing it in the posterior half. Biochemical analysis of the somites has identified glycoproteins expressed only in the posterior half-somite that, when applied to growth cones in culture, cause them to cease advancing and to collapse so that they lose their whole leading-edge structure (see Figure 3.5.10).[45]

Molecules that can cause collapse of the leading edges of neuronal growth cones and migratory cells are widespread in the embryo. One of the best-studied families consists of the Ephrins, which are important in many places. One of these places is the optic tectum, the region of the brain that will receive axons coming from the eyes (as described in more detail in Chapter 3.6). The optic tectum expresses a number of guidance molecules, among which is the glycosylphosphatidylinositol (GPI)-linked membrane protein Ephrin-A5 (synonyms: AL1/RAGS). This molecule is expressed in a gradient in the tectum, being strongest in the posterior and weakest in the anterior.[46] The growth cones that grow into the tectum from the eye, to lay down axons that will eventually carry visual information, express at least five different receptors for Ephrin-A5[47,48]; these are EphA3, EphA4, EphA5, EphA6, and EphA7.♣ Interfering with the expression of Ephrin-A5, either by overexpression or mutation, causes retinal axons to

♣Please note that Eph is *NOT* an abbreviation for "Ephrin"; Ephrins and Ephs are two distinct families of receptors, members of each family bring ligands for certain members of the other family. The natural, but

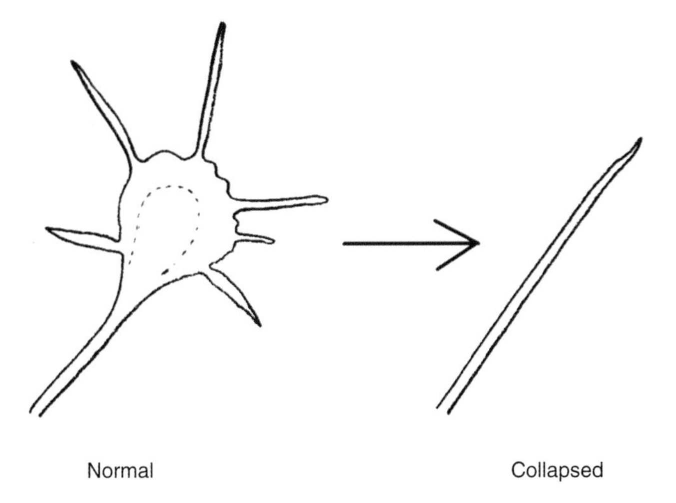

FIGURE 3.5.10 Collapse of growth cones in culture elicited by treatment with repulsive molecules from the posterior half-somite.

lose their way, to fail to produce a proper retinotectal map, and in some cases to end up in the wrong part of the brain altogether.[49,50]

Simple culture experiments in which growing axons are challenged with exogenously applied Ephrin A5 show that the molecule is a powerful inhibitor of axon elongation: contact with it causes growth cones to "collapse" so that they lose their fanned-out appearance and dwindle to a blunt-ended axon[46] (see Figure 3.5.10). This collapse is mediated by a signal transduction system that interferes directly with cell locomotion. The receptor EphA4 interacts directly with a guanidine nucleotide exchange factor (GEF) called Ephexin. Ephexin has the ability to activate small GTPases of the Rho family by exchanging their GDP for GTP, but its substrate specificity is altered by binding to activated EphA4. When no Ephrin-A is bound by EphA4, Ephexin activates Rac and cdc42, each of which encourages leading-edge protrusion (see Chapter 3.2). When EphA4 binds an Ephrin-A ligand, though, Ephexin strongly activates Rho, leading to the formation of myosin-associated stress fibres, and it no longer activates Rac and cdc42.[51] The importance of this pathway is illustrated by the fact that loading neurons with an inhibitor of Rho, such as the bacterial enzyme C-3 transferase, or by inhibiting the Rho's downstream effector, the Rho-associated kinase ROCK, renders them relatively insensitive to Ephrin As.[52] Together, these results suggest that inhibitory Ephrins cause growth-cone collapse by switching the organization of actin from that characteristic of an active leading edge to that characteristic of the trailing edge of a cell, or the main shaft of a growth cone (see Figure 3.5.11).

Migratory cells have no choice but to interact with their substrate (because, except for swimming cells like spermatozoa, they have no other way of moving). The molecules they find there can affect their migration either by direct mechanical means or by signalling to components of the locomotory machinery. The targets of these signals are the same as those of chemotactic signalling, so that integration of the various guidance

incorrect, assumption that one name is just an abbreviation of the other is a common source of confusion in lectures and conferences.

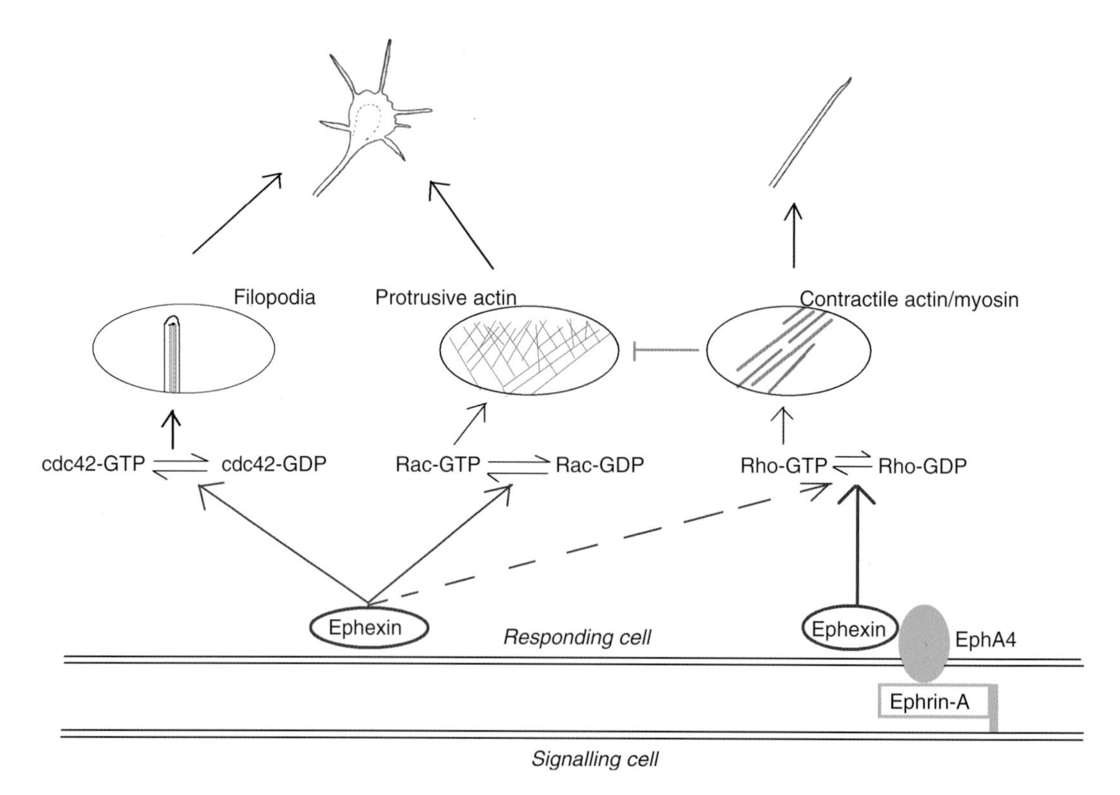

FIGURE 3.5.11 Control of the machinery of locomotion by the Ephrin-A-EphA signalling system.

cues impinging on a cell takes place at the level of the locomotive machinery itself. The integration is also local; the decision whether to migrate at all may be taken at the level of gene expression by the whole cell, but the decision about where to go is taken very much by micro-domains in and near the leading edge. The following chapter will describe how the ability of cells to perform nanometer-scale integration of signals can be used to allow them to perform highly complex feats of navigation, and even to help each other as they do.

Reference List

1. Cantat, I, Misbah, C, and Saito, Y (2004) Vesicle propulsion in haptotaxis: A local model. *Eur. Phys. J. E.* **3**: 403–412
2. Letourneau, PC (1975) Cell-to-substratum adhesion and guidance of axonal elongation. *Dev. Biol.* **44**: 92–101
3. O'Connor, TP, Duerr, JS, and Bentley, D (1990) Pioneer growth cone steering decisions mediated by single filopodial contacts in situ. *J. Neurosci.* **10**: 3935–3946
4. Albrecht-Buehler, G (1976) Filopodia of spreading 3T3 cells. Do they have a substrate-exploring function? *J. Cell Biol.* **69**: 275–286
5. Hammarback, JA, Palm, SL, Furcht, LT, and Letourneau, PC (1985) Guidance of neurite outgrowth by pathways of substratum-adsorbed laminin. *J. Neurosci. Res.* **13**: 213–220
6. Wessells, NK and Nuttall, RP (1978) Normal branching, induced branching, and steering of cultured parasympathetic motor neurons. *Exp. Cell Res.* **115**: 111–122

7. Hammarback, JA and Letourneau, PC (1986) Neurite extension across regions of low cell-substratum adhesivity: Implications for the guidepost hypothesis of axonal pathfinding. *Dev. Biol.* **117**: 655–662
8. Suter, DM, Errante, LD, Belotserkovsky, V, and Forscher, P (6-4-1998) The Ig superfamily cell adhesion molecule, apCAM, mediates growth cone steering by substrate-cytoskeletal coupling. *J. Cell Biol.* **141**: 227–240
9. Lo, CM, Wang, HB, Dembo, M, and Wang, YL (2000) Cell movement is guided by the rigidity of the substrate. *Biophys. J.* **79**: 144–152
10. Pelham, RJ, Jr. and Wang, Y (9-12-1997) Cell locomotion and focal adhesions are regulated by substrate flexibility. *Proc. Natl. Acad. Sci. U.S.A.* **94**: 13661–13665
11. Hanks, SK, Calalb, MB, Harper, MC, and Patel, SK (15-9-1992) Focal adhesion protein-tyrosine kinase phosphorylated in response to cell attachment to fibronectin. *Proc. Natl. Acad. Sci. U.S.A.* **89**: 8487–8491
12. Wang, HB, Dembo, M, Hanks, SK, and Wang, Y (25-9-2001) Focal adhesion kinase is involved in mechanosensing during fibroblast migration. *Proc. Natl. Acad. Sci. U.S.A.* **98**: 11295–11300
13. Ren, XD, Kiosses, WB, Sieg, DJ, Otey, CA, Schlaepfer, DD, and Schwartz, MA (2000) Focal adhesion kinase suppresses Rho activity to promote focal adhesion turnover. *J. Cell Sci.* **113 (Pt 20)**: 3673–3678
14. Chikumi, H, Barac, A, Behbahani, B, Gao, Y, Teramoto, H, Zheng, Y, and Gutkind, JS (8-1-2004) Homo- and hetero-oligomerization of PDZ-RhoGEF, LARG and p115RhoGEF by their C-terminal region regulates their in vivo Rho GEF activity and transforming potential. *Oncogene.* **23**: 233–240
15. Saffell, JL, Williams, EJ, Doherty, P, and Walsh, FW (1995) Axonal growth mediated by cell adhesion molecules requires activation of fibroblast growth factor receptors. *Biochem. Soc. Trans.* **23**: 469–470
16. Saffell, JL, Williams, EJ, Mason, IJ, Walsh, FS, and Doherty, P (1997) Expression of a dominant negative FGF receptor inhibits axonal growth and FGF receptor phosphorylation stimulated by CAMs. *Neuron.* **18**: 231–242
17. Doherty, P and Walsh, FS (1994) Signal transduction events underlying neurite outgrowth stimulated by cell adhesion molecules. *Curr. Opin. Neurobiol.* **4**: 49–55
18. Schaechter, JD and Benowitz, LI (1993) Activation of protein kinase C by arachidonic acid selectively enhances the phosphorylation of GAP-43 in nerve terminal membranes. *J. Neurosci.* **13**: 4361–4371
19. He, Q, Dent, EW, and Meiri, KF (15-5-1997) Modulation of actin filament behavior by GAP-43 (neuromodulin) is dependent on the phosphorylation status of serine 41, the protein kinase C site. *J. Neurosci.* **17**: 3515–3524
20. Meiri, KF, Saffell, JL, Walsh, FS, and Doherty, P (15-12-1998) Neurite outgrowth stimulated by neural cell adhesion molecules requires growth-associated protein-43 (GAP-43) function and is associated with GAP-43 phosphorylation in growth cones. *J. Neurosci.* **18**: 10429–10437
21. Strittmatter, SM, Fankhauser, C, Huang, PL, Mashimo, H, and Fishman, MC (10-2-1995) Neuronal pathfinding is abnormal in mice lacking the neuronal growth cone protein GAP-43. *Cell.* **80**: 445–452
22. Perris, R and Perissinotto, D (2000) Role of the extracellular matrix during neural crest cell migration. *Mech. Dev.* **95**: 3–21
23. Bronner-Fraser, M (1985) Alterations in neural crest migration by a monoclonal antibody that affects cell adhesion. *J. Cell Biol.* **101**: 610–617
24. Lallier, T and Bronner-Fraser, M (29-1-1993) Inhibition of neural crest cell attachment by integrin antisense oligonucleotides. *Science.* **259**: 692–695
25. Lofberg, J, Nynas-McCoy, A, Olsson, C, Jonsson, L, and Perris, R (1985) Stimulation of initial neural crest cell migration in the axolotl embryo by tissue grafts and extracellular matrix transplanted on microcarriers. *Dev. Biol.* **107**: 442–459

26. Rutishauser, U (1985) Influences of the neural cell adhesion molecule on axon growth and guidance. *J. Neurosci. Res.* **13**: 123–131
27. Fischer, G, Kunemund, V, and Schachner, M (1986) Neurite outgrowth patterns in cerebellar microexplant cultures are affected by antibodies to the cell surface glycoprotein L1. *J. Neurosci.* **6**: 605–612
28. Dahme, M, Bartsch, U, Martini, R, Anliker, B, Schachner, M, and Mantei, N (1997) Disruption of the mouse L1 gene leads to malformations of the nervous system. *Nat. Genet.* **17**: 346–349
29. Cremer, H, Chazal, G, Goridis, C, and Represa, A (1997) NCAM is essential for axonal growth and fasciculation in the hippocampus. *Mol. Cell Neurosci.* **8**: 323–335
30. Bastiani, MJ, Harrelson, AL, Snow, PM, and Goodman, CS (13-3-1987) Expression of fasciclin I and II glycoproteins on subsets of axon pathways during neuronal development in the grasshopper. *Cell.* **48**: 745–755
31. Kolodkin, AL, Matthes, DJ, O'Connor, TP, Patel, NH, Admon, A, Bentley, D, and Goodman, CS (1992) Fasciclin IV: Sequence, expression, and function during growth cone guidance in the grasshopper embryo. *Neuron.* **9**: 831–845
32. Zinn, K, McAllister, L, and Goodman, CS (20-5-1988) Sequence analysis and neuronal expression of fasciclin I in grasshopper and *Drosophila. Cell.* **53**: 577–587
33. Patel, NH, Snow, PM, and Goodman, CS (27-3-1987) Characterization and cloning of fasciclin III: A glycoprotein expressed on a subset of neurons and axon pathways in *Drosophila. Cell.* **48**: 975–988
34. McAllister, L, Goodman, CS, and Zinn, K (1992) Dynamic expression of the cell adhesion molecule fasciclin I during embryonic development in *Drosophila. Development.* **115**: 267–276
35. Jay, DG and Keshishian, H (6-12-1990) Laser inactivation of fasciclin I disrupts axon adhesion of grasshopper pioneer neurons. *Nature.* **348**: 548–550
36. Grenningloh, G, Rehm, EJ, and Goodman, CS (4-10-1991) Genetic analysis of growth cone guidance in *Drosophila*: Fasciclin II functions as a neuronal recognition molecule. *Cell.* **67**: 45–57
37. Lin, DM, Fetter, RD, Kopczynski, C, Grenningloh, G, and Goodman, CS (1994) Genetic analysis of fasciclin II in *Drosophila*: Defasciculation, refasciculation, and altered fasciculation. *Neuron.* **13**: 1055–1069
38. Baines, RA, Seugnet, L, Thompson, A, Salvaterra, PM, and Bate, M (1-8-2002) Regulation of synaptic connectivity: Level of fasciclin II influence synaptic growth in the *Drosophila* CNS. *J Neurosci.* **22**: 6587–6595
39. Boucaut, JC and Darribere, T (1983a) Fibronectin in early amphibian embryos. Migrating mesodermal cells contact fibronectin established prior to gastrulation. *Cell Tissue Res.* **234**: 135–145
40. Boucaut, JC and Darribere, T (1983b) Presence of fibronectin during early embryogenesis in amphibian Pleurodeles waltlii. *Cell Differ.* **12**: 77–83
41. Boucaut, JC, Darribere, T, Boulekbache, H, and Thiery, JP (26-1-1984) Prevention of gastrulation but not neurulation by antibodies to fibronectin in amphibian embryos. *Nature.* **307**: 364–367
42. Boucaut, JC, Johnson, KE, Darribere, T, Shi, DL, Riou, JF, Bache, HB, and Delarue, M (1990) Fibronectin-rich fibrillar extracellular matrix controls cell migration during amphibian gastrulation. *Int. J. Dev. Biol.* **34**: 139–147
43. Winklbauer, R and Nagel, M (1991) Directional mesoderm cell migration in the Xenopus gastrula. *Dev. Biol.* **148**: 573–589
44. Keynes, RJ and Stern, CD (30-8-1984) Segmentation in the vertebrate nervous system. *Nature.* **310**: 786–789
45. Davies, JA, Cook, GM, Stern, CD, and Keynes, RJ (1990) Isolation from chick somites of a glycoprotein fraction that causes collapse of dorsal root ganglion growth cones. *Neuron.* **4**: 11–20

46. Drescher, U, Kremoser, C, Handwerker, C, Loschinger, J, Noda, M, and Bonhoeffer, F (11-8-1995) In vitro guidance of retinal ganglion cell axons by RAGS, a 25 kDa tectal protein related to ligands for Eph receptor tyrosine kinases. *Cell.* **82**: 359–370
47. Connor, RJ, Menzel, P, and Pasquale, EB (1-1-1998) Expression and tyrosine phosphorylation of Eph receptors suggest multiple mechanisms in patterning of the visual system. *Dev. Biol.* **193**: 21–35
48. O'Leary, DD and Wilkinson, DG (1999) Eph receptors and ephrins in neural development. *Curr. Opin. Neurobiol.* **9**: 65–73
49. Frisen, J, Yates, PA, McLaughlin, T, Friedman, GC, O'Leary, DD, and Barbacid, M (1998) Ephrin-A5 (AL-1/RAGS) is essential for proper retinal axon guidance and topographic mapping in the mammalian visual system. *Neuron.* **20**: 235–243
50. Nakamoto, M, Cheng, HJ, Friedman, GC, McLaughlin, T, Hansen, MJ, Yoon, CH, O'Leary, DD, and Flanagan, JG (6-9-1996) Topographically specific effects of ELF-1 on retinal axon guidance in vitro and retinal axon mapping in vivo. *Cell.* **86**: 755–766
51. Shamah, SM, Lin, MZ, Goldberg, JL, Estrach, S, Sahin, M, Hu, L, Bazalakova, M, Neve, RL, Corfas, G, Debant, A, and Greenberg, ME (20-4-2001) EphA receptors regulate growth cone dynamics through the novel guanine nucleotide exchange factor ephexin. *Cell.* **105**: 233–244
52. Wahl, S, Barth, H, Ciossek, T, Aktories, K, and Mueller, BK (17-4-2000) Ephrin-A5 induces collapse of growth cones by activating Rho and Rho kinase. *J. Cell Biol.* **149**: 263–270

CHAPTER 3.6

CELL MIGRATION: Waypoint Navigation in the Embryo

The foregoing chapters on cell migration concentrated on the basic mechanisms of locomotion and how cells direct that motion in response to local cues. Most of the experimental evidence came from simple culture systems in which cells were presented with just one potential guidance cue and in which a response was continuous migration in one direction. Cells migrating in a developing embryo generally face a much more complicated task; they have to make journeys that are seldom simple straight lines and have to respond to, or ignore, many potential guidance cues. The ultimate destination may not exist at the time that a migration is initiated and, even when it does, a beeline straight to it may not be the optimum route as it may conflict with other events that need to take place in the intervening space. For this reason, it is perhaps not surprising that complex embryos have evolved a multistage process of direction finding analogous to what sailors and aviators call "waypoint navigation." In waypoint navigation, a route is split into several phases that run from one distinct location—a "waypoint"—to the next. For each phase of the journey, navigational effort is concentrated on reaching the next waypoint rather than on the ultimate destination; once a particular waypoint has been reached, it is forgotten about and all effort is now invested in the next waypoint in the series. The "turn left at the Rose and Crown, turn right after the hump-backed bridge, and right again at the garage" style of navigation, familiar to motorists, is a homely example of waypoint navigation.

WAYPOINT NAVIGATION BY GERM CELLS IN *DROSOPHILA MELANOGASTER*

All of the gametes produced in the life of a *Drosophila melanogaster* fly derive ultimately form 30 to 40 primordial germ cells that arise very early in embryogenesis. These cells arise at the very posterior of the early embryo, far from the eventual location of the gonads and, like the vertebrate germ cells described in Chapter 3.1, they have to undertake a complex migration. The first part of their journey appears to be entirely passive; they are caught up in a general flow of cells that takes them and their neighbouring

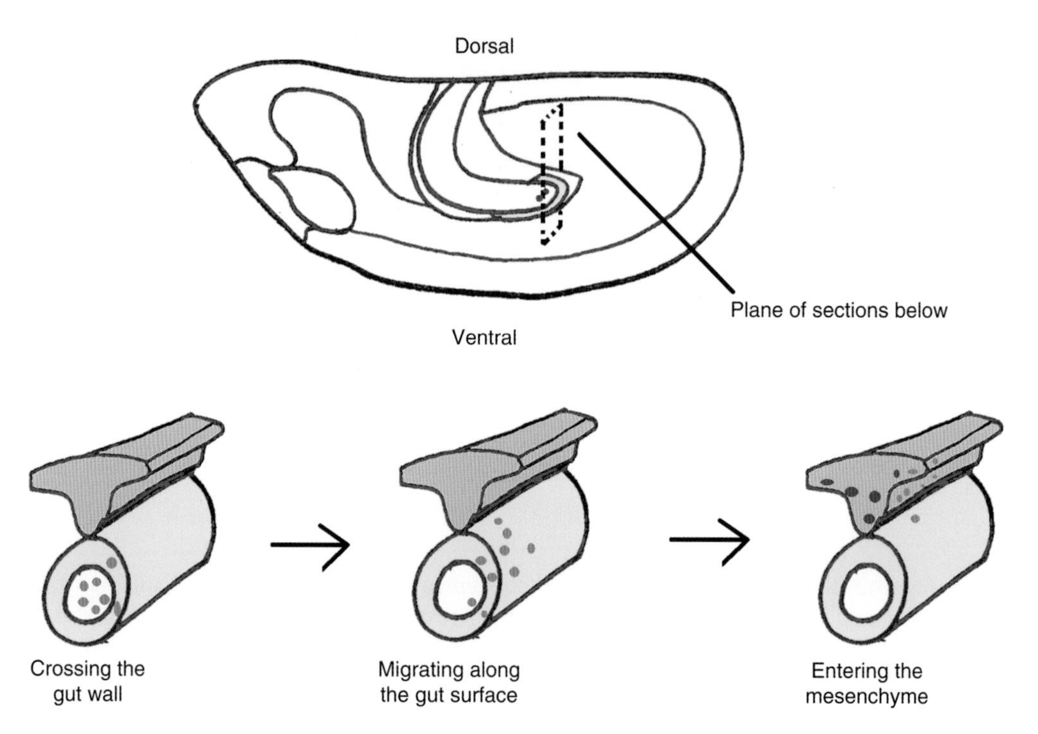

FIGURE 3.6.1 A schematic representation of primordial germ cell migration in *Drosophila melanogaster*. The depiction of the gut and mesenchyme anatomy in the lower figures is simplified a little to make the germ cell movements more clear.

endodermal cells, which will form the posterior midgut, toward the dorsal surface of the embryo quite near its middle (see Figure 3.6.1). From that point, they begin active migration.

During the time that it and the primordial germ cells are being swept along the embryo, the posterior midgut primordium undergoes morphogenesis to form a typical gut-like tube (see Chapter 4.4). The topology of this process presents the primordial germ cells with a problem: they find themselves on the luminal surface of the gut wall, the wrong side to be topologically within the body. The first step of active germ cell migration is therefore to migrate across the gut wall itself. This epithelium is specialized to open up spaces between its cells to allow germ cell migration,[1,2] and it opens up even in mutant flies in which germ cells fail to form. This unusual ability of the posterior midgut to allow the passage of cells is essential, and germ cell migration fails in fate-specification mutants (e.g., *huckebein, serpent*) in which the region of the gut that is supposed to be posterior midgut instead develops hindgut character and will not open up.[3,4] The timing of germ cell migration seems to be set by the gut wall itself. The crossing must also require active participation by the germ cells because other cells transplanted to the same location will not cross. Activation of the wall-crossing phase of germ cell migration requires activation of a specific G-protein coupled receptor of the chemokine receptor family, called Tre (from the gene *trapped in endoderm*).[5] Tre acts autonomously in the germ cells themselves, and although its ligand has not yet been identified, it is probably significant that a related molecule, CXCR4, is the receptor for the gonad-derived chemoattractant CXCL12 in mice.[6,7] The precise pathway of Tre signalling has not yet been elucidated, but it seems to require Rho1 and therefore may affect actin organization.[5]

Once through and on to the correct side of the gut wall, the germ cells have to migrate along its surface toward the dorsal side of the embryo and thence to the mesoderm in which gonads will form (see Figure 3.6.1). Their accurate migration on the gut surface depends on expression of the closely linked genes *wunen1* and *wunen2* by precisely those gut cells that do NOT form a suitable pathway for the germ cells.[8,9] In the absence of any *wunen* expression (i.e., a double mutant), germ cells migrate directionlessly and scatter over the gut surface so that few find their ultimate mesenchymal targets, even though these targets are still present. The suspicion that *wunen* is involved in repulsive signalling is confirmed by the observation that forcing expression of *wunen* in a tissue normally permissive for germ-cell migration keeps that tissue germ-cell free.[9] *Wunen* proteins are expressed on the cell membrane, but they are, at least by homology, phosphatidic acid phosphatases with their catalytic domains outside the cell. The molecular target of the *wunens* has not been identified, but it is generally assumed that they work either by converting a phospholipid into a repulsive signal or by destroying a phospholipid necessary for cell surfaces to permit germ cell migration.[10] As long as *wunen* expression is normal, primordial germ cells will migrate across the gut surface in the right direction and to the right local destination, even if their ultimate destination, the gonad-forming mesoderm, is absent.[11,12,10]

Once they have completed their migration across the surface of the gut, the primordial germ cells leave it and enter the mesoderm, where they split into two streams to colonize the gonads, one on each side of the body. This phase depends on the expression of an enzyme, HMGCoA reductase, by the somatic cells of the gonad.[13] Forcing the expression of HMGCoA in other tissues makes these tissues attractive to primordial germ cells as well, suggesting that the enzyme is involved in making a chemotactic signal. In mammals, HMGCoA is important in the synthesis of cholesterol and a number of other small compounds. Cholesterol cannot itself be the molecule involved in *D. melanogaster* because this species lacks homologues of other mammalian enzymes essential for using the HMGCoA pathway for cholesterol biosynthesis, but there is genetic evidence that farnesyl-diphosphate synthase and geranylgeranyl-diphosphate synthase are required downstream of HMGCoA for germ cell migration.[14] These enzymes are involved in the synthesis of isoprenoids, suggesting that the still-unidentified chemoattractant is an isoprenylated protein. Pharmacological evidence suggests that vertebrates, too, use HMGCoA and isoprenylated proteins to attract primordial germ cells to their gonads.[15] Once in the gonads, germ cells form tight associations with somatic cells and begin the morphogenesis of the mature organ (see Chapter 3.7).

The reason for considering the migration of primordial cells to be an example of waypoint navigation is not simply the fact that it proceeds by a series of discrete phases. Rather, it is because absence of the final destination does not prevent the cells beginning even the active part of their journey normally. They cross the gut normally, even in mutants in which the *wunen*-based guidance system of the next part of their journey is missing, and in flies with normal *wunen* cues, they navigate the gut surface normally, even when there is no prospective gonad to invade at the end of their journey.[11,12,10] They cannot, however, complete normal journeys if any of the intermediate navigation systems are missing, even when the final destination is intact and normal.

WAYPOINT NAVIGATION BY GROWTH CONES

Peripheral axons of insect nervous systems often make sharp turns as they navigate from the cell body, in the tissues, to the central nervous system. In developing grasshopper prothoracic legs, an early-migrating axon from the cell body of a neuron called Ti1, near the end of the leg, grows in a stereotyped series of straight lines. The corners of these lines coincide with a set of large and easily identifiable immature neuron cell bodies called F1, F2, and CT1 (see Figure 3.6.2). The CT1 cell, in particular, is associated with a very sharp turn. This course naturally suggests that the F1, F2, and CT1 cells may act as waypoints, and are often called "guidepost cells." The hypothesis that they do act as waypoints can be tested by obliterating the CT1 cell by UV irradiation; without CT1, the Ti1 axon fails to make its turn correctly and it becomes lost, sending out multiple branches and failing to enter the central nervous system correctly.[16] Irradiation of other, nearby cells does not have any effect.

The developing central nervous system is probably the area of the body in which cell migration pathways are at their most complex and the need for accuracy is at its greatest. The early development of the central nervous system of *Drosophila melanogaster* is a widely used system for analysis of mechanisms of growth-cone guidance because it shows a rich variety of cell type-specific migratory pathways without being so complicated that it is beyond experimental analysis. The central nervous system of this organism follows the usual arthropod pattern, consisting of a chain of ganglia just inside the ventral surface of the organism (see Figure 3.6.3). Neurons develop each side of the midline, the nervous system being approximately symmetrical about this midline, and project axons to their targets. Approximately 10 percent of neurons project axons only to their own side of the midline.[17] Others cross the midline to meet targets on the opposite side and form the axon bundle commissures that link the left and right halves of the nervous system.

The growth cones of axons that cross the midline use the population of glial cells, at the midline itself, as an important waypoint. If these cells are obliterated using

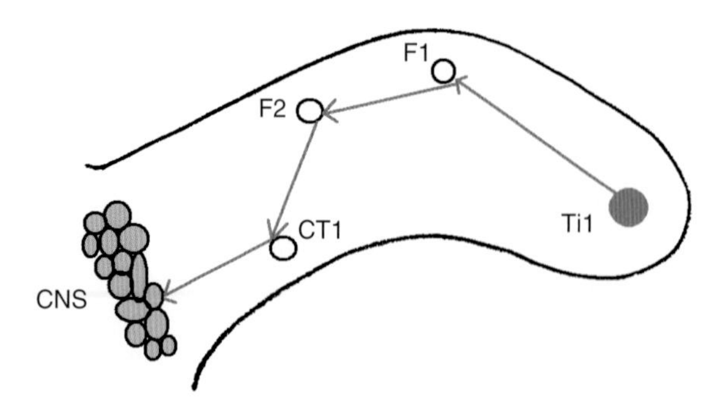

FIGURE 3.6.2 The normal path taken by the Ti1 axon in a developing grasshopper leg. Diagram based on dots in Ref. 16.

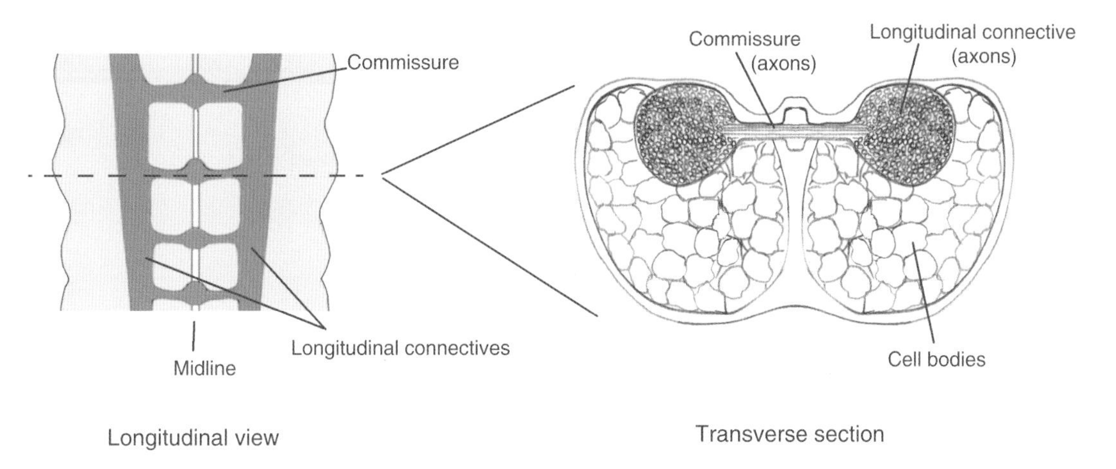

FIGURE 3.6.3 The structure of a section of the developing ventral central nervous system of *D. melanogaster.*

the ricin toxin, axons fail to cross the midline even though their distant targets are still present.[18] The midline cells express the diffusible chemoattractants Netrin 1 and 2, and in the absence of these molecules, commissures fail to form properly,[19,20] being absent or much thinner than usual.[21] The receptor for Netrin, Frazzled (the homologue of the mammalian protein DCC), is expressed by and required by growth cones of axons destined to cross the commissures[22]; there is some evidence that Frazzled operates not as a conventional receptor but rather by presenting Netrins to other receptor molecules.[23] In *frazzled* mutant flies, axons typically fail to cross the midline.

Attraction to the midline is a double-edged sword in promoting the crossing of the central nervous system by commissural axons because, like any other waypoint, it creates a potential trap. Growth cones attracted to Netrin will never leave the midline unless their navigational preferences can be changed on making contact with their midline waypoint. Some parts of the mechanism that alter these preferences have now been identified.

As well as producing the chemoattractants Netrin 1 and 2, the midline cells express the diffusible chemorepellant Slit. Slit signals through the receptor Robo, which antagonizes production of protrusive leading edges. Growth cones of axons that are never meant to cross the central nervous system express Robo and are sensitive to chemorepulsion, so they stay away from the midline. Mutations that block expression of functional Robo cause even these axons to cross.[24] Conversely, if Robo expression is forced on even in what should be commissural axons, these avoid crossing the midline.[25]

Surprisingly, even neurons whose axons are destined to cross the central nervous system express Robo protein. The transport of Robo to the membrane is, however, thwarted by the activity of another protein called Comm (from the gene *commissureless*),[26] which targets nascent Robo peptide to the late endocytic pathway and destroys it.[17] Without Comm, Robo is active, and axons cannot cross the midline, preventing the formation of commissures.[26] Comm is inactivated when growth cones reach the midline, by a mechanism that is not understood, so Robo becomes active and the midline is perceived as a repulsive place (see Figure 3.6.4); the growth cones

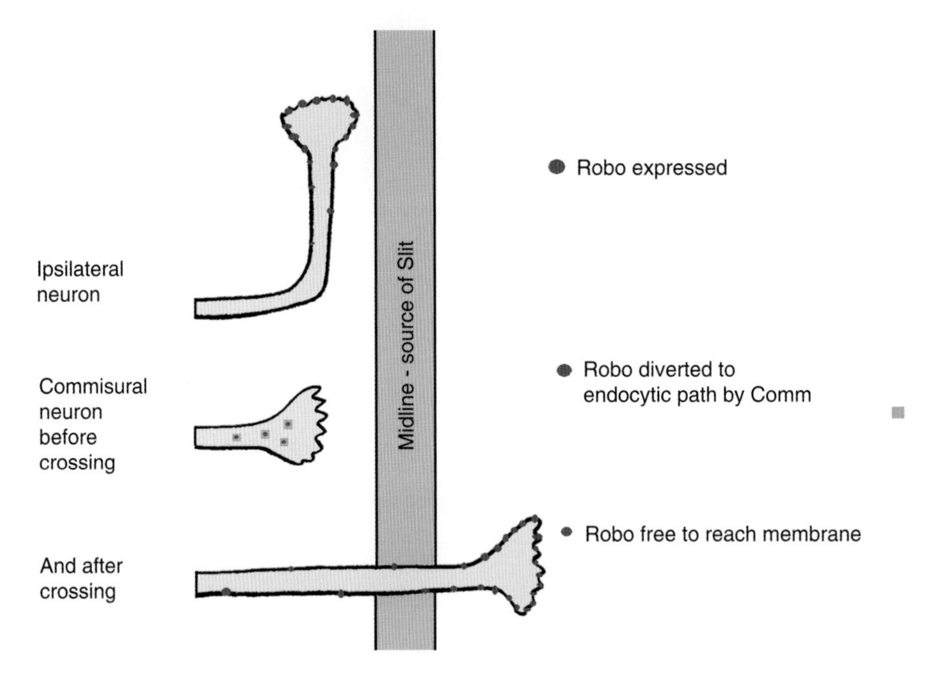

FIGURE 3.6.4 The change in receptor expression in commissural axons once they have crossed the midline.

therefore head away from it and are prevented from crossing again.[27] If the expression of Robo is blocked, growth cones cannot perceive the midline as repulsive, even when they have crossed it, so they keep re-crossing in pointless circles.[24] (This phenotype is how *robo*, an abbreviation for "roundabout," acquired its name.)

Although midline crossing shows several attributes of waypoint navigation, careful analysis of mutants reveals that cells heading for their midline waypoint are capable of detecting at least one subsequent waypoint, too. The basic symmetry of metazoan bodies means that most crossings from left to right have balancing crossings from right to left, and most targets present on the left side are also present on the right side. When mutations in the *comm* and *netrin* genes are used to block the formation of commissures, axons now incapable of crossing to the contralateral side of the central nervous system frequently find their "correct" targets on the ipsilateral♣ half of the central nervous system.[21,28]

The use of Netrin, Slit, and Robo proteins in regulating crossing of the midline is strongly conserved phylogenetically. Although the central nervous system of vertebrates has a quite different morphology to that of insects, and vertebrate genomes contain families of *netrin*, *slit*, and *robo* genes, they use them in essentially the same way and suffer similar errors of axon guidance if the genes are mutated.[29]

A problem related to that of ignoring the influence of an already-reached waypoint is that of becoming newly responsive to the next waypoint in the sequence. In the vertebrate spinal cord, commissural axons grow down toward the floorplate of the spinal cord, cross its midline, and then grow up the other side until they meet the correct

♣ipsilateral = on the same side (as).

longitudinal fibre tract. Once they have found this tract, they turn and grow along it. In a normal spinal cord, they ignore the presence of a similar longitudinal tract on the side of the spinal cord from which they originate, suggesting that they gain interest only once they have passed the midline waypoint. In chick embryos, commissural axons heading toward their midline crossing express no EphA2, but they do express it as soon as they have passed the midline and continue to do so when they are in the longitudinal fibre tract (the ventral funiculus). The expression of the protein is quite specific to those parts of the axon beyond the midline crossing and is not seen on the parts of the axon between the cell body and the midline, suggesting that EphA2 is being inserted into the membrane by the growth cone only after it has passed the midline.[30] There is now good biochemical evidence that protein synthesis can take place locally at growth cones[30] and a highly conserved 3′ untranslated sequence from EphA2 mRNA can drive expression of reporter proteins in the same growth cone-specific, post-midline-specific manner. The signalling events that take place at the growth cone to switch on synthesis of specific new mRNAs have not yet been worked out, but in principle local protein synthesis may prove to be a key mechanism in waypoint navigation. Being local, it would also allow multiple growth cones emanating from the same cell—for example, those of a dendritic arbour—to make their own independent decisions about where to go.

DENSE ARRAYS OF WAYPOINTS MAKE PATHWAYS

When waypoints are arrayed densely enough, they form a continuous pathway for guidance of cells or growth cones from source to destination. An example of such a pathway is important in guiding axons that will carry olfactory information to their targets deep in the brain. Mitral cells in the olfactory bulb of the brain, the part concerned with the sensation of smell, project axons to connect with the amygdala, part of the limbic system that is connected with emotion and sexual arousal. The axons follow a rather indirect route, bundling together along the side of the telencephalon (forebrain) before making a sharp turn toward the amygdala. While in the telencephalon, the mitral axons form a bundle distinct enough to be given an anatomical name, the lateral olfactory tract. If the telencephalon, which includes the olfactory bulb and the amygdala, is removed from an embryonic mouse and placed in organ culture, mitral cells still send their axons precisely through the correct region of the telencephalon to form a lateral olfactory tract. The axons grow across a specific set of neurons already present in the telencephalon, all of which express a distinct antigen apparently unique to that area, which is detectable by the monoclonal antibody "mAblot1."[31] These neurons form a dense array that marks out the whole lateral olfactory tract, even before mitral axons have arrived. If these neurons are ablated in organ culture, mitral axons fail to follow the lateral olfactory tract; indeed they fail to go anywhere.[31] This is true even though the amygdala, the ultimate targets, will still be present in cultured telencephalon. If just some of the mABlot1-binding neurons are ablated, creating an interrupted chain of cells, mitral axons extend normally as far as the break in the chain and then stall. This suggests that, for mitral axons, hopping between waypoints is not enough and that they require a continuous pathway of positive guidance cues.

THE MANY WAYPOINTS OF THE VERTEBRATE VISUAL SYSTEM

The visual system of vertebrates depends on light being detected by retinal cells at the back of a simple eye, and information about the levels of light being relayed, after some local processing and analysis at the back of the eye, to appropriate regions of the brain. The system requires a high precision of connections so that the spatial arrangement of photoreceptors in the retina is represented in a similar spatial arrangement of the cells that receive information from them, in the brain. In mice, each eye produces about 50,000 axons, each of which has to find its appropriate target; in humans, there are over 1,000,000.[ref32] These axons are produced by retinal ganglion cells, which collect information from photoreceptors and pass it on to the brain. The path between retina and brain is complex and involves many intermediate cues.

Retinal ganglion cells do not all differentiate at once but are produced in a centrifugal sequence as the retina develops. The first retinal ganglion cells, and therefore the axons that first form the optic nerve, represent the centre of the retina. More peripheral parts of the retina are represented by retinal ganglion cells that differentiate later, as a "wave" of maturation sweeps out from the centre. This growth pattern effectively re-codes spatial information relating to the radial axis of the retina as differences in the time domain. Axons from peripheral retinal ganglion cells arrive at the optic nerve later and therefore travel on the outside of the axon bundle already formed.[33] The optic nerve is automatically structured by this process so that its centre represents the centre of the retina and its periphery represents the periphery. In "lower" vertebrates such as fish, retinal growth is continuous through adult life and new axons keep being added to the optic nerve; in mammals, the process stops during late foetal development.

The first navigational problem faced by the axonal growth cone produced by a newly differentiated peripheral retinal ganglion cell is turning the right way to find the optic nerve. They are probably aided in this by cells even further to the periphery that have not yet differentiated to produce axons. These cells secrete large amounts of chondroitin sulphate proteoglycan,[34] which is known to be repulsive to a variety of growth cones.[35,36,37] Axons mistakenly attempting to migrate centripetally will therefore be repelled and turned back toward the centre of the retina, where the optic nerve is found.

Axons migrating toward the optic nerve migrate across scattered cells that express Slit1 and many cells that express Slit2.[ref38] In many other contexts, such as control of midline crossing described above, Slit proteins are repulsive, but in the retina they seem to be attractive to growing axons: interfering with their production, by mutating a regulator of Slit1 expression (only), inhibits the migration of retinal ganglion cell growth cones to the optic nerve[38] and those that do penetrate the area fasciculate unusually much, as if they now strongly prefer each other's company to any other available substrate. Forcing some cells of the retina to express Slit1 strongly, using electroporation of an expression construct, causes axons to associate mainly with these cells. It is not clear why Slit1 is attractive in the retina, although in other systems Slit proteins are cleaved proteolytically,[39] and N-terminal fragments of Slit proteins are attractive even when the whole protein is repulsive.[40] Also, again in other systems, some extracellular matrix components or cAMP can persuade cells to view fragments of Slit proteins as attractive rather than repulsive.[41] Together, these observations suggest that the scattered Slit1-expressing cells of the retina form a pathway of attractive waypoints leading toward the centre of the retina and the optic nerve. It is possible that the Slit2 expressed by other cells remains repulsive, and that the opposing effects of the two proteins help to confine the axons to their pathway.

The centrifugal sequence of retinal ganglion cell differentiation means that growth cones from new cells heading toward the optic nerve will soon come across the axons already laid down by cells older than themselves. Faced with the choice between continuing to migrate on the Slit1-expressing cells or fasciculating with existing axons, growth cones tend to chose fasciculation. This depends on specific adhesion molecules expressed on the axon surface. The axons of retinal ganglion cells in the goldfish *Carrassius auratus*, for example, express the adhesion molecules neuropilin and L1[42]; interfering with either of these molecules using monoclonal antibodies interferes with normal guidance. Antibodies against L1 disrupt fasciculation without preventing the axons eventually reaching the optic nerve, while those against neuropilin disrupt fasciculation and cause the axons to become seriously lost and to wander about in pointless circles.[42,43] Similar results have been obtained using anti-L1 in rat retina.[44] L1 is not just an adhesion molecule; it is part of the Semaphorin 3A receptor complex,[45] and it is also a ligand for certain integrins,[46] so it can participate in several aspects of signal transduction. The role of L1 in fasciculation in the retina may therefore be either by direct adhesion or by modulation of responses to other factors by intracellular signalling, or some combination of these effects.

The centripetal chronotopic♣ relationship is not the only large-scale pattern in the retina; there are also orthogonal gradients of Ephs and Ephrins (see Chapter 3.5). Ephrin B proteins and their receptors follow dorso-ventral gradients, Ephrin B proteins being highest dorsally and lowest ventrally, while their Eph B receptors are lowest dorsally and highest ventrally (see Figure 3.6.5). Similarly, Ephrin A proteins are highest nasally and lowest temporally, and their EphA receptors lowest nasally and highest temporally (see Figure 3.6.5).[47,48,49,50] This effectively sets up a coordinate system independent (at the time of reading) from the chronotopic one. In some ways, the use of four additional gradients to create a system for specifying positional information seems surprising since it would be possible to convert the radial coordinate chronotopic gradient into a complete position-specifying system by adding just only one extra component, a circumferential one, to make a polar coordinate system like the global system of latitude and longitude as seen from above the North Pole. There is, however, a problem with polar coordinate systems: they require a sharp discontinuity in an otherwise steady circumferential gradient, like the 23h59min→0h0min transition on a clock face, and it may have just been biologically easier for embryos to avoid this issue by using gradients across opposite diameters; besides, evolution is not bound by any selection pressure for mathematical elegance.

The Eph/Ephrin gradients are an important component of the guidance system that brings retinal ganglion axons to the optic nerve. Mutant mice that lack EphB function (double knockouts for EphB2 and EphB3) begin to develop normally, but some axons coming from the dorsal retina defasciculate close to the beginning of the optic nerve, then bypass it and extend right past it into the opposite side of the eye. In a wild-type embryo, these axons would be running into a rising gradient of EphB that is no longer there; perhaps carrying on until they perceive enough EphB is an important part of the growth cones' developmental program. Or, more realistically, the *ratio* of EphrinB/EphB may be important; a ratio would provide a better scale-independence for

♣chronotopic = linking time and place.

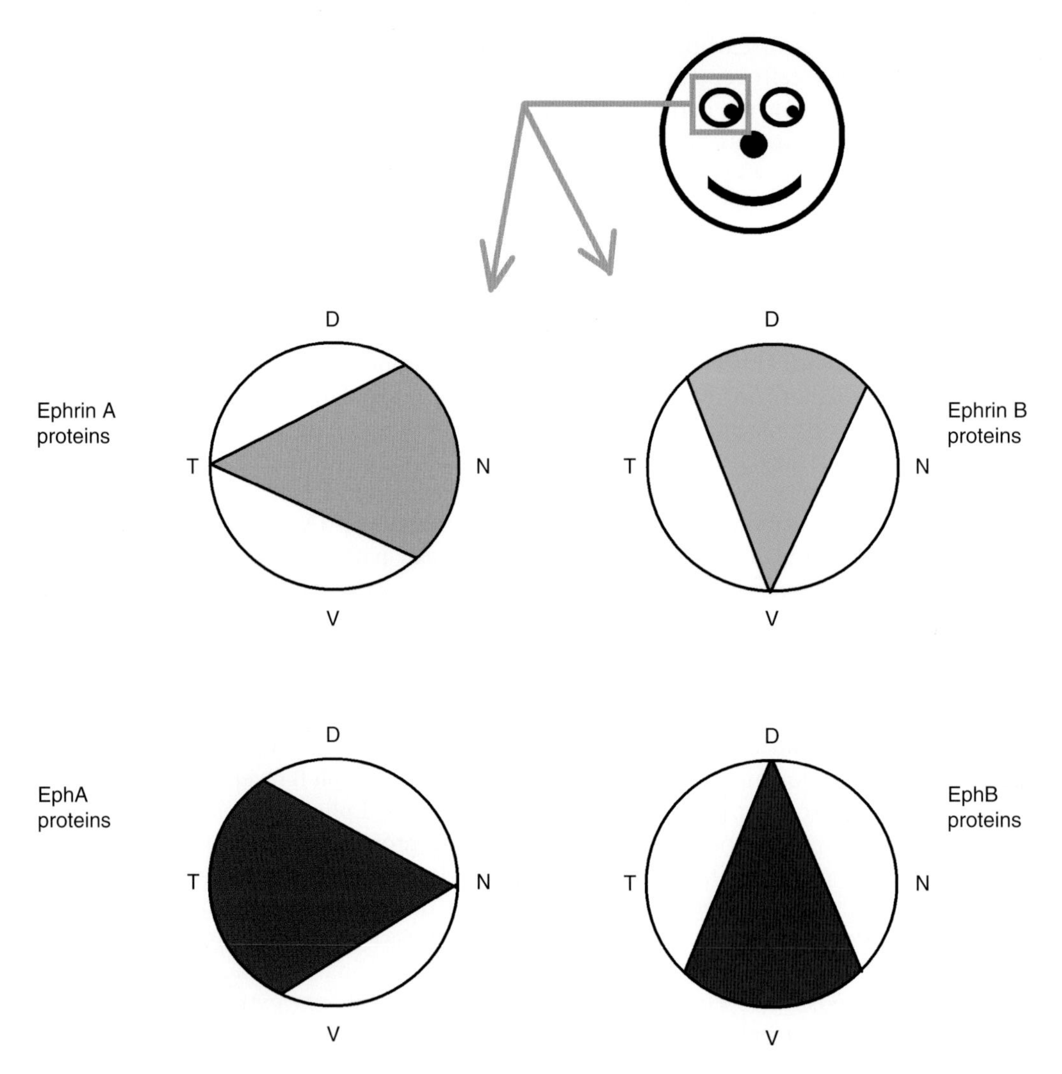

FIGURE 3.6.5 Orthogonal, reciprocal gradients of EphrinA-EphA and EphrinB-EphB proteins in the retina (in each case, several proteins from the families are involved). The cartoon face shows the orientation of retinas, which all represent the right eye; they have been "flattened" for ease of drawing. The triangles represent changing concentrations and are not meant to imply that the expression pattern is actually triangular (it extends across the retina perpendicular to the gradient vector).

navigating retinas that may be large or small. Alternatively, EphB may simply be acting as an inhibitory molecule, as it certainly does to retinal growth cones in culture.[51] Axons coming from the dorsal retina will, as they meet more and more EphB on the cells there, be less and less inclined to migrate on the cells and more inclined to fasciculate with other axons and therefore enter the nerve with them. In this model, fasciculation will be driven both by attraction to other axons (L1, Neuropilin, and so on) and repulsion from alternative pathways.

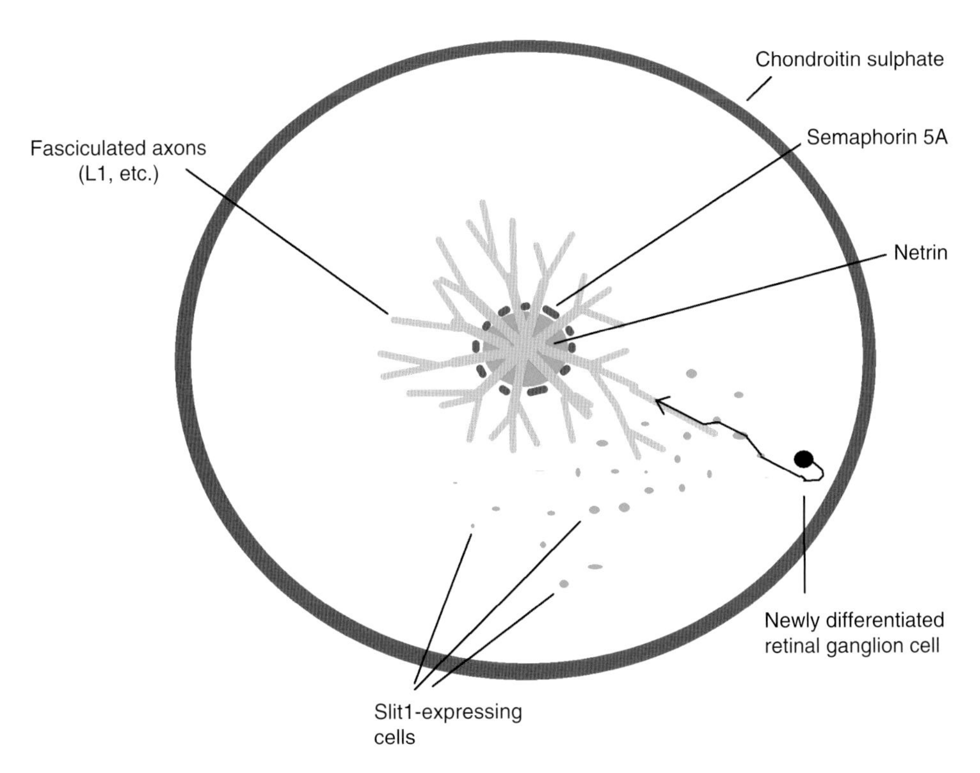

FIGURE 3.6.6 A schematic summary of the cues guiding retinal ganglion axons within the retina. Red indicates repulsive molecules; green and grey indicate attractive molecules.

Neuropepithelial cells at the point where the optic nerve leaves the eye are a source of Netrin, whereas axons of retinal ganglion cells express its receptor, DCC.[52] Deletion of Netrin or DCC function inhibits the ability of axons from all parts of the retina to leave the retina and follow the optic nerve. In mice, Netrin is expressed along the optic nerve itself, which presumably makes it attractive to growth cones.[52] Additional guidance is provided by the repulsive molecule Semaphorin 5A, which is expressed in a halo around the Netrin-rich area at the base of the optic nerve and which is assumed to prevent any would-be escapees from leaving the Netrin-rich routes to the nerve[32] (see Figure 3.6.6).

The optic nerves from both eyes come together at the midline of the body in an area called the "optic chiasm." The chiasm is important in combining signals from each eye that pertain to the same area in the visual field. In animals without binocular vision, in which there is no overlap between visual field, the optic nerves simply cross at the chiasm and project contralaterally (that is, to the opposite side of the brain). In mice, the eyes of which show hardly any overlap in visual field, over 97 percent of axons do this.[53,54] In animals with binocular vision, however, axons representing areas of overlap segregate out and project ipsilaterally (that is, to the same side of the brain). Even in animals with strong binocular vision, the very periphery of the visual field is perceived by only one eye (as anyone with two working eyes can verify by holding a hand beside their face as far back toward the ear as it is visible, and then closing the nearest eye).

Because of the inverting action of a simple lens such as the one in the eye, the edge of the human visual field projects to the *nasal* retina, whereas the central parts of the field are perceived by the *temporal* retinas of both eyes. It is therefore the temporal axons that can sort out at the chiasm to project ipsilaterally. The placing of the eyes on the face and the shape of the face in different species means that binocular vision may also vary along the dorsoventral axis.

The decision between contralateral or ipsilateral projection at the optic chiasm boils down, once again, to a decision about whether to cross the midline of the central nervous system. The molecular mechanisms that operate there can be regarded as analogous to those operating in the midline of the spinal cord or of the insect central nervous system, in that they seem to use an inhibitory signal at the midline to which only some growth cones are sensitive, but the molecules involved are different. The midline of the mouse optic chiasm expresses Ephrin B2 at the time that axons growing through it have to make their contralateral/ipsilateral choice. Axons coming from the dorsal retina around day 15 of mouse development represent parts of the eye that will not produce binocular information. They express little or no EphB1, the receptor for Ephrin B2, and are therefore able to cross the midline unimpeded. Those coming from the ventral temporal zone express high amounts of EphB1, however (see Figure 3.6.5), and are repelled by the midline and forced to follow an ipsilateral pathway[55] (see Figure 3.6.7). The importance of the Ephrin B2/EphB1 system is demonstrated by the phenotype of EphB1 mutant mice, which show very little ipsilateral projection.[55] Choice of the ipsilateral pathway by axons from the left eye is aided by the presence of axons from the right eye that have already crossed to follow the left side of the body, and vice versa. Animals in which one eye is missing show a reduced number of ipsilateral axons even when EphB1 is being expressed normally.[56] The influence of axons from

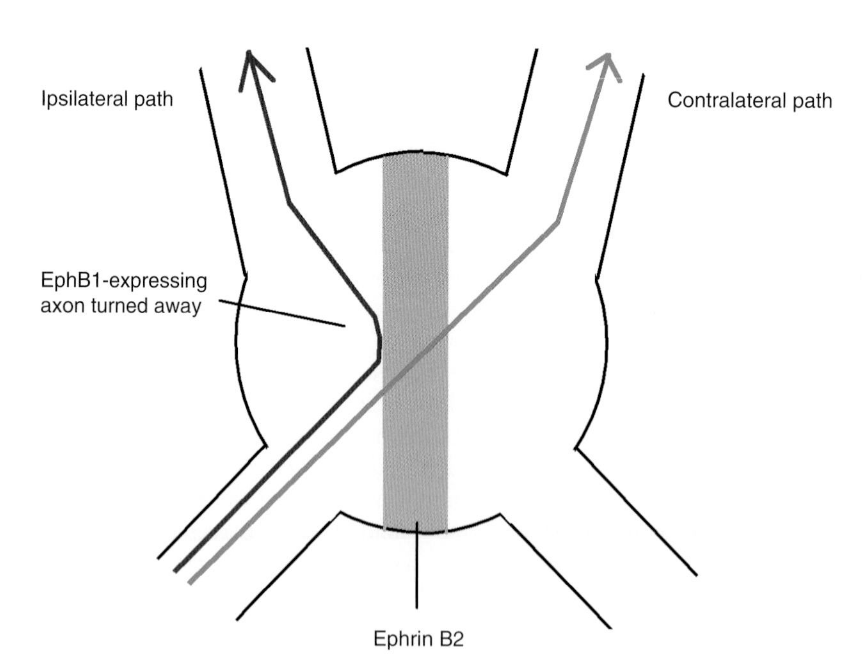

FIGURE 3.6.7 Mechanism for deciding whether to cross the midline at the optic chiasm.

the contralateral eye may be due to general attractive surface molecules such as L1, without which they are less good at supporting ipsilateral choices by axons from the other eye.[57] The need for the opposite eye for ipsilateral projection to take place creates an interesting adaptive response in the embryo, given that it makes sense for it not to bother creating pathways for binocular vision when one eye is missing. It is probably an "accidental" adaptive ability, though, as it is hard to see how it would be selected for directly in evolution.

Once they are through the optic chiasm, axons have to find their appropriate target cells in the optic tectum, an area of the midbrain. The arrangement of axons that terminate on the tectum transforms the continuous spatial map of the retina to a corresponding continuous (albeit differently shaped) spatial map on the tectum. It has been known for over 40 years that this mapping depends on specific biochemical cues borne by axons and tectum.[58] One set of cues is provided, once again, by the Eph/Ephrin signalling system. The optic tectum expresses Ephrins A2 and A5 in a gradient with high expression in the posterior of the tectum and lower expression in the anterior, the Ephrin A5 gradient being steeper[59,60] (see Figure 3.6.8). The nasal retina projects to the posterior tectum and the temporal retina projects to the anterior tectum, intermediate positions on the retinal naso-temporal axis projecting to correspondingly intermediate positions on the tectal antero-posterior axis. There is therefore a correspondence between the gradients of Ephrins A2 and A5 expressed by the retina and those gradients in the tectum: axons from a zone of retina with high Ephrin A2 project to regions of tectum with high Ephrin A2, and vice versa.

Retinal axons express EphA3, a receptor for Ephrins A2 and A5, in a gradient that is high in the temporal region and low nasally. Since binding of A-type Ephrins by EphAs exerts a repulsive influence on growth cones (see Chapter 3.5), these expression patterns strongly suggest a mechanism in which retinotectal mapping is mediated by repulsion. Repulsion is maximal when expression of both EphA3 and EphrinA2/5 are highest, so temporal axons are repelled most strongly from posterior retina. They will still be repelled, but less strongly, from regions of tectum less posterior. Similarly, axons originating from intermediate locations on the naso-temporal axis will express less EphA3 and will be repelled less than temporal axons are by the same level of Ephrin A2/5. If space on the tectum were infinite, the repulsion would steer almost all axons to the anterior, since almost all would be repelled to some extent by posterior tectum. Space is not infinite, however, and axons have to compete with each other. The mapping is therefore caused by a combination of graded repulsion with competition. The model presented above predicts that the antero-posterior map of the tectum will be lost if expression of Ephrins A2 and A5 is prevented, and this is exactly what has been observed.[61]

Competition implies that the "decision" made by an axon about when to terminate is not a once and for all affair, and detailed observations of the developing visual system have shown that the initial places that axons seem to decide to stop growing are often inaccurate. Generally they overshoot their final locations and later correct themselves either by turning back or by extending branches and allowing their original ends to die back.[62] Even the branching is often initially inaccurate and is sharpened up by the destruction of branches that connect with the wrong places.[63] This iterative approach to correct innervation is probably a universal feature of complex neural systems, although others have not been studied in enough detail for this to be clear yet.

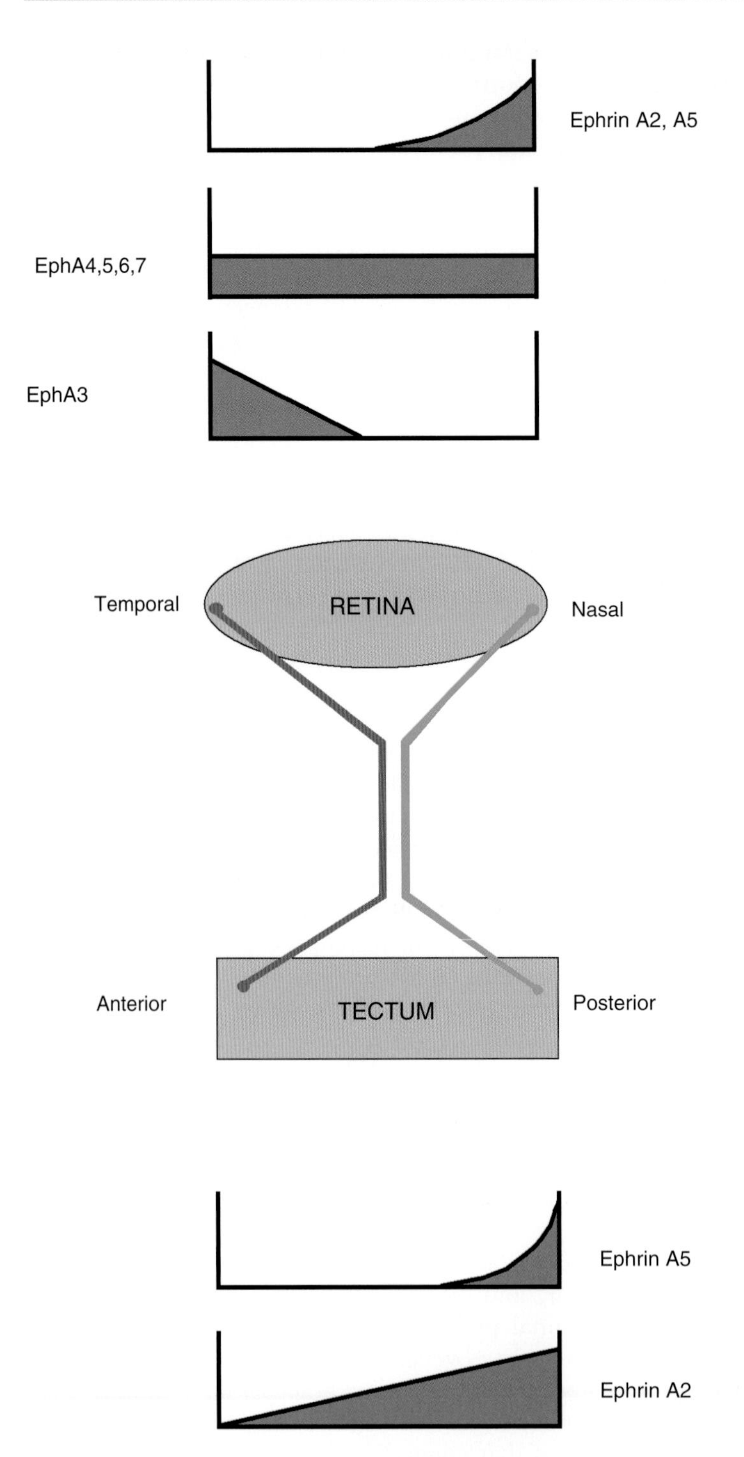

FIGURE 3.6.8 A schematic of the retinotectal projection system: the upper half of the drawing represents the retina, the source of the axons and the Ephs and Ephrins expressed there; the lower half represents the destination, the optic tectum, and the Ephrins expressed there.

The developing visual system, then, depends on a long sequence of guidance cues and on the ability of migrating growth cones to integrate the effects of several cues simultaneously. Some known cues have been described here; there are more cues known already, and probably many more to find. Some cues are "absolute" (for example, repulsion by the chondroitin sulphate at the edge of the developing retina), whereas others are relative (for example, the repulsive gradients of the tectum). Interference with any of the critical cues inhibits normal navigation, even when the ultimate destination is present, but intermediate cues perform their tasks normally, even in the absence of the final destination (many experiments on guidance within the retina were performed on cultured eye tissue isolated from brain). The intermediate cues therefore behave as bona fide waypoints.

Reference List

1. Jaglarz, MK and Howard, KR (1995) The active migration of *Drosophila* primordial germ cells. *Development.* **121**: 3495–3503
2. Jaglarz, MK and Howard, KR (1994) Primordial germ cell migration in *Drosophila melanogaster* is controlled by somatic tissue. *Development.* **120**: 83–89
3. Reuter, R (1994) The gene serpent has homeotic properties and specifies endoderm versus ectoderm within the *Drosophila* gut. *Development.* **120**: 1123–1135
4. Moore, LA, Broihier, HT, Van Doren, M, Lunsford, LB, and Lehmann, R (1998) Identification of genes controlling germ cell migration and embryonic gonad formation in *Drosophila*. *Development.* **125**: 667–678
5. Kunwar, PS, Starz-Gaiano, M, Bainton, RJ, Heberlein, U, and Lehmann, R (2003) Tre1, a g protein-coupled receptor, directs transepithelial migration of *Drosophila* germ cells. *PLoS. Biol.* **1**: E80
6. Molyneaux, KA, Zinszner, H, Kunwar, PS, Schaible, K, Stebler, J, Sunshine, MJ, O'Brien, W, Raz, E, Littman, D, Wylie, C, and Lehmann, R (2003) The chemokine SDF1/CXCL12 and its receptor CXCR4 regulate mouse germ cell migration and survival. *Development.* **130**: 4279–4286
7. Ara, T, Nakamura, Y, Egawa, T, Sugiyama, T, Abe, K, Kishimoto, T, Matsui, Y, and Nagasawa, T (29-4-2003) Impaired colonization of the gonads by primordial germ cells in mice lacking a chemokine, stromal cell-derived factor-1 (SDF-1). *Proc. Natl. Acad. Sci. U.S.A.* **100**: 5319–5323
8. Zhang, N, Zhang, J, Cheng, Y, and Howard, K (1996) Identification and genetic analysis of *wunen*, a gene guiding *Drosophila melanogaster* germ cell migration. *Genetics.* **143**: 1231–1241
9. Zhang, N, Zhang, J, Purcell, KJ, Cheng, Y, and Howard, K (2-1-1997) The *Drosophila* protein *wunen* repels migrating germ cells. *Nature.* **385**: 64–67
10. Starz-Gaiano, M, Cho, NK, Forbes, A, and Lehmann, R (2001) Spatially restricted activity of a *Drosophila* lipid phosphatase guides migrating germ cells. *Development.* **128**: 983–991
11. Rongo, C, Broihier, HT, Moore, L, Van Doren, M, Forbes, A, and Lehmann, R (1997) Germ plasm assembly and germ cell migration in *Drosophila*. *Cold Spring Harb. Symp. Quant. Biol.* **62**: 1–11
12. Warrior, R (1994) Primordial germ cell migration and the assembly of the *Drosophila* embryonic gonad. *Dev. Biol.* **166**: 180–194
13. Van Doren, M, Broihier, HT, Moore, LA, and Lehmann, R (3-12-1998) HMG-CoA reductase guides migrating primordial germ cells. *Nature.* **396**: 466–469
14. Santos, AC and Lehmann, R (2004) Isoprenoids control germ cell migration downstream of HMGCoA reductase. *Dev. Cell.* **6**: 283–293

15. Thorpe, JL, Doitsidou, M, Ho, SY, Raz, E, and Farber, SA (2004) Germ cell migration in zebrafish is dependent on HMGCoA reductase activity and prenylation. *Dev. Cell.* **6**: 295–302
16. Bentley, D and Caudy, M (7-7-1983) Pioneer axons lose directed growth after selective killing of guidepost cells. *Nature.* **304**: 62–65
17. Keleman, K, Rajagopalan, S, Cleppien, D, Teis, D, Paiha, K, Huber, LA, Technau, GM, and Dickson, BJ (23-8-2002) Comm sorts robo to control axon guidance at the *Drosophila* midline. *Cell.* **110**: 415–427
18. Hidalgo, A (2003) Neuron-glia interactions during axon guidance in *Drosophila. Biochem. Soc. Trans.* **31**: 50–55
19. Mitchell, KJ, Doyle, JL, Serafini, T, Kennedy, TE, Tessier-Lavigne, M, Goodman, CS, and Dickson, BJ (1996) Genetic analysis of Netrin genes in *Drosophila*: Netrins guide CNS commissural axons and peripheral motor axons. *Neuron.* **17**: 203–215
20. Harris, R, Sabatelli, LM, and Seeger, MA (1996) Guidance cues at the *Drosophila* CNS midline: identification and characterization of two *Drosophila* Netrin/UNC-6 homologs. *Neuron.* **17**: 217–228
21. Furrer, MP, Kim, S, Wolf, B, and Chiba, A (2003) Robo and Frazzled/DCC mediate dendritic guidance at the CNS midline. *Nat. Neurosci.* **6**: 223–230
22. Kolodziej, PA, Timpe, LC, Mitchell, KJ, Fried, SR, Goodman, CS, Jan, LY, and Jan, YN (1996) Frazzled encodes a *Drosophila* member of the DCC immunoglobulin subfamily and is required for CNS and motor axon guidance. *Cell.* **87**: 197–204
23. Hiramoto, M, Hiromi, Y, Giniger, E, and Hotta, Y (24-8-2000) The *Drosophila* Netrin receptor Frazzled guides axons by controlling Netrin distribution. *Nature.* **406**: 886–889
24. Seeger, M, Tear, G, Ferres-Marco, D, and Goodman, CS (1993) Mutations affecting growth cone guidance in *Drosophila*: Genes necessary for guidance toward or away from the midline. *Neuron.* **10**: 409–426
25. Kidd, T, Bland, KS, and Goodman, CS (19-3-1999) Slit is the midline repellent for the robo receptor in *Drosophila. Cell.* **96**: 785–794
26. Tear, G, Harris, R, Sutaria, S, Kilomanski, K, Goodman, CS, and Seeger, MA (1996) Commissureless controls growth cone guidance across the CNS midline in *Drosophila* and encodes a novel membrane protein. *Neuron.* **16**: 501–514
27. Kidd, T, Brose, K, Mitchell, KJ, Fetter, RD, Tessier-Lavigne, M, Goodman, CS, and Tear, G (23-1-1998) Roundabout controls axon crossing of the CNS midline and defines a novel subfamily of evolutionarily conserved guidance receptors. *Cell.* **92**: 205–215
28. Wolf, BD and Chiba, A (2000) Axon pathfinding proceeds normally despite disrupted growth cone decisions at CNS midline. *Development.* **127**: 2001–2009
29. Long, H, Sabatier, C, Le, Ma, Plump, A, Yuan, W, Ornitz, DM, Tamada, A, Murakami, F, Goodman, CS, and Tessier-Lavigne, M (22-4-2004) Conserved roles for slit and robo proteins in midline commissural axon guidance. *Neuron.* **42**: 213–223
30. Brittis, PA, Lu, Q, and Flanagan, JG (26-7-2002) Axonal protein synthesis provides a mechanism for localized regulation at an intermediate target. *Cell.* **110**: 223–235
31. Sato, Y, Hirata, T, Ogawa, M, and Fujisawa, H (1-10-1998) Requirement for early-generated neurons recognized by monoclonal antibody lot1 in the formation of lateral olfactory tract. *J. Neurosci.* **18**: 7800–7810
32. Oster, SF, Deiner, M, Birgbauer, E, and Sretavan, DW (2004) Ganglion cell axon pathfinding in the retina and optic nerve. *Semin. Cell Dev. Biol.* **15**: 125–136
33. Bignell, G, Micklem, G, Stratton, MR, Ashworth, A, and Wooster, R (1997) The BRC repeats are conserved in mammalian BRCA2 proteins. *Hum. Mol. Genet.* **6**: 53–58
34. Brittis, PA, Canning, DR, and Silver, J (7-2-1992) Chondroitin sulfate as a regulator of neuronal patterning in the retina. *Science.* **255**: 733–736
35. Yu, X and Bellamkonda, RV (15-10-2001) Dorsal root ganglia neurite extension is inhibited by mechanical and chondroitin sulfate-rich interfaces. *J. Neurosci. Res.* **66**: 303–310

36. Snow, DM, Mullins, N, and Hynds, DL (1-9-2001) Nervous system-derived chondroitin sulfate proteoglycans regulate growth cone morphology and inhibit neurite outgrowth: A light, epifluorescence, and electron microscopy study. *Microsc. Res. Tech.* **54**: 273–286
37. Hynds, DL and Snow, DM (1999) Neurite outgrowth inhibition by chondroitin sulfate proteoglycan: Stalling/stopping exceeds turning in human neuroblastoma growth cones. *Exp. Neurol.* **160**: 244–255
38. Jin, Z, Zhang, J, Klar, A, Chedotal, A, Rao, Y, Cepko, CL, and Bao, ZZ (2003) Irx4-mediated regulation of Slit1 expression contributes to the definition of early axonal paths inside the retina. *Development.* **130**: 1037–1048
39. Brose, K, Bland, KS, Wang, KH, Arnott, D, Henzel, W, Goodman, CS, Tessier-Lavigne, M, and Kidd, T (19-3-1999) Slit proteins bind Robo receptors and have an evolutionarily conserved role in repulsive axon guidance. *Cell.* **96**: 795–806
40. Wang, KH, Brose, K, Arnott, D, Kidd, T, Goodman, CS, Henzel, W, and Tessier-Lavigne, M (19-3-1999) Biochemical purification of a mammalian slit protein as a positive regulator of sensory axon elongation and branching. *Cell.* **96**: 771–784
41. Nguyen-Ba-Charvet, KT, Brose, K, Marillat, V, Sotelo, C, Tessier-Lavigne, M, and Chedotal, A (2001) Sensory axon response to substrate-bound Slit2 is modulated by laminin and cyclic GMP. *Mol. Cell Neurosci.* **17**: 1048–1058
42. Ott, H, Bastmeyer, M, and Stuermer, CA (1-5-1998) Neurolin, the goldfish homolog of DM-GRASP, is involved in retinal axon pathfinding to the optic disk. *J. Neurosci.* **18**: 3363–3372
43. Leppert, CA, Diekmann, H, Paul, C, Laessing, U, Marx, M, Bastmeyer, M, and Stuermer, CA (25-1-1999) Neurolin Ig domain 2 participates in retinal axon guidance and Ig domains 1 and 3 in fasciculation. *J. Cell Biol.* **144**: 339–349
44. Brittis, PA, Lemmon, V, Rutishauser, U, and Silver, J (1995) Unique changes of ganglion cell growth cone behavior following cell adhesion molecule perturbations: A time-lapse study of the living retina. *Mol. Cell Neurosci.* **6**: 433–449
45. Castellani, V, Chedotal, A, Schachner, M, Faivre-Sarrailh, C, and Rougon, G (2000) Analysis of the L1-deficient mouse phenotype reveals cross-talk between Sema3A and L1 signalling pathways in axonal guidance. *Neuron.* **27**: 237–249
46. Montgomery, AM, Becker, JC, Siu, CH, Lemmon, VP, Cheresh, DA, Pancook, JD, Zhao, X, and Reisfeld, RA (1996) Human neural cell adhesion molecule L1 and rat homologue NILE are ligands for integrin alpha v beta 3. *J. Cell Biol.* **132**: 475–485
47. Marcus, RC, Gale, NW, Morrison, ME, Mason, CA, and Yancopoulos, GD (15-12-1996) Eph family receptors and their ligands distribute in opposing gradients in the developing mouse retina. *Dev. Biol.* **180**: 786–789
48. Braisted, JE, McLaughlin, T, Wang, HU, Friedman, GC, Anderson, DJ, and O'Leary, DD (1-11-1997) Graded and lamina-specific distributions of ligands of EphB receptor tyrosine kinases in the developing retinotectal system. *Dev. Biol.* **191**: 14–28
49. Holash, JA, Soans, C, Chong, LD, Shao, H, Dixit, VM, and Pasquale, EB (15-2-1997) Reciprocal expression of the Eph receptor Cek5 and its ligand(s) in the early retina. *Dev. Biol.* **182**: 256–269
50. Connor, RJ, Menzel, P, and Pasquale, EB (1-1-1998) Expression and tyrosine phosphorylation of Eph receptors suggest multiple mechanisms in patterning of the visual system. *Dev. Biol.* **193**: 21–35
51. Birgbauer, E, Oster, SF, Severin, CG, and Sretavan, DW (2001) Retinal axon growth cones respond to EphB extracellular domains as inhibitory axon guidance cues. *Development.* **128**: 3041–3048
52. Deiner, MS, Kennedy, TE, Fazeli, A, Serafini, T, Tessier-Lavigne, M, and Sretavan, DW (1997) Netrin-1 and DCC mediate axon guidance locally at the optic disc: Loss of function leads to optic nerve hypoplasia. *Neuron.* **19**: 575–589
53. Jeffery, G (2001) Architecture of the optic chiasm and the mechanisms that sculpt its development. *Physiol. Rev.* **81**: 1393–1414

54. Williams, SE, Mason, CA, and Herrera, E (2004) The optic chiasm as a midline choice point. *Curr. Opin. Neurobiol.* **14**: 51–60
55. Williams, SE, Mann, F, Erskine, L, Sakurai, T, Wei, S, Rossi, DJ, Gale, NW, Holt, CE, Mason, CA, and Henkemeyer, M (11-9-2003) Ephrin-B2 and EphB1 mediate retinal axon divergence at the optic chiasm. *Neuron.* **39**: 919–935
56. Chan, SO, Chung, KY, and Taylor, JS (1999) The effects of early prenatal monocular enucleation on the routing of uncrossed retinofugal axons and the cellular environment at the chiasm of mouse embryos. *Eur. J. Neurosci.* **11**: 3225–3235
57. Cohen, NR, Taylor, JS, Scott, LB, Guillery, RW, Soriano, P, and Furley, AJ (1-1-1998) Errors in corticospinal axon guidance in mice lacking the neural cell adhesion molecule L1. *Curr. Biol.* **8**: 26–33
58. Sperry, RW (1963) Chemoaffinity in the orderly growth of nerve fiber patterns and connections. *Proc. Natl. Acad. Sci. U.S.A.* **50**: 703–710
59. Drescher, U, Kremoser, C, Handwerker, C, Loschinger, J, Noda, M, and Bonhoeffer, F (11-8-1995) In vitro guidance of retinal ganglion cell axons by RAGS, a 25 kDa tectal protein related to ligands for Eph receptor tyrosine kinases. *Cell.* **82**: 359–370
60. Nakamoto, M, Cheng, HJ, Friedman, GC, McLaughlin, T, Hansen, MJ, Yoon, CH, O'Leary, DD, and Flanagan, JG (6-9-1996) Topographically specific effects of ELF-1 on retinal axon guidance in vitro and retinal axon mapping in vivo. *Cell.* **86**: 755–766
61. Feldheim, DA, Kim, YI, Bergemann, AD, Frisen, J, Barbacid, M, and Flanagan, JG (2000) Genetic analysis of Ephrin-A2 and Ephrin-A5 shows their requirement in multiple aspects of retinocollicular mapping. *Neuron.* **25**: 563–574
62. Nakamura, H and O'Leary, DD (1989) Inaccuracies in initial growth and arborization of chick retinotectal axons followed by course corrections and axon remodeling to develop topographic order. *J. Neurosci.* **9**: 3776–3795
63. Simon, DK and O'Leary, DD (1992) Development of topographic order in the mammalian retinocollicular projection. *J. Neurosci.* **12**: 1212–1232

CHAPTER 3.7

CELL MIGRATION: Condensation

Once migrating cells have completed their journey, their next step is often to "condense" together to form a compact mass of cells that subsequently forms a new structure such as a bone or a hair follicle. There are several mechanisms for condensation and these may work separately or in concert, depending on the system concerned. They include directed migration, local proliferation, changes of cell adhesion, and elimination of interstitial matrix. Migration, such as that used by *D. discoideum*, and proliferation are covered in Chapters 3.1 and 5.1. This chapter concentrates instead on the mechanisms that commonly drive cell condensation in developing animals: enhanced cell adhesion and local elimination of matrix.

CONDENSATION THROUGH ENHANCED CELL ADHESION

One simple way in which cells can be induced to aggregate together is for them to express cell-cell adhesion molecules. Provided that the cells move enough for chance encounters between them to take place, switching on the expression of a cell-cell adhesion molecule may be all that is required to cause them to form a single mass, which will grow as more cells expressing the cell arrive and are added.

Developing limb bones provide a good example of condensation that is driven, at least in part, by changes in the expression of cell-cell adhesion molecules. Limb bones develop from cartilaginous predecessors, and these develop from condensations of mesenchymal cells.[1] A local increase in cell density is achieved without cell proliferation in this system,[2] which makes the role of other mechanisms particularly clear. Before condensation begins, cells of the mesenchyme are separated by matrix-rich spaces and show little interest in making contact with each other. As they condense, however, they display an increased tendency to make contact so that the membranes of adjacent cells in the condensate touch over much of their area.[3] This change in adhesive properties is due primarily to the cells activating the expression of the homophilic cell-cell adhesion molecule, N-cadherin. Shortly before condensation begins, N-cadherin is expressed by scattered cells in the region that will become the developing long bones. As condensation proceeds, expression of N-cadherin rises markedly in the condensing cells.[4] Limb mesenchyme can be removed and cultured in small clumps ("micromass culture"), and

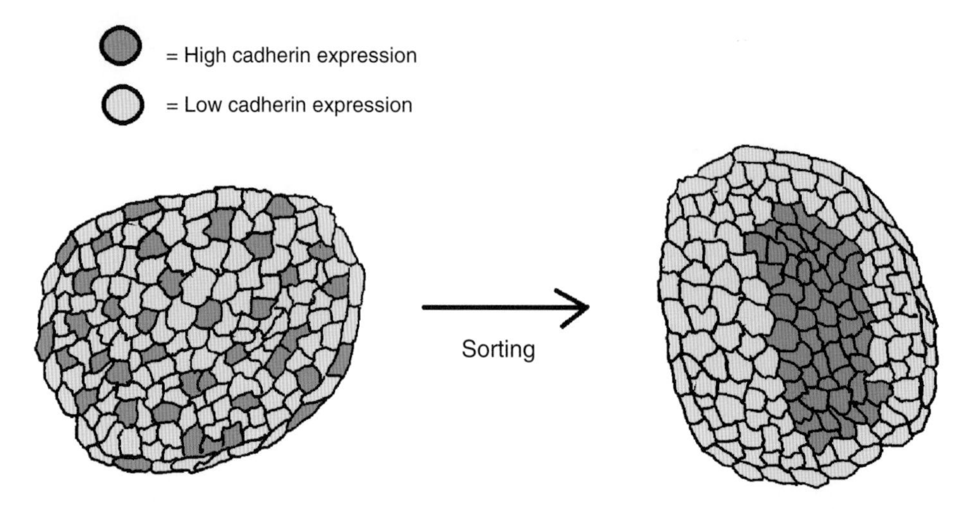

FIGURE 3.7.1 Spontaneous sorting of cells expressing high and low levels of N- (or P-) cadherin.

condensation will occur within these clumps as it does within an intact limb.[5] Treating cultured mesenchymes with function-blocking anti-N-cadherin inhibits condensation[4,6] and expressing dominant negative mutants of N-cadherin has the same effect *in vivo*.[7] These observations suggest strongly that condensation is driven by N-cadherin-mediated cell-cell adhesion.

The appearance of N-cadherin on some of the mesenchyme cells has the effect not only of sticking them together, but also of sorting them out from cells not expressing the molecule. Mixtures of cells that express different amounts of cadherin have an inherent ability to sort out so that those expressing more cadherin cluster together in the inside. The ability of N-cadherin, and of its close relative P-cadherin, to mediate this sorting has been tested by mixing cells that express high levels of cadherin with those that express lower levels of the same cadherin; the cells sort out spontaneously so that the high-expressers form an aggregate inside the low-expressers (see Figure 3.7.1).[8,9] That different levels of cadherin expression can cause cells to sort is therefore in no doubt, but the precise mechanism that achieves this is not yet clear and two rival theories have been proposed.

The first theory, sometimes called the differential adhesion hypothesis or the Steinberg hypothesis after its first champion,[10,11] is based on simple physics and involves no "biological" features beyond the requirement that cells move about. According to the Steinberg hypothesis, sorting is driven by the second law of thermodynamics. This law describes the tendency of systems that are capable of rearrangement to reach a state of minimum free energy. The free energy of the binding domains of an adhesion molecule is minimized when it is adhering to its ligand (that fact is essential to its being an "adhesion" molecule). The free energy of a cell expressing a homophilic adhesion molecule is therefore minimized when it is completely surrounded by other cells that express the same molecule. A mixture of cells that express the adhesion molecule and similar cells that do not would therefore be expected to sort out so that adhesive cells amass in the middle, where the number completely surrounded would be maximized. Less adhesive cells would be left on the outside (see Figure 3.7.2). The

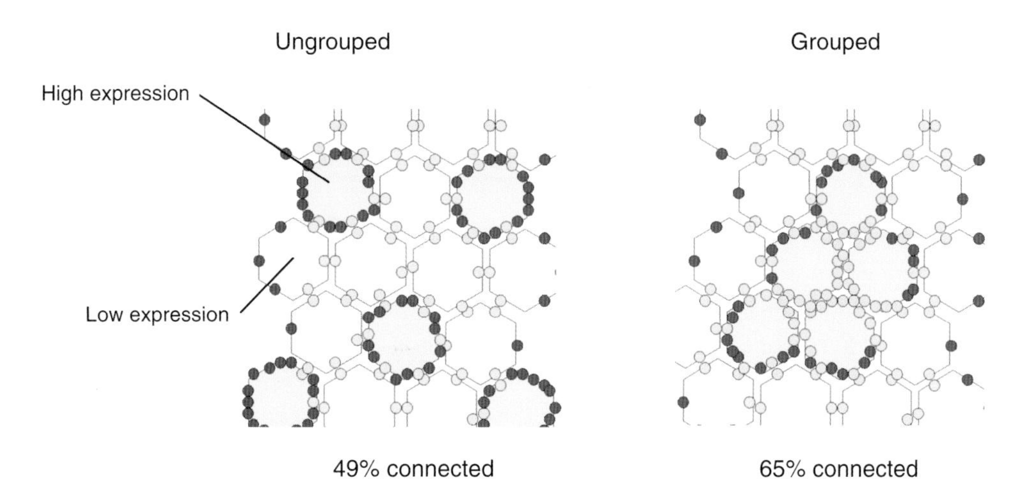

FIGURE 3.7.2 Minimization of free energy by grouping more adhesive cells. In the diagram, low-expressing cells have six molecules of cadherin each (one per face of the hexagonal cells) and high-expressers have 18 molecules each (three per face). Cadherins connected homophilically to partners on other cells are shown in green and ones still unconnected are shown in red. Grouping high-expressers together in the middle reduces the overall fraction of unconnected cadherins, and thus reduces the overall free energy. In reality, cells will not be hexagons, and cadherins will be free to move within the membrane to maximize their chances of finding partners on any neighbouring cell.

explanation is broadly similar to that which accounts for the separation of immiscible liquids. Computer simulations of cell sorting by the Steinberg method have been constructed, and they do result in sorting that has an appearance remarkably similar to that seen in mixtures of real cells, although sorting can be surprisingly slow.[12]

The purely thermodynamic model presented above ought not to depend on "biological" features of the cells. The observation that inhibiting the formation of an actin cytoskeleton using cytochalasin D inhibits cell sorting[9] is therefore worrying. It may reflect the incapability of cadherins to adhere properly without an underlying cytoskeleton, or it may reflect a direct role for the cytoskeleton in cell sorting. A variation on the Steinberg hypothesis, first proposed by A.K. Harris[13] and subsequently refined by others,[14,15] makes great use of contractile forces under the cell membrane. Formation of cadherin-mediated adhesions recruits contractile microfilaments to the membrane (see Chapter 2.2). These pull on the junctions, and the forces they create have components normal to the membrane, directed into the cell, and also tangential to the membrane, directed away from non-adhering regions of membrane to adhering ones (see Figure 3.7.3). Both of these forces are resisted by the cadherin adhesion systems. The source of resistance to inward forces is obvious (cadherins keep hold of their partners in the other cell). The source of resistance to tangential forces is the energy that is released when further adhesions are made in virgin membrane,[14] which appears as an outward-directed tangential force (see Figure 3.7.3).

The total tangential force in the area of contact is called the "interfacial force γ." For two identical cells of a type we shall call A for the purposes of writing an equation,

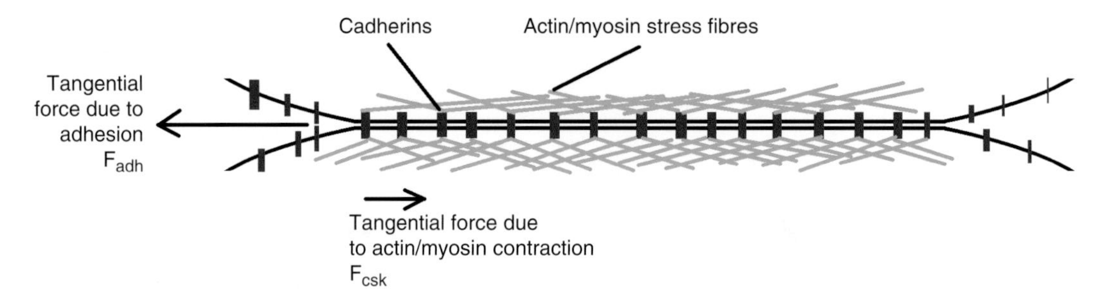

FIGURE 3.7.3 Tangential forces (those actin along the plane of the membrane) due to adhesion and actin/myosin contraction.

the interfacial force γ^{AA} is given by:

$$\gamma^{AA} = 2F^{A}_{csk} - F^{A}_{adh}$$

where the "A" superscripts are used to keep track of the cell type involved (the "2" arises because each cell contributes its own cytoskeletal force). If the interface is between two different cells, A and B, that exert different contractile forces, the interface force will be:

$$\gamma^{AB} = F^{A}_{csk} + F^{B}_{csk} - F^{AB}_{adh}$$

If the interfacial force γ^{AA} is identical to γ^{AB}, cells will show no inclination to sort. If the interfacial forces are different, though, junctions between cells of different types will be subjected to forces that will encourage cells of one type to slip between cells of the other type. A simple illustration of this, for an idealized junction between four cells, is shown in Figure 3.7.4. The position of the four-way junction is controlled by the balance of the interfacial forces. If $\gamma^{BB} > \gamma^{AA}$, the interface will be drawn downward until the angle at which the A cells penetrate the B cells is acute enough for the forces to balance. If the structure of the A cells will not permit an acute enough angle ϕ to form, the envelopment of A cells in type B cells will continue. It does not matter to this model whether the differences in interfacial force are due to forces of adhesion, cytoskeletal forces modulated by adhesion, or both.

The idealized geometry illustrated in Figure 3.7.3 does not do justice to the complete model, but detailed computer simulations have been carried out and they produce biologically sensible results.[14] In simulations of mixed cells, those with higher interfacial tensions envelop those with lower interfacial tensions. Reconciling this result with observations on the behaviour of cells expressing different amounts of N-cadherin would require that higher levels of cadherin adhesion correlate (cause) lower interfacial tensions. This may be counterintuitive, considering the close associations between adhesions and the contractile cytoskeleton, but is not impossible given the many ways in which adhesion complexes can initiate signals that modulate myosin activity. Unfortunately, no measurements on either interfacial tensions or absolute adhesiveness, which would settle the matter, have yet been made.

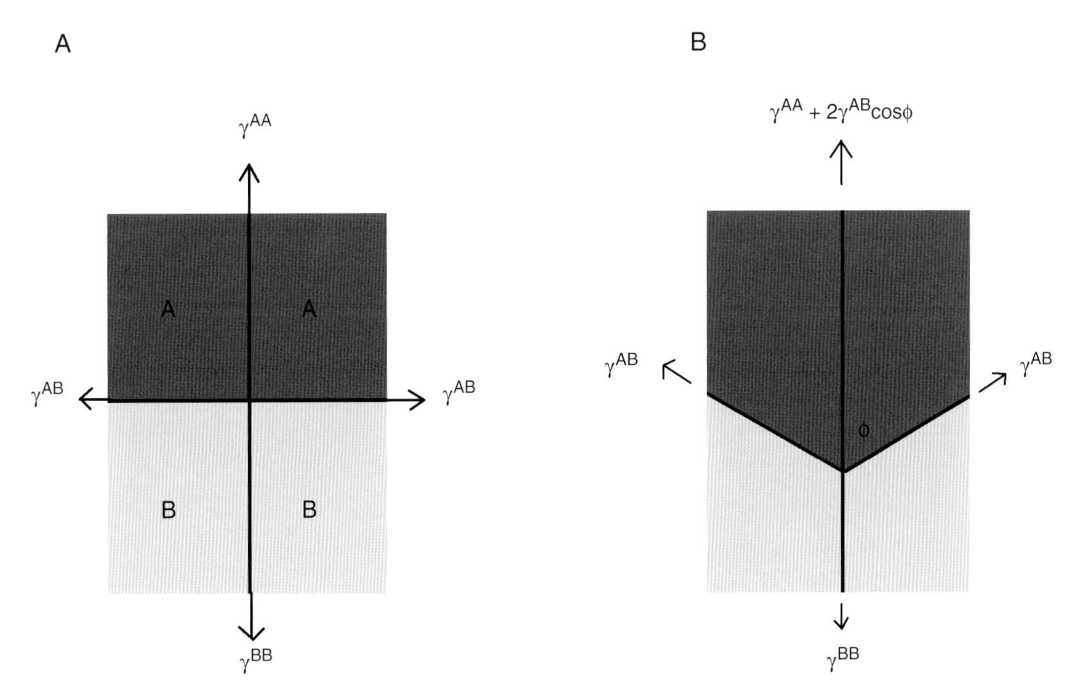

FIGURE 3.7.4 Interfacial tensions at an interface between two A cells and two B cells. (This diagram is based loosely on the discussion in G.W. Brodland and H.H. Chen).[14]

Condensation of limb mesenchyme, driven as it is by N-cadherin, probably also requires some other mechanisms described in this chapter; for example, the interstitial matrix has to be cleared out somehow before new matrix of the cartilage is laid down. The patterning of the limb, which sets up positional signals to ensure that condensation takes place at appropriate sites and times, is outside the scope of this book (which restricts its focus to morphogenesis itself), but good reviews can be found elsewhere.[16,17]

Adhesion-driven coalescence may involve more than one cell type, as occurs in the developing gonads of *D. melanogaster*. The cells of the gonads are of two broad types: somatic cells, which develop from mesoderm local to parasegments 10–12, and germ cells that arrive at the somatic cells after a long migration through the embryo (the migration is described in Chapter 3.6). During normal development, primordial germ cells arrive at and adhere to the presumptive gonad when it is still quite an elongated structure. The somatic cells then extend around the primordial germ cells and condense together to make a more rounded and compact organ[18,19] (see Figure 3.7.5).

Like the mesenchyme of tetrapod limbs, the coalescence of cells in the gonad depends on homophilic adhesion molecules of the cadherin family, in this case the E-cadherin homologue, DE-cadherin. Somatic gonad cell precursors and primordial germ cells both express this molecule. In mutant embryos that lack DE-cadherin function, the primordial germ cells fail to become ensheathed by the somatic cells, and the presumptive gonad fails to round up and remains in its elongated state.[20,21] The cells also express the transmembrane protein Fear of Intimacy.[20] It seems not to be needed by

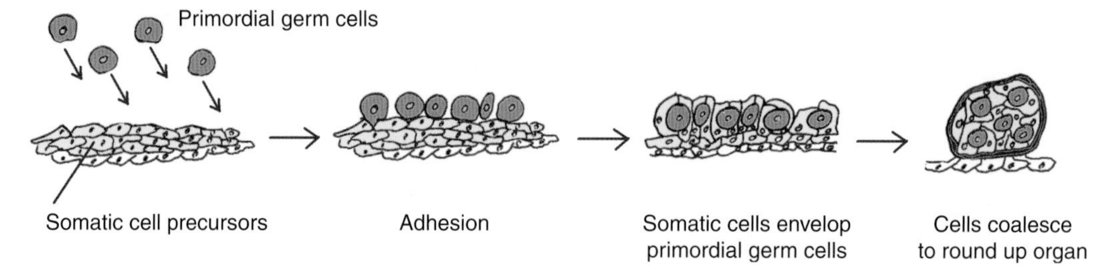

FIGURE 3.7.5 Schematic diagram of gonad formation in *D. melanogaster,* based on D. Godt and U. Teppass.[18] In reality there are more cells than depicted here (about 30 somatic cells per gonad).

the germ cells, but if somatic cells lack it then their levels of DE-cadherin are abnormally low and envelopment and rounding again fail.

The *D. melanogaster* gonad contains an important remaining mystery: since both primordial germ cells and somatic cells express DE-cadherin, why do the germ cells not bind each other and form a clump instead of being enveloped by the somatic cells? If germ cells were more adhesive than the somatic cells, they would be expected to segregate to form a clump in the centre of the organ, as N-cadherin-expressing mesenchyme cells segregate to the middle of the limb. Indeed, if primordial germ cells are engineered to overexpress DE-cadherin, this is precisely what happens.[21] If, on the other hand, somatic cells were more adhesive, they would be expected to segregate to the inside. The observed pattern of cells, with germ cells enveloped by somatic cells, would be expected only if adhesions between heterologous cells were stronger than those between homologous cells (or if they featured lower interfacial tension). Possibly there are additional heterotypic cell-cell adhesion molecules that ensure an especially strong interaction between germ cells and somatic cells.

CONDENSATION BY ELIMINATION OF INTERSTITIAL MATRIX

Mesenchyme cells are not really separated by "space" as is implied by conventional histological diagrams, including many in this book, but are instead embedded in and separated by extracellular matrix. One way of bringing cells together into a tight condensate is to eliminate the matrix between them while maintaining it between uninvolved cells. An example of this mechanism can be found in the mammalian dermis, where mesenchyme cells form condensates as part of the early development of hair follicles.

Mammalian skin is a layered structure in which a stratified epithelial epidermis (see Chapter 4.1) is supported by a matrix-rich, mesenchymal dermis (see Figure 3.7.6a). During foetal development, the dermis secretes signals that induce the overlying epidermis to form a series of evenly spaced placodes.[22] The size and spacing of the placodes is under the control of the dermis itself, so that dermis from a region of the body characterized by densely spaced, thick dark hair will induce formation of this type of hair even when it is transplanted under epidermis at a site characterized by sparse, fine

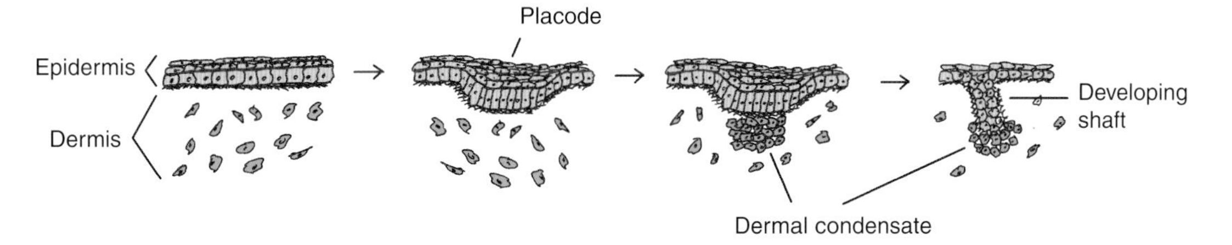

FIGURE 3.7.6 Cell movements in the early development of a single hair follicle.

hair. This is true even in adults, as has been shown by one scientist who transplanted dermis from his own scalp into the arm of a female colleage![23] The mechanism that controls spacing of placodes has not been elucidated yet, but it is assumed that there is competition and lateral inhibition between placode-promoting and placode-repressing signals, probably based on Wnt proteins.[24] Once formed, each of these placodes then signals back to the dermis, probably using a combination of PDGF[25] and Wnt proteins,[26] and causes the dermal cells to condense together The condensates then signal back to the epidermis to induce its invasive growth to form the shaft of a hair follicle. Further signalling events cause the differentiation of the hair follicle into its different functional components. (These signalling systems are beyond the scope of this chapter, but have been reviewed well elsewhere.)[24]

Before their condensation, cells of the dermis are separated by a matrix that consists of proteins such as interstitial collagens, and of large glycosaminoglycans, notably hyaluronic acid. Hyaluronic acid is an uncharged glycosaminoglycan that consists of alternating residues of N-acetylglucosamine and glucuronic acid. The carbohydrate chains tend to adopt and extended conformation and make complexes with water to produce a space-filling, gel-like structure with a viscosity about 5,000 times that of water alone.[27] Long chains of hyaluronic acid are a feature of extracellular matrix molecules such as aggrecan. Before condensation, the interstitial matrix of the dermis is rich in hyaluronic acid, but as condensation takes place, hyaluronic acid disappears between the condensing cells and is absent from areas of condensation once they have formed.[28] This is probably because condensing cells, but not neighbouring cells uninvolved in condensation, express the cell-surface hyaluronidase CD44.

The expression of the CD44 hyaluronidase by condensing cells suggests that one mechanism of condensation may simply be destruction of the most space-filling component of the matrix that normally keeps cells apart. As long as there were to be positive pressure from neighbouring tissues, this mechanism alone would be expected to bring cells closer together (see Figure 3.7.7). The importance of hyaluronic acid destruction to condensation can be demonstrated by injecting a soluble bacterial hyaluronidase into the dermis; mesenchymal cells at the injection site clump together as if making a condensate.[28] This observation strongly supports the model of condensation by elimination of matrix, but does not prove it conclusively because it remains possible that the effect of destroying hyaluronic acid is biochemical rather than mechanical. CD44 is a "receptor" for hyaluronic acid and other glycosaminoglycans as well as being a hyaluronidase, and signalling from the matrix via CD44 modulates cellular activity.[29] Also, hyaluronic acid can itself modulate signalling by growth factors such as TGFβ.[30] It is therefore possible that loss of the molecule due to bacterial hyaluronidase causes

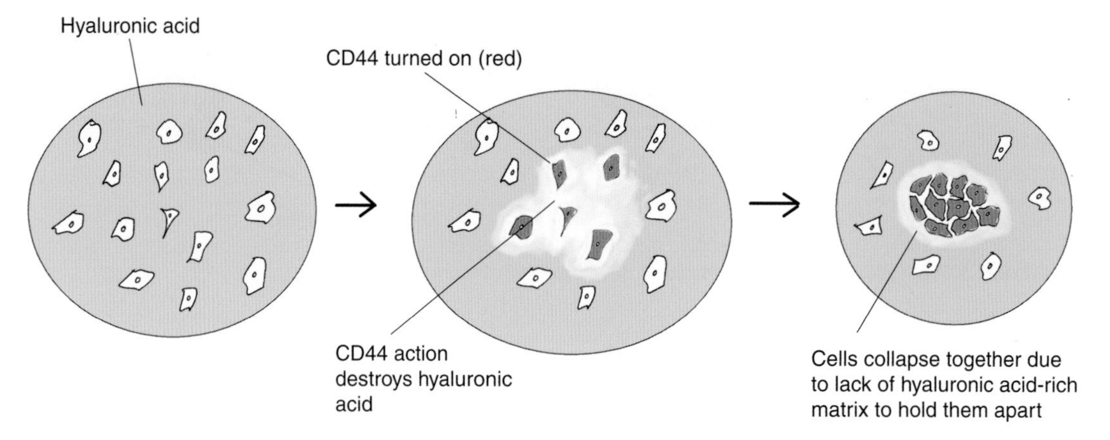

FIGURE 3.7.7 Model for condensation of dermal cells by elimination of interstitial hyaluronic acid. The model depends on there being external pressure present always, to drive the eventual coalescence of cells no longer being kept apart by matrix.

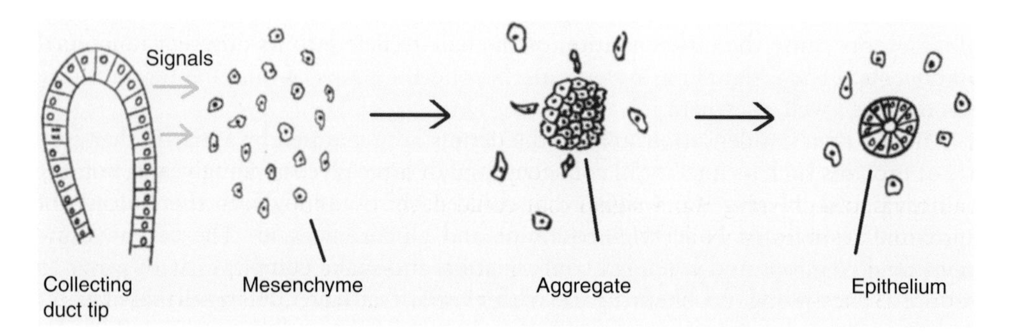

FIGURE 3.7.8 Condensation in kidney development: signals from the developing collecting duct induce groups of mesenchymal cells to form tight aggregates of cells that, when large enough, undergo a mesenchyme-to-epithelial transition.

condensation because of the effect on signalling rather than by purely mechanical means. The simple mechanism depicted in Figure 3.7.8, though attractive, should therefore be viewed with appropriate caution.

It must be stressed that the mechanisms of adhesion and of elimination of interstitial matrix are not mutually exclusive and both are likely to operate together in most instances.

CONTROL OF THE SIZES OF CONDENSATES AND AGGREGATES: Quorum Sensing

Cells form condensates for a purpose, usually to bring together a critical mass for some subsequent developmental event. If the cells attempt to rush on with the next event before this critical mass has built up, the event will generally fail, or at least proceed abnormally. If, on the other hand, they never detect that the critical mass has been reached, the subsequent events will never begin, and cells will keep flowing into an ever-growing condensate. To avoid either of these problems, aggregating cells must possess a mechanism for asking "are we yet a critical mass?" or, in the language of committees that has already been applied to analogous biological problems by bacteriologists, "do we constitute a quorum?"[31] The problem of quorum sensing in animal development has received rather little attention to date, but one system in which the critical components may have been identified is provided by the mammalian kidney.

During development of the mammalian metanephric (permanent) kidney, mesenchymal cells condense together to form tight aggregates.[32,33] These subsequently undergo a mesenchymal-to-epithelial transition to form excretory nephrons (see Figure 3.7.8; see also see Chapter 4.1).

Formation of the condensates is triggered by signals that emanate from the tips of the developing collecting duct system as it ramifies through the mesenchyme to form an epithelial tree (see Chapter 4.6).[34] These signals induce changes in gene expression, and cells begin to secrete the signalling protein Wnt4 as they condense.[35] Wnt4 synthesis is driven even harder by the presence of Wnt4 so that as cells condense, the local levels of the protein become high. These levels might, therefore, be a useful proxy for the concentration of condensing cells at a particular location, and crossing a threshold in Wnt4 levels may drive the decision to undergo the mesenchymal-to-epithelial transition. This idea is supported by the phenotype of $Wnt4^{-/-}$ mice, in which kidney mesenchyme cells form large, diffuse groups of mesenchyme cells but fail to undergo mesenchymal to epithelial transition.[35] This phenotype would be expected if they were waiting for a threshold-crossing Wnt4 signal that could never arrive. Conversely, treating the cells with lithium ions, which stimulate the canonical Wnt signal transduction pathway downstream of the receptor,[36] results in an earlier mesenchyme-to-epithelial transition and the formation of very small epithelial nephrons that have abnormally few cells.[37] This is exactly what would be expected if cells use their perception of Wnt4 levels to judge whether they are yet part of a quorate group (see Figure 3.7.9).

If the system were as simple as that described above, it would be in danger of recruiting all available mesenchymal cells to condense and become nephrons. This would be bad for the kidney because the organ also requires supportive stromal cells that differentiate from the same stock of mesenchyme that gives rise to nephrons. There seem to be at least two solutions to this. One is provided by cells that have already been in a potentially nephron-producing area but have not become part of a condensate and have instead taken the alternative fate of becoming stroma (the oldest of these will have made that decision before Wnt4 levels would have become overwhelming). These cells produce sFRP1, a secreted inhibitor of Wnt signalling. Experimental application of sFRP1 to developing kidneys inhibits nephron formation,[38] and its secretion by stroma probably provides the kidney with a suitable antagonist of Wnt4 signalling so that cells away from a forming condensate are not affected inappropriately by small amounts of Wnt4 diffusing from it. To prevent the sFRP1 from blocking Wnt4 function in the condensate, condensing cells that are already responding to Wnt4 produce sFRP2; this antagonizes sFRP1 and therefore releases Wnt4 signalling from inhibition. The diffusion constants of these proteins have not been measured, but if sFRP2 has a shorter diffusion range, this mechanism would work well in terms of the range of Wnt4 signalling. The other main system that kidneys use to avoid runaway condensation is a

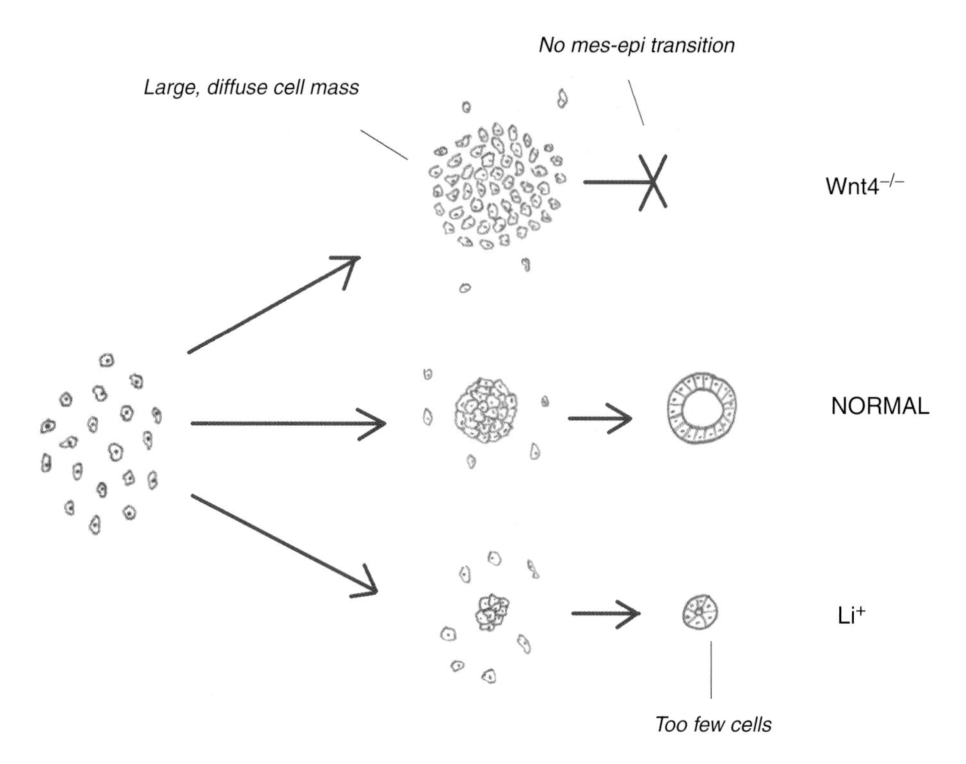

FIGURE 3.7.9 Removing the Wnt4 signal from developing kidneys prevents cells from every undergoing post-condensation mesenchyme-to-epithelial transition, while mimicking a strong Wnt signal pharmacologically causes precocious differentiation before the normal number of cells has been reached.

feedback in which developing nephrons produce BMP-7, which biases the fate choice of uncommitted mesenchyme toward becoming stroma; nephrons therefore effectively supply themselves with appropriate amounts of stroma.[39]

Once condensation has taken place, a condensate has to separate itself from the tissues that surround it; it has to make a boundary. In most of the systems that have received intensive study, condensation is followed almost immediately by differentiation that alters the adhesion systems of the cells. This is most obvious in tissues such as the kidney and developing somites, where condensation is followed almost immediately by differentiation of the condensed cells into epithelia, which, because of their quite different adhesion systems (see Chapter 4.1), will quite naturally segregate from mesenchyme. Even in areas such as the limb, condensed mesenchyme quickly differentiates to become cartilage, and its expression of adhesion and matrix molecules changes.

Reference List

1. Fell, HB (1925) The histogenesis of cartilage and bone in the long bones of the embryonic fowl. *J. Morphol.* 40: 417–451
2. Janners, MY and Searls, RL (1970) Changes in rate of cellular proliferation during the differentiation of cartilage and muscle in the mesenchyme of the embryonic chick wing.

Dev. Biol. **23**: 136–165
3. Thorogood, PV and Hinchliffe, JR (1975) An analysis of the condensation process during chondrogenesis in the embryonic chick hind limb. *J. Embryol. Exp. Morphol.* **33**: 581–606
4. Oberlender, SA and Tuan, RS (1994) Expression and functional involvement of N-cadherin in embryonic limb chondrogenesis. *Development.* **120**: 177–187
5. Ahrens, PB, Solursh, M, and Reiter, RS (1-10-1977) Stage-related capacity for limb chondrogenesis in cell culture. *Dev. Biol.* **60**: 69–82
6. Delise, AM and Tuan, RS (2002) Analysis of N-cadherin function in limb mesenchymal chondrogenesis in vitro. *Dev. Dyn.* **225**: 195–204
7. Delise, AM and Tuan, RS (2002) Alterations in the spatiotemporal expression pattern and function of N-cadherin inhibit cellular condensation and chondrogenesis of limb mesenchymal cells in vitro. *J. Cell Biochem.* **87**: 342–359
8. Steinberg, MS and Takeichi, M (4-1-1994) Experimental specification of cell sorting, tissue spreading, and specific spatial patterning by quantitative differences in cadherin expression. *Proc. Natl. Acad. Sci. U.S.A.* **91**: 206–209
9. Friedlander, DR, Mege, RM, Cunningham, BA, and Edelman, GM (1989) Cell sorting-out is modulated by both the specificity and amount of different cell adhesion molecules (CAMs) expressed on cell surfaces. *Proc. Natl. Acad. Sci. U.S.A.* **86**: 7043–7047
10. Steinberg, MS (15-9-1962) On the mechanism of tissue reconstruction by dissociated cells. I. Population kinetics, differential adhesiveness, and the absence of directed migration. *Proc. Natl. Acad. Sci. U.S.A.* **48**: 1577–1582
11. Steinberg, MS (1970) Does differential adhesion govern self-assembly processes in histogenesis? Equilibrium configurations and the emergence of a hierarchy among populations of embryonic cells. *J. Exp. Zool.* **173**: 395–433
12. Glazier, JA and Graner F (2004) Simulation of the differential adhesion rearrangement of biological cells. *Physical Review E.* **47**: 2128–2154
13. Harris, AK (21-9-1976) Is cell sorting caused by differences in the work of intercellular adhesion? A critique of the Steinberg hypothesis. *J. Theor. Biol.* **61**: 267–285
14. Brodland, WG, and Chen, HH (2000) The mechanics of cell sorting and envelopment. *J. Biomech. Eng.* **33**: 845–851
15. Brodland, GW and Chen, HH (2000) The mechanics of heterotypic cell aggregates: Insights from computer simulations. *J. Biomech. Eng.* **122**: 402–407
16. Tickle, C (2003) Patterning systems–from one end of the limb to the other. *Dev. Cell.* **4**: 449–458
17. Tickle, C (15-10-2002) Molecular basis of vertebrate limb patterning. *Am. J. Med. Genet.* **112**: 250–255
18. Godt, D and Tepass, U (2-9-2003) Organogenesis: Keeping in touch with the germ cells. *Curr. Biol.* **13**: R683–R685
19. Jenkins, AB, McCaffery, JM, and Van Doren, M (2003) *Drosophila* E-cadherin is essential for proper germ cell-soma interaction during gonad morphogenesis. *Development.* **130**: 4417–4426
20. Van Doren, M, Mathews, WR, Samuels, M, Moore, LA, Broihier, HT, and Lehmann, R (2003) Fear of Intimacy encodes a novel transmembrane protein required for gonad morphogenesis in *Drosophila. Development.* **130**: 2355–2364
21. Jenkins, AB, McCaffery, JM, and Van Doren, M (2003) *Drosophila* E-cadherin is essential for proper germ cell-soma interaction during gonad morphogenesis. *Development.* **130**: 4417–4426
22. Hardy, MH (1992) The secret life of the hair follicle. *Trends Genet.* **8**: 55–61
23. Reynolds, AJ, Lawrence, C, Cserhalmi-Friedman, PB, Christiano, AM, and Jahoda, CA (4-11-1999) Trans-gender induction of hair follicles. *Nature.* **402**: 33–34
24. Millar, SE (2002) Molecular mechanisms regulating hair follicle development. *J. Invest. Dermatol.* **118**: 216–225
25. Karlsson, L, Bondjers, C, and Betsholtz, C (1999) Roles for PDGF-A and sonic hedgehog in development of mesenchymal components of the hair follicle. *Development.* **126**: 2611–2621

26. DasGupta, R and Fuchs, E (1999) Multiple roles for activated LEF/TCF transcription complexes during hair follicle development and differentiation. *Development.* **126**: 4557–4568
27. Varki, A, Cummings, R, Esko, J, Freeze, H, Hart, G, and Marth, J (1999) Essentials of glycobiology. (Cold Spring Harbor Laboratory Press)
28. Underhill, CB (1993) Hyaluronan is inversely correlated with the expression of CD44 in the dermal condensation of the embryonic hair follicle. *J. Invest. Dermatol.* **101**: 820–826
29. Singleton, PA and Bourguignon, LY (15-4-2004) CD44 interaction with ankyrin and IP3 receptor in lipid rafts promotes hyaluronan-mediated Ca2+ signalling leading to nitric oxide production and endothelial cell adhesion and proliferation. *Exp. Cell Res.* **295**: 102–118
30. Ito, T, Williams, JD, Fraser, DJ, and Phillips, AO (11-6-2004) Hyaluronan regulates transforming growth factor-beta1 receptor compartmentalization. *J. Biol. Chem.* **279**: 25326–25332
31. Fuqua, WC, Winans, SC, and Greenberg, EP (1994) Quorum sensing in bacteria: the LuxR-LuxI family of cell density-responsive transcriptional regulators. *J. Bacteriol.* **176**: 269–275
32. Davies, JA and Bard, JB (1998) The development of the kidney. *Curr. Top. Dev. Biol.* **39**: 245–301
33. Davies, JA (1996) Mesenchyme to epithelium transition during development of the mammalian kidney tubule. *Acta. Anat. (Basel).* **156**: 187–201
34. Davies, J (2001) Intracellular and extracellular regulation of ureteric bud morphogenesis. *J. Anat.* **198**: 257–264
35. Stark, K, Vainio, S, Vassileva, G, and McMahon, AP (1994) Epithelial transformation of metanephric mesenchyme in the developing kidney regulated by Wnt-4. *Nature.* **372**: 679–683
36. Klein, PS and Melton, DA (6-8-1996) A molecular mechanism for the effect of lithium on development. *Proc. Natl. Acad. Sci. U.S.A.* **93**: 8455–8459
37. Davies, JA and Garrod, DR (1995) Induction of early stages of kidney tubule differentiation by lithium ions. *Dev. Biol.* **167**: 50–60
38. Yoshino, K, Rubin, JS, Higinbotham, KG, Uren, A, Anest, V, Plisov, SY, and Perantoni, AO (2001) Secreted Frizzled-related proteins can regulate metanephric development. *Mech. Dev.* **102**: 45–55
39. Davies, JA and Fisher, CE (2002) Genes and proteins in renal development. *Exp. Nephrol.* **10**: 102–13

4

Epithelial Morphogenesis

CHAPTER 4.1

THE EPITHELIAL STATE: A Brief Overview

Most of the cell types that were discussed in previous chapters can demonstrate their main characteristics individually. Even if they use cooperation and swarm intelligence in a real embryo, mesenchymal cells retain their chief physiological characteristics when isolated in culture. Epithelia and endothelia are different. Their most important characteristic is that they adhere to one another to build almost impermeable sheets that separate the inside of the body from the outside, and separate different compartments of the body from each other.

Epithelia are of great importance to the metazoan body plan, and epithelium was probably the first true tissue type to evolve. One of the simplest of all multicellular animal body plans is provided by the phylum placozoa, in which *Trichoplax adhaerens* is the only known species. *T. adhaerens* has a flat and very flexible body about 3 mm in diameter (see Figure 4.1.1a). It consists mainly of two discs of epithelium: an upper one that consists of a single cell type and a lower one that consists of ciliated cells and glandular cells. These discs are joined at their edges, where cells expressing the animal's only Hox gene, *Trox2*, may act as stem cells capable of becoming either upper-edge or lower-edge epithelium.[1] Between the discs are contractile star cells that can change the shape of the whole multicellular body in a manner reminiscent of unicellular amoebae. That, apparently, is it. There is no body axis, no nervous system, no internal organs; animal life is, at its simplest, almost completely epithelial.

The primacy of epithelium in phylogenetic history is reflected in developmental history. The embryos of even "advanced" metazoa such as ourselves form an outer epithelial covering before making any other type of specialized tissue (see Figure 4.1.1b). The stage of early embryogenesis at which the body consists of an epithelial "cyst," perhaps with some cells inside it, is called the "blastula" and is universal in metazoan development; indeed, passing through a blastula stage is, according to some definitions, the essential qualification for membership of the animal kingdom.[2]

Although both evolutionary and developmental progress add many other cell types to the body, the internal anatomy of "higher" animals is still dominated by epithelial tissues. Most substance exchange between compartments happens across epithelia and endothelia, and their net area in some organs, such as the mammalian lung, is often many orders of magnitude larger than the area of the outer surface of the organ itself.

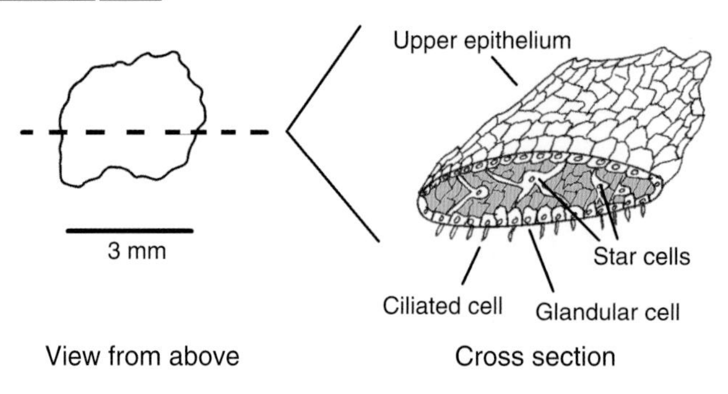

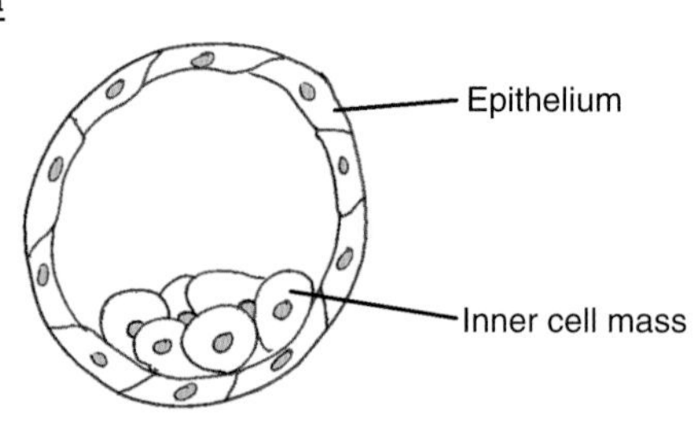

FIGURE 4.1.1 The dominance of epithelia in very simple animals, such as *Trichoplax adhaerens*, and in the early stages of even complex animals such as ourselves.

To achieve a large area, epithelia may be folded, for example in the mammalian large intestine, or they may form highly branched tubes as they do in the lung. The histological appearance of most organs is dominated by the epithelial structures that they contain; epithelial morphogenesis is therefore very important to the organogenesis as a whole.

The epithelial state is relatively easy to define in anatomical terms, implying a sheet of closely apposed cells that have an innate apico-basal polarization across the plane of the sheet. Defining the epithelial state in molecular terms is, however, more difficult as there seem to be no genes that are expressed in epithelia but nowhere else.[3] There seem to be two reasons for this. The first is that there is a great variety of different epithelial tissues, at least 200 in an adult human,[4] each of which is specialized for different functions and expresses different subsets of genes. The second is that "typically epithelial" proteins tend to be expressed in some non-epithelial sites as well and cannot therefore be used to define epithelia; examples of such non-epithelial expression of "epithelial" proteins include E-cadherin in the central nervous system,[5] desmosomal proteins in cardiac muscle,[6] cytokeratin in smooth muscle,[7] ZO1 in astrocytes,[8] α6 integrin in lymphocytes,[9] and collagen IV in embryonic lung mesenchyme.[10]

It is perhaps fitting that the epithelial state is best defined in anatomical rather than molecular terms, for the intimate association between epithelial cells is so central to the character of the tissue that an isolated epithelial cell is arguably not an epithelial cell at all. The communal nature of an epithelium (/ endothelium) has deep implications for morphogenetic mechanisms because it prevents cells from undergoing many of the

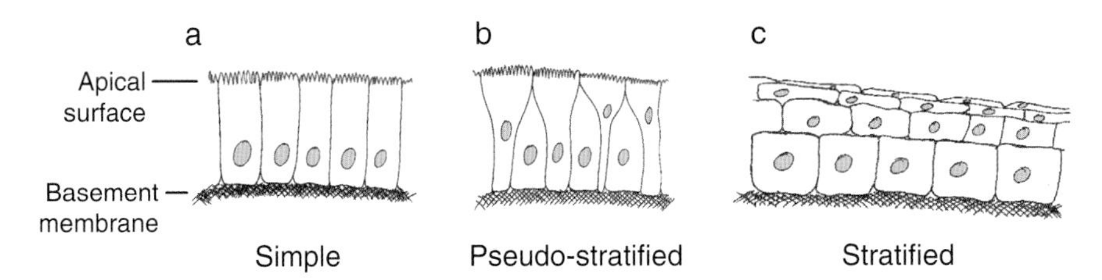

FIGURE 4.1.2 The three basic types of epithelium.

processes typical of mesenchymal cells. Instead, the epithelial state allows a different set of morphogenetic processes that are unavailable to other cell types to occur. To understand how some morphogenetic mechanisms are restricted while others are allowed, it is necessary first to understand the structure of a typical epithelium.

Most epithelia are "simple" and consist of a single layer of cells attached to a basement membrane (see Figure 4.1.2a). Some, such as skin, are "stratified" and have several layers that typically include a basal layer of multiplying cells and overlying layers of differentiating cells (see Figure 4.1.2c). Others are "pseudo-stratified," with all of their cells making contact with the basement membrane but arranged so that not all reach the top (see Figure 4.1.2b).

Epithelial cells connect to each other and to the underlying basement membrane via specific types of junctions. In a typical vertebrate epithelium, the junction type found closest to the apical surface is the tight junction, a complex of proteins that forms an effective seal between cells so that the fluid bathing the apical domains of cells is kept separate from that bathing the basolateral domains (see Figure 4.1.3). The tight junction also plays a role in preventing the mixing of membrane components across the boundary between the apical and basal domains.[11] This is important because these domains can possess quite distinct sets of proteins that arrive by specific targeting of secretory vesicles from the ER/Golgi pathway. Intercellular contacts in the tight junction occur between proteins of the claudin family, which line up side by side in the membrane to make impermeable belts that surround the cell. Tight junctions are important for sealing and also originate signals,[12] but in most systems they do not seem to play a major role in the mechanics of morphogenesis; they are not, for example, usually responsible for carrying substantial mechanical forces.

Just below the tight junctions are adherens junctions,[13,14] typically arranged as a "belt" running right around the cell. Adherens junctions use cadherin molecules, usually E-cadherin, to mediate intercellular adhesion. Their cytoplasmic faces are connected, via adaptor proteins, to the actin cytoskeleton. This allows tension generated in the cytoskeleton by actin-myosin contraction to be transmitted between neighbouring cells by means of adherens junctions. Some of the proteins of the complex on the cytoplasmic face of adherens junctions can trigger intracellular signalling cascades according to the adhesion state, and perhaps the tension in, the junction.[15,16]

Below the zone of adherens junctions are desmsomes,[17] typically more scattered than are adherens junctions. Desmosomes also use cadherins, though these are a specialized set (desmogleins, desmocollins). Their cytoplasmic faces can also originate signals, and they connect to the cytokeratin cytoskeleton so that mechanical loads on an epithelium can be shared by all cells in the tissue.

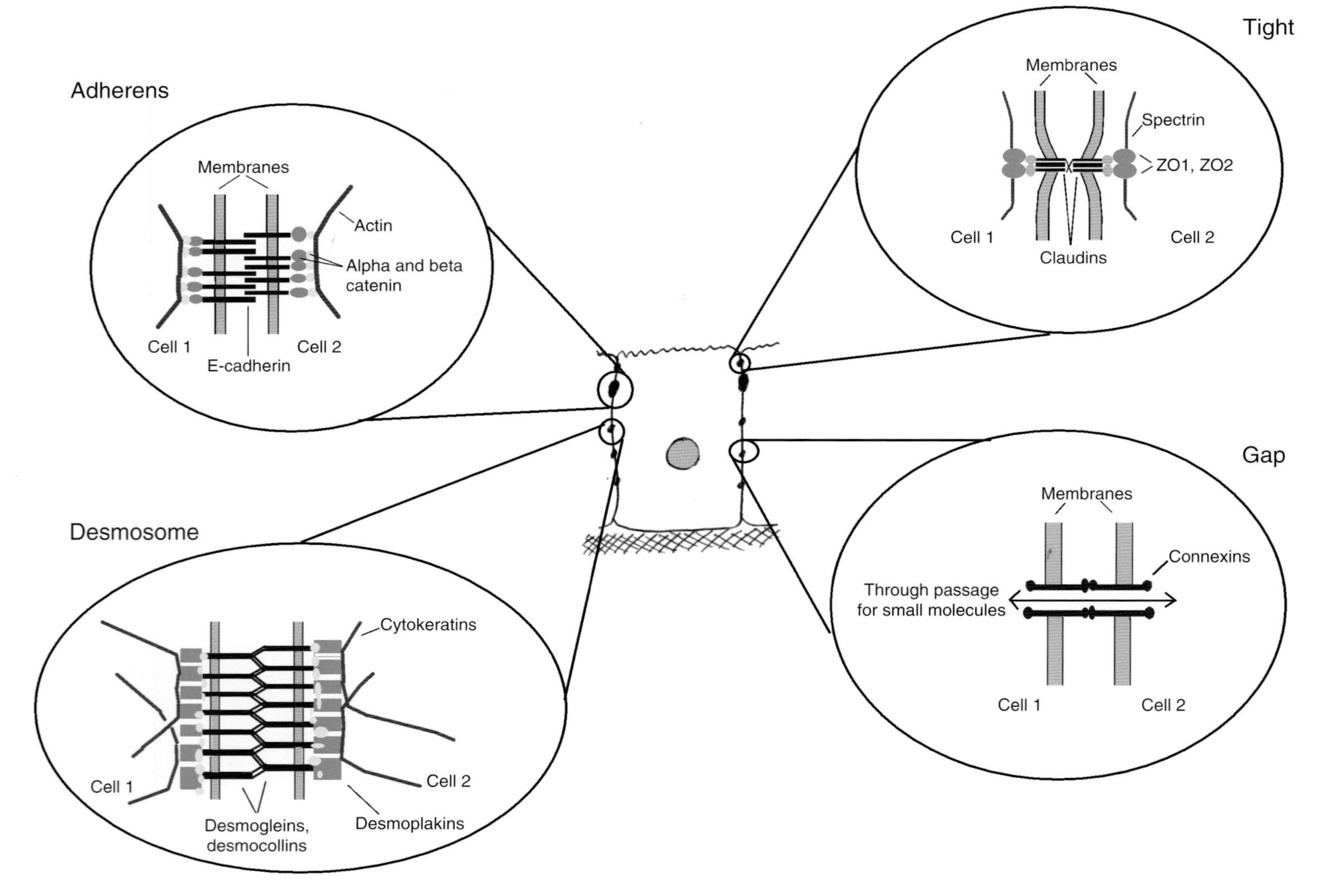

FIGURE 4.1.3 The junctions of a typical epithelial cell. The diagrams of individual junction show only a few of the many proteins that are present at the cytoplasmic face; more information can be gleaned from the reviews cited in the text.

Cells may also be connected by gap junctions,[18] in which connexin protein complexes form a pore that is continuous between the cytoplasms of adjacent cells, although separate from the intercellular medium. Gap junctions allow the passage of small molecules, which can include signalling molecules that regulate morphogenesis. For example, some of the signals that set up left-right asymmetry in vertebrates depend on gap junctions.[19] Other cells, particularly neurons, make great use of gap junction communication.

The interface between the basal surface of the cell and the basement membrane is rich in integrin-containing junctions such as hemidesmosomes,[20] which connect with the cytokeratin system, and other contacts that connect with actin. These, too, originate intercellular signals[21] and are important in preventing programmed cell death due to anoikis (see Chapter 5.3). In addition to these stable junctions, epithelial cells can form integrin-containing contacts similar to those of mesenchyme. For example, the laminin receptor $\alpha6\beta1$ integrin can recruit motile actin structures to allow cell migration and can signal via PI-3-kinase and small GTPases of the Rho family to organize the cytoskeleton,[22] as described in Chapter 3.2.

In general, the structure of an epithelium sketched above is maintained during morphogenesis so that it remains tightly sealed. This means that many of the morphogenetic processes used by mesenchyme cells, such as migration and condensation, are unavailable. Instead, morphogenesis has to take place by growth, folding, and invagination of the sheet. These distortions usually require cells to stream past each other to alter their neighbour relationships, and this in turn requires that the junctions between epithelial cells and between cells and the matrix are labile and can be released when required. This release cannot be global—that would destroy the epithelium♥—but must be restricted to only the junctions that have to move. The pattern of junction modification is discussed further in Chapter 4.2. The molecular mechanisms that control it are not yet understood, although internal regulators of cell-cell adhesion, such as the small GTPase Rap1, have been identified.[24]

THE MAKING OF AN EPITHELIUM

Most new epithelia arise from the morphogenesis of an existing epithelium, but the formation of the very first epithelium of the embryo, that of the blastula, has to be an exception. In the mouse, the development of the first epithelium begins when the embryo contains a mere eight cells (see Figure 4.1.4). At this time, E-cadherin already present in the embryo is phosphorylated and becomes functional.[25] As it reaches the cell membranes of the blastomeres that produce it, their mutual adhesion increases, and they maximize their contact. This adhesion, and the compaction of the embryo that results, is inhibited by function-blocking anti-E-cadherin.[25]

Between the 8 and 32 cell stages, other junction components are transcribed in a sequence. For example, gap junctions components are made at the 8 cell stage,[26] the tight junction protein ZO1α- is transcribed from the 8 cell stage, starting about an hour after E-cadherin-mediated compaction, cingulin is made at the 16 cell stage, but ZO1α+ is not made until the 32 cell stage. Each protein heads to sites of cell contact

♥Some examples of morphogenesis do involve complete breakdown of the epithelium (e.g. during the epithelial-to-mesenchymal transitions required to supply mesenchyme cells to the developing heart[23]), but these are not being discussed here.

immediately, but only when the ZO1α+ protein is present is the intercellular path (the path that allows medium to pass through spaces between cells) sealed.[27] Desmosomal proteins are expressed from the 16 cell stage. Their expression is independent of E-cadherin function[28] but does depend on the cell concerned not being surrounded by other cells and therefore not being deep inside the embryo.[28] Deep cells will become the inner cell mass and are not, therefore, destined to be epithelial yet. In a similar manner, only cells that are not surrounded by others assemble tight junctions.[29] Presumably, free edges generate a permissive signal that is suppressed when they make contact with other cells. As adhesion proceeds, E-cadherin-rich contact sites recruit the actin cytoskeleton and its associated proteins to the developing junctions. Of these actin-associated proteins, spectrin, ankyrin, and fodrin are particularly important because they form a network that can bind membrane proteins, such as Na^+/K^+ ATPase, which are localized to the basolateral membranes in mature epithelia.[30] This may be one mechanism by which these proteins are recruited away from membranes that are not making E-cadherin-mediated adhesions and toward those that are.[31] Membranes that cannot form any E-cadherin-mediated adhesions will be those facing the outside of the embryo, which will form the apical domain. Adhering E-cadherin also recruits a complex of the proteins Sec6 and Sec8. The Sec6/8 complex acts as a docking receptor for vesicles carrying basolateral proteins from the Golgi complex[32] and acts as another mechanism to drive polarization of the cells.[33] This connection between cell adhesion and polarization means that the forming epithelium can organize its own polarity according to the environment in which it finds itself—another example of adaptive self-organization.

The polarization of pumps such as the Na^+/K^+ ATPase means that, once the tight junctions have formed, cells can pump fluid inward and inflate the blastocoel cavity. The epithelial cells of the trophectoderm, the outer layer of the embryo, continue to divide as development proceeds. If they divide tangentially, both daughters are epithelial, whereas if they divide radially, the daughter still facing the free edge remains epithelial but the inner one loses its epithelial nature and joins the inner cell mass.

As well as forming from blastomeres early in development, epithelia sometimes form from mesenchymal cells during organogenesis. In the examples of mesenchyme-to-epithelium transition most closely studied, such as formation of epithelial nephrons from the mesenchyme of kidneys in vertebrates, the process is similar to that found in early embryos. Cells down-regulate their expression of typically mesenchymal proteins and express the adhesion and cytoskeletal systems characteristic of epithelia, again in a temporal succession.[34] These proteins, particularly cadherins, promote cell-cell

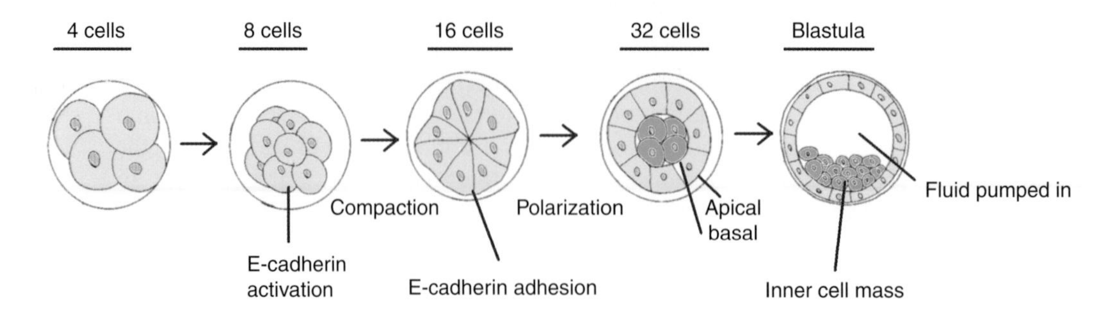

FIGURE 4.1.4 Formation of the first epithelium of a mouse.

adhesion and polarization so that cells in contact with each other organize themselves into a polarized epithelial cyst.[35]

The growth of an epithelium is generally achieved by proliferation of cells within that epithelium, and the role of controlled proliferation in driving morphogenesis is described in Chapter 5.2. Some epithelia, however, also grow by recruitment of mesenchymal cells in a process of continuous mesenchyme-to-epithelial transition. An example is the nephric duct of at least some vertebrates. The nephric duct is the drainage system for the pronephric and mesonephric kidneys, which are temporary embryonic structures in reptiles, mammals, and birds but are retained as permanent kidneys in fish and amphibians. Experiments using chimaeric embryos that include tissues from the frogs *Xenopus laevis* and *Xenopus borealis* demonstrate that a nephric duct derived from one species will, when migrating through the mesenchyme derived from the other species, take up cells from that other species.[36] These cells appear to become fully epithelial. Whether this mode of development takes place in the nephric ducts of other species, and also in the development of the drainage ducts of metanephric (permanent) kidneys of mammals, is currently a matter of debate.

THE FORCES THAT SHAPE AN EPITHELIUM

Epithelia are shaped by a combination of the forces they generate internally and the various pressures that derive from the rest of the embryo and its environment. The attachment of the cytoskeleton to the cell-cell junctions of epithelia has the consequence that forces generated by or experienced in one cell are communicated to the rest of the epithelium, and the tissue therefore behaves as a mechanically integrated whole. The actin-myosin stress fibres attached to cell-cell and cell-matrix junctions are, like all stress fibres, in tension (see Chapter 3.2). The net effect of the stress fibres and junctions of all of the cells is to impart a tension to the complete epithelial sheet that is analogous to the surface tension of soap bubbles. Except when some cells of the epithelium behave differently to others to generate specific additional forces—for example, by the cell wedging that will be described in Chapter 4.4—the general shape of an epithelium might therefore be expected to follow the same rules as soap bubbles, whose behaviour can be explained completely by surface tension.

Surfaces under tension take up forms that minimize their surface area (see Chapter 2.2). If a one-dimensional object, such as an idealized length of wire, is stretched over a curve of radius R and is under tension T (see Figure 4.1.5), the inwards pressure per unit length that it exerts is given by:

$$\rho = \frac{T}{R}$$

Similarly, if a two-dimensional surface is stretched over a two-dimensional surface of radii R_1 and R_2:

$$\rho = \frac{T}{R_1} + \frac{T}{R_2} \qquad \text{refs}^{37,38}$$

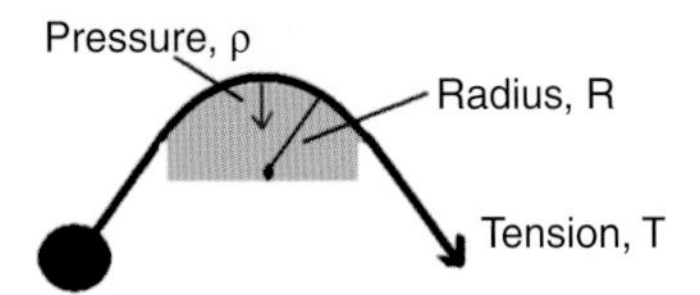

FIGURE 4.1.5 When a frictionless wire is stretched with tension T over a curve with radius R, it exerts a pressure per unit length $\rho = T/R$.

In equilibrium, this inward pressure must equal whatever outward pressure is present in the system (for example, that which is due to fluid contained inside an epithelium). If the surface is everywhere the same (no special zones of cell wedging and so on), then T will be everywhere the same, so $(1/R_1 + 1/R_2)$, and therefore the mean curvature,♣ will be the same everywhere. Two well-known forms that satisfy the condition of constant mean curvature are the plane (zero curvature) and the hollow sphere ("cyst"), with $R_1 = R_2$. Other shapes are also possible and can be explored by dipping an arbitrarily distorted wire loop into a solution of detergent. The resulting shapes will be of minimum surface area and, although they may include complex curves of convex (R positive) and concave (R negative) types, the mean curvature will be constant everywhere.

Modelling epithelia as tense films does, however, reveal a serious problem with the stability of one of the most common of all epithelial configurations: the hollow tube. At first sight, there is nothing wrong with a tense film taking up a cylindrical configuration; it will, after all, have a constant curvature throughout its length. Unlike a plane or sphere, though, a cylinder is inherently unstable to small disturbances. This fact may be demonstrated by drawing detergent solution between two wire rings (see Figure 4.1.6). Once the ratio of length to diameter increases beyond about π, the middle section of the cylinder constricts so that its radius diminishes. Constancy of curvature is maintained because the increasing positive curvature around the circumference of the cylinder is balanced by increasing negative curvature along its length (see Figure 4.1.6). In extremis, the process continues until the cylinder breaks up completely to leave isolated spherical bubbles.[38]

This instability of long cylinders under "surface tension"ℵ creates a real problem for organisms that need to organize epithelia as long tubules. This fact was pointed out at least as long ago as 1917 by D'Arcy Thompson.[38] There are several possible solutions to the problem. Most epithelia are embedded in surrounding mesenchyme, and it is possible that the mesenchyme itself would resist the collapse of an underlying epithelium. However, since mesenchymes are generally flexible enough, through literal flexing and also through remodelling, to permit morphogenetic movements of epithelia, there would have to be a very subtle sensing system to allow them to tell the difference between morphogenesis and collapse. Alternatively, it is possible that the epithelia themselves possess the ability to resist collapse. This cannot be simply by there being a maximum radius of curvature through which an epithelial cell can bend, because real tubules vary in diameter from microns to centimeters, so there must be more

♣The mean curvature of a surface is defined as $0.5\ (1/R_1 + 1/R_2)$.[ref39]

ℵThe phrase is in quotation marks because the tension shown by epithelia is not strictly "surface tension," but it is mechanically analogous to it.

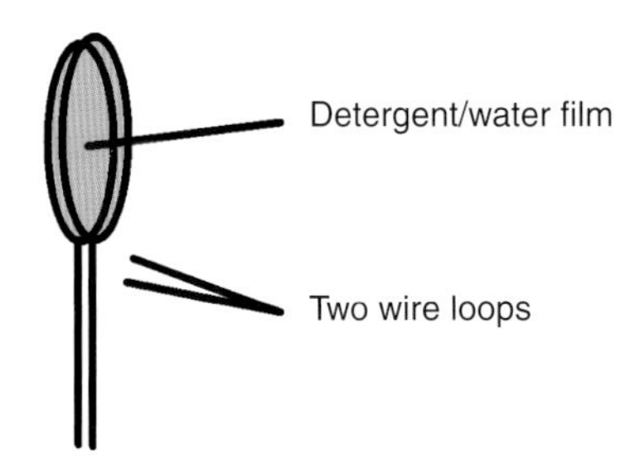

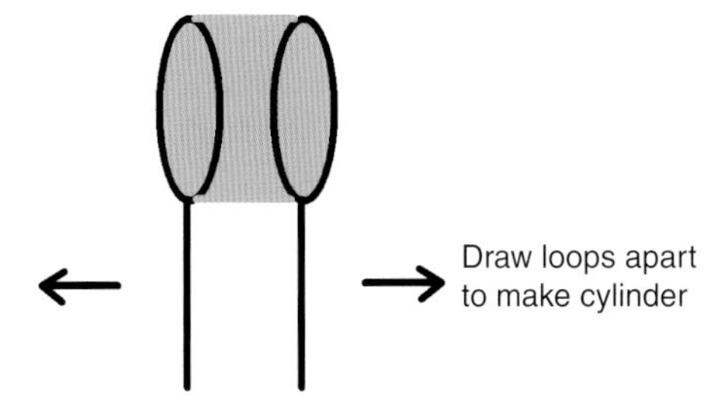

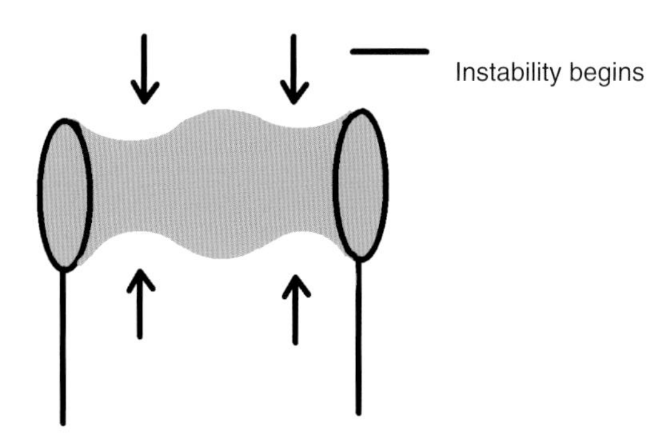

FIGURE 4.1.6 The inherent instability of a long cylinder made of a surface under tension, as demonstrated with a child's bubble-blowing apparatus.

specific mechanisms. A recent set of experiments in *D. melanogaster* has identified a protein without which tubules "collapse" into disconnected cysts. Although the researchers describing the abnormality do not mention it, this phenotype is so reminiscent of the catastrophic collapse predicted by Thompson that it may lead directly to the systems that epithelia have evolved to avoid that collapse.

The molecule concerned is a cadherin of the Fat family, of which there are two members (Fat and Fat-like) in *D. melanogaster*[40,41] and at least three (Fat1,2,3) in mammals.[42,43,44] Fat proteins are unusually large by the standards of the cadherin family and have 34 cadherin-type repeats, and also laminin-like domains, in their extracellular portions. By comparison, *D. melanogaster's* E-cadherin has just six cadherin-type repeats.[45] Fat-like is expressed in tubular epithelia such as the tracheal trunks,[41] while in mammals Fat1 is expressed in lung and renal epithelia, as well as in some non-epithelial sites.[42,46,47,48] When expression of *Fat-like* is prevented by interfering RNAs, the tracheal system is severely malformed and collapses in places.[41] Significantly, perhaps, Fat1 is expressed in the slit diaphragms that form the filters of mammalian kidneys: without the protein, the small spaces of the slit diaphragm collapse and the filter is complete blocked.[49] Fat1 does, therefore, seem to have a function in keeping epithelia apart.

How might Fat function? The Fat cadherins are so large, reaching about 0.16 μm from the membrane, that it is unlikely that they would be directly involved in cell-cell adhesion, and attempts to demonstrate an adhesive function directly have failed.[41] It is possible that Fat-like is involved in keeping epithelial apart rather than holding them together, although the diameter of tubules is significantly larger than the length of Fat-like so it cannot work simply by being a compression strut running across the tubule. Fat1 is known to regulate the arrangement of the cytoskeleton at cell junctions,[50] and it may be that it is part of a feedback system that detects either a close approach or too small a radius of curvature, and reduces tension accordingly.

The problem of the stability of tubules emphasizes that much remains to be learned about how epithelia maintain the shape they have, let alone about the more difficult problem of how they acquire that shape in the first place. Nevertheless, recent years have brought a much-improved understanding about the generation and modification of epithelial form. The following chapters will focus on the main modes of epithelial morphogenesis—movement of sheets, closure of holes, convergent extension, invagination, fusion, and branching—and will describe what is known about how they work.

Reference List

1. Jakob, W, Sagasser, S, Dellaporta, S, Holland, P, Kuhn, K, and Schierwater, B (2004) The Trox-2 Hox/ParaHox gene of Trichoplax (Placozoa) marks an epithelial boundary. *Dev. Genes Evol.* **214**: 170–175
2. Barnes, RSK (1998) The diversity of living organisms. (Blackwell)
3. Davies, JA and Garrod, DR (1997) Molecular aspects of the epithelial phenotype. *Bioessays.* **19**: 699–704
4. Alberts, B, Bray, D, Lewis, J, Raff, M, Roberts, K, and Watson, JD (1994) Molecular biology of the cell. **3rd**: (Garland)
5. Babb, SG, Barnett, J, Doedens, AL, Cobb, N, Liu, Q, Sorkin, BC, Yelick, PC, Raymond, PA, and Marrs, JA (2001) Zebrafish E-cadherin: Expression during early embryogenesis and regulation during brain development. *Dev. Dyn.* **221**: 231–237

6. Schafer, S, Stumpp, S, and Franke, WW (1996) Immunological identification and characterization of the desmosomal cadherin Dsg2 in coupled and uncoupled epithelial cells and in human tissues. *Differentiation.* **60**: 99–108
7. Jahn, L, Fouquet, B, Rohe, K, and Franke, WW (1987) Cytokeratins in certain endothelial and smooth muscle cells of two taxonomically distant vertebrate species, Xenopus laevis and man. *Differentiation.* **36**: 234–254
8. Howarth, AG and Stevenson, BR (1995) Molecular environment of ZO-1 in epithelial and non-epithelial cells. *Cell Motil. Cytoskeleton.* **31**: 323–332
9. Botling, J, Oberg, F, and Nilsson, K (1995) CD49f (alpha 6 integrin) and CD66a (BGP) are specifically induced by retinoids during human monocytic differentiation. *Leukemia.* **9**: 2034–2041
10. Thomas, T and Dziadek, M (1993) Genes coding for basement membrane glycoproteins laminin, nidogen, and collagen IV are differentially expressed in the nervous system and by epithelial, endothelial, and mesenchymal cells of the mouse embryo. *Exp. Cell Res.* **208**: 54–67
11. Matter, K and Balda, MS (2003a) Functional analysis of tight junctions. *Methods.* **30**: 228–234
12. Matter, K and Balda, MS (2003b) Signalling to and from tight junctions. *Nat. Rev. Mol. Cell Biol.* **4**: 225–236
13. Tepass, U (2002) Adherens junctions: New insight into assembly, modulation and function. *Bioessays.* **24**: 690–695
14. Nagafuchi, A (2001) Molecular architecture of adherens junctions. *Curr. Opin. Cell Biol.* **13**: 600–603
15. Perez-Moreno, M, Jamora, C, and Fuchs, E (21-2-2003) Sticky business: Orchestrating cellular signals at adherens junctions. *Cell.* **112**: 535–548
16. Jamora, C and Fuchs, E (2002) Intercellular adhesion, signalling and the cytoskeleton. *Nat. Cell Biol.* **4**: E101–E108
17. Garrod, DR, Merritt, AJ, and Nie, Z (2002) Desmosomal adhesion: Structural basis, molecular mechanism and regulation (Review). *Mol. Membr. Biol.* **19**: 81–94
18. Nicholson, BJ (15-11-2003) Gap junctions: From cell to molecule. *J. Cell Sci.* **116**: 4479–4481
19. Mercola, M (15-8-2003) Left-right asymmetry: nodal points. *J. Cell Sci.* **116**: 3251–3257
20. Jones, JC, Hopkinson, SB, and Goldfinger, LE (1998) Structure and assembly of hemidesmosomes. *Bioessays.* **20**: 488–494
21. Nievers, MG, Schaapveld, RQ, and Sonnenberg, A (1999) Biology and function of hemidesmosomes. *Matrix Biol.* **18**: 5–17
22. Mercurio, AM, Rabinovitz, I, and Shaw, LM (2001) The alpha 6 beta 4 integrin and epithelial cell migration. *Curr. Opin. Cell Biol.* **13**: 541–545
23. Munoz-Chapuli, R, Perez-Pomares, JM, Macias, D, Garcia-Garrido, L, Carmona, R, and Gonzalez-Iriarte, M (2001) The epicardium as a source of mesenchyme for the developing heart. *Ital. J. Anat. Embryol.* **106**: 187–196
24. Price, LS, Hajdo-Milasinovic, A, Zhao, J, Zwartkruis, FJ, Collard, JG, and Bos, JL (27-5-2004) Rap1 regulates E-cadherin-mediated cell-cell adhesion. *J. Biol. Chem.* **279**: 35127–32
25. Hyafil, F, Morello, D, Babinet, C, and Jacob, F (1980) A cell surface glycoprotein involved in the compaction of embryonal carcinoma cells and cleavage stage embryos. *Cell.* **21**: 927–934
26. Aghion, J, Gueth-Hallonet, C, Antony, C, Gros, D, and Maro, B (1994) Cell adhesion and gap junction formation in the early mouse embryo are induced prematurely by 6-DMAP in the absence of E-cadherin phosphorylation. *J Cell Sci.* **107 (Pt 5)**: 1369–1379
27. Sheth, B, Fesenko, I, Collins, JE, Moran, B, Wild, AE, Anderson, JM, and Fleming, TP (1997) Tight junction assembly during mouse blastocyst formation is regulated by late expression of ZO-1 alpha+ isoform. *Development.* **124**: 2027–2037

28. Collins, JE, Lorimer, JE, Garrod, DR, Pidsley, SC, Buxton, RS, and Fleming, TP (1995) Regulation of desmocollin transcription in mouse preimplantation embryos. *Development.* **121**: 743–753
29. Collins, JE and Fleming, TP (1995) Epithelial differentiation in the mouse preimplantation embryo: Making adhesive cell contacts for the first time. *Trends Biochem. Sci.* **20**: 307–312
30. Bennett, V (1990) Spectrin-based membrane skeleton: A multipotential adaptor between plasma membrane and cytoplasm. *Physiol. Rev.* **70**: 1029–1065
31. Yeaman, C, Grindstaff, KK, and Nelson, WJ (1999) New perspectives on mechanisms involved in generating epithelial cell polarity. *Physiol. Rev.* **79**: 73–98
32. Grindstaff, KK, Yeaman, C, Anandasabapathy, N, Hsu, SC, Rodriguez-Boulan, E, Scheller, RH, and Nelson, WJ (29-5-1998) Sec6/8 complex is recruited to cell-cell contacts and specifies transport vesicle delivery to the basal-lateral membrane in epithelial cells. *Cell.* **93**: 731–740
33. Yeaman, C, Grindstaff, KK, and Nelson, WJ (1-2-2004) Mechanism of recruiting Sec6/8 (exocyst) complex to the apical junctional complex during polarization of epithelial cells. *J. Cell Sci.* **117**: 559–570
34. Davies, JA and Garrod, DR (1995) Induction of early stages of kidney tubule differentiation by lithium ions. *Dev. Biol.* **167**: 50–60
35. Davies, JA (1996) Mesenchyme to epithelium transition during development of the mammalian kidney tubule. *Acta. Anat. (Basel).* **156**: 187–201
36. Cornish, JA and Etkin, LD (1993) The formation of the pronephric duct in Xenopus involves recruitment of posterior cells by migrating pronephric duct cells. *Dev. Biol.* **159**: 338–345
37. Laplace, P-S (1806) Theorie de l'action capillaire. (Mechanique Celeste)
38. Thompson, DW (1961) On growth and form. Abridged by Bonner JT (Cambridge University Press)
39. Weisstein, EW (2002) CRC Concise encyclopaedia of mathematics. **2nd**: (Chapman & Hall/CRC Press)
40. Mahoney, PA, Weber, U, Onofrechuk, P, Biessmann, H, Bryant, PJ, and Goodman, CS (29-11-1991) The fat tumor suppressor gene in *Drosophila* encodes a novel member of the cadherin gene superfamily. *Cell.* **67**: 853–868
41. Castillejo-Lopez, C, Arias, WM, and Baumgartner, S (4-6-2004) The fat-like gene of *Drosophila* is the true orthologue of vertebrate fat cadherins and is involved in the formation of tubular organs. *J. Biol. Chem.* **279**: 24034–24043
42. Dunne, J, Hanby, AM, Poulsom, R, Jones, TA, Sheer, D, Chin, WG, Da, SM, Zhao, Q, Beverley, PC, and Owen, MJ (20-11-1995) Molecular cloning and tissue expression of FAT, the human homologue of the *Drosophila* fat gene that is located on chromosome 4q34-q35 and encodes a putative adhesion molecule. *Genomics.* **30**: 207–223
43. Nakayama, M, Nakajima, D, Yoshimura, R, Endo, Y, and Ohara, O (2002) MEGF1/fat2 proteins containing extraordinarily large extracellular domains are localized to thin parallel fibres of cerebellar granule cells. *Mol. Cell Neurosci.* **20**: 563–578
44. Mitsui, K, Nakajima, D, Ohara, O, and Nakayama, M (1-2-2002) Mammalian Fat3: A large protein that contains multiple cadherin and EGF-like motifs. *Biochem. Biophys. Res. Commun.* **290**: 1260–1266
45. Oda, H, Uemura, T, Harada, Y, Iwai, Y, and Takeichi, M (1994) A *Drosophila* homolog of cadherin associated with armadillo and essential for embryonic cell-cell adhesion. *Dev. Biol.* **165**: 716–726
46. Cox, B, Hadjantonakis, AK, Collins, JE, and Magee, AI (2000) Cloning and expression throughout mouse development of mfat1, a homologue of the *Drosophila* tumour suppressor gene fat. *Dev. Dyn.* **217**: 233–240
47. Ponassi, M, Jacques, TS, Ciani, L, and ffrench-Constant, C (1999) Expression of the rat homologue of the *Drosophila* fat tumour suppressor gene. *Mech. Dev.* **80**: 207–212

48. Inoue, T, Yaoita, E, Kurihara, H, Shimizu, F, Sakai, T, Kobayashi, T, Ohshiro, K, Kawachi, H, Okada, H, Suzuki, H, Kihara, I, and Yamamoto, T (2001) FAT is a component of glomerular slit diaphragms. *Kidney Int.* **59**: 1003–1012
49. Ciani, L, Patel, A, Allen, ND, and ffrench-Constant, C (2003) Mice lacking the giant protocadherin mFAT1 exhibit renal slit junction abnormalities and a partially penetrant cyclopia and anophthalmia phenotype. *Mol. Cell Biol.* **23**: 3575–3582
50. Tanoue, T and Takeichi, M (24-5-2004) Mammalian Fat1 cadherin regulates actin dynamics and cell-cell contact. *J. Cell Biol.* **165**: 517–528

CHAPTER 4.2

EPITHELIAL MORPHOGENESIS: Neighbour Exchange and Convergent Extension

It is a feature of animal development, especially of early development, that epithelial tissues have to change shape dramatically and quickly. In many instances, epithelial sheets appear to "flow" as if they were fluid membranes able to stretch and bend as new structures form, and they do so rapidly enough that there is no time for cell division to play a role in this process. It could, in principle, be achieved either by a change in the shapes of individual cells or by a reshuffling of cells so that they exchange neighbours. Deformation of individual cells is a common mechanism of morphogenesis in plant tissues (see Chapter 2.4), but in many animal systems, the shapes of individual epithelial cells are "normal" (not unusually stretched in any direction) both before and after the change of tissue shape. This is only possible if epithelial cells are able to change neighbours (see Figure 4.2.1). It used to be thought that the change in epithelial shape was driven by some hidden mechanism and that cells' ability to exchange their old neighbours for new ones was an essentially passive process in response to that mechanism. It is now becoming clear, however, that controlled neighbour exchange may be the morphogenetic mechanism itself.

Convergent extension of epithelia driven by neighbour exchange is a key process during the formation of an elongated axis in many animal phyla. It is particularly easy to observe in ascidians due to the transparency of their embryos and the low numbers of cells involved. In embryos of the ascidian *Corella inflata*, an epithelial monolayer of 40 cells transforms over 6 hours to produce a long stack of cells—the notochord—that is the main skeletal element in the forming tadpole.[1,2,3] Three-dimensional time-lapse recording of this transformation in *C. inflata* shows the cells undergo convergent extension so complete that a patch of epithelium about six cells along the left-right axis by about seven cells along the antero-posterior axis become a single row of cells[3] (see Figure 4.2.2). As convergent extension tales place, the area also invaginates so that the developing notochord becomes internalized (invagination is discussed in Chapter 4.4).

Convergent extension can also happen in epithelial tubes. For example, after the gut of the sea urchin forms by invagination, it elongates markedly. During this process,

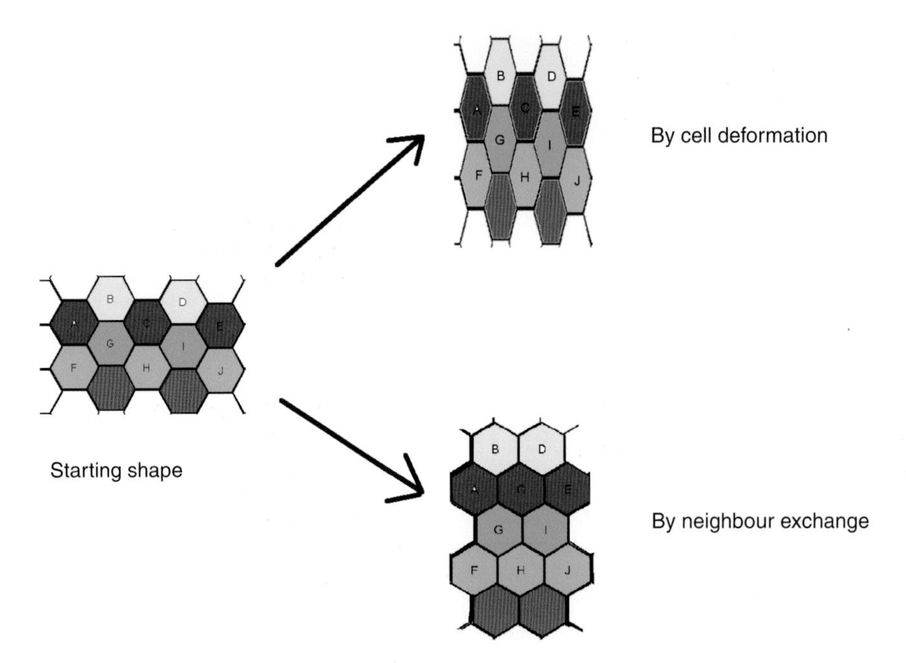

FIGURE 4.2.1 The difference between two strategies for deforming a sheet to make it tall and thin. In the cell deformation model, cells maintain their neighbour relationships but alter their shapes. This method is seen mainly in plants (see Chapter 2.4). In the neighbour exchange model, cells have the same shape at the end of the process as they do at the beginning, but they change neighbours; this can be seen in the schematic by the coming together of cells of like colour in rows.

the numbers of cells around its circumference decreases, while the number along its length increases, the change being brought about by cells changing their neighbour relationships[4] (see Figure 4.2.3). Rearrangement of neighbour relationships is also seen during the evagination of insect imaginal discs[5] and also during the closure of holes (see Chapter 4.3).

One of the best-studied examples of epithelial convergent extension is the elongation of the germ band in *Drosophila melanogaster*. The germ band is the part of the embryo that will form the segmented trunk of the future larva, and during development it thins and extends to curl around the inside of the egg and to bend back on itself (see Figure 4.2.4). As it does so, cells converge toward the ventral midline and intercalate between one another so that the tissue as a whole extends along the midline. The integrity of the epithelium is retained as the cells rearrange,[6] which means that they never acquire "free" edges and cannot move by the action of lamellipodia and filopodia in the way that converging mesenchyme cells do.

Before convergent extension begins, cells are packed in a hexagonal array. This array is arranged so that the centres of the cells form fairly ordered rows along the antero-posterior axis; this means that one-third of cell-cell boundaries are at 90 degrees to the antero-posterior axis, one-third at +30 degrees, and the remaining third at –30 degrees (see Figure 4.2.5).

The different types of cell boundaries, those perpendicular to the antero-posterior axis and those at an angle to it, behave differently, and this is critical to the movement

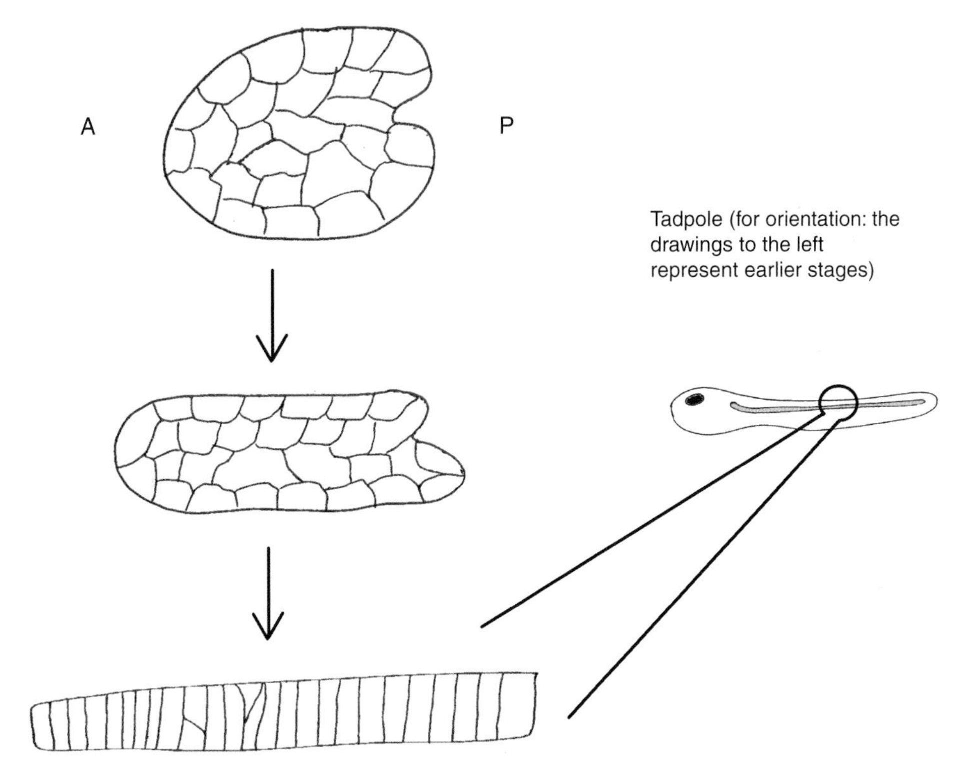

FIGURE 4.2.2 Convergent extension in *C. inflata*, based on detailed drawings made from three-dimensional time-lapse images.[3] "A" and "P" refer to the antero-posterior axis. The schematic shows the epithelium flattened out throughout and does not depict invagination that is happening at the same time. The drawing of the tadpole is to indicate the eventual position of the notochord at a later stage of development.

of the cells. Cell boundaries that are perpendicular to the antero-posterior axis—and *only* those boundaries—shorten. Where diagonal boundaries meet, a new boundary begins to form parallel to the antero-posterior axis, and this boundary lengthens. The lengthening of this new boundary is along the axis in which the tissue as a whole lengthens. A schematic of the process is shown in Figure 4.2.6. In practice, the shapes of the cells are a good deal more variable in the intermediate stages, as indeed they have to be when all of their neighbours, not shown in Figure 4.2.6, are taken into account, but the end result is another hexagonal array orientated with its hexagons at 90 degrees to the original array. This can be seen clearly in a movie provided on *Nature*'s website as information supplementary to a paper,[6] and readers are strongly recommended to view it for themselves.

Epithelial cells involved in germ-band extension are held together mainly by adherens junctions that include the transmembrane adhesion molecule *Drosophila* DE-cadherin and the β-catenin *armadillo*. Also present at the apical-most junctions only are the myosin II heavy chain encoded by *zipper*, the myosin II regulatory light chain encoded by *spaghetti squash*, and the myosin-associated protein Slam (here, "apical" refers as always to the apico-basal polarity of the epithelium and has nothing to do

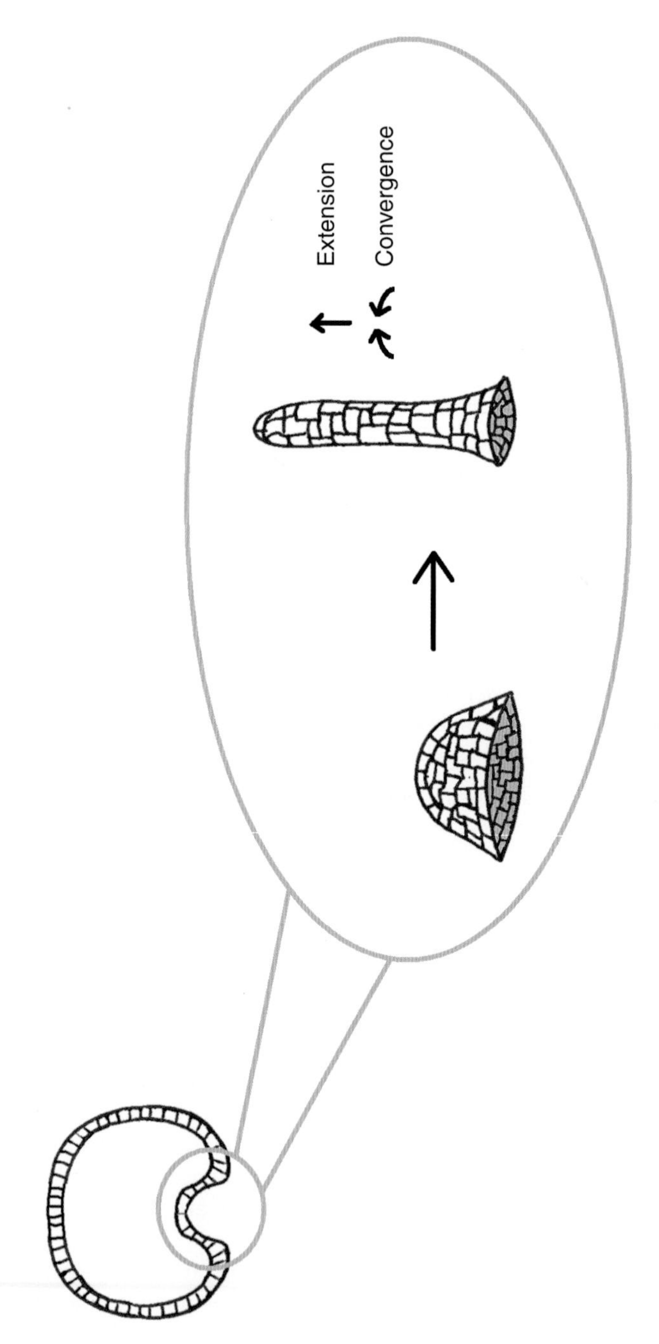

FIGURE 4.2.3 Convergent extension in the elongation of the gut primordium in sea urchin embryos.

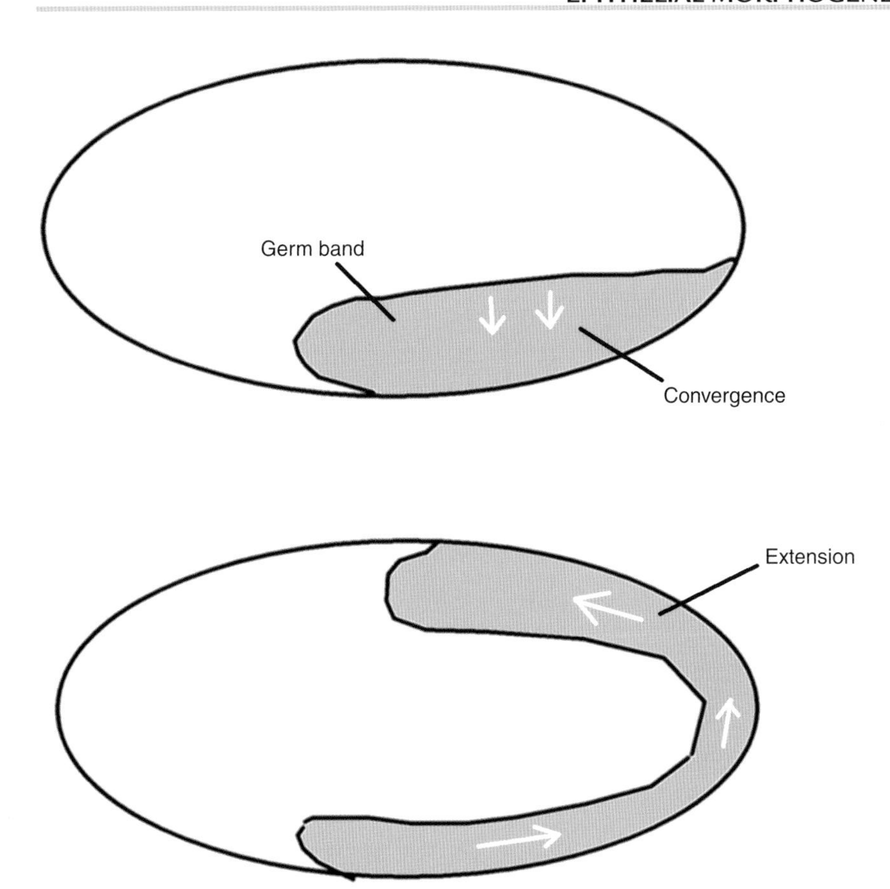

FIGURE 4.2.4 Convergent extension in the early embryo of *D. melanogaster*. The processes of convergence and extension happen together and are separated on the diagram simply for clarity of labelling.

with vertices of hexagons and so on). The distribution of Spaghetti squash protein is not uniform within the plane of the epithelium, however, and it locates preferentially along the shrinking side boundaries of the cells. Embryos carrying mutants in *zipper* or *spaghetti squash* show severe disruption to germ-band extension because junction remodelling fails and the cell boundaries that are perpendicular to the antero-posterior axis do not shorten.[6] It is not yet clear how myosin physically mediates the shortening of the boundaries at which it is active. It may be that, by acting on the actin filaments present at and linking between adherens junctions, it simply draws the junctions together and therefore shortens that face of the membrane, although some additional mechanism would still be required to take the junctions apart, as must happen for cells to let go of one another as they exchange neighbours.

The targeting of active myosin and its associated proteins to specific cell boundaries may explain why only those boundaries shrink, but it leaves open the question of how myosin is activated in, or targeted to, the correct location in the first place. This targeting of myosin requires the cells to have an internal representation of their orientation within the plane of the epithelium, usually referred to as "planar polarity." The fact that

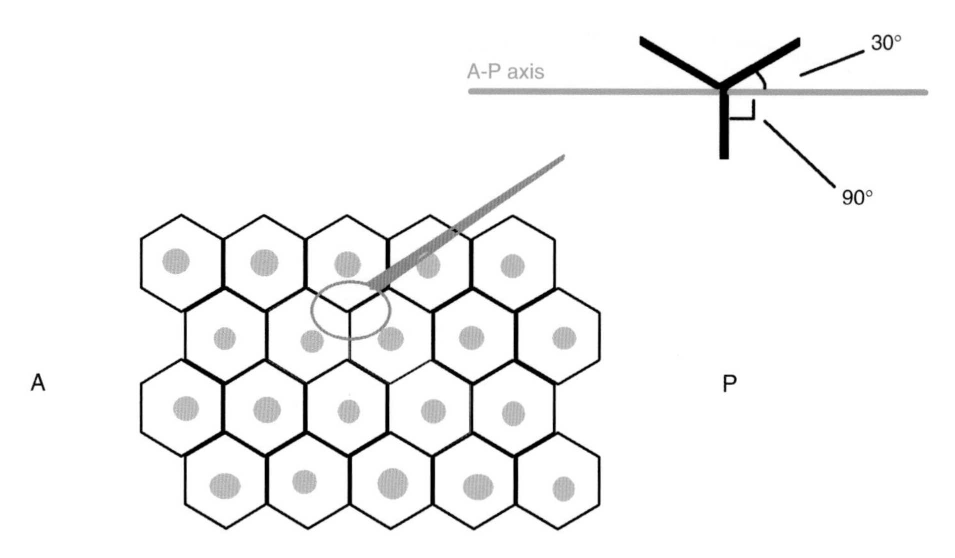

FIGURE 4.2.5 The alignment of the pre-convergent-extension hexagonal array of epithelial cells with the antero-posterior axis prior to germ band extension in *D. melanogaster*.

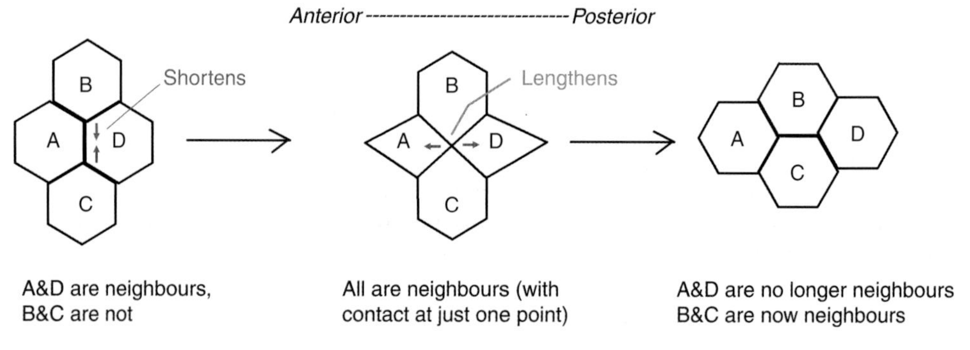

FIGURE 4.2.6 A schematic diagram to show how applying different length-change processes to different borders can drive convergent extension. Because many cells are undergoing these changes at the same time, the shapes of intermediate stages are irregular and include triangles, diamonds, pentagons, and hexagons (all can be seen in the movie[6] referred to in the text).

opposite sides of a cell show the same behaviour in terms of boundary shortening or lengthening suggests at first sight that the response of the cell to its orientation must be bipolar (see Figure 4.2.7a). This may not be so, however, due to the special nature of epithelial tissues. Adherens junctions can only be made and maintained through the cooperation of the cells they link, and if cell A "decides" to destroy the junctions between itself and cell B, there may be nothing that cell B can do about it. The shrinkage of the border between cells A and B might therefore be achieved through the action of cell A alone; all that cell B has to do is shrink the border between it and C and so on (see Figure 4.2.7b); in this way, monopolar cells would appear to show bipolar behaviour, although their own efforts would really be focused in only one direction.

Bipolar model

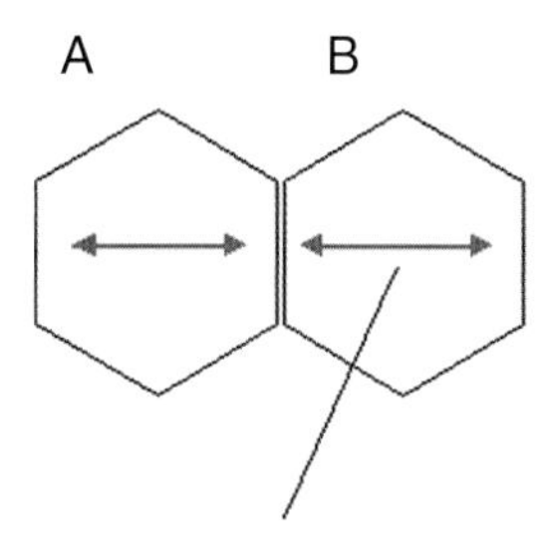

Each cell targets both sides for shortening

Monopolar model

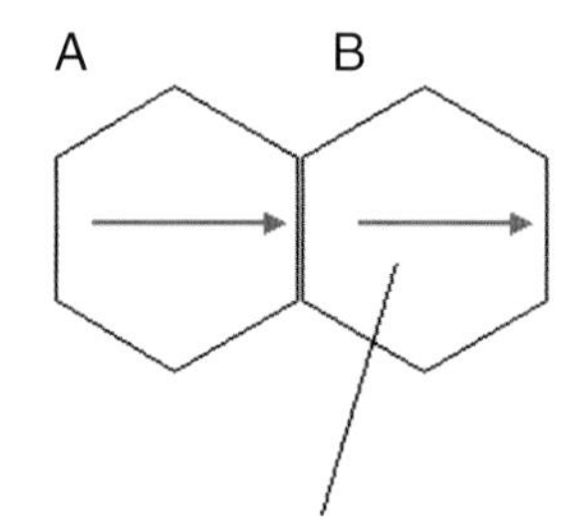

Each cell targets only one side for shortening, and has its other side shortened passively

FIGURE 4.2.7 Cells may have a bipolar orientation in which each actively targets the shortening of both of its sides, or they may have a monopolar orientation in which one side shortens passively due to the actions of neighbouring cells.

Whether this is true in the real embryo might be tested using mosaics in which some marked cells unable to shorten junctions themselves (e.g. *zipper* mutants) are scattered among wild-type cells, and the behaviour of their common junctions with wild-type cells is observed.

The monopolar mechanism could, in principle at least, be the means of communicating polarity through the tissue. If having a border shortened by a neighbour could initiate a polarization signal within a cell that points away from that border, then as long as one row of cells could be polarized by an external organizer (for example, a neighbouring tissue secreting a short-range signal), the polarization would sweep as a wave through the epithelium. There are good reasons to view this model with suspicion, however. One is that it is difficult to see how the polarization would be accurate enough to keep pointing directly across the epithelium, as any errors would be cumulative. The other is that movies of epithelial convergent extension[6] suggest that cells commence the process more or less simultaneously, and not in a sequence.

The near simultaneity of border rearrangements implies that cells acquire their planar polarity by some means independent of their morphogenetic movements. The evidence available to date, although patchy, suggests that components of the same planar polarity pathway that controls mesodermal convergent extension in *X. laevis* and bristle orientation on the wings of *D. melanogaster* may be involved.

The activation of myosin II in the shrinking borders of the epithelial cells is controlled by the Zipper/Spaghetti squash complex, and this complex can in turn be activated by *D. melanogaster's* Rho-dependent kinase, Drok.[7] Inhibition of Drok using the drug Y27632 strongly reduces association of myosin-associated proteins with the cell boundaries and inhibits cell junction remodelling so that convergent extension fails.[6] The involvement of Drok in convergent extension is particularly interesting because it is already implicated in the translation of planar polarity signals in the wing primordium of *D. melanogaster* via the non-canonical Wnt signalling pathway shown in Figure 4.2.8.

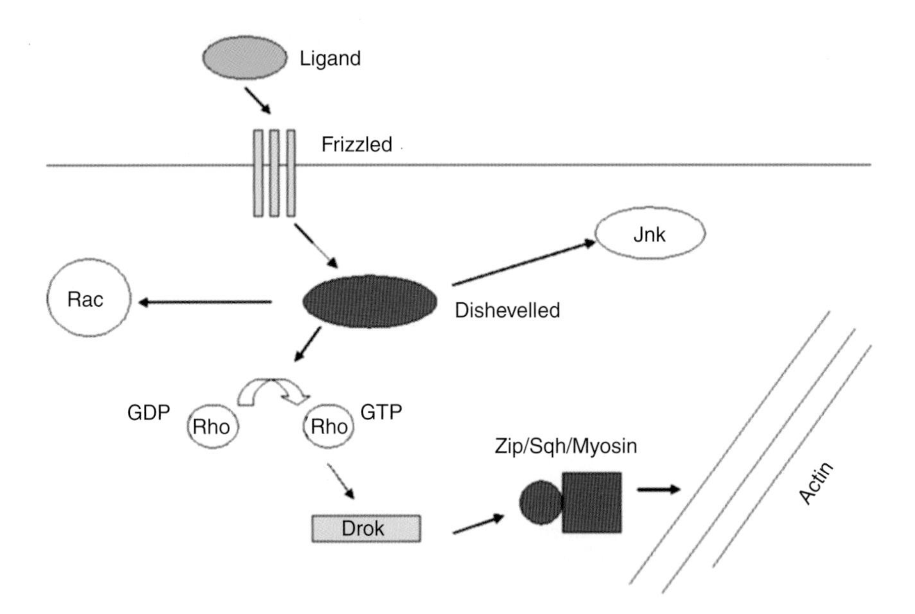

FIGURE 4.2.8 The signalling system implicated in passing planar polarity information via Drok to the myosin/actin system in wing bristle polarization.

In the wing system, the expression of the Frizzled and Dishevelled proteins is inhomogeneous so that they are expressed only at the end of the cell pointing toward the distal wing.[8] This is the point of the cell from which the wing bristle, produced by actin ordered by the effects of Drok on myosin, emerges.[7,9] The reason for the inhomogeneity of Frizzled and Dishevelled location is not yet understood, but it may be the result of a gradient-amplifying mechanism (as discussed in Chapter 3.2). If Frizzled and Dishevelled were expressed in a similarly polarized manner in epithelia undergoing convergent extension, they may be expected to activate Drok and therefore myosin-mediated border shrinkage on just that side of the cell. Convergent extension of mesenchyme in *X. laevis*, used at a similar stage of development for the same purpose of elongating the main axis of the body, does involve the Frizzled-Dishevelled pathway, and the convergent extension seen in the overlying epithelium presumably does too. This pathway orientates protrusive activity in those parts of the mesenchymal cells orientated along the direction of convergence. There are no data available on the distributions of Frizzled, Dishevelled, or Drok in the epithelium, but in a spirit of fun one can propose the pathway suggested in Figure 4.2.9, and wait to see how the data come out.

Dishevelled signalling is certainly involved in regulating convergent extension during notochord formation in the ascidian *Ciona intestinalis*, and expression of mutant Dishevelled protein in this region of the embryo blocks the process without blocking normal assignation of cell fate.[10] In mosaic embryos, the effect of mutant Dishevelled is cell-autonomous, and wild-type cells that surround those expressing the mutant protein behave normally.[10] In this species at least, polarity must be sensed by cells independently and does not depend on being passed from cell to cell.

Epithelial convergent extension, depending as it does on the polarized, boundary-changing activity of individual cells, is another example of the self-organizing of

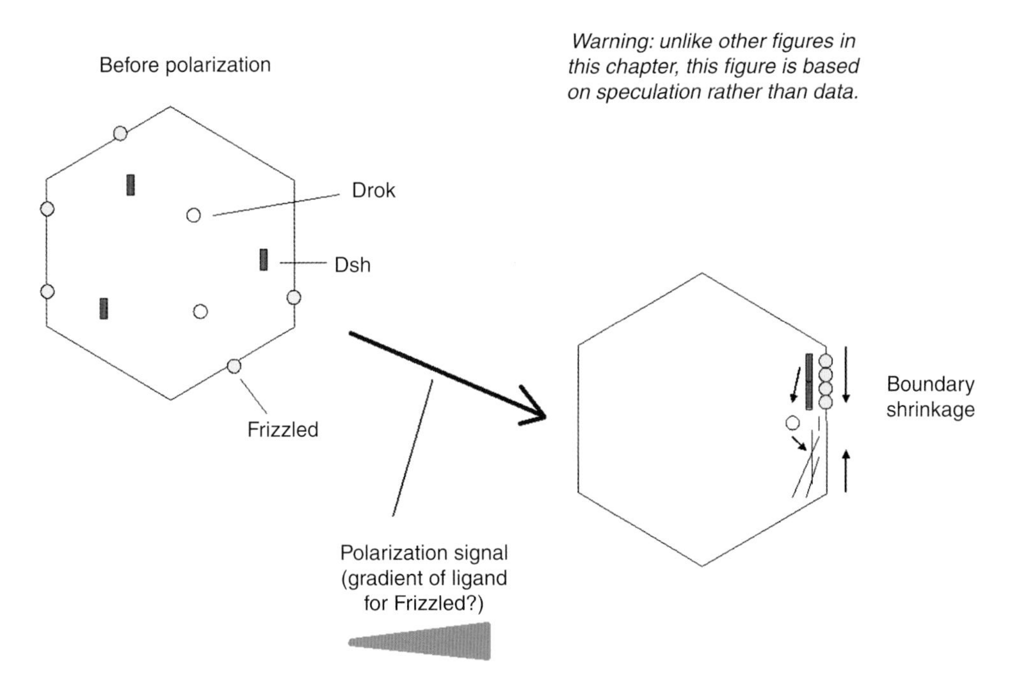

FIGURE 4.2.9 One way in which the "planar polarity" pathway triggered by Frizzled might be coupled to the boundary shrinkage required for epithelial convergent extension.

morphogenetic change by a multitude of autonomous agents, each of which follows simple rules. What is slightly unusual is that the morphological change of each cell does not reflect the resulting change of the tissue in microcosm, as is the case when tissues extend because their constituent cells do (see Figure 4.2.1 top) or when each cell moves in the direction that the population as a whole moves during cell migration. It therefore serves as a reminder to researchers attempting to find a cellular explanation of a morphogenetic change at the tissue level that the relationship between events at the two levels may involve complex geometrical transformations rather than a simple scaling up of the same change of shape.

Reference List

1. Conklin, EG (1905) The organization and cell lineage of the ascidian egg. *J. Acad. Nat. Sci.* **13**: 1–119
2. Miyamoto, DM and Crowther, RJ (1985) Formation of the notochord in living ascidian embryos. *J. Embryol. Exp. Morphol.* **86**: 1–17
3. Munro, EM and Odell, GM (2002) Polarized basolateral cell motility underlies invagination and convergent extension of the ascidian notochord. *Development.* **129**: 13–24
4. Ettensohn, CA (1985) Gastrulation in the sea urchin embryo is accompanied by the rearrangement of invaginating epithelial cells. *Dev. Biol.* **112**: 383–390
5. Fristrom, D (1976) The mechanism of evagination of imaginal discs of *Drosophila melanogaster*. III. Evidence for cell rearrangement. *Dev. Biol.* **54**: 163–171
6. Bertet, C, Sulak, L, and Lecuit, T (10-6-2004) Myosin-dependent junction remodelling controls planar cell intercalation and axis elongation. *Nature.* **429**: 667–671

7. Winter, CG, Wang, B, Ballew, A, Royou, A, Karess, R, Axelrod, JD, and Luo, L (6-4-2001) *Drosophila* Rho-associated kinase (Drok) links Frizzled-mediated planar cell polarity signalling to the actin cytoskeleton. *Cell.* **105**: 81–91
8. Strutt, DI (2001) Asymmetric localization of Frizzled and the establishment of cell polarity in the *Drosophila* wing. *Mol. Cell.* 7: 367–375
9. Wong, LL and Adler, PN (1993) Tissue polarity genes of *Drosophila* regulate the subcellular location for prehair initiation in pupal wing cells. *J. Cell Biol.* **123**: 209–221
10. Keys, DN, Levine, M, Harland, RM, and Wallingford, JB (15-6-2002) Control of intercalation is cell-autonomous in the notochord of Ciona intestinalis. *Dev. Biol.* **246**: 329–340

CHAPTER 4.3

EPITHELIAL MORPHOGENESIS: Closure of Holes

The closure of holes within an embryonic or foetal epithelium is an essential part of normal development in many animals. It is also important in the repair of wounds that are not part of normal development but may occur, either because of accidents or predators, in the case of free-living embryos, or because of surgical intervention in the case of mammals.

DORSAL CLOSURE IN *DROSOPHILA MELANOGASTER*

The morphogenetic movements of the *D. melanogaster* embryo leave a gap on the dorsal side of the egg in which the embryonic epidermis fails to cover the yolky tissue. This gap, the "dorsal hole," is later covered by an upward spreading of the epidermis from the flanks of the embryo (see Figure 4.3.1).

Dorsal closure begins about 9 hours after the egg is laid (in standard laboratory conditions). In principle, such a movement might be driven either by active advance of the edge cells into the space of the "hole" or by their passive advance under pressure from cell expansion or proliferation in the epidermis behind them. Ablation of areas of epidermis near but not at the leading edge, using an ultraviolet laser, fails to prevent dorsal closure.[1] Indeed it accelerates it a little. This shows that the closure of the hole cannot be driven by cells being pushed from behind, but rather by mechanisms that must be local to the area of the hole itself. Again, it turns out that the process relies heavily on changes in the actin cytoskeleton that are brought about via the Rho family of GTPases.

The cells at the edge of the epidermis, which border the "hole," acquire thick bands of actin that run across them, adjacent to and parallel to the hole's edge.[1] Formation of this actin band is dependent on Rho GTPase, which is also responsible for causing formation of stress fibres in fibroblasts (see Chapters 2.2 and 3.2), and it fails to form around the dorsal hole of embryos lacking Rho activity.[2] The actin bands terminate at cell-cell junctions and are attached to them. The alternating fibre-junction–fibre-junction system therefore constitutes a band that runs right around the edge of the hole (see Figure 4.3.2). Myosin-mediated sliding of oppositely polarized actin filaments pulls the sides of each cell together and therefore contracts the entire band running around the hole. The importance of myosin-mediated contraction is underlined

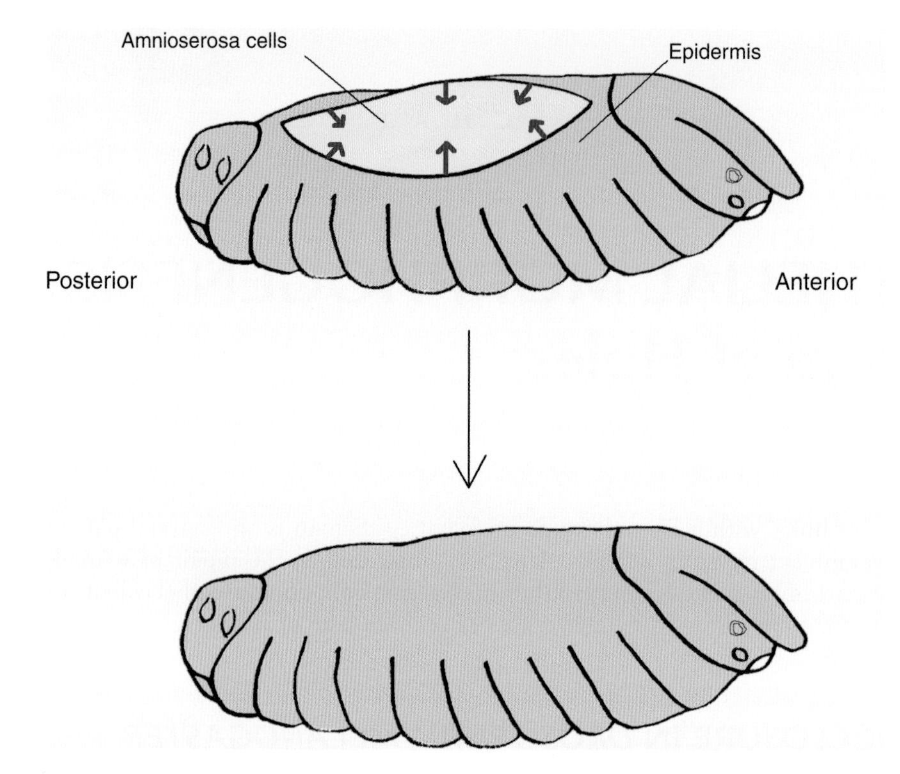

FIGURE 4.3.1 Dorsal closure in the embryo of *D. melanogaster*. The embryo is covered by an epithelial epidermis except at the dorsal "hole," shown white in the diagram. This hole in the epidermis is covered by amnioserosa cells, also epithelial, which disappear as the epidermal edges advance and seal the hole up.

by the observation that mutant embryos that have no myosin II heavy chain fail to complete dorsal closure,[3] although that fact does not in itself prove that contraction of the leading edge is the essential event for which myosin II is needed. Laser ablation of a small number (two to three) cells at the leading edge of the hole causes the cells adjacent to their flanks to spring away sideways, providing further evidence that the alternating band of actin and cell junctions around the hole really is under tension.[1] Laser ablation of leading edge cells does not, however, block dorsal closure, even if multiple sites are ablated (see Figure 4.3.3b). Even though the advance of a site of ablation is slowed down temporarily, it catches up before long. Other mechanisms must therefore be capable of drawing the sides of the hole together.

The cells of the amnioserosa are also under tension, which can again be revealed by their springing away from a site of laser ablation. What is more, ablation of amnioserosa cells near the edge of the hole causes nearby epidermis to spring away too (see Figure 4.3.3c). If the epidermis were advancing solely by means of its circumferential tension, there is no reason that it should spring back; indeed, its progress may be expected to accelerate with less amnioserosa in the way. That it springs back therefore indicates that it was being pulled in by forces in the amnioserosa (but Newton's third law♣ makes it

♣Action and reaction are equal and opposite: if you pull on a rope, the rope pulls as hard on you.

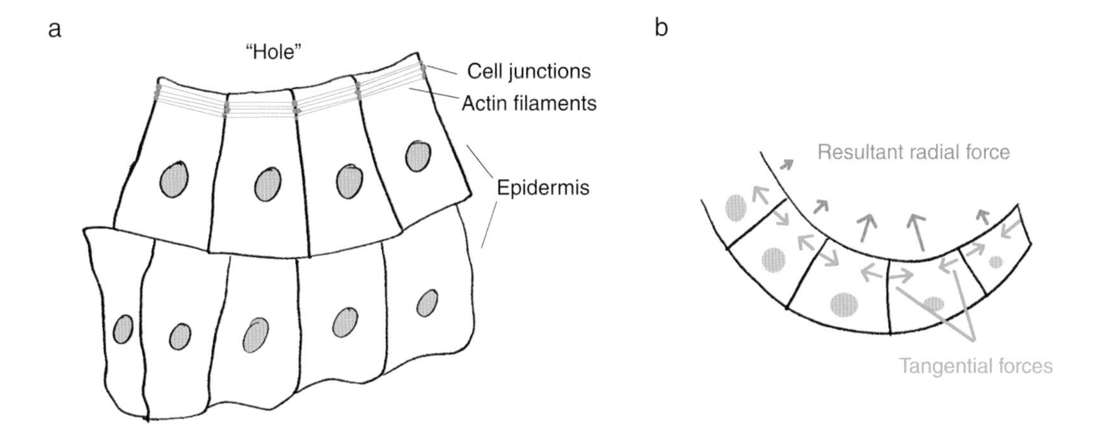

FIGURE 4.3.2 Leading edge actin bands in dorsal closure. (a) Alternation of actin filaments (green) and cell junctions (red) connects adjacent cells mechanically to form a tissue-scale contractile apparatus. (b) The circumferential contractile forces (green) produce a resultant radial force (red).

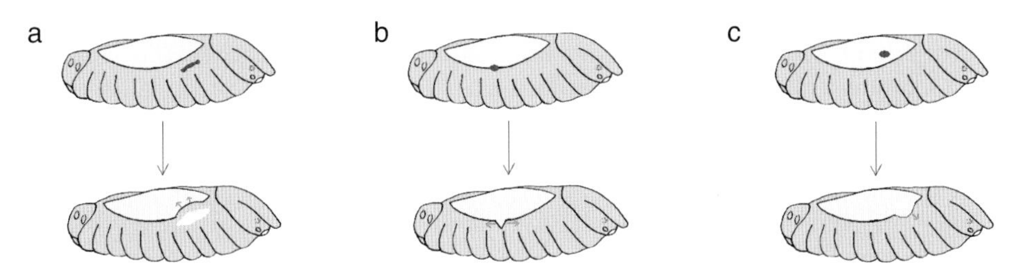

FIGURE 4.3.3 Effects of laser ablation on dorsal closure in *D. melanogaster*. The site of ablation is shown as a red spot, and resulting movements by green arrows. (a) Ablation of a region of epidermis away from the edge increases, rather than decreases, progress towards the midline, suggesting that progress cannot be driven by pushing from behind. (b) Ablation of cells at the leading edge causes their neighbours to spring apart, as if under tension. (c) Ablation of an area of amnioserosa delays progress, as if the amnioserosa is pulling on the epidermis.

impossible to deduce from the force whether this pulling is generated by activity in the epidermis, the amnioserosa, or both). Repeated ablation of the amnioserosa does not abolish dorsal closure, so while the amnioserosa can assist closure process, it seems not to be essential for it. Ablation of both the amnioserosa and tension-bearing cells at the edge of the epidermis does block closure, suggesting that while closure can be driven by either one of these cell types, it cannot take place in the absence of both.[1] The possibility remains, however, that the double-ablation causes such damage that closure may have been prevented by non-specific effects.

The leading edge of the advancing epidermis is atypical, by the usual standards of epithelial tissue, in that it represents a "free" lateral edge (the lateral edges of most epithelial cells are in contact with neighbouring epithelial cells). This free edge has a propensity to behave like the leading edge of a migratory cell and to produce lamellipodia and filopodia. In fibroblasts, Rho antagonizes the production of these motile structures (see Chapter 3.2), and the same thing seems to happen in epithelial cells taking part in dorsal closure: wild-type cells show restrained motility but *Rho* mutants show much more lamellipodial and filipodial acitvity.[2] This may provide the embryo with an automatic backup system for dorsal closure; if the Rho-driven actin belt system is working normally, cells will move by their circumferential tension, but if Rho-activated actin belt formation fails, the cells produce machinery to migrate forward by direct cell locomotion. Mosaic experiments, in which dominant negative *Rho* is expressed by groups of cells surrounded by wild-type neighbours, show that the dorsal migration of the mutants is actually faster than that of wild-type cells, but the coherence of the migration is lost so that the mutant cells spread throughout the dorsal midline. The actin cable system seems not to have evolved for optimum efficiency of migration, therefore, but as a restraining mechanism that allows migration to take place without cells' spatial relationships to one another to become too randomized. (Orderliness is important because the cells concerned have already been assigned a segment identity, and segment identity needs to be conserved without stagger at the midline.)

The asymmetry of cells at the leading edge, which are surrounded by epithelial cells except in the direction in which they have to head, may itself provide a sufficient polarizing cue to direct migration and actin cable formation. These cells do, however, show an asymmetric localization of proteins involved in the "planar polarity" pathway discussed in Chapter 4.2. Before dorsal closure begins, the locations of the transmembrane receptor protein Frizzled, the signal transduction protein Dishevelled, and the cadherin Flamingo show no obvious polarity. As closure begins, all three associate with adherens junctions and with associated actin nucleating centres each side of the leading edge, but they eschew the leading edge itself.[4] This polarized expression is lost once the two sides of the closing hole meet. In the planar polarity pathway, Dishevelled signals through JNK (see Figure 4.2.8) and, in other examples of planar polarity such as that found in wing bristle development, correct polarization of pathway components depends on pathway function, suggesting feedback. When embryos lacking JNK function begin dorsal closure, Frizzled protein is cleared away from the leading edge, but it fails to accumulate at the adherens junctions to the sides of that edge, and actin nucleation is weak and not well organized. Embryos mutant for Dishevelled show incomplete closure of the hole. These observations suggest that the planar polarity pathway is active in leading edge cells and that it is used to establish actin nucleating centres in the correct place to make the actin cable.[4] Frizzled is a receptor for the Wnt family protein Wingless, and possibly for other yet-to-be-identified ligands as well. Embryos lacking Wingless protein show very delayed and disrupted dorsal closure with weak actin cables. Wingless seems not to set up the direction of the gradient (expressing the protein ubiquitously does not interfere with orientation of dorsal closure), but it may be required to keep Frizzled active enough to allow signalling through the Dishevelled pathway, which is polarized either by a different diffusible factor or by the inherent asymmetry of cells at the leading edge.

The closure of the dorsal hole presents a formidable problem of accurate navigation because the epithelium on each side of the hole has already established its segmental identity (using Hox genes and so on). When the two sides come together, it is essential that the stripe of epithelium from the left side of the embryo representing abdominal segment 1 fuses with the abdominal segment 1 stripe from the right side, and only with that stripe. The same applies to all of the other segments—any other outcome would throw the identities of the left and right sides of the dorsal body out of registration with each other, create unexpected segment boundaries, and make the creation of

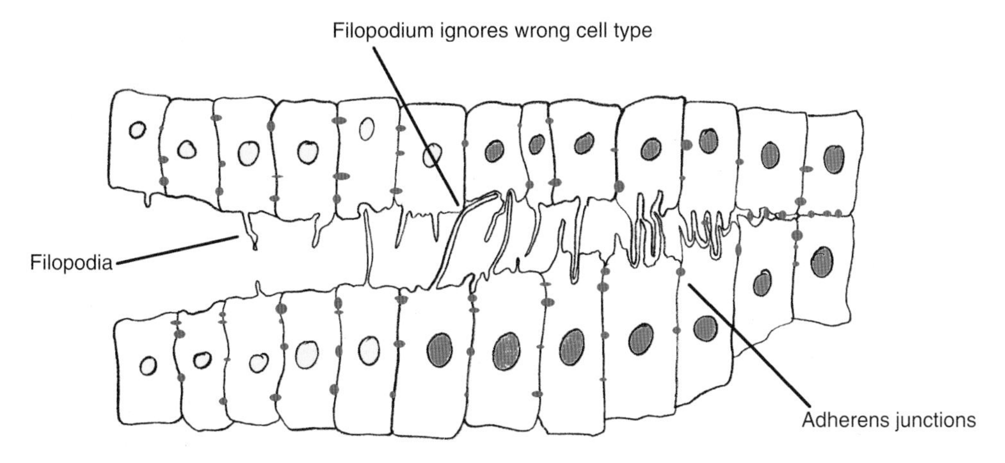

FIGURE 4.3.4 Filopodial activity at the end of dorsal closure. As the opposing sides near each other, cells produce filopodia with which they explore one another. If they find a cell of the wrong type (represented here by nuclear colour) they ignore it, but when they find a cell of the correct type they push into it and form adhesive contacts. As more contacts are made, the parts of the cells between filopodia also advance in a lamellipodium-like manner, meet, and adhesive contacts mature to become "normal."

normal anatomical septa between segments impossible. Relying on mechanical perfection in a complex and ever-changing embryo is not enough, so before they meet and fuse, the cells from approaching sides actively seek suitable partners. As the sides come close, leading edge cells become much more active in production of very long filopodia, which reach out across the gap and make contact with the other side.[5] The filopodia seem to be capable of recognizing stripes of their own sort; for example, filopodia from cells expressing *engrailed*, i.e. from the anterior portion of each para-segment, will adhere to other cells expressing *engrailed* but will pass right over non-*engrailed*-expressing cells to find the nearest ones that do express the gene.[5] Since Engrailed is a nuclear protein, the filopodia cannot be detecting it directly but presumably detect a cell surface marker that is expressed under Engrailed control.

Once filopodia find a cell they recognize, they push against its plasma membrane and cause it to yield around them to create a large surface area of contact between the filopodial membrane and that of the other cell (see Figure 4.3.4). Junction molecules are recruited to the sites of contact, and the fleeting filopodial kiss is transformed into a committed partnership as the contacts mature to become conventional adherens junctions.[6] If formation of filopodia is inhibited by dominant negative cdc42 (see Chapter 2.1), cells do not form proper adhesive contacts, and dorsal closure fails.

WOUND HEALING IN THE EMBRYO

Wounds in the skin or cornea of mammalian and avian embryos heal by a process similar to dorsal closure in *D. melanogaster*. While wound healing might be seen to be of only peripheral relevance to a book centred on mechanisms of normal morphogenesis, it provides a very useful model precisely because wounds do not take place at stereotyped, predictable times and places. The construction and activation of the healing mechanism therefore has to be by local adaptive self-organization.

When a small wound is made in embryonic skin, the undamaged tissue moves across the hole and seals it up perfectly and scarlessly. Shortly after wound formation, a band of actin filaments, connected between cells using adherens junctions, forms just under the edges of the "free" lateral membranes that border the wound.[6] The cable is similar to that seen in *D. melanogaster*'s dorsal closure, and it seems to perform a similar function; if actin polymerization is blocked using cytochalasin D, actin cables fail to form and the wound fails to close.[7] As in dorsal closure, the actin cable is associated with myosin and is formed under the action of Rho GTPase.[8] The meeting of cells and their "zipping up" uses exploratory filopodia whose action is very similar to that seen in dorsal closure.[9]

The "unplanned" nature of embryonic wounds makes the self-organizing properties of the actin cable system easier to study than in situations where "pre-programming" of cells may be involved. There are at least two necessary aspects to the self-organization of the wound-closing actin cable: its expression in the correct parts of cells that appose the wound edge, and its disappearance from cells that have already zipped up. The regulation of assembly and disassembly has not been studied in complete detail in the wound system, but findings from that system can be combined with studies in simple cell culture to suggest a simple and automatic mechanism for cable regulation.

Cells in the unwounded epithelium are in tension; that is obvious from the way that a wound made by a fine needle or blade gapes open. Because of the positive relationship between the stability of actin/myosin fibres and the tension they carry (see Chapter 2.2), contractile actin bundles concentrate along lines of tension that, in an unwounded epithelium, tend to be spread evenly around the adhesive junctions of the cell. If the epithelium is wounded, tension becomes greater at the edge of the wound and greatest where the radius of curvature of the tissue is smallest.♣ Epithelial cells subjected to increased tension in culture respond by laying down thick contractile actin cables parallel to the direction of the tension.[10] There is evidence from a number of systems that this response is mediated by local activation of Rho.[11] Contractile actin cables have two effects: they provide the cell with a means to resist being stretched by tensile forces, and they block the formation of lamellipodial structures on the newly free edge of the cell, just as stress fibres at the lateral margins of a migrating fibroblast antagonize the production of a lamellipodium there (see Chapter 3.2). Because tension is always greatest parallel to the local wound edge, actin cables are automatically positioned in the right place. The contractile actin cable can function only if it is connected by adherens junctions to the cable in adjacent cells. If adhesion at these junctions is inhibited using a function-blocking anti-E-cadherin antibody, actin filaments disappear, and the free edge of the cell makes a lamellipodium instead.[12] The lamellipodium crawls forward, and healing still takes place by a secondary migratory mechanism that has been selected automatically and quite directly by the failure of the principal, actin-cable-based system.

When the actin cables are being assembled, it is clearly important that a substantial supply of E-cadherin is made available in precisely the correct place to connect the

♣At the end of a cut in a tissue being stressed at right angles to the cut, application of Inglis' Law states that stress on the tissue at the tip of the cut is approximately $s \cdot (1 + 2\sqrt{(L/r)})$, where s is the stress in the uncut tissue, L the half length of the cut, and r the radius of its end.

new actin cables of adjacent cells. This supply is assisted by the dependence of adherens junction stability on actin filaments. If the actin of cultured cells is removed by cytochalasin, junction components are internalized for re-use.[13] Thus, reorganization of actin to a tense band at the wound edge automatically assists reorganization of junctions, and well-organized junctions assist the stability of the actin that terminates in them. At least part of this mechanism probably rests on the fact that binding of the extracellular domain of E-cadherin to E cadherin on adjacent cells allows its intracellular domain to recruit the actin nucleating molecule Arp2/3.[14] Adjacent cells therefore adjust their arrangements of actin and junctions continuously and iteratively, and since all cells connect to neighbours on both sides, the actin band is always arranged appropriately at the tissue scale by events at the molecular scale. As the hole finally zips up, in a manner similar to that described for *D. melanogaster*, new adhesions that form between cells of opposing sides encourage actin filaments to end in them, where tension is now concentrated, and the thick band that was at the wound edge disintegrates to give rise to a "normal" ring of cortical actin filaments running inside the adherens junctions around the cell.

It is interesting to observe that the healing of wounds in embryonic epithelium uses mechanisms very similar to the healing of wounds in the membrane of a single cell. The zone immediately underneath the plasma membrane of most cells is richly supplied with actin fibres. If a plasma membrane is breached, for example with a fine needle, external Ca^{2+} ions flood into it and trigger a healing response.[15] The Ca^{2+} causes an almost immediate fusion of the membranes of cytoplasmic vesicles to form a vesicle membrane bilayer that plugs the damaged site. This action plugs the hole but does not reinstate the normal plasma membrane or the normal cell cortex underneath it. To reinstate the proper sub-membrane environment, the undamaged cell cortex from around the hole moves inward radially until it has completely filled in the damaged site. Within 30 seconds or so of wounding, a cable of actin and myosin forms around the circumference of the hole, and that cable contracts, pinching off the hole and dragging the normal cell cortex with it.[16] Yet again, activity of Rho GTPase is essential for the formation of this healing contractile ring.[16] A similar contractile ring of actin is formed during normal cytokinesis. It is therefore possible that this key means to close over a hole was invented just once in evolution, probably for cytokinesis, was then connected to cell wound-detection apparatus to become part of the cell wound-healing machinery, and was then generalized at tissue level to become part of the machinery used to close off tissue-scale holes. It is another illustration of an important theme in morphogenesis: "new" morphogenetic events generally co-opt existing simple semi-autonomous mechanisms and adapt them to a new purpose instead of inventing new ones.

Like many of the morphogenetic events described in this book, closure of holes depends on the concerted action of cells that perform essentially simple tasks (cable building and contraction, lamellipodial locomotion) directed by local cues. No cell has to "know" the shape of a hole or where it is along the hole's circumference; as long as each cell performs its simple, local tasks properly, the assembly of cells will organize itself automatically to close a small hole of any shape.

It is important to note that the wound-healing response of adult skin is very different from that seen in embryos. It uses a rapid deposition of matrix (clotting and scarring) and cell migration of the lamellipodial rather than contractile ring (purse-string) type. There is considerable research effort being expended on finding ways to persuade adult skin to heal by the scarless embryonic route, which would allow post-operative wounds to heal into skin as good as new.[17,18]

Reference List

1. Kiehart, DP, Galbraith, CG, Edwards, KA, Rickoll, WL, and Montague, RA (17-4-2000) Multiple forces contribute to cell sheet morphogenesis for dorsal closure in *Drosophila*. *J. Cell Biol.* **149**: 471–490
2. Jacinto, A, Wood, W, Woolner, S, Hiley, C, Turner, L, Wilson, C, Martinez-Arias, A, and Martin, P (23-7-2002) Dynamic analysis of actin cable function during *Drosophila* dorsal closure. *Curr. Biol.* **12**: 1245–1250
3. Young, PE, Richman, AM, Ketchum, AS, and Kiehart, DP (1993) Morphogenesis in *Drosophila* requires nonmuscle myosin heavy chain function. *Genes Dev.* **7**: 29–41
4. Kaltschmidt, JA, Lawrence, N, Morel, V, Balayo, T, Fernandez, BG, Pelissier, A, Jacinto, A, and Martinez-Arias, A (2002) Planar polarity and actin dynamics in the epidermis of *Drosophila*. *Nat. Cell Biol.* **4**: 937–944
5. Jacinto, A, Wood, W, Balayo, T, Turmaine, M, Martinez-Arias, A, and Martin, P (16-11-2000) Dynamic actin-based epithelial adhesion and cell matching during *Drosophila* dorsal closure. *Curr. Biol.* **10**: 1420–1426
6. Martin, P and Lewis, J (12-11-1992) Actin cables and epidermal movement in embryonic wound healing. *Nature.* **360**: 179–183
7. McCluskey, J and Martin, P (1995) Analysis of the tissue movements of embryonic wound healing–DiI studies in the limb bud stage mouse embryo. *Dev. Biol.* **170**: 102–114
8. Brock, J, Midwinter, K, Lewis, J, and Martin, P (1996) Healing of incisional wounds in the embryonic chick wing bud: Characterization of the actin purse-string and demonstration of a requirement for Rho activation. *J. Cell Biol.* **135**: 1097–1107
9. Wood, W, Jacinto, A, Grose, R, Woolner, S, Gale, J, Wilson, C, and Martin, P (2002) Wound healing recapitulates morphogenesis in *Drosophila* embryos. *Nat. Cell Biol.* **4**: 907–912
10. Kolega, J (1986) Effects of mechanical tension on protrusive activity and microfilament and intermediate filament organization in an epidermal epithelium moving in culture. *J. Cell Biol.* **102**: 1400–1411
11. Olson, MF (2004) Contraction reaction: Mechanical regulation of Rho GTPase. *Trends Cell Biol.* **14**: 111–114
12. Danjo, Y and Gipson, IK (1998) Actin 'purse string' filaments are anchored by E-cadherin-mediated adherens junctions at the leading edge of the epithelial wound, providing coordinated cell movement. *J. Cell Sci.* **111 (Pt 22)**: 3323–3332
13. Quinlan, MP and Hyatt, JL (1999) Establishment of the circumferential actin filament network is a prerequisite for localization of the cadherin-catenin complex in epithelial cells. *Cell Growth Differ.* **10**: 839–854
14. Kovacs, EM, Goodwin, M, Ali, RG, Paterson, AD, and Yap, AS (5-3-2002) Cadherin-directed actin assembly: E-cadherin physically associates with the Arp2/3 complex to direct actin assembly in nascent adhesive contacts. *Curr. Biol.* **12**: 379–382
15. Steinhardt, RA, Bi, G, and Alderton, JM (21-1-1994) Cell membrane resealing by a vesicular mechanism similar to neurotransmitter release. *Science.* **263**: 390–393
16. Bement, WM, Mandato, CA, and Kirsch, MN (3-6-1999) Wound-induced assembly and closure of an actomyosin purse string in Xenopus oocytes. *Curr. Biol.* **9**: 579–587
17. O'Kane, S and Ferguson, MW (1997) Transforming growth factor beta S and wound healing. *Int. J. Biochem. Cell Biol.* **29**: 63–78
18. Shah, M, Foreman, DM, and Ferguson, MW (25-1-1992) Control of scarring in adult wounds by neutralising antibody to transforming growth factor beta. *Lancet.* **339**: 213–214

CHAPTER 4.4

INVAGINATION AND EVAGINATION: The Making and Shaping of Folds and Tubes

Epithelial sheets are essentially two-dimensional structures. They come to occupy and frequently to dominate three-dimensional space by a process of controlled bending and deformation, a biological origami made far more powerful than its paper-based equivalent by the fact that living sheets are not limited to folding but can stretch and shrink within their planes as well. This feature allows two-dimensional sheets to be transformed into tubes by invagination or evagination; these transformations are the basis of three of the five ways in which tubes can be made (see Figure 4.4.1).

INVAGINATION

Invagination♣ is the production of a tube by local in-pushing of a surface, as when an extended finger is pressed radially against a partially inflated balloon. There are two main forms of invagination: axial and orthogonal. Axial invagination occurs at a point and can only produce a dent or a tube; the surface pushes inward directly down the axis of this. Orthogonal invagination occurs along a line rather than at a single point and produces a trough, the axis of which is parallel to the original surface and therefore at right angles to the direction of invagination.

AXIAL INVAGINATION: The Sea Urchin Archenteron

One of the anatomically simplest examples of invagination is provided by the early embryos of sea urchins. The epithelium of the vegetal plate flattens and then invaginates so that a hollow column of epithelium, the archenteron, invades the blastocoelic cavity

♣"Invagination" derives directly from the Latin *in*, meaning "in," and *vagina*, meaning "sheath," and does not imply any analogy with development of the vagina (birth canal). Ironically, the vagina is an example of a tube that does *not* develop by invagination but rather by a complex series of fusions in the cloacal area. It gains its name directly from the Latin (sheath), probably because Roman soldiers' slang for penis was *gladius*, meaning sword.

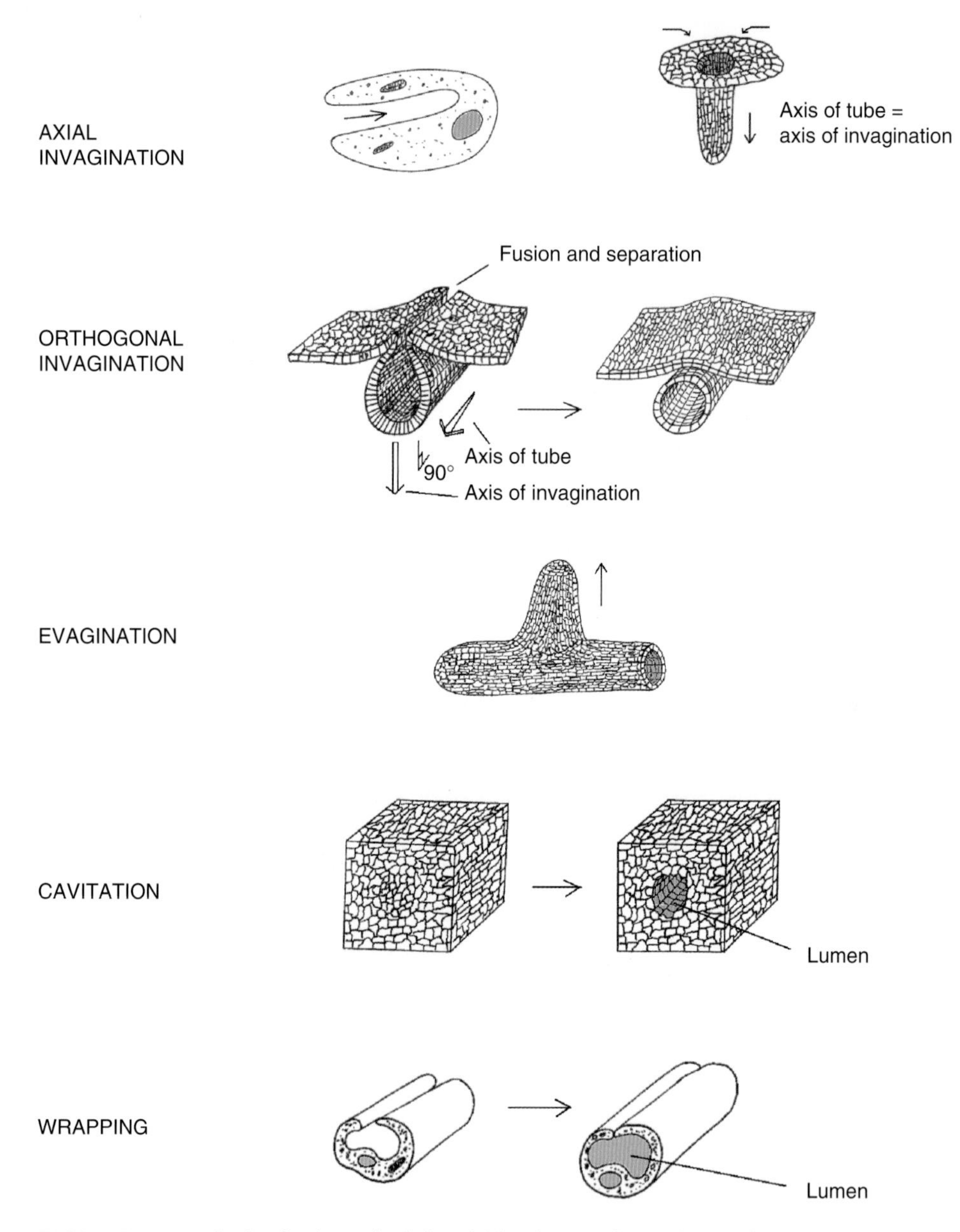

FIGURE 4.4.1 The five basic methods by which tubes can be made. Axial invagination can take place at the scale of an individual cell or of an epithelial sheet and proceeds by infolding of the cell or sheet surface along what will be the axis of the tube. Orthogonal invagination proceeds by infolding of a line of epithelium in a direction at right angles to what will be the axis of the tube to form a trough, the lips of which then fuse and separate from their parent epithelium. Evagination is similar to invagination but involves outfolding rather than infolding. Cavitation creates a tube by clearing cells away from a line in an initially solid tissue, either by cell rearrangement or cell death. Wrapping consists of a cell curling up on itself; it is topologically distinct from orthogonal evagination because wrapping involves adhesion of the opposite sides of the cell but does not involve their fusion.

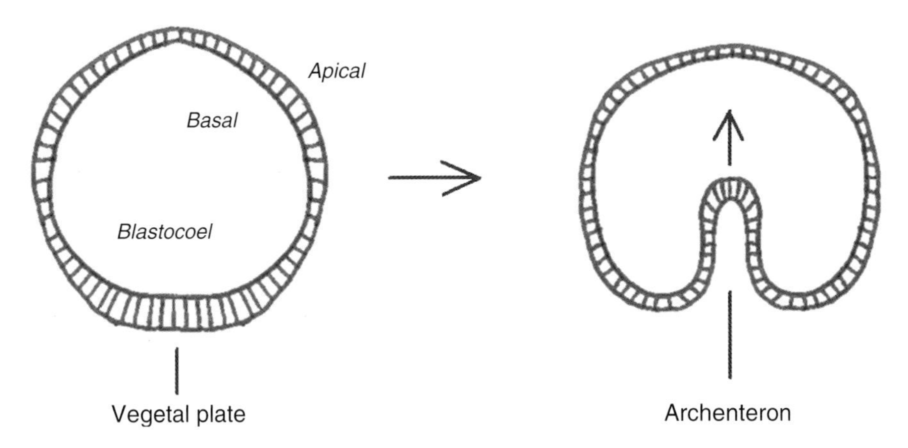

FIGURE 4.4.2 Invagination of the archenteron from the vegetal plate of a sea urchin.

of the embryo (see Figure 4.4.2). The invagination is driven locally and will take place in isolated vegetal plates, a fact that rules out any mechanisms based on pressure in the blastocoel (i.e., the archenteron cannot simply be "sucked in"). It also rules out mechanisms based on tangential compression in the growing epithelium.[1] There is a local increase in cell division in the area around the time of invagination, but treatment of embryos with the DNA synthesis inhibitor aphidicollin does not prevent invagination from taking place, suggesting that cell multiplication (see Chapter 5.2) cannot be the primary mechanism.[2]

Cells in the vegetal plate show very strong expression of actin/myosin filaments that run mainly circumferentially under their junctions. In principle, contraction of these filaments could squeeze cytoplasm from the apical to the basal end of each cell and therefore expand it, forcing the basal surface of the epithelium to bow inward (see Figure 4.4.3). Finite element♣ computer models of this hypothesis produce very realistic behaviour.[1] Despite this, direct tests of the role of actin filaments in invagination fail to show any need for actin at all. Invagination begins normally in the presence of the actin-depolymerizing drug cytochalasin D.[ref3] As well as arguing against an actin-based mechanism for invagination, this observation underlines an important warning: the fact that a hypothetical model works well on a computer does not prove that it is the actual mechanism used in life.

The most likely mechanism for invagination, based on data to hand at the time of writing, relies more on the mechanics of extracellular matrix than of the cells themselves. The apical (external) surface of the epithelial cells is covered in a thick extracellular matrix consisting of an inner "apical lamina" and an outer, clear, hyaline layer. The matrix seems to be important in driving invagination, and invagination is blocked by monoclonal antibodies raised against uncharacterized components of the apical lamina.[4] As invagination begins, cells of the vegetal plate secrete a large chondroitin sulphate proteoglycan into the apical matrix but not into the hyaline layer. Chondroitin sulphates can form gels that swell as they absorb water, so that the apical layer of the

♣Finite element models consider a complex surface to be made up of a large number of small and simple elements such as triangles, the deformations of which, and the effect of those deformations on the shape of the entire surface, are comparatively simple to compute.

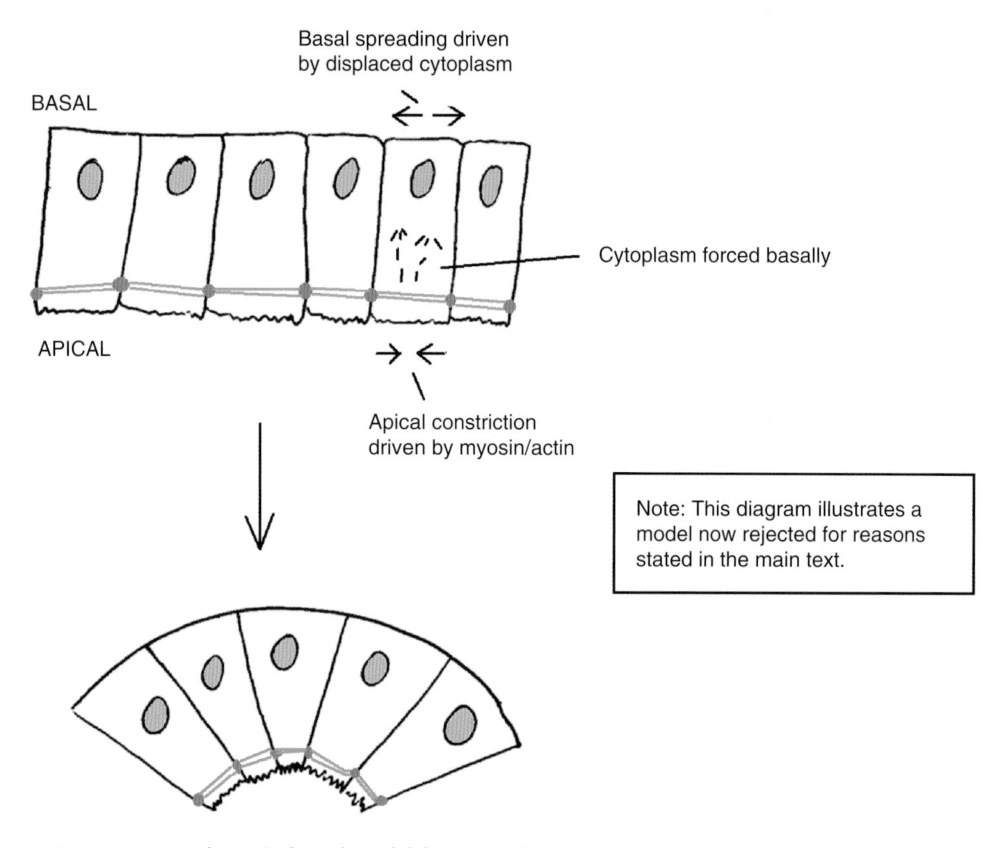

FIGURE 4.4.3 The actin-based model for sea urchin invagination. The ability of sea urchins to invaginate even in the presence of cytochalasins argues against acceptance of this model.

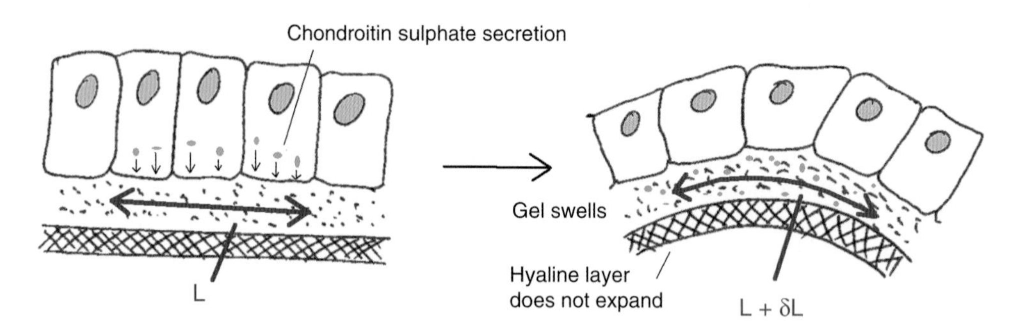

FIGURE 4.4.4 The gel swelling model for sea urchin invagination. Secretion and swelling of a chondroitin sulphate gel expands the apical matrix without expanding the hyaline layer. The extra length, marked δL in the diagram, forces the combination to bend in much the way that the bimetallic strip in a thermostat is forced to bend by the unequal expansion of its two layers.

matrix expands while the overlying hyaline layer does not. This differential expansion then forces the matrix to buckle inwards (see Figure 4.4.4). Finite-element modelling confirms that this mechanism could work in principle, but only if the epithelium itself deforms significantly more easily than (has a $\geq$60 times lower elastic modulus than) the matrix overlying it.[1] Indirect testing of the hypothesis in living embryos shows that blocking the secretion of chondroitin sulphate blocks invagination. On the other hand, inducing the chondroitin sulphate earlier, by pharmacological manipulation of Ca^{2+} signalling, causes a corresponding advance of the timing of invagination.[3] There has not yet been any direct evidence of gel swelling in this system, though, and the elastic moduli of the layers involved have not been measured.

ORTHOGONAL INVAGINATION: Neural Tube Formation in Vertebrates

The neural tube of vertebrates, which forms the brain, spinal cord, and neural crest, derives from an area of the dorsal ectoderm epithelium called the "neural plate." The events that shape it first into a trough, by orthogonal invagination, and then into a neural tube (see Figure 4.4.5) are broadly similar in all vertebrate classes, although there are inevitable differences in detail caused by different embryonic geometries.

The first event of neural tube formation is an apico-basal thickening of the epithelium; such a thickening is characteristic of regions of ectoderm that are about to invaginate, and the thickened area is usually called a "placode." In birds, at least, this thickening will take place even when the presumptive neural plate is isolated from its surrounding tissues, so it must rely on strictly local mechanisms.[5] The apico-basal elongation of neural plate cells is associated with the presence of arrays of microtubules extending from base to apex of the cell, and it can be inhibited by microtubule-depolymerizing agents.[6,7,8,9] For a further discussion of this, see Chapter 2.3.

The next event is the formation of curvature in the neural plate, a process sometimes referred to as "kinking."[10] This curving is localized to specific "hinge points" and generates convex or concave curvature depending on where it takes place. At the edges of the neural plate, curvature is convex (from the point of view of the outside of the embryo), whereas at the middle of the plate it is concave (see Figure 4.4.6). A key feature of the hinge points is that cells within them acquire a distinct wedge shape, being constricted apically in the case of the median hinge point and basally in the case of the dorsolateral hinge points.

Cells in the region of the presumptive concave hinge points, red in Figure 4.4.6, express the apically located actin-binding protein Shroom just before and during neural tube folding.[12,13] Several experimental results suggest that Shroom is a key regulator of the folding process. Neural folding fails in *Xenopus tropicalis* embryos that express a dominant-negative version of the protein Shroom754–1108 or in embryos treated with morpholino antisense oligonucleotides that target the endogenous Shroom message.[12] On the other hand, forcing the expression of Shroom by transfection of epithelial cells is sufficient to cause them to accumulate actin fibres in their apical regions and to undergo pronounced apical constriction.[12] This apical constriction causes cells transfected with Shroom to take on an obvious wedge shape, even when in a flat culture dish surrounded by other epithelial cells. This key observation indicates that formation of a wedge shape by cells in the embryo is probably an active consequence of Shroom expression and not just a passive accommodation of cell shape to being in a curved tissue. It suggests that acquisition of a wedge shape may be a mechanism for,

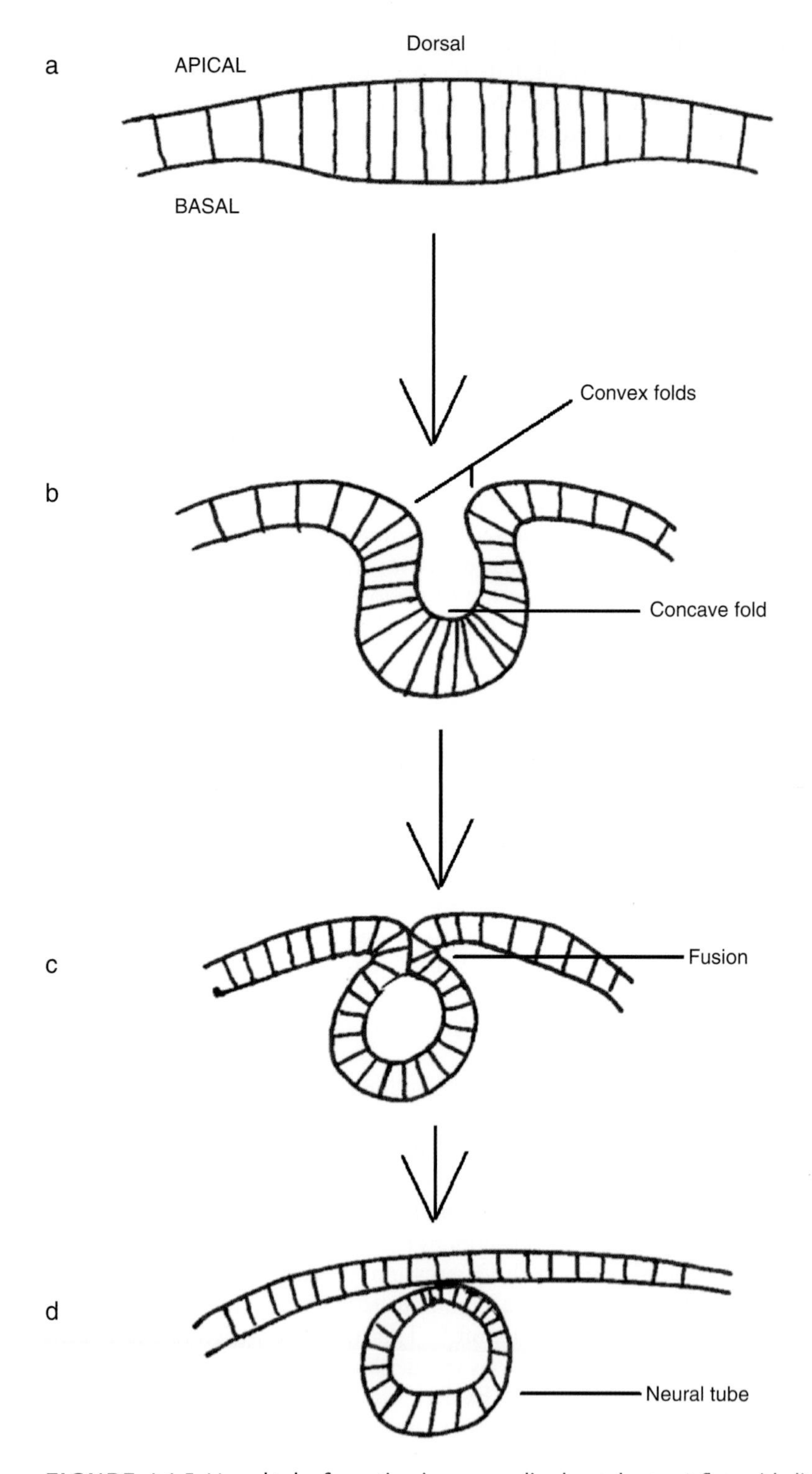

FIGURE 4.4.5 Neural tube formation in a generalized vertebrate. A flat epithelium folds to create a trough, the lips of which then fuse to form a tube. In chicks, mice, and salamanders, the epithelium is a monolayer as depicted here, but in some species, such as the frog *Xenopus laevis*, it has more layers.

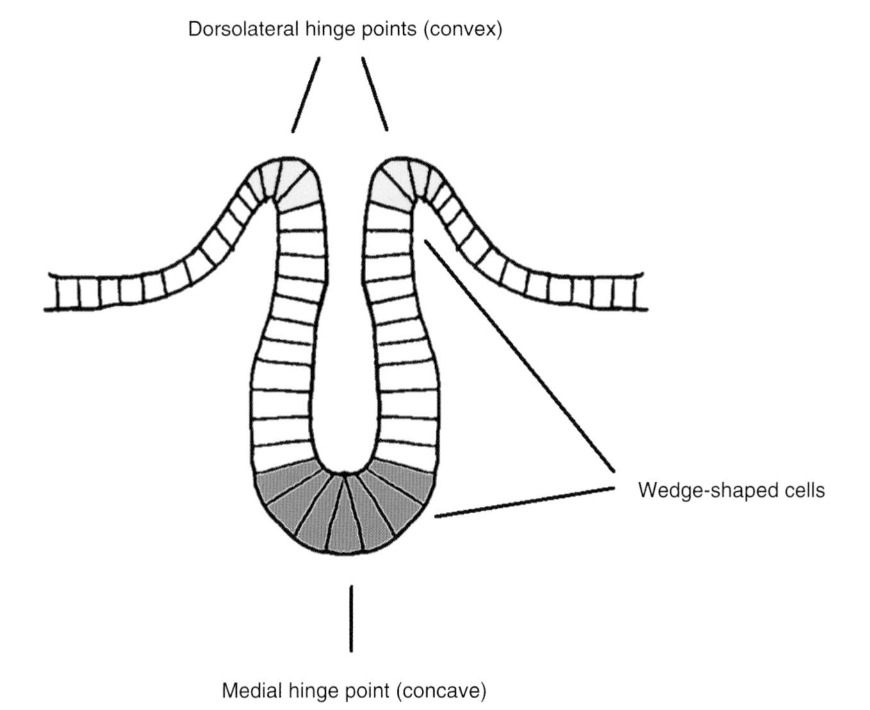

FIGURE 4.4.6 The neural plate folds mainly at hinge points, drawn here from a micrograph of avian neurulation.[11] Note that the wedge-shaped cells marked in red and yellow have opposite directions of wedging with respect to their apico-basal polarity. For clarity, this diagram exaggerates the size of cells, and in reality there are many more of them, densely packed but still in a single layer.

rather than a consequence of, tissue bending. This conclusion is supported by the observation that cell wedging takes place even in isolated midline strips of neural plate so cannot be the result of bending pressures from tissues that flank that region.[14]

The accumulation of actin fibres in the apical domains of Shroom-expressing cells, and the fact that Shroom can bind to F-actin,[13] suggests that myosin-mediated actin contraction of the apical end of the cell may be the mechanism for wedging (see Figure 4.4.7); this is the view presented in most textbooks.[15,16] Actin filaments are certainly required for neural tube formation in a variety of species, including amphibians, chicks, and rats, and inhibition of actin filament formation using cytochalasins prevents neural tube formation.[17,18] Careful studies of the events involved, though, show that cytochalasins do not prevent the acquisition of a wedge shape at the hinge points but rather interfere only with later development.[19] In mammals, formation of the cranial neural tube is inhibited by cytochalasin D, but formation of the more posterior neural tube is apparently unaffected by the drug.[20] This variation probably warns us that actin-based contraction is not the fundamental mechanism for wedge formation. Also, where actin-myosin contraction is clearly an agent of epithelial morphogenesis—for example, in the closure of holes (see Chapter 4.3)—Rho-mediated organization of actin-myosin filaments is essential. Shroom-induced cell wedging has no need of Rho,[21] though, and neural tube formation fails in 30 percent of mice that are missing a

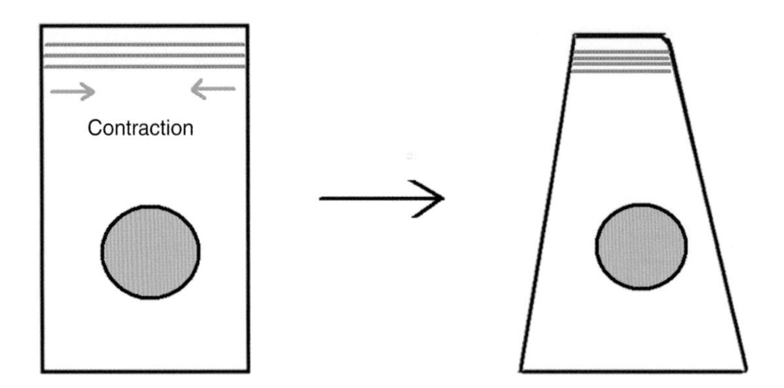

FIGURE 4.4.7 The popular actin-myosin contraction model for cell wedging. For reasons discussed in the main text, this should be treated with great caution until more data are available.

p190RhoGAP, a *deactivator* of Rho.[22] This is precisely the opposite of what one would expect if the prevalent view of wedging by actin-myosin contraction were correct.

Although Shroom does not seem to act via Rho, it does require a different small GTPase of the Ras family, Rap1.[ref12] The connection between Shroom and Rap1 and the events downstream of Rap1 have not yet been worked out in the neural tube, but data from other systems suggest that cell adhesion systems may be important. Integrin-mediated adhesion of some cell types, such as T cells, is enhanced by Rap1 through "inside-out" signalling (control of adhesion by signals on the inside of the cell) and fails in the absence of the molecule.[23,24] In *Drosophila melanogaster*, Rap1 is involved in reassembling adherens junctions after cell division,[25] and it is known to bind the protein Canoe,[26,27] which is located on the inside of adherens junctions, where it binds to the junctional protein Tamou (homologous to vertebrate ZO1).[28] Rap1 activity is important in determining the way that adherens junctions space out around a cell, and cells mutant for Rap1 show an abnormal grouping of junctions to one part of the cell instead of an even distribution around the circumference of the apical end.[29]

This last-mentioned activity of Rap1, control of the spacing of adherens junctions, might be very important in the control of cell wedging since a change in the spacing of existing cell-cell junctions might be expected to expand or constrict the apical end of the cell. Unfortunately, the mechanism by which Rap1 regulates spacing between junctions is not yet known. It may not involve the usual actin-myosin contraction system because, in at least some systems, adhesions promoted by Rap1 are stable and functional even in the absence of actin microfilament systems.[30]

Until more data are available, it is important to note that the observation that Shroom requires Rap1 to drive cell wedging does not prove that Rap1 drives wedging. It may simply permit cell junctions to be rearranged in response to the real driver of apical constriction, in the way that it is required to permit junctional rearrangements during mitosis in *D. melanogaster*.

The *convex* hinge-point cells of the neural tube do not express Shroom, presumably because they have to contract at their basal rather than apical ends, and no gene specific to them has yet been identified. In mice, Shroom is expressed under the regulation of the secreted signalling molecule Sonic Hedgehog (Shh) and in *shh*$^{-/-}$ mutant embryos the medial, concave hinge point fails to form, but the neural tube still forms and it closes

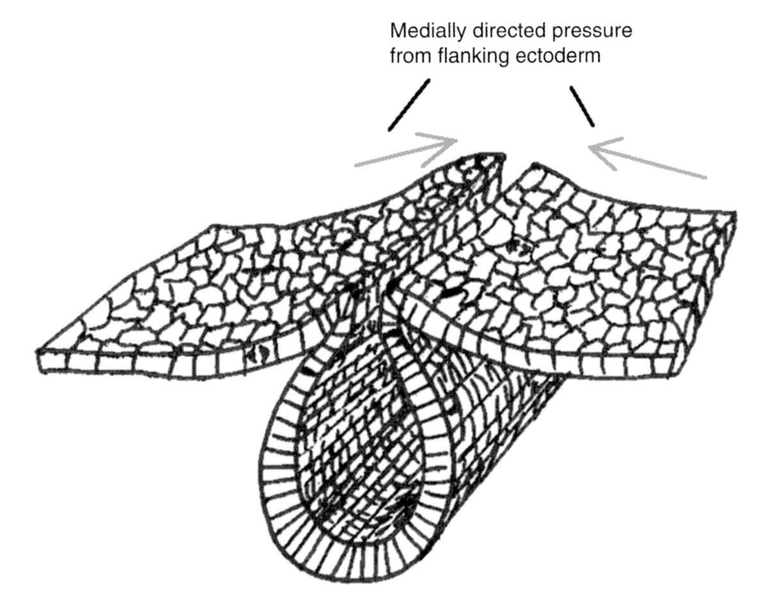

FIGURE 4.4.8 Meeting of the dorsal parts of the neural tube (the neural folds) is driven at least in part by medial expansion of the surrounding ectoderm, driven by cell flattening and orientated mitosis.

under the action of the dorsolateral, convex hinge points alone.[31] This strongly suggests that basal constriction by the dorsolateral hinge cells is an active process rather than being a passive consequence of bending by the medial hinge point.

The elevation of the neural folds and their subsequent meeting contrasts with plate thickening and cell wedging in that it seems not to be driven by purely local mechanisms and it will not take place if the neural plate is isolated from its flanking ectoderm.[32] The critical non-local influence may be simple mechanical pressure from nearby ectoderm that expands medially and pushes the edges of the neural folds upward and together (see Figure 4.4.8). The medial expansion of the ectoderm seems to be driven by cells contracting apico-basally so that the once-cuboidal epithelium becomes squamous, and its surface area increases.[33] Its medial advance may also involve orientated mitoses (see Chapter 5.2) in the medio-lateral axis.[34]

INVAGINATION WITHOUT TUBE FORMATION

Epithelial folding is often used as a means of increasing the surface area for substance exchange within an organ, with no aim of subsequent tube formation. The mammalian colon develops from an initially smooth-walled tube of stratified (multi-layered) epithelium to become highly folded. The process of colon folding will take place if sections of hindgut are removed from the foetus and placed in culture. The mechanism driving folding must therefore be endogenous to the organ rather than relying, for example, on pressure from elsewhere in the embryo. As in neurulation, the process seems to begin when stripes of F-actin fibres appear at putative hinge points, spaced irregularly along the epithelium. These zones fold downwards, creating the folded structure of the mature epithelium. The process is dependent on actin and will not take place in the presence of cytochalasin D.[35] Although Shroom is known to be expressed in the gut during early development, no studies of its expression during late development, around the time of colon folding, have yet been published, so it is not clear whether folding in the colon and the neural tube are driven by exactly the same mechanisms.

EVAGINATION OF IMAGINAL DISCS

Evagination is the morphological reverse of invagination and results in the outpushing of a dome or a tube. Whereas invagination has the basal ends of cells leading the advance of a fold or tube, evagination is led by the apical ends of cells.

One of the most striking and best-studied examples of evagination is responsible for the limb imaginal discs of *D. melanogaster* transforming from almost flat plates of epithelia to limb-shaped tubes. Imaginal discs are epithelial sheets that develop inside the body during embryogenesis and remain hidden inside the larval stages of the life cycle. When the hormone ecdysone reaches a high enough concentration to trigger metamorphosis, the imaginal discs undergo spectacular morphogenetic movements to form structures such as wings and legs, characteristic of the adult body plan.

The leg imaginal disc is a typical epithelium in that it consists of a sheet of cells linked to each other by adherens junctions towards the apical end of the cell and linked to a basement membrane by integrin-containing junctions on the basal surface. Both the adherens cell-cell junctions and the integrin cell-matrix junctions are associated with large arrays of actin microfilaments.[36] Early in the development of a disc, the morphology of the epithelial cells is unremarkable. As the disc grows, however, the cells change shape, particularly in the regions that will give rise to the proximal and dorsal regions of the leg. The previously columnar cells become highly anisotropic, showing a pronounced elongation along an axis tangential to the edge of the disc and shortening along the radial axis (see Figure 4.4.9). Furthermore, this anisotropy is more extreme at each cell's apical end than at its basal end, which has the result that the set of cells with which the apical end of a given cell makes contact overlaps but is not identical with the set of cells with which its basal end makes contact.[36] As these shape changes take place, the epithelium begins to fold, although it does not yet evert.

At the onset of metamorphosis, the shape of the anisotropic cells changes once more, and the circumferential elongation disappears so that the cells are again isotropic. This has the effect of elongating the tissue in a radial direction and, because the disc has already begun to evert so that its centre is now at the tip of what will be the leg, elongation in a radial direction with respect to the disc becomes elongation along the axis of what is now an emerging leg (see Figure 4.4.9). There must also presumably be a degree of neighbour exchange because without it the number of cells around the circumference of the leg at different points of its length would vary with the number around the corresponding radius of the original disc, which would be proportional to the radius itself.

Unlike the axial invagination of sea urchins and the orthogonal invagination of the frog neural tube, evagination in the leg disc does seem to depend on myosin-mediated contraction of actin fibre bundles. Actin is expressed strongly in cables under the adherens junctions of the disc and is required for morphogenesis; evagination fails

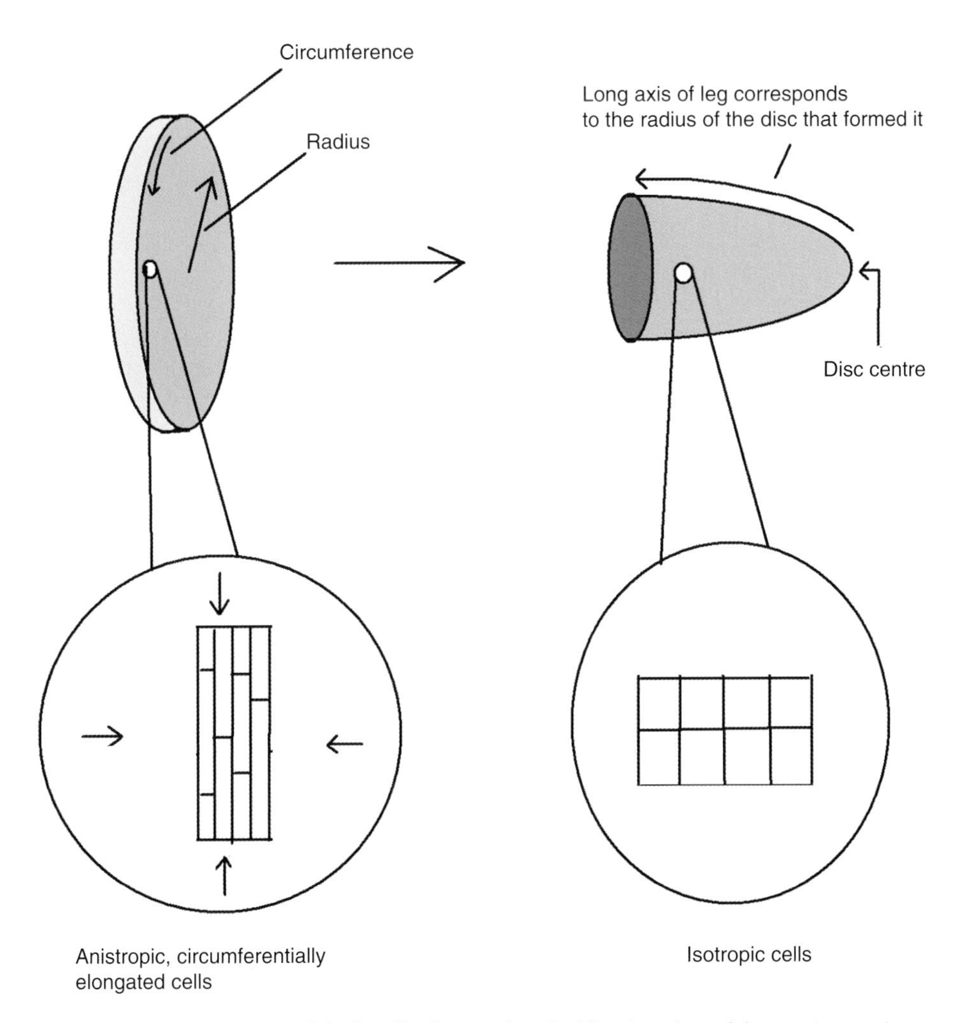

FIGURE 4.4.9 Eversion of the leg disc is associated with relaxation of the previous anisotropy of cells so that they contract in a circumferential direction and elongate in a direction corresponding to the radius of the original disc, a direction that now corresponds to the long axis of the leg. The shapes of the disc and leg have been simplified in this diagram in order to make the correspondence of old and new axes clear; both are in reality rather folded.

if discs are first treated with the actin-depolymerizing drug cytochalasin B.[37] Myosin is also required, and this has been demonstrated by an ingenious technique. The requirement for myosin in the discs cannot be tested by a simple gene knockout because embryos carrying the *spaghetti-squash* (*sqh*) mutation, which disrupts the gene encoding the myosin regulatory light chain, fail to develop far enough for imaginal disc development to be studied. The mutants can be rescued, however, by a construct encoding wild-type myosin regulatory light chain driven by a heat-shock promoter as long as they are subjected to a daily heat shock. If embryos are grown with daily myosin-transcription-inducing heat shocks up to a particular stage of development and then the heat shock is omitted, myosin is made available for early development but can be taken away by the time discs evaginate. When the cessation of heat shock is timed to coincide with leg disc eversion, eversion fails. This suggests that myosin regulatory light chain is important to the process.[38] The small GTPase Rho, which activates myosin regulatory light chain via Rho kinase in many systems, is also required for leg eversion.[39] This suggests that the evaginating leg may assemble its contractile machinery through the "classical" Rho pathway used by so many cells.

Activation of the Rho pathway in the discs seems to be under the control of a multi-functional protein, Stubbled-stubbloid. When disc cells receive the metamorphosis-inducing hormone ecdysone, they activate the expression of the *stubble-stubbloid* (*sb-sbd*) gene, which encodes a transmembrane serine protease with an extracellular protease domain and receptor activity.[40] The receptor activity seems, through genetic evidence at least, to control Rho activation and therefore to control the formation of a contractile actin-myosin system (see Figure 4.4.10), although the ligand that acts on Stubbled-stubbloid has not yet been identified.[41]

The proteolytic activity of the Stubbled-stubbloid may play a direct role in morphogenesis as well. Cells of the imaginal disc are attached by their basal surfaces to extracellular matrix, which will resist the deformation of the disc and the exchange of neighbours. Cleavage of basement membrane components such as collagen IV is a feature of disc eversion,[43] and inhibitors of serine proteases block disc eversion,[42] although it is not clear whether this reflects an essential requirement for matrix degradation or for serine protease activity in signalling events. The Stubbled-stubbloid protein is a protease that can, in principle at least, activate itself by proteolytic cleavage and,

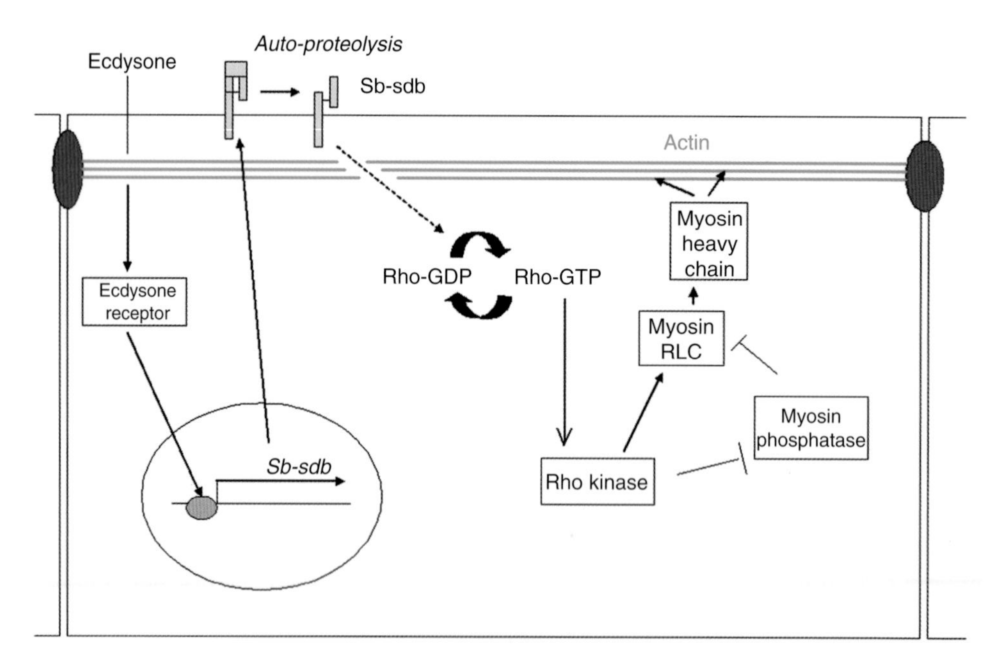

FIGURE 4.4.10 The current view of the pathway leading from ecdysone reception to contraction of the apical ends of leg disc cells (adapted from the discussions in C.A. Bayer et al.[41] and R.E. Ward et al[42]).

once activated, be involved in the degradation of extracellular matrix and a consequent increase in the flexibility of the disc. It should be borne in mind, though, that the protein is located mainly on the apices of cells, at which place it is arguably an unlikely candidate for being a major modifier of the basement membrane. Other proteases may therefore be involved.

There are many other examples of invagination and evagination, but these have not received as much close study as the systems described here. In each case, complex and large-scale changes in the shape of tissue are again the result of simple changes at the cell level, which do not require cells to have any special "knowledge" of the grand plan of morphogenesis.

Reference List

1. Davidson, LA, Koehl, MA, Keller, R, and Oster, GF (1995) How do sea urchins invaginate? Using biomechanics to distinguish between mechanisms of primary invagination. *Development.* **121**: 2005–2018
2. Stephens, L, Hardin, J, Keller, R, and Wilt, F (1986) The effects of aphidicolin on morphogenesis and differentiation in the sea urchin embryo. *Dev. Biol.* **118**: 64–69
3. Lane, MC, Koehl, MA, Wilt, F, and Keller, R (1993) A role for regulated secretion of apical extracellular matrix during epithelial invagination in the sea urchin. *Development.* **117**: 1049–1060
4. Burke, RD, Myers, RL, Sexton, TL, and Jackson, C (1991) Cell movements during the initial phase of gastrulation in the sea urchin embryo. *Dev. Biol.* **146**: 542–557
5. Schoenwolf, GC (22-10-1988) Microsurgical analyses of avian neurulation: Separation of medial and lateral tissues. *J. Comp. Neurol.* **276**: 498–507
6. Karfunkel, P (1971) The role of microtubules and microfilaments in neurulation in Xenopus. *Dev. Biol.* **25**: 30–56
7. Handel, MA and Roth, LE (1971) Cell shape and morphology of the neural tube: Implications for microtubule function. *Dev. Biol.* **25**: 78–95
8. Karfunkel, P (1972) The activity of microtubules and microfilaments in neurulation in the chick. *J. Exp. Zool.* **181**: 289–301
9. Ferm, VH (1963) Colchicine teratogenesis in hamster embryos. *Proc. Soc. Exp. Biol. Med.* **112**: 775–778
10. Lawson, A, Anderson, H, and Schoenwolf, GC (1-2-2001) Cellular mechanisms of neural fold formation and morphogenesis in the chick embryo. *Anat. Rec.* **262**: 153–168
11. Colas, JF and Schoenwolf, GC (2001) Towards a cellular and molecular understanding of neurulation. *Dev. Dyn.* **221**: 117–145
12. Haigo, SL, Hildebrand, JD, Harland, RM, and Wallingford, JB (16-12-2003) Shroom induces apical constriction and is required for hingepoint formation during neural tube closure. *Curr. Biol.* **13**: 2125–2137
13. Hildebrand, JD and Soriano, P (24-11-1999) Shroom, a PDZ domain-containing actin-binding protein, is required for neural tube morphogenesis in mice. *Cell.* **99**: 485–497
14. Schoenwolf, GC (22-10-1988) Microsurgical analyses of avian neurulation: Separation of medial and lateral tissues. *J. Comp. Neurol.* **276**: 498–507
15. Alberts, B, Bray, D, Lewis, J, Raff, M, Roberts, K, and Watson, JD (1994) Molecular biology of the cell. **3rd**: (Garland)
16. Slack, JMW (2001) Essential developmental biology. (Blackwell)
17. Lee, HY and Kalmus, GW (1976) Effects of cytochalasin B on the morphogenesis of explanted early chick embryos. *Growth.* **40**: 153–162
18. Morriss-Kay, G and Tuckett, F (1985) The role of microfilaments in cranial neurulation in rat embryos: Effects of short-term exposure to cytochalasin D. *J. Embryol. Exp. Morphol.* **88**: 333–348

19. Schoenwolf, GC, Folsom, D, and Moe, A (1988) A reexamination of the role of microfilaments in neurulation in the chick embryo. *Anat. Rec.* **220**: 87–102
20. Ybot-Gonzalez, P and Copp, AJ (1999) Bending of the neural plate during mouse spinal neurulation is independent of actin microfilaments. *Dev. Dyn.* **215**: 273–283
21. Haigo, SL, Hildebrand, JD, Harland, RM, and Wallingford, JB (16-12-2003) Shroom induces apical constriction and is required for hingepoint formation during neural tube closure. *Curr. Biol.* **13**: 2125–2137
22. Brouns, MR, Matheson, SF, Hu, KQ, Delalle, I, Caviness, VS, Silver, J, Bronson, RT, and Settleman, J (2000) The adhesion signalling molecule p190 RhoGAP is required for morphogenetic processes in neural development. *Development.* **127**: 4891–4903
23. Bos, JL, de Bruyn, K, Enserink, J, Kuiperij, B, Rangarajan, S, Rehmann, H, Riedl, J, de Rooij, J, van Mansfeld, F, and Zwartkruis, F (2003) The role of Rap1 in integrin-mediated cell adhesion. *Biochem. Soc. Trans.* **31**: 83–86
24. McLeod, SJ, Shum, AJ, Lee, RL, Takei, F, and Gold, MR (26-3-2004) The Rap GTPases regulate integrin-mediated adhesion, cell spreading, actin polymerization, and Pyk2 tyrosine phosphorylation in B lymphocytes. *J. Biol. Chem.* **279**: 12009–12019
25. Tepass, U (2002) Adherens junctions: New insight into assembly, modulation and function. *Bioessays.* **24**: 690–695
26. Linnemann, T, Geyer, M, Jaitner, BK, Block, C, Kalbitzer, HR, Wittinghofer, A, and Herrmann, C (7-5-1999) Thermodynamic and kinetic characterization of the interaction between the Ras binding domain of AF6 and members of the Ras subfamily. *J. Biol. Chem.* **274**: 13556–13562
27. Boettner, B, Harjes, P, Ishimaru, S, Heke, M, Fan, HQ, Qin, Y, Van Aelst, L, and Gaul, U (2003) The AF-6 homolog canoe acts as a Rap1 effector during dorsal closure of the *Drosophila* embryo. *Genetics.* **165**: 159–169
28. Takahashi, K, Matsuo, T, Katsube, T, Ueda, R, and Yamamoto, D (1998) Direct binding between two PDZ domain proteins Canoe and ZO-1 and their roles in regulation of the jun N-terminal kinase pathway in *Drosophila* morphogenesis. *Mech. Dev.* **78**: 97–111
29. Knox, AL and Brown, NH (15-2-2002) Rap1 GTPase regulation of adherens junction positioning and cell adhesion. *Science.* **295**: 1285–1288
30. Bos, JL, de Rooij, J, and Reedquist, KA (2001) Rap1 signalling: Adhering to new models. *Nat. Rev. Mol. Cell Biol.* **2**: 369–377
31. Ybot-Gonzalez, P, Cogram, P, Gerrelli, D, and Copp, AJ (2002) Sonic hedgehog and the molecular regulation of mouse neural tube closure. *Development.* **129**: 2507–2517
32. Schoenwolf, GC (22-10-1988) Microsurgical analyses of avian neurulation: Separation of medial and lateral tissues. *J. Comp. Neurol.* **276**: 498–507
33. Schoenwolf, GC and Alvarez, IS (1991) Specification of neurepithelium and surface epithelium in avian transplantation chimeras. *Development.* **112**: 713–722
34. Sausedo, RA, Smith, JL, and Schoenwolf, GC (19-5-1997) Role of nonrandomly oriented cell division in shaping and bending of the neural plate. *J. Comp. Neurol.* **381**: 473–488
35. Colony, PC and Conforti, JC (1993) Morphogenesis in the fetal rat proximal colon: Effects of cytochalasin D. *Anat. Rec.* **235**: 241–252
36. Condic, ML, Fristrom, D, and Fristrom, JW (1991) Apical cell shape changes during Drosophila imaginal leg disc elongation: A novel morphogenetic mechanism. *Development.* **111**: 23–33
37. Fristom, D and Fristom, JW (1975) The mechanism of evagination of imaginal discs of *Drosophila melanogaster*. 1. General considerations. *Dev. Biol.* **43**: 1–23
38. Edwards, KA and Kiehart, DP (1996) *Drosophila* nonmuscle myosin II has multiple essential roles in imaginal disc and egg chamber morphogenesis. *Development.* **122**: 1499–1511
39. Halsell, SR, Chu, BI, and Kiehart, DP (2000) Genetic analysis demonstrates a direct link between rho signalling and nonmuscle myosin function during Drosophila morphogenesis. *Genetics.* **155**: 1253–1265

40. Appel, LF, Prout, M, Abu-Shumays, R, Hammonds, A, Garbe, JC, Fristrom, D, and Fristrom, J (1-6-1993) The *Drosophila* Stubble-stubbloid gene encodes an apparent transmembrane serine protease required for epithelial morphogenesis. *Proc. Natl. Acad. Sci. U.S.A.* **90**: 4937–4941
41. Bayer, CA, Halsell, SR, Fristrom, JW, Kiehart, DP, and von Kalm, L (2003) Genetic interactions between the RhoA and Stubble-stubbloid loci suggest a role for a type II transmembrane serine protease in intracellular signalling during *Drosophila* imaginal disc morphogenesis. *Genetics.* **165**: 1417–1432
42. Pino-Heiss, S and Schubiger, G (1989) Extracellular protease production by *Drosophila* imaginal discs. *Dev. Biol.* **132**: 282–291
43. Fessler, LI, Condic, ML, Nelson, RE, Fessler, JH, and Fristrom, JW (1993) Site-specific cleavage of basement membrane collagen IV during *Drosophila* metamorphosis. *Development.* **117**: 1061–1069

CHAPTER 4.5

EPITHELIAL MORPHOGENESIS: Fusion of Epithelia

Most of the epithelial tissues of the body connect to at least one other epithelial tissue to become part of a system. Frequently, these tissues form by growth and differentiation of regions of a continuous embryonic epithelium so that they are automatically connected in the right way. For example, the Bowman's capsule, proximal convoluted tubule, loop of Henlé, and distal convoluted tubule of a nephron in the human kidney form by differentiation of successive zones of a common progenitor tube so that no additional morphogenetic tricks are needed to ensure that one leads into the other. Similarly, the respiratory alveoli, bronchioles, bronchi, and trachea of the human lung form by differentiation of zones of a branching epithelium and are therefore connected automatically. Not all connections can be made this way, however, and some connections are made by the fusion of epithelial components that formed separately from one another.

In many cases, the need to form an epithelial structure by the fusion of existing epithelial structures is forced on the embryo by the laws of mathematics. The simple epithelial sphere of a typical early embryo is an object of topological genus 0, meaning that it is not pierced by any holes. Body plans such as those of cnidaria (jellyfish, hydroids, and so on) that have a blind-ended gut and no other openings are also of topological genus 0: it would be possible, in principle, to model them from a hollow sphere of flexible clay without having to introduce tears or joins (see Figure 4.5.1a). The metazoan body plan, in which the gut has two openings to the outside world is, however, an object of topological genus 1: it is mathematically impossible to derive it from a hollow sphere without cutting and joining (see Figure 4.5.1b). The foregut and the hindgut therefore have to meet and join together somewhere along their lengths to produce a continuous tube. Many examples of epithelial and endothelial fusion are necessitated by these laws of topology, but others, such as the production of an isolated excretory nephron and its subsequent joining to a urine-collecting duct within the kidney, are not formally required by the laws of mathematics and may instead be quirks of evolutionary history. (One could imagine an alternative way of building a kidney in which nephrons developed as branches from the urinary collecting ducts, in the way that mammary alveoli develop as branches from milk ducts in the mammary gland.)

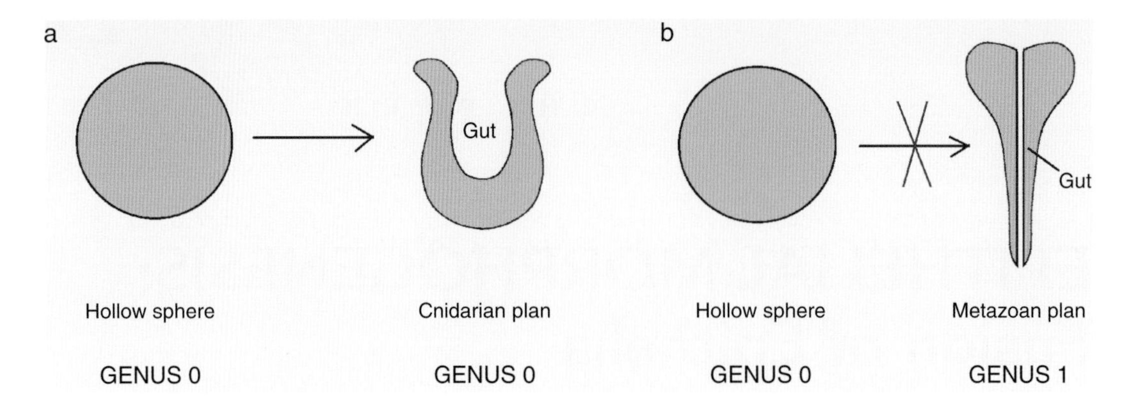

FIGURE 4.5.1 Formation of a single-ended gut from a hollow sphere is topologically possible without cutting and fusing, but forming a double-ended gut from a hollow sphere is not. The bottom line of text indicates topological, rather than biological, genera.

There is an important distinction between the epithelial fusions described in this chapter and the hole closures described in Chapter 4.2. In hole closures, there is a break in the epithelium and cells that meet across it already have the correct apico-basal polarization for the final tissue and meet lateral edge to lateral edge: all that needs to happen is that these edges adhere (see Figure 4.5.2a). In fusions, each interacting epithelium is intact and cells therefore meet basal surface to basal surface, as when tubules join, or apex to apex, as when palatal shelves join. To fuse into a single epithelium, cells have to change polarity, neighbour relationships, or both (see Figure 4.5.2b). Fusions are particularly prone to failure in humans, with cleft palate (failure of fusion of palatal shelves in the mouth) and hypospadias (failure of fusion at the underside of the penis, sometimes accompanied by failure of urethral migration) being among the most common congenital abnormalities detected in children.

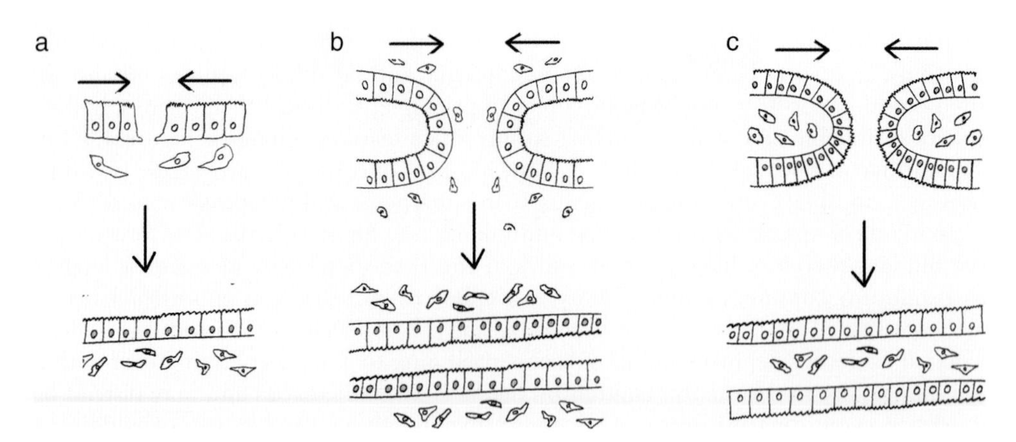

FIGURE 4.5.2 The difference between (a) hole closure, in which an epithelium seals up lateral domain to lateral domain, and fusion, in which epithelia meet either basal domain to basal domain (b), as in fusion of renal nephrons and collecting ducts or apical domain to apical domain (c), as in the secondary palate.

TRACHEAL FUSION IN *DROSOPHILA MELANOGASTER*

The tracheal system of *Drosophila melanogaster*, also discussed in Chapter 4.6, develops as a series of branched epithelial tubules that come in from the outside of the body and ramify through the tissues. At the tips of these tubes are single cells that are specialized for fusion. The tip cells approach and connect with adjacent tracheal systems at stereotyped locations in the intersegmental walls to form continuous trunk connections leading along the body (see Figure 4.5.3). This probably allows air to circulate, its movement being driven by movements in the body of the insect as it moves.

The cells at the tracheal tips are typical epithelial cells in that they are polarized and they express the "epithelial" adhesion molecule DE-cadherin, mainly at the apical end of the lateral domains of the cell. DE-cadherin is *D. melanogaster*'s homologue of E-cadherin and is encoded by the gene *shotgun*. Tip cells are specialized in terms of differentiation state, however, and express genes such as *escargot*, which are not expressed elsewhere in the tracheal tree and which are required for the expression of molecules directly concerned with fusion in these cells, such as cadherins.[1,2] The specialization and subsequent fate of these cells justifies their having a specific name: fusion cells.

The first morphological event of tracheal joining seems to be the production of exploratory filopodia by cells at the tracheal tips. Formation of filopodia requires the

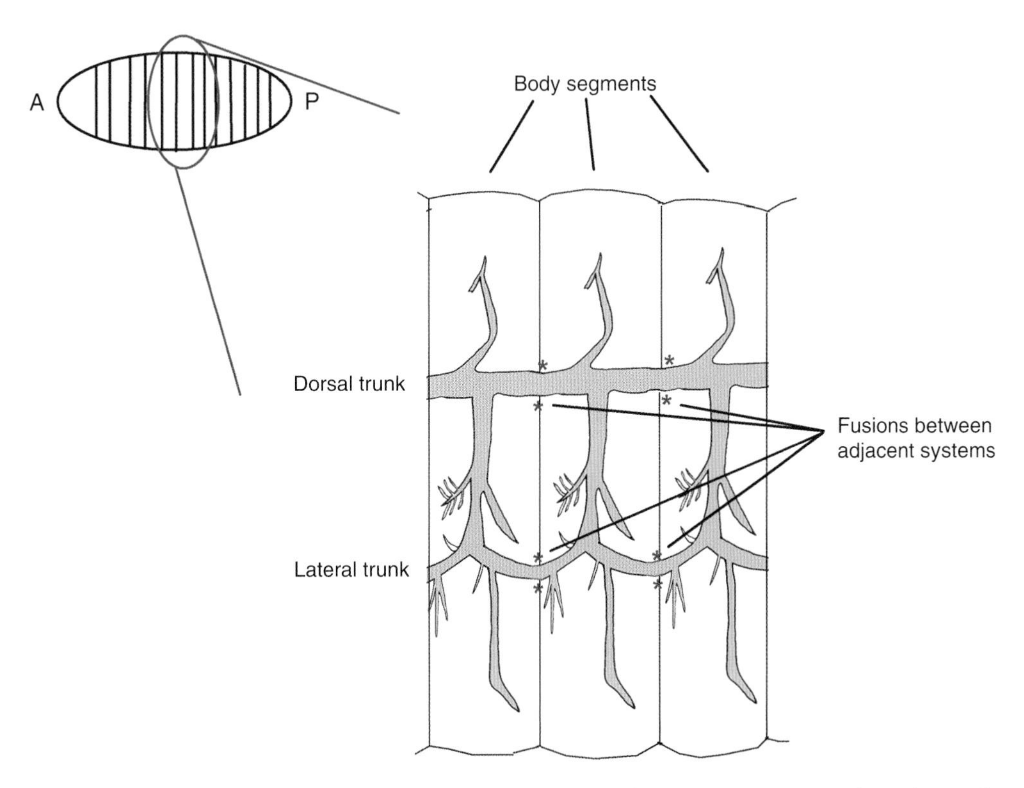

FIGURE 4.5.3 The connections between tracheal systems of adjacent segments in the embryo of *D. melanogaster*. Fusions are flanked by red asterisks. This figure is based on data in Ref. 1.

receipt of FGF-type signalling via Branchless and Breathless (see Chapter 4.6) and depends, as would be expected for filopodia, on activation of the small GTPase cdc42 (see Chapter 3.2).[3] The filopodia project and retract, apparently seeking out a suitable partner—a tracheal tip from an adjacent body segment—for fusion.[4] They migrate on and are apparently guided by a large bridge cell, which makes contact with both adjacent tracheal systems.[3,5]

Once filopodia from approaching tip cells touch, the cell adhesion molecule DE-cadherin accumulates at the sites of contact (see Figure 4.5.4). This accumulation takes about 10 minutes and may require new synthesis since it is one of the few aspects of epithelial morphogenesis for which maternal DE-cadherin RNA is insufficient and for which zygotically encoded DE-cadherin in required.[6] DE-cadherin is normally expressed in the lateral domains of epithelial cells, mainly near the apex. The filopodia form from the basal side of the cell, which is the side more typically associated with integrin-mediated cell-matrix adhesion than with cadherin-mediated cell-cell adhesion. The appearance of DE-cadherin in the filopodial contacts is therefore an early indication that cell polarity changes when contact takes place. DE-cadherin-mediated adhesion between fusing tracheal tips is important to the process, which fails in *shotgun* (i.e., DE-cadherin$^{-/-}$) mutants.[6] It also fails in mutants lacking β-catenin (encoded by the *armadillo* gene), which is an essential internal component on DE-cadherin-containing adherens junctions.[7] In cultured mammalian cells, the direction of cell polarity is determined, at least in part, by the location of E-cadherin-mediated adhesion.[8] If this is also true of *D. melanogaster*, DE-cadherin-mediated adhesion may be responsible for the re-polarization of touching cells so that the contacting surfaces, previously basal, are now regarded as lateral. They will, after all, have to be lateral from the perspective of the fused tube system.

DE-cadherin is not the only protein that is required to mediate fusion: the transmembrane protein Fear of Intimacy is also required, and in *fear of intimacy* mutants, tracheal tip cells seem to recognize one another but they do not adhere.[9] Perhaps the protein is required to activate adhesion by DE-cadherin or to mediate connection between the cadherin and the cytoskeleton. *Fear of intimacy* mutants also show failure of cell adhesion in gonad development.

The protein Kakapo provides an important link between integrins and the actin and microtubule cytoskeletons in *D. melanogaster*.[10] During normal tracheal fusion, Kakapo protein accumulates in the region of the DE-cadherin contacts between fusion cells.[11] This accumulation of Kakapo re-organizes the actin in the cell so that fibres form to link the new DE-cadherin-mediated contacts between tracheae with the existing DE-cadherin-containing adherens junctions around the lumens of each trachea.[11] Once formed, these actin fibres seem to perform the twin roles of contraction to draw the tracheal systems closer together (presumably mediated by myosin) and marking the place for insertion of new membrane and apical determinants.

EPITHELIAL FUSION IN PALATE DEVELOPMENT

The secondary palate of mammals, the bony plate that separates the nose from the mouth, is formed by the in-growth of two opposing shelves, one from the left and one from the right. These palatal shelves commence development as processes that grow downward from the maxillary processes each side of the tongue. They then elevate to a

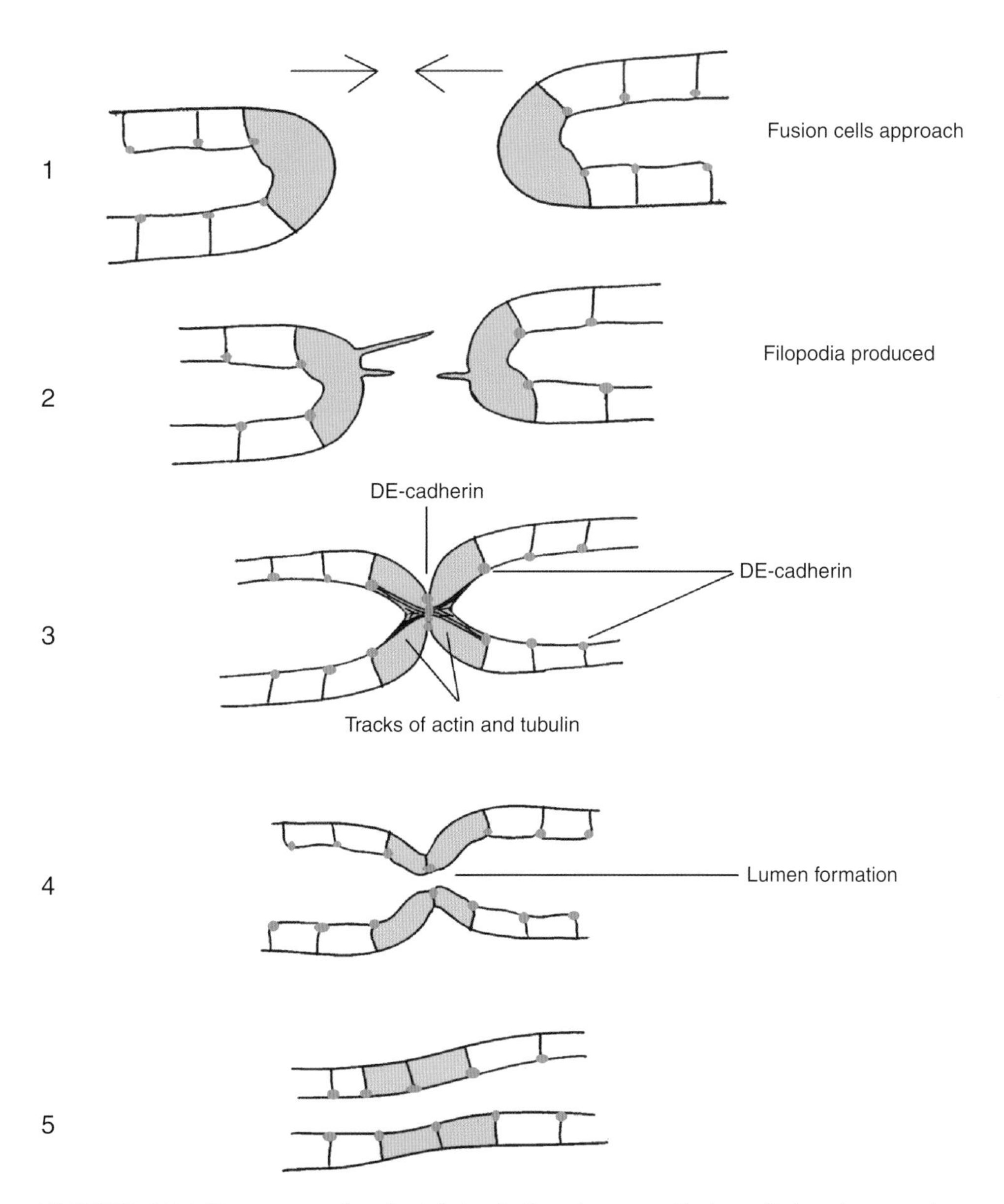

FIGURE 4.5.4 The process of tracheal fusion in *D. melanogaster*. Fusion cells (grey) approach and reach out to one another with filopodia. When the cells make contact, they adhere using DE-cadherin (red), and signalling through DE-cadherin, β-catenin, and Kakapo produces tracks of actin and microtubules. New membrane and apical organizers are then inserted along these tracks, and as apical character spreads, the two lumens approach closer and closer until they fuse. The bridge cell is omitted from the diagram for simplicity.

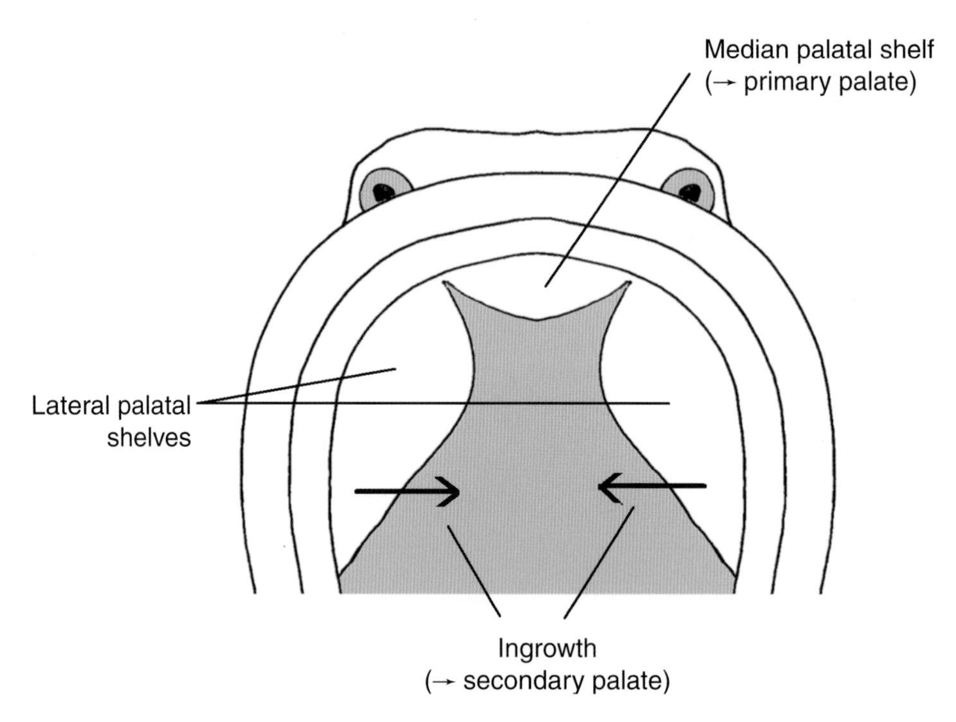

FIGURE 4.5.5 Formation of the secondary palate. This diagram is from a point of view looking up from under the chin, with the lower jaw removed for clarity. The oral cavity is closed off from the nasal cavity by the meeting of three processes. The median palatal shelf gives rise to the primary palate, whereas the secondary palate forms by the fusion of in-growing lateral palatal shelves.

horizontal position and extend across the top of the tongue to meet along the midline (see Figure 4.5.5). Because it is accessible, and because "cleft palate" (see Figure 4.5.6) is a common human abnormality, the process of palate fusion has been studied in more detail than most epithelial fusions. Many of the signals involved in the process have been identified, though there is still much to discover at the level of morphogenetic mechanisms.

The inside of the oral cavity is covered by a multilayered epithelium that consists of a columnar epithelium and an overlying periderm. When growth of the palatal shelves brings the epithelia from opposing shelves together at the midline, the columnar epithelia meet apical domain to apical domain. The first event connected directly with the process of fusion is a change in the behaviour of cells as the epithelia approach. Soon after the palatal shelves elevate to grow across the top of the tongue, cells at their medial edges begin to produce large numbers of microvilli[12] and then produce filopodia that project toward the oncoming opposite shelf (see Figure 4.5.7).[10] This presumably reflects activity of small GTPases such as cdc42, although this assumption has not been investigated experimentally. Formation of the filopodia requires the presence of the signalling molecule TGFβ3 in the epithelium: filopodia do not form in *tgfβ3*$^{-/-}$ embryos unless these animals are treated with exogenous TGFβ3.

The filopodia are coated in a thick covering of proteoglycans,[13] particularly chondroitin sulphate.[14] This glycosaminoglycan seems to be critical for the initial adhesion

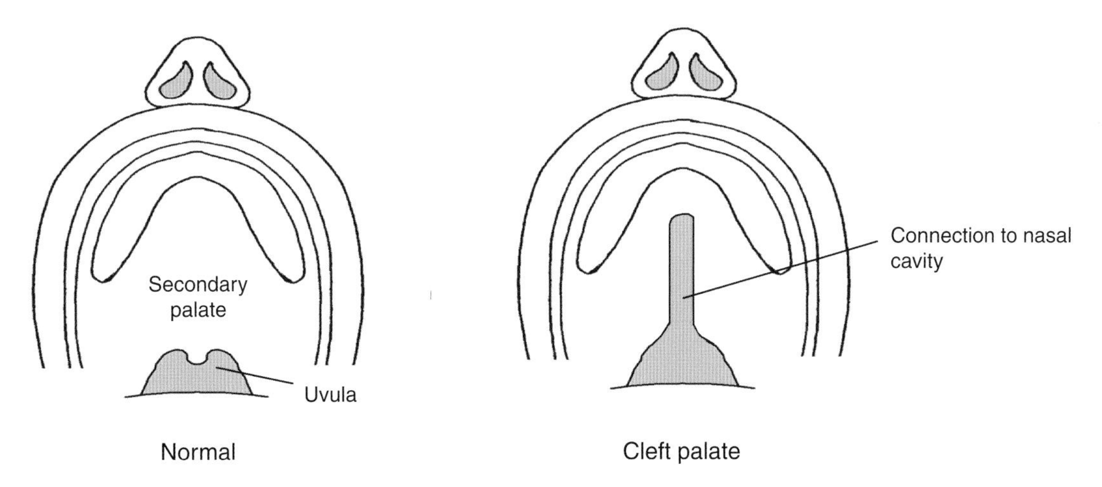

FIGURE 4.5.6 Cleft palate in the human mouth. The sketch is taken from a viewpoint directly below the mouth, with the lower jaw "removed" for clarity. The nasal cavity and passages to it are shaded gray.

of converging palate shelves, and if it is removed by chondroitinase enzymes, or if its synthesis is blocked with β-D-xylosides, adhesion fails.[15] The precise nature of the adhesive mechanisms is not yet understood, but the use of chondroitin sulphate here and in neural tube closure (below) suggests that one basic mechanism is being re-used in several places. Once adhesion has been achieved, the periderm cells move away to the top and bottom of the medial edge, where they die,[16,17] and the underlying columnar epithelia meet.

The line of adhering medial epithelia, needed for protection of the underlying stroma before fusion, is now an obstruction to the fusion of this stroma to make a continuous tissue, and it has to be removed. Two processes seem to be involved in this: removal by apoptosis and removal by epithelial-to-mesenchymal transition and subsequent migration. TUNEL staining of the fusing palate reveals a high frequency of apoptosis (see Chapter 5.3) in the epithelial cells between apposed palatal shelves, but only once the shelves have met.[17] This apoptosis is important to the process of fusion. Blocking apoptosis with inhibitors of caspases prevents proper destruction of the medial epithelium and its associated basement membrane, and, although they adhere, the palate shelves never fuse.[17] The medial epithelium, like any other, has a basement membrane at its basal surface, and this, too, needs to be cleared away. Matrix metalloproteinases are activated locally where cells are undergoing apoptosis and presumably account for the disappearance of basement membrane components such as collagen IV. The production of metalloproteinases fails if apoptosis is blocked but is encouraged if apoptosis is increased by the addition of retinoic acid, which is a powerful inducer of cell death in this system.[17,16] If adhering palates are treated with inhibitors of matrix metalloproteinases, the basement membrane remains, and, although adhesion between shelves and apoptosis of medial epithelial cells still takes place normally, fusion fails.[16]

The main evidence for clearance of medial-edge epithelial cells by epithelial-mesenchymal transition comes from retroviral cell-marking experiments in which cells

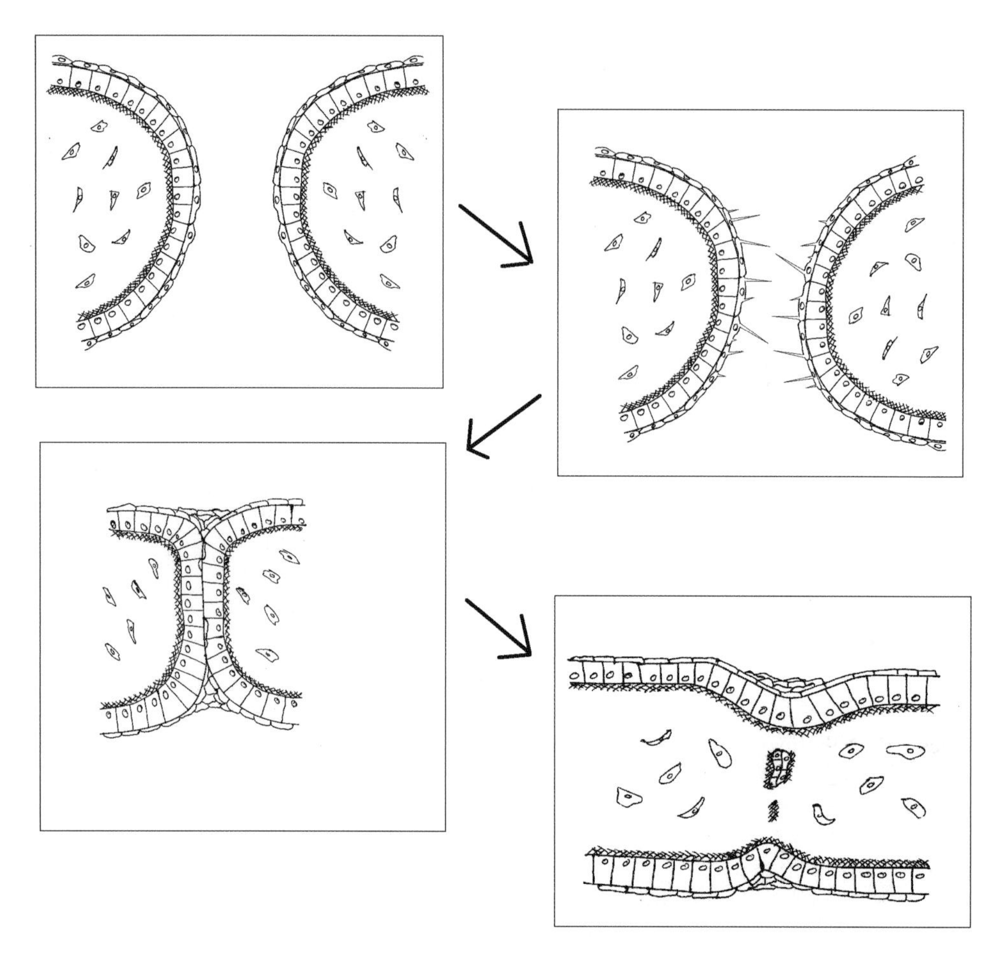

FIGURE 4.5.7 Fusion of the secondary palate. As the palate shelves approach, filopodia rich in chondroitin sulphate are produced, and the cells meet and adhere. Periderm cells move away from the site of adhesion to form a triangular collection of dying cells above and below the seam. Cells of the medial epithelium disappear by cell death and by epithelium-mesenchymal transition.

in the palatal epithelium are labelled permanently by retroviral transformation. After fusion, these labelled cells turn up in the mesenchyme and have an apparently completely mesenchymal phenotype.[14] This process of differentiation is again dependent on signalling by TGFβ3, which signals via Smads to the transcription factor LEF1.[18] Epithelial-to-mesenchymal transition also seems to involve the protein Snail, which is associated with epithelium-mesenchyme transitions in other systems.[19] In these other systems, Snail requires the PI-3-kinase signalling pathway,[20] although the precise linkages between the TGFβ3 pathway and PI-3-kinase-Snail is unclear.

Epithelial fusion is a necessary step in the formation of a tube by orthogonal invagination (see Chapter 4.4), as the folds above an invaginating trough have to fuse in order to produce a tube from it (see Figure 4.5.8). The neural tube is one example of this, but there are also others.

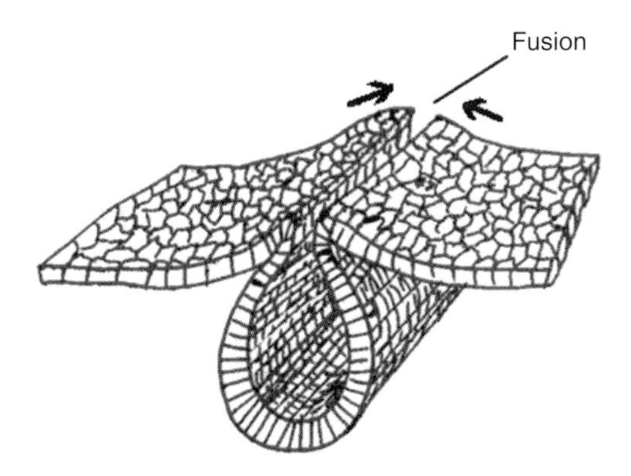

FIGURE 4.5.8 Fusion in the orthogonal invagination that forms the neural tube.

One of these other examples is associated with a very common congenital abnormality of boys, hypospadias, with an incidence of 1:250.[ref21] In this condition, the urethra opens not at the end of the penis, where it is supposed to, but rather somewhere along its underside or even at the junction between the penis and the scrotum. During early genital development, both sexes form a phallus just anterior to the urethral opening. In females, the phallus matures to form a clitoris, and the urethral opening remains posterior to it. In males, the phallus matures to form a penis, and the urethral opening "migrates" along a ventral groove to its end, the hole being closed by fusion at its proximal end as it advances distally (see Figure 4.5.9).

The precise anatomical details are somewhat complex because of the many different tissues that contribute to genital development, but the overall pattern is that of the

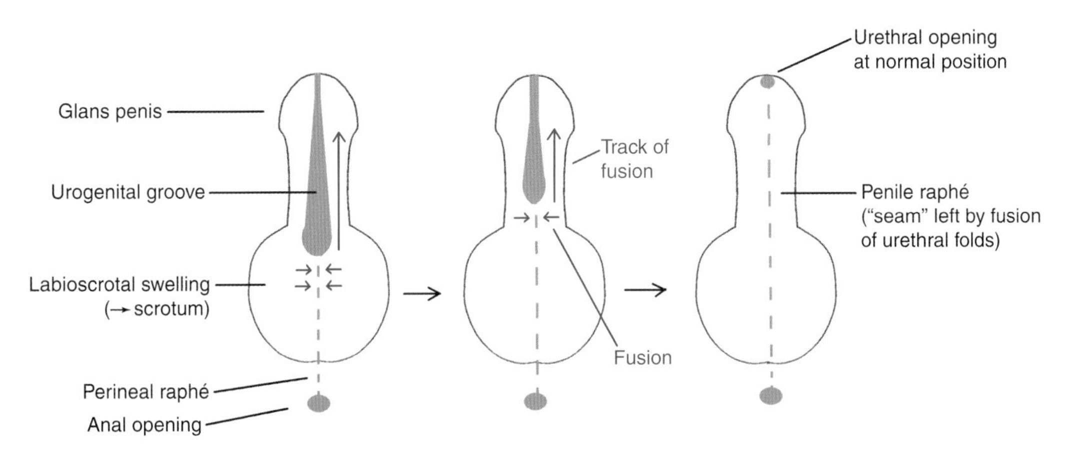

FIGURE 4.5.9 The "zipping up" of the urethral groove to move the urethral opening to its normal position at the end of the penis. These diagrams are drawn from a viewpoint between the legs, looking up. In reality, the shape of the labioscrotal folds and shaft change somewhat during the zipping up of the groove, but this is not represented in the diagram for the sake of simplicity; the shape drawn is correct for the final stage.

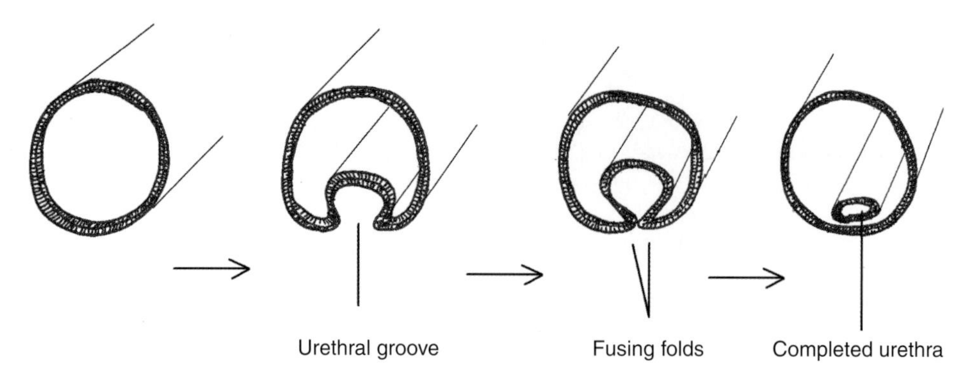

FIGURE 4.5.10 Fusion in formation of the penile urethra, drawn from sections in L. S. Baskin et al.[21]

formation of a groove and a distally moving fusion between folds from the left and right sides of the underside of the penis to cover over the groove (see Figure 4.5.10).[22,21]

The urethral groove is but one site of fusion in the perineal region of the mammalian embryo. Early in development, there is only one opening, the cloaca, in which the urinary, genital, and gut systems unite. The cloacal cavity is then divided by the in-growth of three folds from its walls, the Tourneau fold and the two Rathke folds, which fuse and separate the anal canal from the vagina (or from the urethra in males). Failure of this fusion leads to the human congenital abnormality of rectovaginal fistula, in which the rectum shares a common opening (and, alas, its bacteria) with the vagina.

There is little detailed information about the mechanisms of epithelial fusion in these other sites, although ultrastructural studies have again suggested the presence of filopodia on the surfaces of approaching cells. The frequent association of filopodia and fusion suggests that a conserved mechanism operates across phyla and across different sites within the same animal. Verifying whether this is so will depend on further studies of fusing systems.

Fusion is a common enough process in development to suggest that it is generally reliable, despite the above-mentioned problems that affect higher mammals. The ability to break and fuse epithelia and endothelia freed embryos from the tyranny of topological impossibility and gaining this ability was one of the key steps toward body plans as rich and complex as ours.

Reference List

1. Samakovlis, C, Hacohen, N, Manning, G, Sutherland, DC, Guillemin, K, and Krasnow, MA (1996) Development of the *Drosophila* tracheal system occurs by a series of morphologically distinct but genetically coupled branching events. *Development.* **122**: 1395–1407
2. Samakovlis, C, Manning, G, Steneberg, P, Hacohen, N, Cantera, R, and Krasnow, MA (1996) Genetic control of epithelial tube fusion during *Drosophila* tracheal development. *Development.* **122**: 3531–3536
3. Wolf, C, Gerlach, N, and Schuh, R (2002) *Drosophila* tracheal system formation involves FGF-dependent cell extensions contacting bridge-cells. *EMBO Rep.* **3**: 563–568
4. Tanaka-Matakatsu, M, Uemura, T, Oda, H, Takeichi, M, and Hayashi, S (1996) Cadherin-mediated cell adhesion and cell motility in *Drosophila* trachea regulated by the transcription factor Escargot. *Development.* **122**: 3697–3705

5. Wolf, C and Schuh, R (1-9-2000) Single mesodermal cells guide outgrowth of ectodermal tubular structures in *Drosophila*. *Genes Dev.* **14**: 2140–2145
6. Uemura, T, Oda, H, Kraut, R, Hayashi, S, Kotaoka, Y, and Takeichi, M (15-3-1996) Zygotic Drosophila E-cadherin expression is required for processes of dynamic epithelial cell rearrangement in the *Drosophila* embryo. *Genes Dev.* **10**: 659–671
7. Lee, M, Lee, S, Zadeh, AD, and Kolodziej, PA (2003) Distinct sites in E-cadherin regulate different steps in *Drosophila* tracheal tube fusion. *Development.* **130**: 5989–5999
8. McNeill, H, Ozawa, M, Kemler, R, and Nelson, WJ (27-7-1990) Novel function of the cell adhesion molecule uvomorulin as an inducer of cell surface polarity. *Cell.* **62**: 309–316
9. Van Doren, M, Mathews, WR, Samuels, M, Moore, LA, Broihier, HT, and Lehmann, R (2003) Fear of intimacy encodes a novel transmembrane protein required for gonad morphogenesis in *Drosophila*. *Development.* **130**: 2355–2364
10. Gregory, SL and Brown, NH (30-11-1998) Kakapo, a gene required for adhesion between and within cell layers in *Drosophila*, encodes a large cytoskeletal linker protein related to plectin and dystrophin. *J. Cell Biol.* **143**: 1271–1282
11. Lee, S and Kolodziej, PA (2002) The plakin Short Stop and the RhoA GTPase are required for E-cadherin-dependent apical surface remodeling during tracheal tube fusion. *Development.* **129**: 1509–1520
12. DeAngelis, V and Nalbandian, J (1968) Ultrastructure of mouse and rat palatal processes prior to and during secondary palate formation. *Arch. Oral Biol.* **13**: 601–608
13. Taya, Y, O'Kane, S, and Ferguson, MW (1999) Pathogenesis of cleft palate in TGF-beta3 knockout mice. *Development.* **126**: 3869–3879
14. Martinez-Alvarez, C, Tudela, C, Perez-Miguelsanz, J, O'Kane, S, Puerta, J, and Ferguson, MW (15-4-2000) Medial edge epithelial cell fate during palatal fusion. *Dev. Biol.* **220**: 343–357
15. Gato, A, Martinez, ML, Tudela, C, Alonso, I, Moro, JA, Formoso, MA, Ferguson, MW, and Martinez-Alvarez, C (15-10-2002) TGF-beta(3)-induced chondroitin sulphate proteoglycan mediates palatal shelf adhesion. *Dev. Biol.* **250**: 393–405
16. Cuervo, R and Covarrubias, L (2004) Death is the major fate of medial edge epithelial cells and the cause of basal lamina degradation during palatogenesis. *Development.* **131**: 15–24
17. Cuervo, R, Valencia, C, Chandraratna, RA, and Covarrubias, L (1-5-2002) Programmed cell death is required for palate shelf fusion and is regulated by retinoic acid. *Dev. Biol.* **245**: 145–156
18. Nawshad, A and Hay, ED (22-12-2003) TGFbeta3 signalling activates transcription of the LEF1 gene to induce epithelial mesenchymal transformation during mouse palate development. *J. Cell Biol.* **163**: 1291–1301
19. Martinez-Alvarez, C, Blanco, MJ, Perez, R, Rabadan, MA, Aparicio, M, Resel, E, Martinez, T, and Nieto, MA (1-1-2004) Snail family members and cell survival in physiological and pathological cleft palates. *Dev. Biol.* **265**: 207–218
20. Kang, P and Svoboda, KK (2002) PI-3 kinase activity is required for epithelial-mesenchymal transformation during palate fusion. *Dev. Dyn.* **225**: 316–321
21. Baskin, LS, Erol, A, Jegatheesan, P, Li, Y, Liu, W, and Cunha, GR (2001) Urethral seam formation and hypospadias. *Cell Tissue Res.* **305**: 379–387
22. Hynes, PJ and Fraher, JP (2004) The development of the male genitourinary system: III. The formation of the spongiose and glandar urethra. *Br. J. Plast. Surg.* **57**: 203–214

CHAPTER 4.6

EPITHELIAL MORPHOGENESIS: Branching

Efficient transport of resources and of waste products is a problem faced by all organisms, particularly by large ones. Critical resources such as oxygen have to be brought close to all cells that require them, while organs that are specialized for secretion have to pack a large amount of glandular epithelium into a small volume and channel its secretions away. Branched architectures provide one solution to this type of problem; they can be found at scales varying from the finest capillaries of the microvasculature to the body plans of the largest organisms on earth (trees and soil fungi).[1] The internal anatomy of many animals contains branched epithelia and endothelia, and the histology of most mammalian organs is dominated by branched tubes.

There are four basic methods of constructing a system of branched tubes. These are confluence of separate tubes, clefting of an ampulla, sprouting of new tips, and intussusceptive division of an existing tube (see Figure 4.6.1).

Confluence, used for example in the construction of pro- and meta-nephric kidneys and also in the formation of blood vessels by vasculogenesis, is not normally considered to be "branching morphogenesis," and it will not therefore be described in detail here; mechanisms of epithelial fusion, required for confluence, are discussed in Chapter 4.5. Branching morphogenesis proper is usually achieved by sprouting, clefting, or intussusception.

BRANCHING BY SPROUTING

Sprouting is the general name given for a very common mode of branching in which new branch tips grow out from an existing tip or tubule. There are two basic forms: monopodial, in which side branches form from a principal trunk, and dipodial, in which tips divide dichotomously so that there is no one dominant main stem (see Figure 4.6.2). The two forms are not exclusive, and it is common for organs to use dipodial branching to set up their basic architecture and then to use monopodial branching to add large numbers of side branches. The mammary gland is an example of such an organ, dipodial branching being used to create the tree of milk ducts and monopodial branching to add alveoli (where milk is actually secreted) to the branches of that tree.[2] There are also

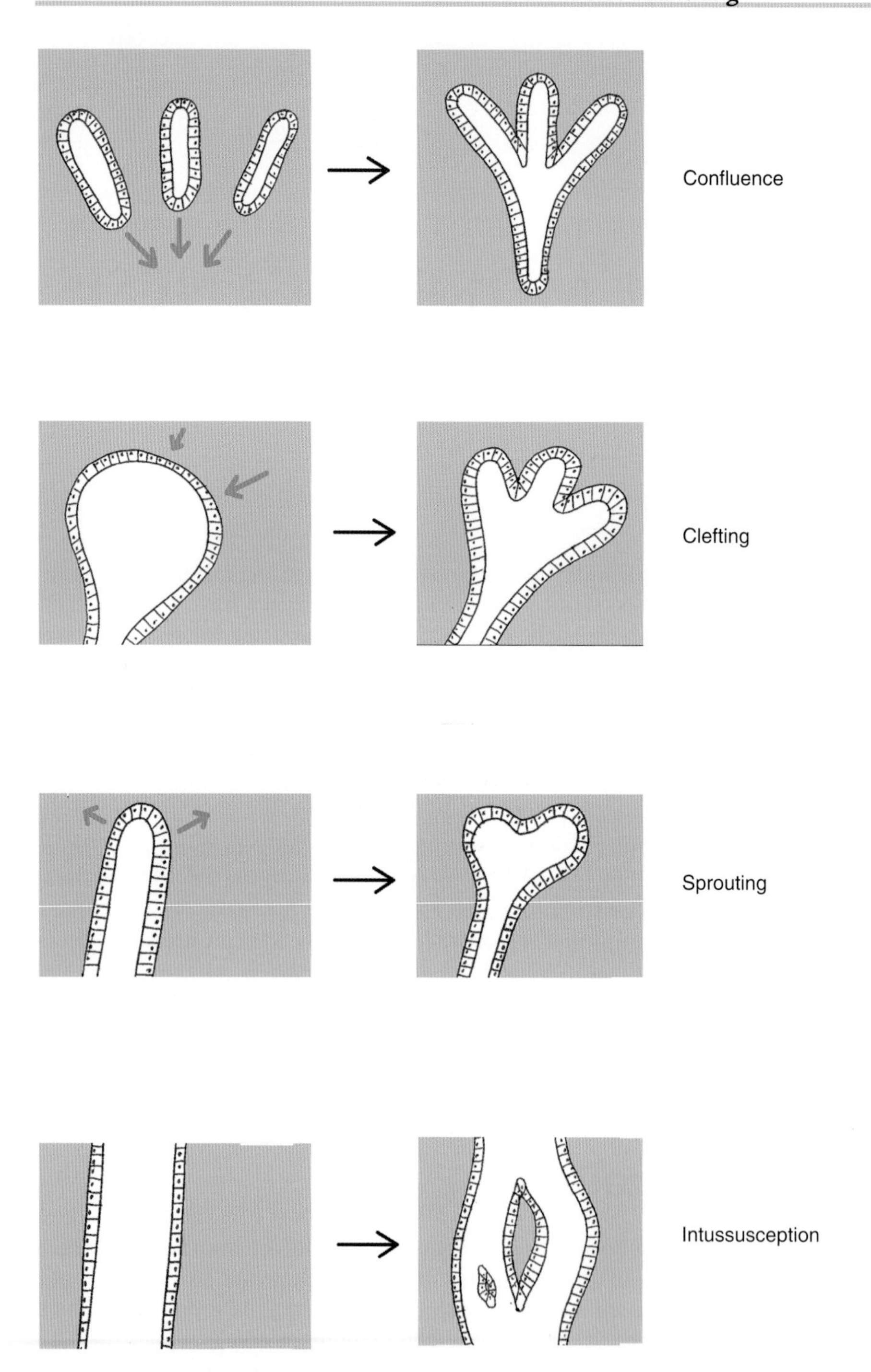

FIGURE 4.6.1 The four main ways of building a branched tube. The lumen is shown in white and surrounding mesenchyme in grey.

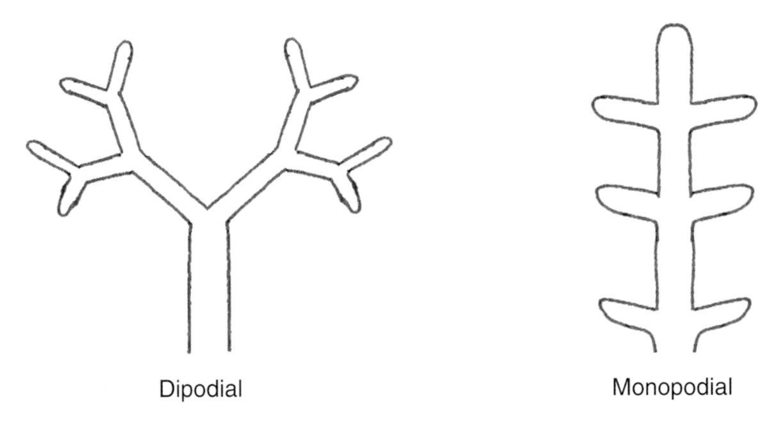

FIGURE 4.6.2 Dipodial and monopodial patterns of branching.

variations on the dipodial form, with trifurcations (and so on) replacing bifurcations, and there is some evidence for this in the kidney.[3]

Branching morphogenesis by sprouting is not restricted to the biological world, and a physical analogy (or a "physical example," depending on one's taste in semantics) is provided by viscous fingering. It is worth considering this simple physical example briefly in order to highlight the differences between it and the mechanisms found in living embryos, especially because this model is widely used by those interested in the physics of embryonic morphogenesis. Viscous fingering occurs when one fluid is forced, under pressure, to penetrate a more viscous fluid, for example when air is injected into oil in a thin space between coverslips.[4,5] The less viscous fluid tends to break up repeatedly into fingers to form a fractal♣ tree. In such a system, the interface between the fluids moves forward at a velocity proportional to the local pressure gradient:

$$v = -k\nabla P$$

where v is velocity of the advancing boundary, P the pressure, and k a constant.♥ At the interface between the fluids:

$$R = \sigma/(P_{outside} - P_{inside})$$

where R is the local radius of curvature of the boundary and σ the surface tension. The pressure field in the system satisfies the Laplace equation:

$$\delta^2P/\delta^2x + \delta^2P/\delta^2y = 0$$

These equations contain an inherent instability, the Mullins-Sekerka instability,[6] which arises because of a positive feedback in the system.[7] If, by some random fluctuation

♣Fractal trees have the property of self-similarity: they look so similar at different magnifications that one cannot deduce the scale of a photograph from the morphology alone. Real "fractal" trees, in the biological and physical worlds, only approximate to this ideal and are self-similar only over a range of magnification that shows less than the whole structure and more than just one or two branches.

♥where the ∇ symbol denotes the gradient; $\nabla f(x, y) = f_x(x, y)\mathbf{i} + f_y(x, y)\mathbf{j}$ for the two-dimensional thin cell being discussed.

(thermal jostling and so on) one part of the interface progresses forward a little further, it will create a sharper gradient of pressure at the front of the air as it pushes further into the surrounding medium. This means that it will tend to progress forward faster (by the first of the equations above). The advance of thin fingers is therefore favoured over that of a broad front. Too thin a finger, however, will "waste" some of the pressure difference because of the surface tension in bending the interface (see the middle equation above). The interface therefore arrives at a compromise between being pointed enough to advance rapidly and smooth enough not to have too sharp a radius of curvature.[7] The precise formula for the thickness of each finger that pushes forward depends on the geometry of the system,[7] but qualitatively the result is that broad fronts break up into fingers, the sides of which break up again as they grow, so that a branched pattern is created. The system therefore organizes itself quite automatically for branching growth.

In a hypothetical biological system that branched by viscous fingering, the "pressure" would probably come from cell proliferation within the branching epithelium, and the viscosity of the surroundings would be provided by the cells and matrix of the mesenchyme. "Pure" viscous fingering would require no local modification of these variables (as might be achieved by modulation of the cytoskeleton, for example). How likely is it that viscous fingering is the mechanism used for branching of real epithelia?

The simplest systems in which this question can be asked are culture models in which the only tissue present is the one that is branching. When cultured in simple media, renal cell lines such as MDCK and mIMCD3s form approximately spherical cysts that are composed of epithelial cells with conventional polarity, the apical side facing the lumen (see Figure 4.6.3). When the medium is enriched with certain ramogens♣ that would normally be synthesized by the mesenchyme that surrounds branching epithelia in the embryo, processes sprout from the cysts.[8,9,10] Many authors consider this to be a good model for branching *in vivo*.[10,11,12]

In a system in which branching were to be explained by viscous fingering, the function of the ramogens would be expected either to increase the "pressure" within the epithelium or to reduce its "viscosity" (its ability to rearrange in response to small forces). MDCK cells are induced to branch by HGF, its name suggests, HGF was

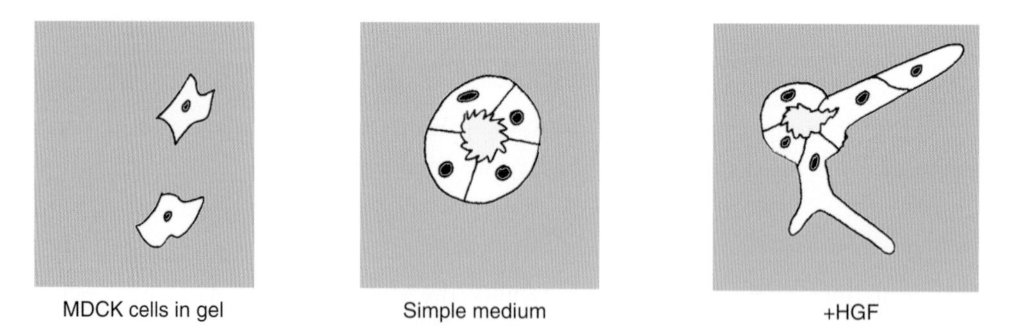

FIGURE 4.6.3 MDCK or mIMCD-3 renal-collecting duct cell lines form cysts when cultured in three-dimensional collagen-rich gels (shown in grey). These cysts sprout processes when treated with molecules such as hepatocyte growth factor (HGF).

♣Ramogens = molecules that promote branching, from the Latin *Ramus*, meaning "branch," and the Greek *Genesis*, meaning "creation."

first identified as a mitogen that stimulates cell division in hepatocytes.[14,15] Its effect on proliferation of MDCK cells is somewhat complicated; in sub-confluent♠ cultures, it seems to have no mitogenic effect,[16,17,18] but in confluent cultures, it seems to enable the cells to ignore contact inhibition of proliferation (see Chapter 5.1) and to proliferate as if they were still subconfluent.[19,20] Critically, in the three-dimensional culture system used for tubulogenesis, it does *not* appear to act as a significant mitogen.[21] This suggests that HGF does not act by increasing pressure by increasing the number of cells in the cyst. Comparison of micrographs of HGF-treated and untreated cysts[22] supports the idea that HGF-treated cysts do not have significantly more cells. Pressure could also be increased by direct pressurization of the lumen by ion and water pumps and channels in the cells. Direct measurements of lumen pressure seem not to have been published (making them would be technically difficult), but again, the "relaxed" morphology of the apical surface of cysts seen in micrographs[22] does not give an impression of great pressure within. There is an increased amount of space between lateral domains of cells, and there are fewer apical actin filaments so that "viscosity" of the epithelium may change somewhat. Cells do not separate completely, though.[23]

There have been no measurements of the effects of growing MDCK cell cysts in matrices of different known viscosities, although there have been some reports of the effects of different concentrations of collagen in the gel, and viscosity will presumably correlate with collagen concentration (although probably not in a linear manner). Inconveniently for the viscous fingering hypothesis, branching works better when the concentration of collagen is low and less well when it is high.[24] In the absence of direct measurements of pressure or viscosity, there is no strong evidence to suggest that "pure" viscous fingering does drive biological branching, and there is some circumstantial evidence against it. This does not rule out the principle altogether, but it makes it likely that local modifications of cell and matrix behaviour play such an important role in regulating branching that it is sensible to consider the alternative hypothesis, that sprouting is an "active" process driven by activities local to the cells involved, rather than global variables such as pressure that are distributed across the system. The simple MDCK cyst model provides a valuable system for identification of "active" processes in sprouting cells. This chapter will therefore continue with this model for a while longer before considering true *in vivo* examples.

HGF, signalling through its Met receptor tyrosine kinase,[25] affects multiple systems in MDCK cells, including the cytoskeleton, cell-cell adhesions, cell-matrix adhesions, polarity, and the composition of the matrix itself. The general pattern of these changes is to give sprouting cells characteristics that are associated with migration, although the essentially "epithelial" character of the cells remains. The cytoplasmic domain of Met contains a single multifunctional docking site, with the (phospho)tyrosine-containing sequence YVHVNATYVNV, which binds a number of proteins, including Shc,[ref26] Grb2,[ref27] Gab1,[ref28] and Stat3.[ref29] On stimulation with HGF, Met activates the PI-3-kinase, MAP-kinase pathway, and Stat3 pathways, all of which are required for branching morphogenesis.[30,29]

PI-3-kinase has two main effects: it protects the cells from anoikis (see Chapter 5.3) when their matrix adhesions weaken in response to other pathways, and it promotes the activation of N-WASP, probably via cdc42. N-WASP is an activator of the Arp2/3

♠Sub-confluent means that cells have not yet covered the surface of a culture dish so that a significant proportion of them border free space.

complex, which drives the formation of actin networks characteristic of cell locomotion, especially filopodia (see Chapter 3.2).[31,32] As well as promoting the formation of motile "leading edge" structures, PI-3-kinase represses the formation of stress fibres and cell-matrix adhesions.[33] Activation of the PI-3-kinase pathway alone, using constitutively active mutants, can induce MDCK cell cysts to sprout processes, but these processes do not develop into long, lumen-containing tubules.[30]

The MAP-kinase pathway also has multiple effects: it increases cell multiplication[27] and alters cell-cell and cell-matrix adhesion,[22] decreasing some adhesions but increasing others, such as those mediated by α2β1 and α3β1 integrins, which are important to the sprouting of tubules.[34] The basal surfaces of cells in MDCK cell tubules secrete copious quantities of the basement membrane protein fibronectin, which binds directly to integrins, including α3β1, and can organize collagen deposition.[35] Interfering with the binding of fibronectin to its integrin receptors using disintegrins that contain the RGD peptide that forms the integrin-binding site of fibronectin disrupts branching morphogenesis.[36] Including fibronectin in the three-dimensional matrix facilitates branching, as does inclusion of laminin and entactin, but inclusion of collagen IV, characteristic of normal basement membrane, inhibits it.[21] Sprouting of tubules also requires the activation of matrix metalloproteinase enzymes that cleave molecules of the matrix, possibly simply to clear a path, and possibly also to expose new epitopes on the matrix molecules. The metalloproteinases are secreted as inactive pro-enzymes, which are themselves activated by proteolysis in a cascade usually triggered by membrane-tethered metalloproteinases. One of these, MT1-MMP, is expressed by the sprouting tubules; if MT1-MMP is inhibited, sprouting fails.[37]

Signaling via Smad proteins, typically activated by members of the TGFβ family, acts as a potent inhibitor of branching morphogenesis. TGFβ itself activates Smads 2 and 3 and inhibits both cell proliferation and the formation of branching tubules in the MDCK cell system.[38] Activin A, a member of the TGFβ superfamily, also signals via Smads 2 and 3 and inhibits branching morphogenesis. Activin is synthesized by MDCK cells and seems to act as an endogenous inhibitor of branching. Its synthesis is reduced by treatment with HGF. Treatment of MDCK cells with the activin antagonist follistatin is sufficient to induce branching morphogenesis even in the absence of HGF.[39] This suggests that sprouting is the "default" behaviour of these cells and that it is normally held in check by autocrine activin signalling.

It seems, then, that branching morphogenesis requires a balance between main pathways that are activated by HGF (see Figure 4.6.4). Activating PI-3-kinase alone induces sprouting but not production of long, branched tubules.[30] Activating MAP-kinase alone simply causes vigorous but disorganized growth and a breakdown of cell-cell adhesion.[29] Activation of both PI-3-kinase and MAP-kinase pathways simultaneously, using a mutant Ras that can trigger both constitutively, produces effects identical to that of activating the MAP-kinase pathway alone, but if the MAP-kinase path is *partially* inhibited using the drug PD98059, morphogenesis is better than with PI-3-kinase alone. This supports the idea that a balance of activation is important.[29] Not surprisingly, activation of Stat3 does not induce formation of branching tubules, but without it the growth and scattering of cells seems to overwhelm morphogenesis, and tubules do not form.[29] Presumably, the balance of activation is normally maintained by feedback systems operating inside the cells.

The picture obtained from the MDCK cell model is therefore one of very active cells that sprout from their cysts by loosening their bonds to their existing location

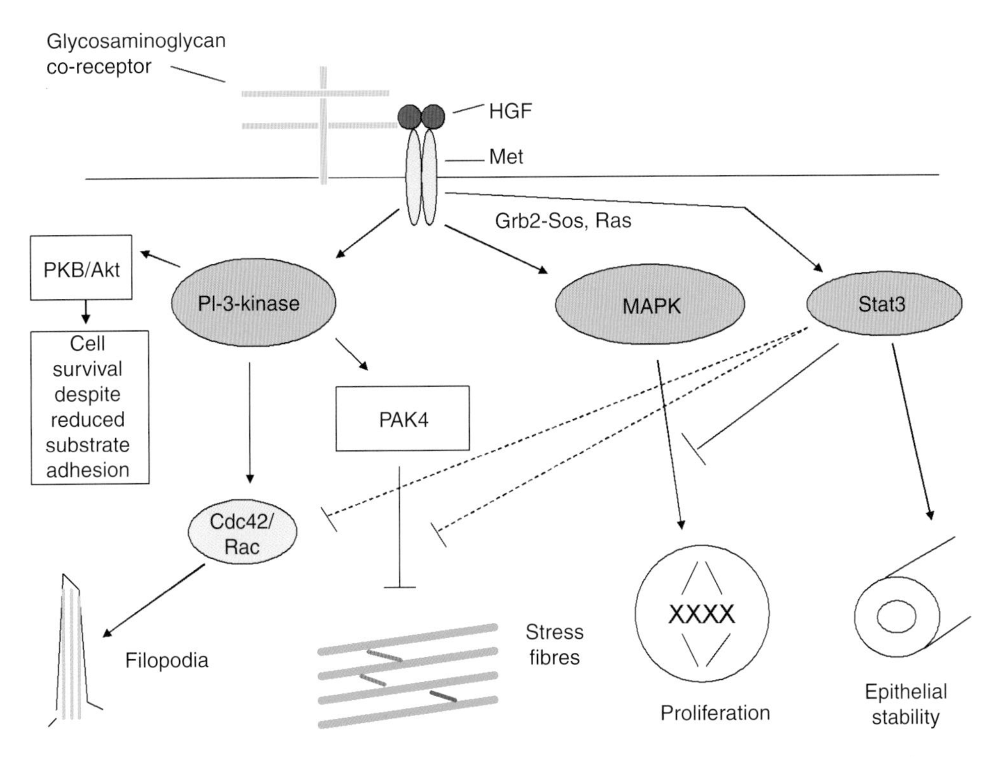

FIGURE 4.6.4 Diagram of the main pathways by which HGF induces MDCK cells to produce branching tubules in three-dimensional collagen matrices. The arrows do not necessarily imply direct connections, and in most cases symbolize cascades of molecules, many of which have been shown in more detailed schematics in other chapters. The dotted lines indicate an antagonism of the general process to which they point, but the level at which that antagonism operates is not yet known. References for each interaction can be found in the main text.

and migrating actively outward before reorganizing to form a proper tubular structure. This is very different from the "passive," pressure-driven model of viscous fingering. How typical is it of branching systems *in vivo*?

The embryonic branching system that is most similar to the MDCK system, at least of those systems that have been studied closely, is the tracheal system of *Drosophila melanogaster*. This system, which has already been described to some extent in Chapter 4.5, is unusual because it contains very few cells and its sprouting tends to be led by the activity of clearly identifiable individual cells. This makes it morphologically similar to the MDCK cell system and distinct from most mammalian epithelia, the branching tubules of which tend to include dozens to hundreds of cells. That is not to say that tracheae are simple; even though they consist of fewer than 100 cells even when mature, they include different zones with different patterns of cell arrangement. In the main "trunks," the lumen is surrounded by two to five cells. In the principal branches, the tubes are narrower and are surrounded by a single cell that has curved around to make junctions with itself, while the finest tubes are hollow processes from a single cell (see Figure 4.6.5). There are also tubes formed by fusion between adjacent tracheal systems; these have already been described in Chapter 4.5.

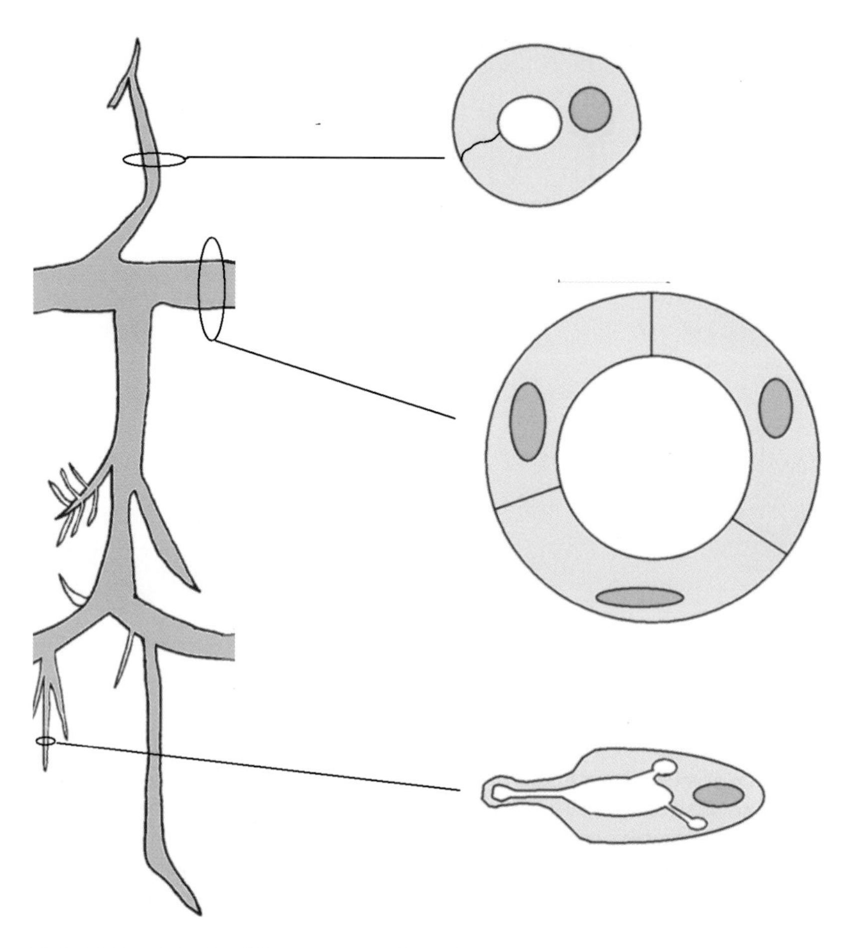

FIGURE 4.6.5 The different types of tubule structure seen in different zones of one body segment's tracheal system in *D. melanogaster*. The diagrams are based on electron micrographs and three-dimensional schematics presented in A. Uv, R. Cantera, and C. Samakovlis's "Drosophila Tracheal Morphogenesis: Intricate Cellular Solutions to Basic Plumbing Problems."[40]

The tracheal system of each body segment forms from an invagination of about 80 cells from body wall. Unlike MDCK cells, and unlike almost every other branching system studied to date, these cells complete their morphogenetic program without any further cell division. This may be an adaptation to the very fast development of *D. melanogaster*, which allows a mere 12 hours for the tracheal system to be completed. Sprouting of branches is driven not by HGF but by the protein Branchless,[41] a homologue of mammalian FGF, which signals via the FGF receptor homologue Breathless.[42] (Both proteins are named after the phenotype caused by their mutation: absence of a proper tracheal system.) Once they receive the Branchless signal, some cells in the invaginations produce filopodia and become the leading cells of sprouting branches.[43] These cells show strongly activated MAP-kinase.[44,45] No information is yet available on activation of PI-3-kinase or Stat3 pathways, but it is known that the downstream effector, Akt, is involved.[46,47] It is also known that formation of tracheal filopodia is under the control of cdc42.[48]

Cell rearrangement necessary for the sprouting of filopodia to lead to the formation of tubules depends on the activity of transcription factors such as Ribbon,[49,50,51] which may act downstream of the MAP-kinase pathway.[40] The cells of embryos mutant for the *ribbon* gene still produce filopodia, and their basal surfaces still extend towards sources of Branchless, but the apical surfaces of the cells remain locked in position so that the trachea cannot form properly, and all that happens is that thin basal processes of cells extend toward the targets that ought to receive complete tracheae. Intriguingly, Ribbon is expressed in other locations associated with cell rearrangement, such as dorsal closure, suggesting that it may be a general factor to allow epithelia to move. The transcription factor Grainyhead is also involved in the apical end of the cell and seems to regulate how much apical membrane is produced.[52] Without Grainyhead, tracheae produce too much apical membrane and their branches become over-elongated. Grainyhead is controlled by the Branchless signalling pathway and, intriguingly, its homologues are expressed in branching epithelia of vertebrates,[53,54] suggesting a conserved system. Cell rearrangements required for the formation of very fine branches are under the control of additional "special" molecules, but study of these processes is too new for a story to have emerged yet.

The branching epithelia of vertebrates (see Figure 4.6.6) tend to be constructed on a larger scale, and the tips of branches are many cells wide (see Figure 4.6.7; sprouting of single cells is, however, common in capillary endothelia). This means that new sprouts are not led by motility of single cells, and it also means that the activities of a group of cells have to be coordinated so that all move forward together. Nevertheless, there is evidence that the same set of control pathways may be used.

These epithelia ramify into mesenchymal cells that surround them, and their growth and branching are controlled by a variety of paracrine signals that include FGF7 and FGF10 (both of which signal via FGFRIIIb), EGF, TGFα, and amphiregulin (all of which signal via RGFR), HGF (which signals via Met), and GDNF (which signals via Ret/GFRα1 and, via GFRα1, via Met too[56]). Most organs use several of these ligands at once, although just one seems to play a dominant role in each case (see Table 4.6.1). Each receptor is capable of triggering both the MAP-kinase and PI-3-kinase pathways,[57,58,59,60,61] and experiments in which these pathways are inhibited by treatment of cultured organ rudiments with specific drugs suggest that these pathways play similar roles to those described for the MDCK cell model. MAP-kinase activation is required for branching morphogenesis in organs in which a requirement has been tested, which include the salivary glands,[62] lungs, and kidneys, where MAP-kinase drives mitosis.[63,64,65] The PI-3-kinase pathway is required for branching in salivary glands[66,67] and kidneys.[68] Smad activation by the TGFβ family of signalling molecules inhibits branching in lungs[69,70] and kidneys.[11] Interactions with the matrix are also required, as they are in the MDCK cell system. Laminins and nidogen are needed for branching in salivary glands[71,72,73] and kidneys,[74] while the branching epithelia of lungs and mammary glands fail to develop properly in animals in which collagen synthesis is inhibited.[75,76] Proteolysis by the matrix is also required, either to clear a passage for the advancing epithelium or to effect more subtle biochemical changes in the matrix.[77]

Unfortunately, the precise cellular mechanisms that allow multicellular epithelial tubes to sprout new tips are still far from clear. Vertebrate epithelial cells tend to proliferate as branching takes place, but while proliferation does tend to be localized to

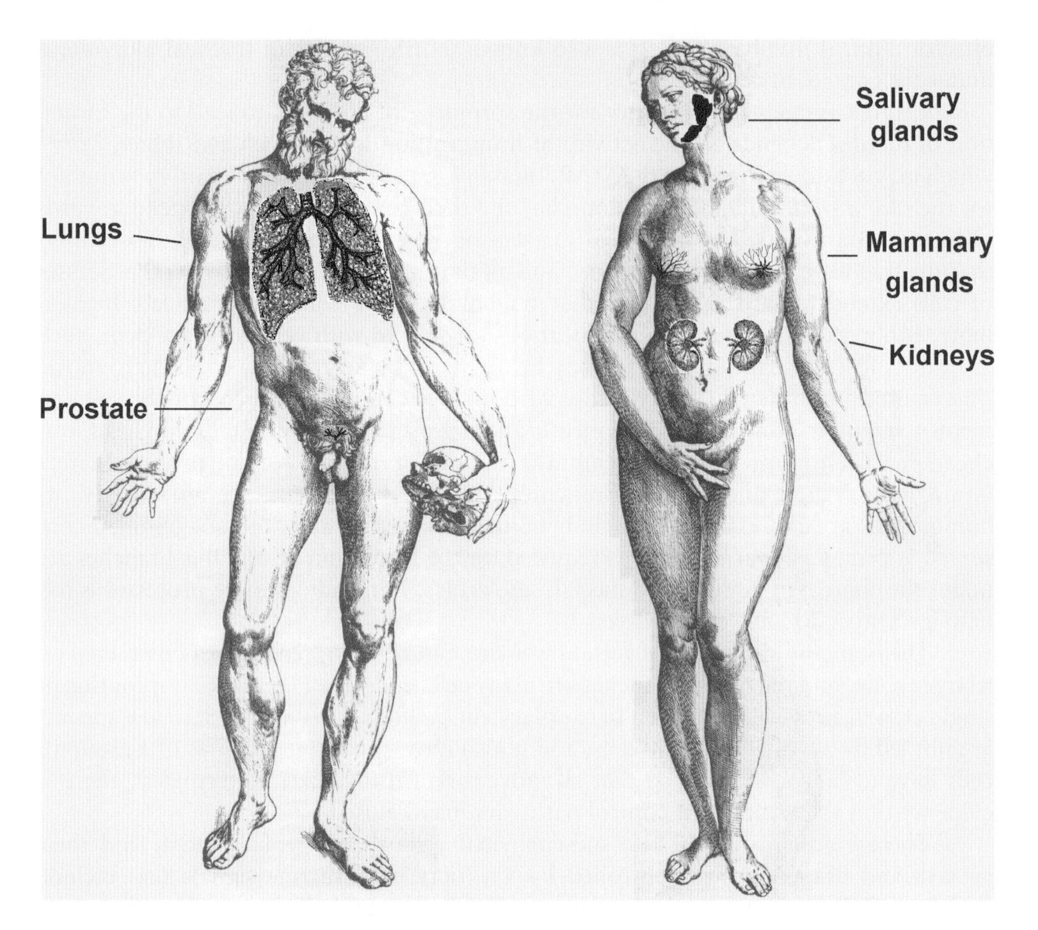

FIGURE 4.6.6 The mammalian branching systems that have received most attention, superimposed on drawings made by the anatomist Versalius in 1543. This diagram is reproduced by kind permission of BioEssays, where it first appeared.

TABLE 4.6.1 Key Ramogens in Branching Organs*

Branching System	Main Paracrine Ramogen
Tracheae in *D. melanogaster*	Branchless (FGF homologue) via Breathless (FGFR homologue)
Lung	FGF10 via FGFRIIIb
Pancreas	FGF10 via FGFRIIIb
Mammary gland	FGF7 via FGFRIIIb
Kidney	GDNF via GFRα1-Ret complex
Prostate	FGF7 via FGFRIIIb

*The species is mouse unless specified otherwise. Original sources for the data can be found in Ref. 78.

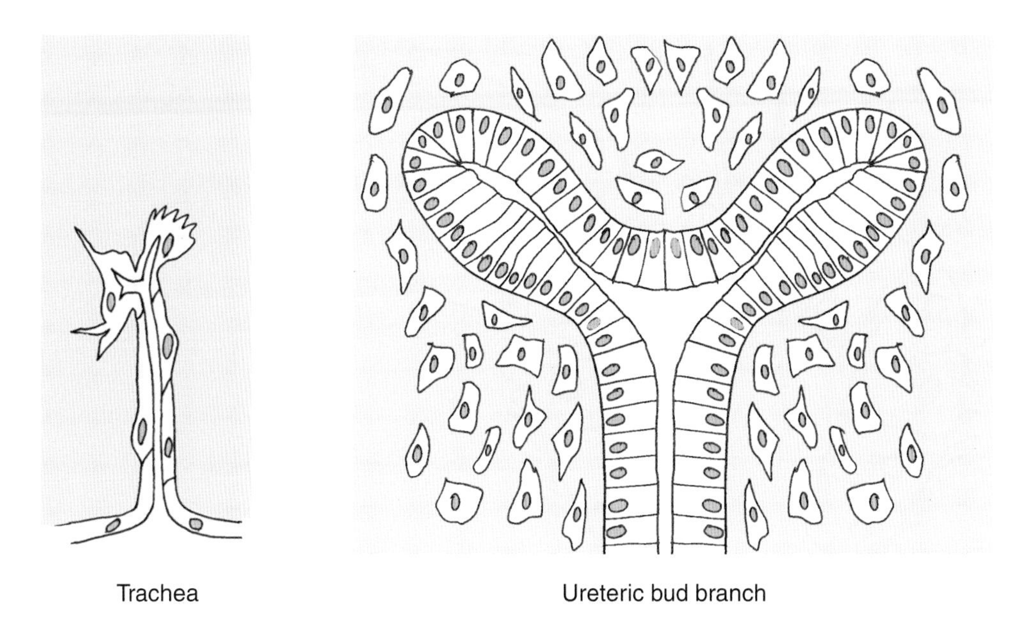

FIGURE 4.6.7 The difference between the morphology of tracheal branching in *D. melanogaster*, where new sprouts tend to be led by single cells, and in the ureteric bud, a typical mammalian branching epithelium in which branch tips are composed of several cells. The tracheal drawing is based on data presented in A. Uv, et al.[40] and the ureteric bud from an electron micrograph in L. Saxen.[55]

the growing tips,[78] there is no evidence that mitoses are orientated to push cells out in any particular direction. The ultrastructure of the epithelium is somewhat relaxed at the tips, at least in salivary glands[79,80] and kidneys,[81] suggesting that cells there are specialized for rearrangement and possibly also for direct locomotion (see Figure 4.6.8). There is little evidence for the construction of a motile leading edge analogous to those seen in MDCK cells and insect tracheae, but at least in kidney it is possible to detect fine matrix-rich processes that reach out from the basal side of the tips of the epithelium[65] and which become very long and obvious when the growth of the tips has been frustrated by depriving them of glycosaminoglycans.[82] It is tempting to speculate that these are filopodia and that the growing front of the epithelium shows "classical" motile character, although at this stage such a conjecture is merely speculation. It is supported, though, by the observation that low concentrations of cytochalasin D, which seem to disrupt cell-cell adhesion at the tip even more than usually, cause tip cells to scatter into the surrounding mesenchyme.[83] Also in the kidney, the apical surfaces of cells at the tip are particularly rich in actin filaments[65] that may drive the sprouting of a new epithelial tip through cell wedging (see Chapter 4.4).

SWITCHING BETWEEN MODES OF BRANCHING

In some branching epithelia, mesenchyme-derived signals can switch the epithelium between different modes of morphogenesis. This is demonstrated clearly by mammary

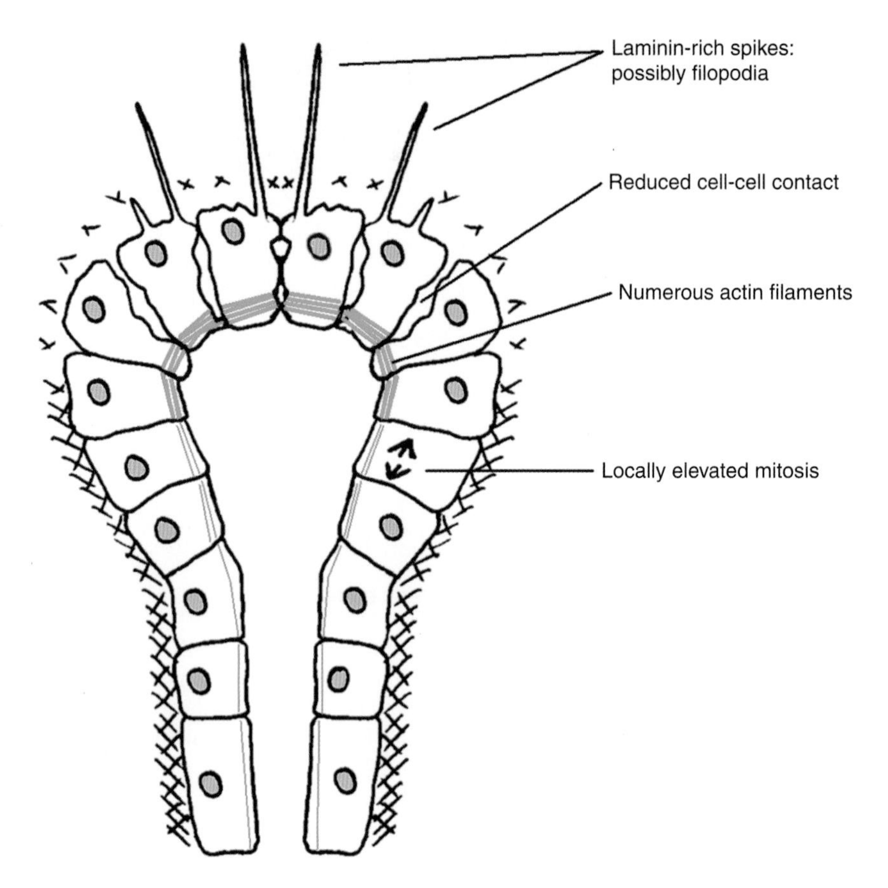

FIGURE 4.6.8 Features of the tips of the branching ureteric bud epithelium of the kidney. This schematic is assembled using information from several sources,[65,83,84] and I know of no real image that shows all features simultaneously. It should therefore be interpreted with appropriate caution.

epithelium, which is capable of developing either a long spindly branched system of milk ducts, which is what happens early in development, or of developing rounded secretory alveoli that actually produce the milk. During early development, the mesenchyme that surrounds the growing ducts produces copious quantities of HGF, but during early pregnancy it produces neuregulin (which signals via ErbB instead). Mammary glands grow well in organ culture. Adding HGF to their medium promotes the production of a branched ductal tree. Neuregulin, on the other hand, induces the same epithelium to produce numerous alveoli, rather than further branches, from the ducts[85] (see Figure 4.6.9). Although both receptors can signal through a variety of pathways, it appears that PI-3-kinase is essential for HGF-induced branching but not for the formation of alveoli, whereas MAP-kinase is essential for alveoli but not branching morphogenesis, at least in a simple culture system in which mammary epithelia are suspended in a three-dimensional matrix.[86]

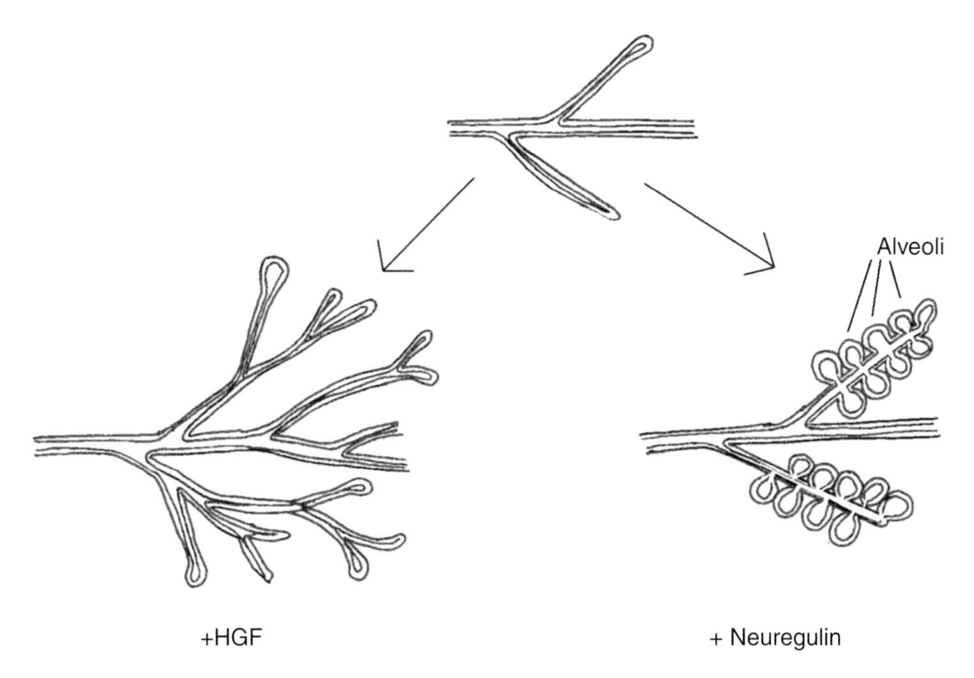

FIGURE 4.6.9 The different types of morphogenesis elicited by HGF and by neuregulin treatment of cultured mammary glands. The drawings are based on original micrographs in Y. Yang et al.[85]

PATTERNING THE BRANCHING TREE

Although the positions and directions of the largest (first-formed) tubules of branched systems tend to be laid down in stereotyped positions under the control of specific patterning molecules, the detailed arrangements of smaller branches are normally more flexible, and the tree organizes itself according to the tissues that it has to serve. This self-organization seems to be achieved by a combination of autocrine and paracrine signalling.

Many mesenchyme-derived ramogens act as chemoattractants to branching epithelia. If, for example, a bead soaked in the renal ramogen GDNF is placed adjacent to a kidney rudiment growing in culture, the nearby parts of the developing ureteric bud tree grow toward it.[87] Indeed, such beads can even induce the formation of new branches from parts of the excretory system that would not normally branch, such as the nephric duct.[87] It is not clear how this chemotaxis works, but if branching tips do show "conventional" locomotory behavior (filopodia, and so on), it may use an adaptation of the chemotactic mechanisms described in Chapter 3.3. Whatever its mechanism, the directional response would have the effect of bringing epithelial branches toward the strongest sources of attractants and would encourage further branching there.

In principle, any negative feedback between the production of chemoattractants by cells and the arrival of branches in their vicinity could be used to organize the branching tree. For example, low oxygen conditions induce cells to produce the ramogenic chemoattractant of endothelia, VEGF.[88,89,90] The VEGF induces sprouting of capillaries and attracts them,[91,92] and, once blood flow through the new capillaries is established, the concentration of available oxygen rises, VEGF synthesis falls, and sprouting ceases. Thus the extent and direction of branching is determined by local need, and the branching system organizes itself accordingly. Similar feedback systems can be found in control of epithelial branching. In the kidney, for example, "virgin" mesenchyme that has not yet been invaded by the branching ureteric bud produces GDNF,[93] which encourages branching and probably sets its direction too. The ureteric bud is a source of short-range signals that cause the differentiation of mesenchyme once it reaches it; the differentiation process is complex because it gives rise to epithelial nephrons,[94] but from the point of view of branching, what is important is that the differentiating mesenchyme ceases to produce GDNF. It is therefore no longer attractive to the ureteric bud. Indeed, after a delay of some hours, the mesenchyme produces negative regulators of branching, such as TGFβ and BMP2 that signal via the Smad pathway.[95] It is reasonable to assume that this pattern of protein expression ensures that the ureteric bud is attracted to areas that it has not yet penetrated and not encouraged to branch into areas it already served (see Figure 4.6.10).

Recent discoveries have shown that it is a mistake, however, to assume that the epithelium is incapable of organizing itself into a reasonably spaced tree using only its own resources. For example, when the ureteric bud of a kidney is isolated and placed in

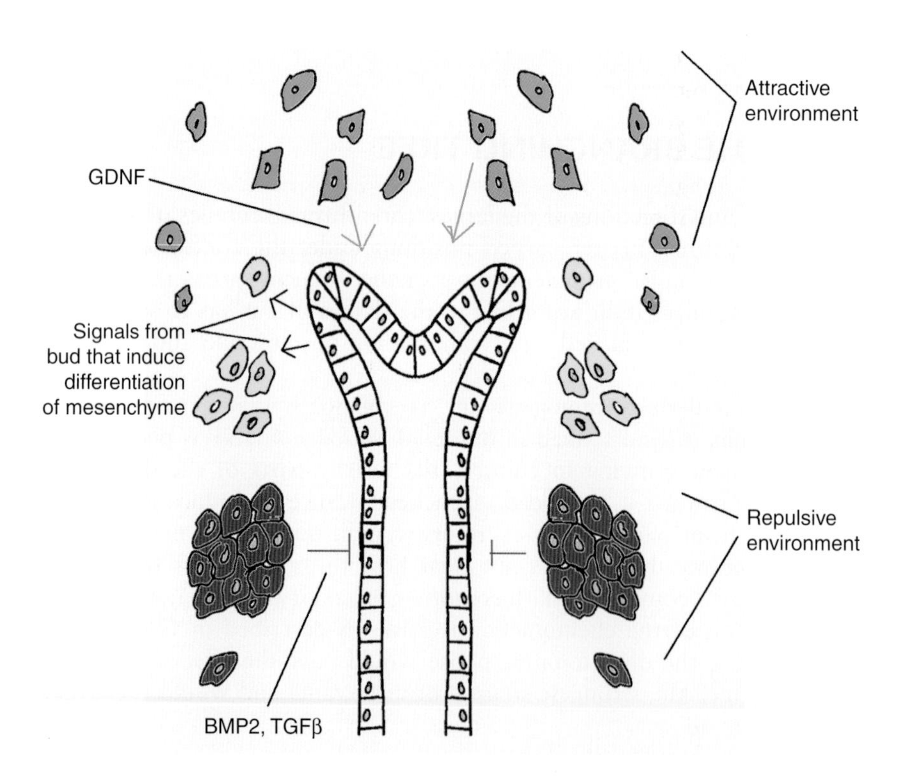

FIGURE 4.6.10 Feedback systems in the developing kidney cause areas of virgin mesenchyme to be attractive to ureteric bud branches (because they produce factors such as GDNF) and areas already penetrated to cease to be attractive or to induce branching.

a three-dimensional matrix, it will still form a tree with an appearance very similar to that found in a real kidney.[96,97] To state the obvious, normal branching requires that some cells initiate the sprouting of new branches while others refrain from this so that there are stable lengths of tubule in the core of the tree. If every cell tried to sprout, one would expect either that the epithelium would grow isotropically, like an inflating balloon, or that it would produce ridiculous numbers of branches, leading to a confused mess. Cells at the sprouting tips must, therefore, be different.

In *D. melanogaster*, the sprouting cells at the branching tips of tracheae show very much higher levels of MAP-kinase activation than do cells in the "shafts" behind the tips.[44,45] This restriction of activation depends on a mechanism of lateral inhibition in which tip cells inhibit the sensitivity of their neighbours to Branchless, the extracellular activator of MAP-kinase during tracheal branching. There are at least two mechanisms for this lateral inhibition. One relies on Notch-Delta signalling. Delta is expressed in the tip cells as a consequence of MAP-kinase activation by Branchless. Delta is a ligand for the protein Notch and activates the Notch pathway in neighbouring cells. Notch has several effects in the cells that express it: it down-regulates Delta and it down-regulates the expression of Breathless, the receptor for Branchless.[44] The net effect is that neighbours of Delta-expressing tip cells are inhibited from being induced, by Branchless, to become tip cells themselves (see Figure 4.6.11).

The other system of lateral inhibition relies on the protein Sprouty, named after the over-branched tracheae of *sprouty* mutants in which too many cells behave as tip cells and show evidence of MAP-kinase activation.[98] Sprouty is produced by the tip cells themselves in response to Branchless signalling. It diffuses from there to the neighbouring cells, where it inhibits the signalling from Breathless to MAP-kinase at the level of Ras proteins.[99] It is not yet clear, however, why the cells that synthesize Sprouty protein are not themselves shut down by it. Sprouty proteins are widespread phylogenetically, and the developing branching organs of mammals express several members of the Sprouty family, which can block signalling from FGF and EGF receptor signals. In lungs, for example, Sprouty2 is expressed in response to signalling by the primary ramogen FGF10, and it acts as an inhibitor of branching.[100]

Feedback systems such as those based on Notch-Delta and Sprouty, and probably other molecules as well, confer on branching epithelia an innate tendency to assign some cells to sprouting and others to being part of a straight tubule; in terms of the sprouting program, all cells are called but few are chosen. The mesenchyme-derived signals superimpose additional information on this basic system, and they bias the choice of which cells will be tip-like because they drive the same basic pathways. Thus the shape of a branched system is formed by interaction of the innate tendencies of the epithelium and signals from its environment—"nature and nurture" in microcosm.

BRANCHING BY CLEFTING

Branching by clefting and branching by sprouting are sometimes difficult to distinguish morphologically, and many authors consider them to be one process, clefting being considered one mechanism for separating newly sprouted tips. I am considering them separately in this book not from a desire to be dogmatic about their being different processes, but rather because there is not yet convincing evidence that they are on a morphological and functional continuum. Epithelia that branch by clefting are

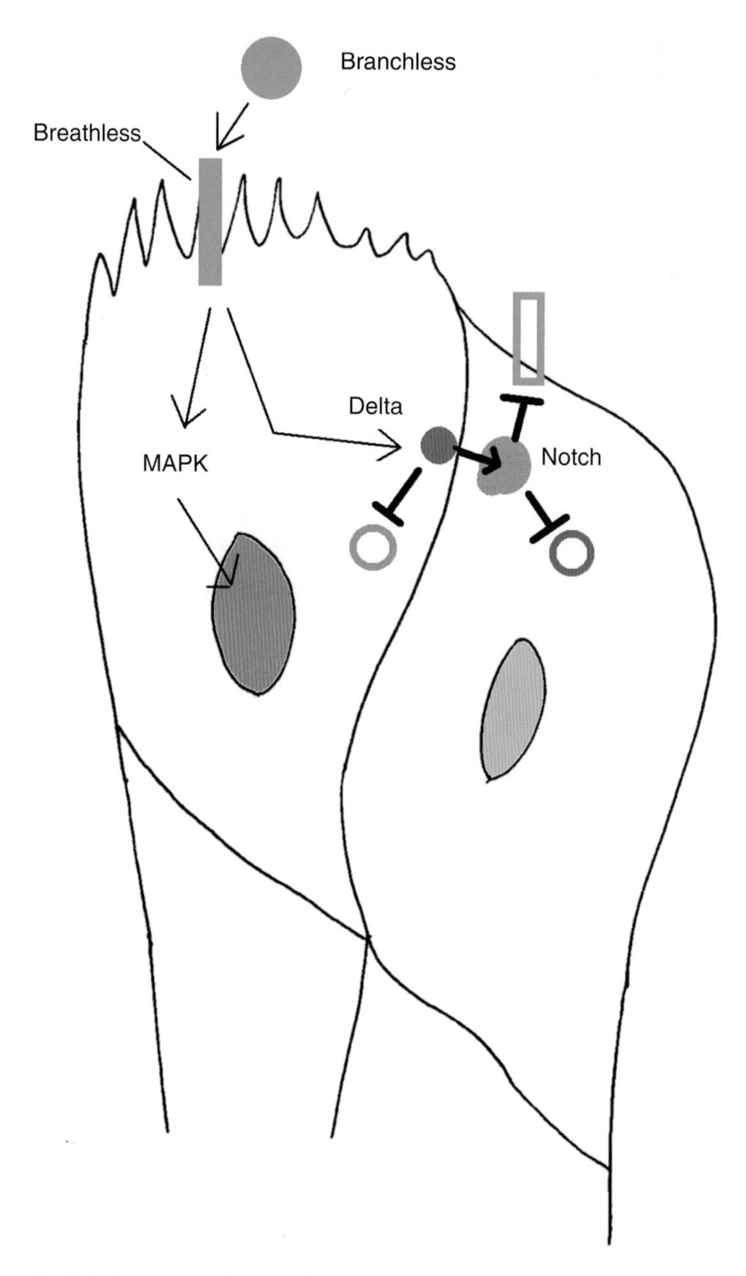

FIGURE 4.6.11 The role of Notch and Delta in restriction of MAP-kinase activation to one tip cell in branching tracheae of *D. melanogaster*.

characterized by having enlarged tips, usually called ampullae because of the way that their shape resembles their namesake Romano-Greek perfume bottles. The first morphological sign of branching occurs when clefts develop in an ampulla so that its leading edge can progress forward only between the clefts, and the old ampulla has therefore effectively divided into two or more new ones.

Branching by clefting has been studied most closely in the submandibular salivary glands of the mouse. These glands form from the epithelium of the oral cavity, which thickens and pushes down into the mesenchyme of the first branchial arch of the embryo. The epithelium forms a solid ampullary structure that undergoes branching morphogenesis and later hollows out by apoptosis.[101] Epithelial growth and branching is controlled by a number of mesenchyme-derived growth factors, including FGF10, without which salivary gland morphogenesis fails;[102] signalling by the sonic hedgehog, EGF, TGF, TNF, and IGF pathways is also probably involved.[101] Rudiments of these glands can be removed from embryos and grown in organ culture, where they can be observed and manipulated easily. Grown in normal media, the epithelial ampullae develop clefts and divide into separate growth fronts, each of which develops an ampulla of its own so that branching continues (see Figure 4.6.12). This branching growth will take place when the epithelium is surrounded by mesenchyme as it would be *in vivo*, and will also take place if it is cultured on its own as long as it is provided with appropriate

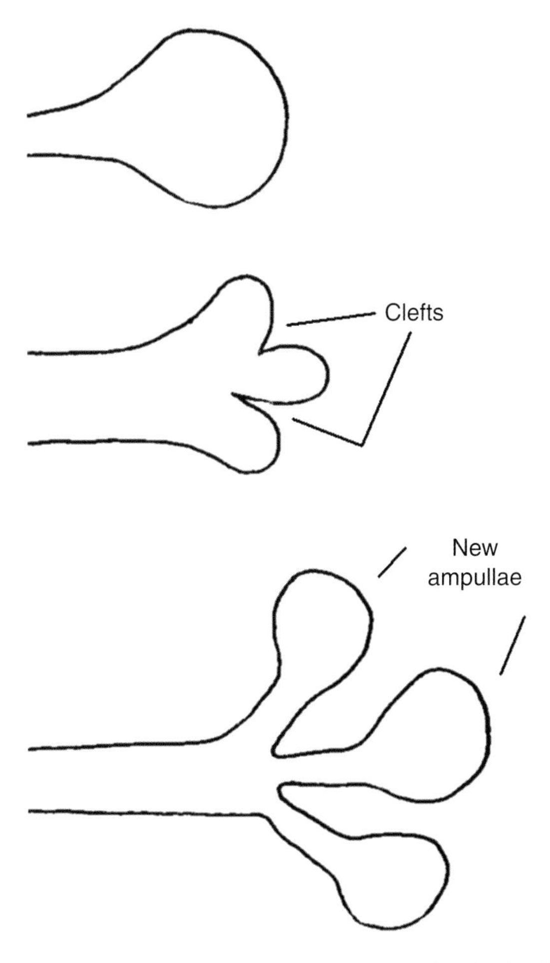

FIGURE 4.6.12 Clefting in the salivary gland. The diagram shows only the epithelium, but in reality it is surrounded by mesenchymal cells.

mesenchyme-derived factors and an artificial matrix.[103] This is an important observation because it indicates that the mechanism for branching morphogenesis lies within the epithelium itself rather than the surrounding cells.

The epithelium grows by cell proliferation as it branches, but proliferation is not required for the development of clefts, and even if rudiments are X-rayed or treated with aphidicollin to reduce greatly both DNA synthesis and mitotic division, clefts still form normally.[104] This emphasizes that clefts are not simply places at which epithelial advance is blocked, but they rather push into the epithelium, invading its space. The clefts, once they have formed, are associated with collagen fibres up to 2.5 μm in diameter, which appear and thicken as the clefts deepen.[105] These contain several types of collagen, but collagen III is particularly strong.[106] Collagen fibres can bear considerable tension so that their presence in the cleft may be responsible for holding back the epithelium in the cleft while the tissue on each side is free to advance. This idea is supported by the observation that treating salivary glands with exogenous collagenase eliminates clefting and branching, whereas treating them with inhibitors of their endogenous collagenases enhances clefting.[107,108]

The organization of type I and III collagens into the type of fibrils found in clefting does not occur by simple self-assembly, at least in *in vivo* conditions, and instead requires a scaffold of the extracellular matrix protein fibronectin.[35] Inhibiting the function of fibronectins using antibodies, or inhibiting fibronectin synthesis using RNA interference, blocks the formation of collagen fibrils and therefore blocks the clefting of salivary gland epithelium.[79] The fibronectin scaffold is itself placed and arranged by cell-substrate adhesion complexes containing integrins such as $\alpha 5\beta 1$, and inhibition of these integrins also blocks branching.[79] This means that, although collagen fibres control epithelial shape, the epithelial cells are still ultimately in control of their morphogenesis since they determine where the collagen fibres will form. Examination of the expression pattern of fibronectin mRNA supports this idea; it is seen only in cells that are about to form or are actually forming a cleft.[79]

The epithelium of the ampulla is unusually loosely connected compared with that of the shafts and of epithelia in general, and their cells show unusually low expression of components of cell-cell junctions such as E-cadherin, desmoplakins, and ZO1.[79,80] This is particularly true in the clefts themselves, and it may be another consequence of fibronectin: addition of fibronectin aggregates to human salivary gland epithelial cells growing in culture causes them to down-regulate their cell-cell junctions locally, where they make contact with the fibronectin. The relaxing of cell-cell adhesion may help the cells, which have not yet formed a true, sealed, lumen-containing epithelium, to "flow" easily past the cleft point. Actin microfilaments are also required for salivary gland morphogenesis. Treatment of cultured rudiments with cytochalasin blocks morphogenesis and clefting, although clefting begins again when the drug is removed, and does so even when the epithelium is not in physical contact with the mesenchyme.[109] It is not clear whether the actin is needed for events such as cell wedging or for another purpose entirely; perhaps myosin-actin contraction, acting via integrin adhesions, helps to pull collagen fibrils tight and drive the cleft hard back into the epithelium.

Clefting, then, seems to be initiated by epithelial cells, which synthesize, secrete, and organize fibronectin proteins that then organize collagens I and III into fibrils capable of making a cleft. The fibronectin also causes epithelial cell-cell adhesions to weaken. What is not at all clear is how the epithelial cells organize themselves so that some initiate clefting while others, between them, do not. The mesenchyme-derived signals mentioned above are required for the stimulation of clefting, but it is highly unlikely that the position of clefts is patterned by local variations in mesenchymal secretion because the epithelium branches normally, even in mesenchyme-free culture systems. It is possible that a lateral-inhibition system operates in the epithelium itself, in which cells already expressing fibronectin inhibit their neighbours from deciding to initiate clefts themselves. Such a lateral inhibition system may rely on chemical messengers, such as the Sprouty protein that mediates lateral inhibition in insect tracheae, or it may rely on the mechanical properties of the epithelium as it is bent. There is some evidence, although no mechanism yet, for the latter possibility, because altering the amount of collagen available to make fibrils, using collagenases or inhibitors of endogenous collagenases, alters the numbers of clefts.[107,108] This suggests that collagen-rich clefts are themselves part of the feedback system that controls spacing, and not just a downstream consequence of that system.

INTUSSUSCEPTIVE BRANCHING

Intussusception♣ is a mode of branching that is seen mainly in the vascular systems of vertebrates. The process contrasts with sprouting, also used in vascular systems, in that it does not involve the formation of blind-ended tubes, and formation of loops does not require tubes to find each other and to join. Instead, ramifications and the increased surface area they bring can be added without any parts of the vessel ceasing to carry moving fluid. This feature makes intussusception ideally suited for the modification and elaboration of blood vessels. The process begins when a vessel wall folds inwards. The infolding may degenerate to leave a "pillar" that divides the stream of blood[112] (see Figure 4.6.13) or it may fuse with an infolding that is approaching from the opposite wall of the vessel (see Figure 4.6.14).

The topology of infolding endothelium ensures that the apical sides of the endothelial cells always face the blood. As the folds form, they carry in with them the extracellular matrix that is associated with their basal surfaces, especially collagen, and they can also carry in with them "endothelial-like" cells that are closely associated

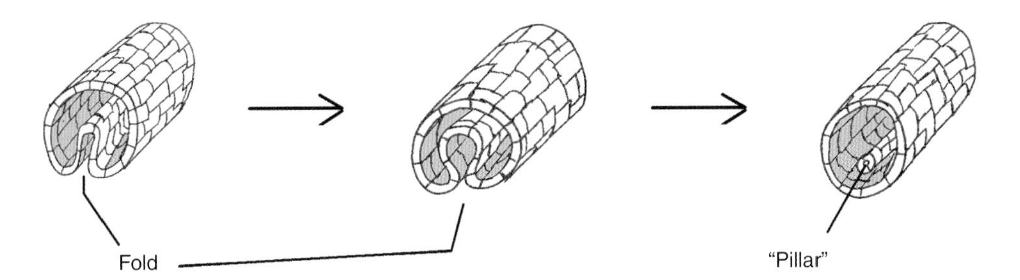

FIGURE 4.6.13 Formation of a pillar by infolding followed by loss of the basal parts of the fold.

♣The term *interssusception* means "a taking within" (*Oxford English Dictionary*, Second Edition, 1989); it has been used to describe this mode of branching morphogenesis by authors such as Patan.[110,111]

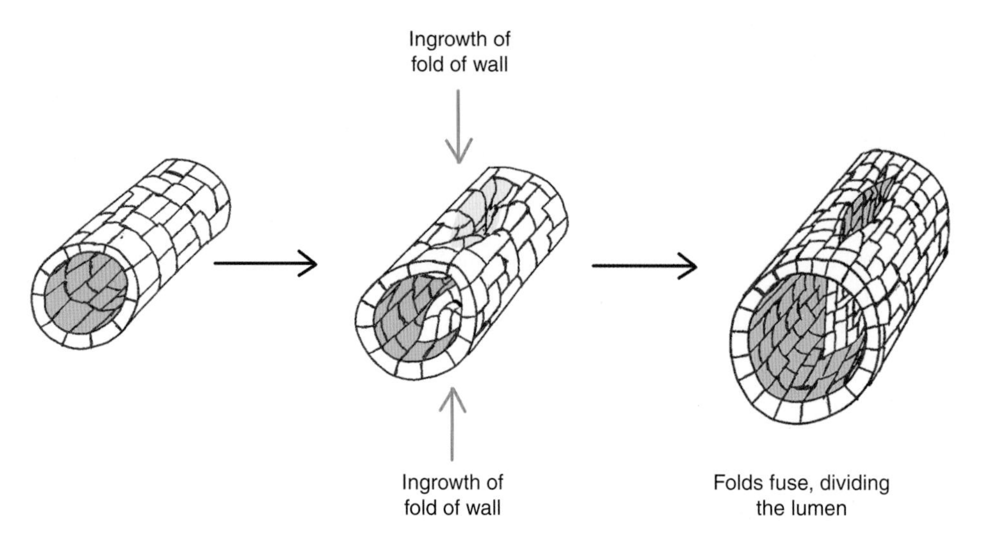

FIGURE 4.6.14 The division of a vessel lumen by fusion of infolding walls. The reality is seldom as geometrically simple as this diagram (see main text).

with the outsides of vessel walls.[112] Where folds completely cross the vessel, the space between the basal surfaces of the endothelial cells can be invaded by cells of the surrounding tissue so that the tissue effectively flows into the new "space" between the two halves of what was once a single vessel. Where folds will form just a pillar, the processes of endothelial-like cells that are trapped within the pillar organize the collagen there so that its fibrils form mechanically strong bundles. The cells and the collagen remain in the pillar core when it is cut off by disappearance of the base of the fold itself (see Figure 4.6.15).

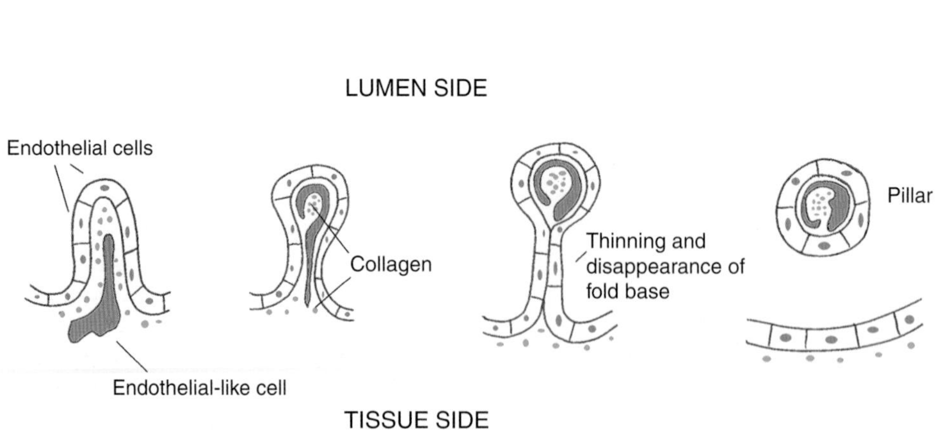

LUMEN SIDE

Endothelial cells

Collagen

Thinning and
disappearance of
fold base

Pillar

Endothelial-like cell

TISSUE SIDE

FIGURE 4.6.15 The role of endothelial-like cells, shown in red, in the organization of collagen during formation of intralumenal pillars. This diagram is based on the discussion of S. Patan et al.[112]

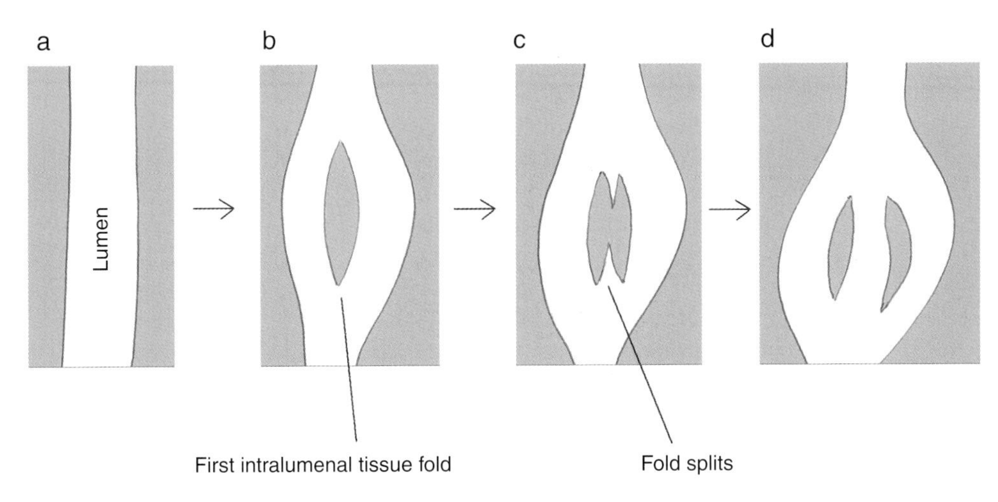

FIGURE 4.6.16 The splitting of an intralumenal tissue fold to create a new passage for blood flow. Again, the lumen is white and the tissue grey.

One they have formed, the tissue barriers that divide a lumen can themselves split so that blood flows through them, thus increasing the surface area for substance exchange even more (see Figure 4.6.16).

The cellular mechanisms of interssusceptive branching are not yet understood in detail, but by drawing analogies with the processes of epithelial folding and fusion, it is possible to make educated guesses that can be tested by experiment. The general decision to undergo branching at all is probably regulated by the tissues that the blood system is there to serve. Tissues that require a better blood supply—for example, those that are anoxic—secrete factors such as VEGF, FGF2, and angiopoietins that encourage both sprouting and interssusceptive branching.[91,110,113,114,115,116] This will presumably create a general stimulus for branching activity but would be unlikely to specify precisely around the edge of a vessel an endothelial fold/pillar should form. The positions of folds and pillars is probably determined at least in part by the conditions of blood flow within the vessel, with the flow conditions either specifying the line of emergence of a fold or determining which of many random attempts to make a fold can be stable enough to succeed.

The idea that blood flow can specify the location of a morphological event is not as far-fetched as it may seem, and there are several ways in which this might happen. Endothelial cells are known to have a number of mechanisms by which they can sense the pressure and shear of the fluid that flows across them and alter their gene expression accordingly.[117,118] It is therefore possible that local measurements of flow act as inputs to the morphogenetic effectors of folding (see below) and at least bias the probability of folds being initiated at particular points. The pressure exerted by a moving fluid decreases as speed of movement increases,♣ and this could, in principle, also induce infolding by a direct sucking in of the walls, which could be stabilized by positive

♣This is the Bernoulli effect: it can be demonstrated very simply in the lab by hanging two Ping-Pong balls on long strings so that they are next to each other, separated by a few centimetres. If a student blows between the balls, to create air whose velocity is greater than that in the rest of the room, the balls draw *together*.

feedback (e.g., cell bending driven by the fluid initiating cytoskeletal changes that extend the fold). Even if folds initiate randomly, those that attempt to fuse and close off a zone carrying a strong flow of blood would be unlikely to succeed because they would be pushed apart by the blood surging through, whereas those growing along lines of slow and laminar blood flow would be much more likely to be able to fuse to make a wall.

The cellular mechanisms of fold formation have not been investigated in detail, but the obvious analogy with epithelial invagination and evagination makes it likely that it is driven either by cell wedging or by matrix swelling. Cell multiplication is also likely to play a role, even if only to provide for the increase in surface area, and this can be driven by local sources of VEGF, FGFs, and so on.[119,120,121] Both the joining of folds and the formation of pillars requires fusion of endothelial sheets. Joining of folds requires apex-apex fusion and seems to involve cell projections that may be similar to the filopodia seen when fusing epithelia approach one another (see Chapter 4.5). Formation of pillars requires some re-polarization of cells so that the stem of the pillar can seal off in both directions as it degenerates; this is probably analogous to the cell re-polarization and fusion that occurs when the neural tube separates from the ectoderm (see Chapter 4.4), except that the apico-basal polarity of the cells is reversed.

AUTOMATIC VERSUS PLANNED ARCHITECTURE IN BRANCHING SYSTEMS

This chapter has placed strong emphasis on the ability of branching systems to grow and to organize themselves by adaptive self-organization, according to the needs of the tissue they serve. This type of development is used widely in developing organs, particularly for the finer details, and is the only one available for the healing of wounded tissue. It is not, however, generally used for the largest-diameter tubes in a branching system; the architecture of the great blood vessels, of the bronchi, and of the largest (earliest formed) tubes of the kidney, pancreas, and so on seems to be programmed differently. A dialog between the developing branched system and its surroundings is still usually involved but, rather than being shaped by signals reflecting the local balance of supply and demand, the pattern is controlled by specific additional cues. In the zebrafish *Brachydanio rerio*, for example, the location of largest arteries (the aortae) just below the notochord is determined by specific cues. The angioblasts, which are precursors of endothelia, arise below and to the sides of the somites and migrate along the medial surfaces of the somites to arrive just below the notochord. They migrate as individual motile cells and, having a VEGF receptor called neuropilin-1, they are attracted by VEGF that is produced by the medial edge of the somites. As the dorsal migration of the angioblasts proceeds, the ventral parts of the somites produce semaphorin-3a, which repels the angioblasts and traps them into the small zone just under the notochord.[122] The notochord itself is also a source of patterning information, and in embryos in which the notochord is absent, the aorta fails to form[123] (see Figure 4.6.17). The main trunks of the blood system are therefore positioned by specific patterning signals before the phase of local adaptive self-organization begins. The largest vessels serving each organ and system also tend to be patterned by specific influences, before local self-organization takes over.

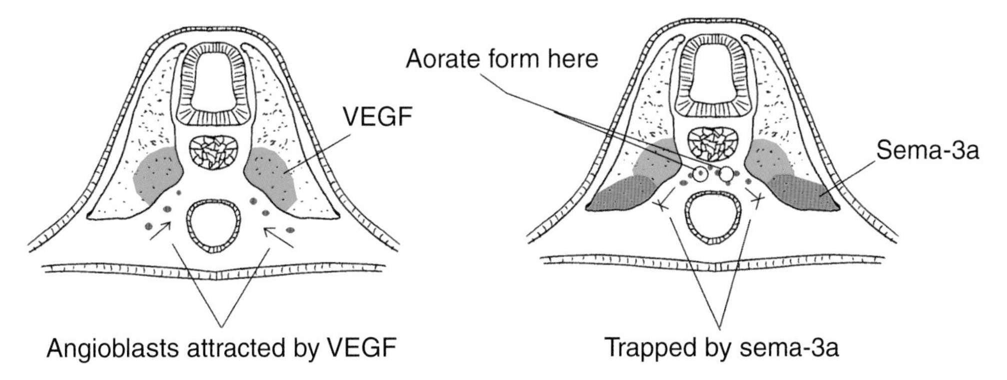

FIGURE 4.6.17 Some of the signals that attract the angioblasts (purple) to the sub-notochord region in which the aortae will form.

The same seems to apply to the "trunk" epithelial tubules in organs such as the lung and kidney. The specifically "early" patterning signals have not been identified yet, but it is clear that the first divisions of the branching system are highly stereotypical. They become more variable, and responsive to vagaries in the tissues around them, only later.[124]

Reference List

1. Davies, JA (2004) Branching morphogenesis. (Landes Biomedical)
2. Radisky, DC, Hirai, Y, and Bissell, MJ (2003) Delivering the message: Epimorphin and mammary epithelial morphogenesis. *Trends Cell Biol.* **13**: 426–434
3. al Awqati, Q and Goldberg, MR (1998) Architectural patterns in branching morphogenesis in the kidney. *Kidney Int.* **54**: 1832–1842
4. Saffman, PG and Taylor, G (1958) The penetration of a fluid into a porous medium or Hele-Shaw cell containing a more viscous liquid. *Proc. Roy. Soc. Lond. Ser. A.* **245**: 312–329
5. Howison, D (1986) Fingering in Hele-Shaw cells. *J. Fluid Mech.* **16**: 439–453
6. Mullins, WW and Sekerka, RF (1963) Morphological stability of a particle growing by diffusion or heat flow. *J. Applied Math.* **34**: 323–329
7. Fleury, V, Watanabe, W, Nguyen, T-H, Unbekandt, M, Warburton, D, Dejmek, M, Nguyen, MB, Lindner, A, and Schwartz, L (2004) Physical mechanisms of branching morphogenesis in animals. In Davies, JA, Branching morphogenesis (Landes Bioscience)
8. Montesano, R, Schaller, G, and Orci, L (23-8-1991) Induction of epithelial tubular morphogenesis in vitro by fibroblast-derived soluble factors. *Cell.* **66**: 697–711
9. Montesano, R, Matsumoto, K, Nakamura, T, and Orci, L (29-11-1991) Identification of a fibroblast-derived epithelial morphogen as hepatocyte growth factor. *Cell.* **67**: 901–908
10. Sakurai, H and Nigam, SK (1997) Transforming growth factor-beta selectively inhibits branching morphogenesis but not tubulogenesis. *Am. J. Physiol.* **272**: F139–F146
11. Gupta, IR, Piscione, TD, Grisaru, S, Phan, T, Macias-Silva, M, Zhou, X, Whiteside, C, Wrana, JL, and Rosenblum, ND (10-9-1999) Protein kinase A is a negative regulator of renal branching morphogenesis and modulates inhibitory and stimulatory bone morphogenetic proteins. *J. Biol. Chem.* **274**: 26305–26314
12. Piscione, TD, Yager, TD, Gupta, IR, Grinfeld, B, Pei, Y, Attisano, L, Wrana, JL, and Rosenblum, ND (1997) BMP-2 and OP-1 exert direct and opposite effects on renal branching morphogenesis. *Am. J. Physiol.* **273**: F961–F975
13. Russell, WE, McGowan, JA, and Bucher, NL (1984) Partial characterization of a hepatocyte growth factor from rat platelets. *J. Cell Physiol.* **119**: 183–192

14. Nakamura, T, Nawa, K, Ichihara, A, Kaise, N, and Nishino, T (30-11-1987) Purification and subunit structure of hepatocyte growth factor from rat platelets. *FEBS Lett.* **224**: 311–316
15. Ishibashi, K, Sasaki, S, Sakamoto, H, Nakamura, Y, Hata, T, Nakamura, T, and Marumo, F (31-1-1992) Hepatocyte growth factor is a paracrine factor for renal epithelial cells: Stimulation of DNA synthesis and NA,K-ATPase activity. *Biochem. Biophys. Res. Commun.* **182**: 960–965
16. Stoker, M and Perryman, M (1985) An epithelial scatter factor released by embryo fibroblasts. *J. Cell Sci.* 77: 209–223
17. Gherardi, E, Gray, J, Stoker, M, Perryman, M, and Furlong, R (1989) Purification of scatter factor, a fibroblast-derived basic protein that modulates epithelial interactions and movement. *Proc. Natl. Acad. Sci. U.S.A.* **86**: 5844–5848
18. Li, S, Gerrard, ER, and Balkovetz, DF (7-4-2004) Evidence for ERK 1/2 Phosphorylation controlling contact inhibition of proliferation in Madin-Darby canine kidney epithelial cells. *Am. J. Physiol. Cell Physiol.* **287**: C432–C439
19. Balkovetz, DF (1999) Evidence that hepatocyte growth factor abrogates contact inhibition of mitosis in Madin-Darby canine kidney cell monolayers. *Life Sci.* **64**: 1393–1401
20. Santos, OF and Nigam, SK (1993) HGF-induced tubulogenesis and branching of epithelial cells is modulated by extracellular matrix and TGF-beta. *Dev. Biol.* **160**: 293–302
21. Williams, MJ and Clark, P (2003) Microscopic analysis of the cellular events during scatter factor/hepatocyte growth factor-induced epithelial tubulogenesis. *J. Anat.* **203**: 483–503
22. Pollack, AL, Apodaca, G, and Mostov, KE (2004) Hepatocyte growth factor induces MDCK cell morphogenesis without causing loss of tight junction functional integrity. *Am. J. Physiol. Cell Physiol.* **286**: C482-C494
23. Jiang, ST, Chiu, SJ, Chen, HC, Chuang, WJ, and Tang, MJ (2001) Role of alpha(3)beta(1) integrin in tubulogenesis of Madin-Darby canine kidney cells. *Kidney Int.* **59**: 1770–1778
24. Sachs, M, Weidner, KM, Brinkmann, V, Walther, I, Obermeier, A, Ullrich, A, and Birchmeier, W (1996) Motogenic and morphogenic activity of epithelial receptor tyrosine kinases. *J. Cell Biol.* **133**: 1095–1107
25. Pelicci, G, Giordano, S, Zhen, Z, Salcini, AE, Lanfrancone, L, Bardelli, A, Panayotou, G, Waterfield, MD, Ponzetto, C, Pelicci, PG (20-4-1995) The motogenic and mitogenic responses to HGF are amplified by the Shc adaptor protein. *Oncogene.* **10**: 1631–1638
26. Ponzetto, C, Zhen, Z, Audero, E, Maina, F, Bardelli, A, Basile, ML, Giordano, S, Narsimhan, R, and Comoglio, P (14-6-1996) Specific uncoupling of GRB2 from the Met receptor. Differential effects on transformation and motility. *J. Biol. Chem.* **271**: 14119–14123
27. Weidner, KM, Di Cesare, S, Sachs, M, Brinkmann, V, Behrens, J, and Birchmeier, W (14-11-1996) Interaction between Gab1 and the c-Met receptor tyrosine kinase is responsible for epithelial morphogenesis. *Nature.* **384**: 173–176
28. Boccaccio, C, Ando, M, Tamagnone, L, Bardelli, A, Michieli, P, Battistini, C, and Comoglio, PM (15-1-1998) Induction of epithelial tubules by growth factor HGF depends on the STAT pathway. *Nature.* **391**: 285–288
29. Khwaja, A, Lehmann, K, Marte, BM, and Downward, J (24-7-1998) Phosphoinositide 3-kinase induces scattering and tubulogenesis in epithelial cells through a novel pathway. *J. Biol. Chem.* **273**: 18793–18801
30. Miki, H, Sasaki, T, Takai, Y, and Takenawa, T (1-1-1998) Induction of filopodium formation by a WASP-related actin-depolymerizing protein N-WASP. *Nature.* **391**: 93–96
31. Yamaguchi, H, Miki, H, and Takenawa, T (1-5-2002) Neural Wiskott-Aldrich syndrome protein is involved in hepatocyte growth factor-induced migration, invasion, and tubulogenesis of epithelial cells. *Cancer Res.* **62**: 2503–2509
32. Wells, CM, Abo, A, and Ridley, AJ (15-10-2002) PAK4 is activated via PI3K in HGF-stimulated epithelial cells. *J. Cell Sci.* **115**: 3947–3956

33. Chiu, SJ, Jiang, ST, Wang, YK, and Tang, MJ (2002) Hepatocyte growth factor upregulates alpha2beta1 integrin in Madin-Darby canine kidney cells: Implications in tubulogenesis. *J. Biomed. Sci.* 9: 261–272
34. Velling, T, Risteli, J, Wennerberg, K, Mosher, DF, and Johansson, S (4-10-2002) Polymerization of type I and III collagens is dependent on fibronectin and enhanced by integrins α 11β 1 and α 2β 1. *J. Biol. Chem.* **277**: 37377–37381
35. Jiang, ST, Chuang, WJ, and Tang, MJ (2000) Role of fibronectin deposition in branching morphogenesis of Madin-Darby canine kidney cells. *Kidney Int.* **57**: 1860–1867
36. Kadono, Y, Shibahara, K, Namiki, M, Watanabe, Y, Seiki, M, and Sato, H (29-10-1998) Membrane type 1-matrix metalloproteinase is involved in the formation of hepatocyte growth factor/scatter factor-induced branching tubules in Madin-Darby canine kidney epithelial cells. *Biochem. Biophys. Res. Commun.* **251**: 681–687
37. Santos, OF and Nigam, SK (1993) HGF-induced tubulogenesis and branching of epithelial cells is modulated by extracellular matrix and TGF-beta. *Dev. Biol.* **160**: 293–302
38. Maeshima, A, Zhang, YQ, Furukawa, M, Naruse, T, and Kojima, I (2000) Hepatocyte growth factor induces branching tubulogenesis in MDCK cells by modulating the activin-follistatin system. *Kidney Int.* **58**: 1511–1522
39. Uv, A, Cantera, R, and Samakovlis, C (2003) *Drosophila* tracheal morphogenesis: Intricate cellular solutions to basic plumbing problems. *Trends Cell Biol.* **13**: 301–309
40. Sutherland, D, Samakovlis, C, and Krasnow, MA (13-12-1996) Branchless encodes a *Drosophila* FGF homolog that controls tracheal cell migration and the pattern of branching. *Cell.* **87**: 1091–1101
41. Reichman-Fried, M and Shilo, BZ (1995) Breathless, a *Drosophila* FGF receptor homolog, is required for the onset of tracheal cell migration and tracheole formation. *Mech. Dev.* **52**: 265–273
42. Ribeiro, C, Ebner, A, and Affolter, M (2002) In vivo imaging reveals different cellular functions for FGF and Dpp signalling in tracheal branching morphogenesis. *Dev. Cell.* **2**: 677–683
43. Ikeya, T and Hayashi, S (1999) Interplay of Notch and FGF signalling restricts cell fate and MAPK activation in the *Drosophila* trachea. *Development.* **126**: 4455–4463
44. Gabay, L, Seger, R, and Shilo, BZ (1997) MAP kinase in situ activation atlas during *Drosophila* embryogenesis. *Development.* **124**: 3535–3541
45. Jin, J, Anthopoulos, N, Wetsch, B, Binari, RC, Isaac, DD, Andrew, DJ, Woodgett, JR, and Manoukian, AS (2001) Regulation of *Drosophila* tracheal system development by protein kinase B. *Dev. Cell.* **1**: 817–827
46. Lavenburg, KR, Ivey, J, Hsu, T, and Muise-Helmericks, RC (2003) Coordinated functions of Akt/PKB and ETS1 in tubule formation. *Faseb J.* **17**: 2278–2280
47. Wolf, C, Gerlach, N, and Schuh, R (2002) *Drosophila* tracheal system formation involves FGF-dependent cell extensions contacting bridge-cells. *EMBO Rep.* **3**: 563–568
48. Jack, J and Myette, G (1997) The genes raw and ribbon are required for proper shape of tubular epithelial tissues in *Drosophila*. *Genetics.* **147**: 243–253
49. Shim, K, Blake, KJ, Jack, J, and Krasnow, MA (2001) The *Drosophila* ribbon gene encodes a nuclear BTB domain protein that promotes epithelial migration and morphogenesis. *Development.* **128**: 4923–4933
50. Bradley, PL and Andrew, DJ (2001) Ribbon encodes a novel BTB/POZ protein required for directed cell migration in *Drosophila melanogaster*. *Development.* **128**: 3001–3015
51. Hemphala, J, Uv, A, Cantera, R, Bray, S, and Samakovlis, C (2003) Grainy head controls apical membrane growth and tube elongation in response to Branchless/FGF signalling. *Development.* **130**: 249–258
52. Wilanowski, T, Tuckfield, A, Cerruti, L, O'Connell, S, Saint, R, Parekh, V, Tao, J, Cunningham, JM, and Jane, SM (2002) A highly conserved novel family of mammalian developmental transcription factors related to *Drosophila* Grainyhead. *Mech. Dev.* **114**: 37–50

53. Ting, SB, Wilanowski, T, Cerruti, L, Zhao, LL, Cunningham, JM, and Jane, SM (15-3-2003) The identification and characterization of human Sister-of-Mammalian Grainyhead (SOM) expands the Grainyhead-like family of developmental transcription factors. *Biochem. J.* **370**: 953–962
54. Saxen, L (1987) Organogenesis of the kidney. (Cambridge University Press)
55. Popsueva, A, Poteryaev, D, Arighi, E, Meng, X, Angers-Loustau, A, Kaplan, D, Saarma, M, and Sariola, H (14-4-2003) GDNF promotes tubulogenesis of GFRalpha1-expressing MDCK cells by Src-mediated phosphorylation of Met receptor tyrosine kinase. *J. Cell Biol.* **161**: 119–129
56. Cantley, LG and Cantley, LC (1995) Signal transduction by the hepatocyte growth factor receptor, c-met. Activation of the phosphatidylinositol 3-kinase. *J. Am. Soc. Nephrol.* **5**: 1872–1881
57. Rosario, M and Birchmeier, W (2003) How to make tubes: Signalling by the Met receptor tyrosine kinase. *Trends Cell Biol.* **13**: 328–335
58. Klint, P and Claesson-Welsh, L (15-2-1999) Signal transduction by fibroblast growth factor receptors. *Front Biosci.* **4**: D165–D177
59. Jorissen, RN, Walker, F, Pouliot, N, Garrett, TP, Ward, CW, and Burgess, AW (10-3-2003) Epidermal growth factor receptor: Mechanisms of activation and signalling. *Exp. Cell Res.* **284**: 31–53
60. Sariola, H and Saarma, M (1-10-2003) Novel functions and signalling pathways for GDNF. *J. Cell Sci.* **116**: 3855–3862
61. Kashimata, M, Sayeed, S, Ka, A, Onetti-Muda, A, Sakagami, H, Faraggiana, T, and Gresik, EW (15-4-2000) The ERK-1/2 signalling pathway is involved in the stimulation of branching morphogenesis of fetal mouse submandibular glands by EGF. *Dev. Biol.* **220**: 183–196
62. Kling, DE, Lorenzo, HK, Trbovich, AM, Kinane, TB, Donahoe, PK, and Schnitzer, JJ (2002) MEK-1/2 inhibition reduces branching morphogenesis and causes mesenchymal cell apoptosis in fetal rat lungs. *Am. J. Physiol. Lung Cell Mol. Physiol.* **282**: L370–L378
63. Watanabe, T and Costantini, F (1-7-2004) Real-time analysis of ureteric bud branching morphogenesis in vitro. *Dev. Biol.* **271**: 98–108
64. Fisher, CE, Michael, L, Barnett, MW, and Davies, JA (2001) Erk MAP kinase regulates branching morphogenesis in the developing mouse kidney. *Development.* **128**: 4329–4338
65. Koyama, N, Kashimata, M, Sakashita, H, Sakagami, H, and Gresik, EW (2003) EGF-stimulated signalling by means of PI3K, PLCgamma1, and PKC isozymes regulates branching morphogenesis of the fetal mouse submandibular gland. *Dev. Dyn.* **227**: 216–226
66. Larsen, M, Hoffman, MP, Sakai, T, Neibaur, JC, Mitchell, JM, and Yamada, KM (1-3-2003) Role of PI 3-kinase and PIP3 in submandibular gland branching morphogenesis. *Dev. Biol.* **255**: 178–191
67. Tang, MJ, Cai, Y, Tsai, SJ, Wang, YK, and Dressler, GR (1-3-2002) Ureteric bud outgrowth in response to RET activation is mediated by phosphatidylinositol 3-kinase. *Dev. Biol.* **243**: 128–136
68. Zhao, J, Crowe, DL, Castillo, C, Wuenschell, C, Chai, Y, and Warburton, D (2000) Smad7 is a TGF-beta-inducible attenuator of Smad2/3-mediated inhibition of embryonic lung morphogenesis. *Mech. Dev.* **93**: 71–81
69. Zhao, J, Lee, M, Smith, S, and Warburton, D (15-2-1998) Abrogation of Smad3 and Smad2 or of Smad4 gene expression positively regulates murine embryonic lung branching morphogenesis in culture. *Dev. Biol.* **194**: 182–195
70. Kadoya, Y, Mochizuki, M, Nomizu, M, Sorokin, L, and Yamashina, S (1-11-2003) Role for laminin-alpha5 chain LG4 module in epithelial branching morphogenesis. *Dev. Biol.* **263**: 153–164

71. Kadoya, Y, Nomizu, M, Sorokin, LM, Yamashina, S, and Yamada, Y (1998) Laminin alpha1 chain G domain peptide, RKRLQVQLSIRT, inhibits epithelial branching morphogenesis of cultured embryonic mouse submandibular gland. *Dev. Dyn.* **212**: 394–402
72. Kadoya, Y, Salmivirta, K, Talts, JF, Kadoya, K, Mayer, U, Timpl, R, and Ekblom, P (1997) Importance of nidogen binding to laminin gamma1 for branching epithelial morphogenesis of the submandibular gland. *Development.* **124**: 683–691
73. Ekblom, P, Ekblom, M, Fecker, L, Klein, G, Zhang, HY, Kadoya, Y, Chu, ML, Mayer, U, and Timpl, R (1994) Role of mesenchymal nidogen for epithelial morphogenesis in vitro. *Development.* **120**: 2003–2014
74. Matsui, R, Thurlbeck, WM, Shehata, EI, and Sekhon, HS (1996) Two different patterns of airway branching regulated by different components of the extracellular matrix in vitro. *Exp. Lung Res.* **22**: 593–611
75. Miyake, T, Komura, H, Tokuhira, A, Yamamoto, T, Miyake, A, Tanizawa, O, Terada, N, Yamamoto, R, Yoshida, S, Tsuji, M (1990) Impaired development of mammary glands in scorbutic rats unable to synthesize ascorbic acid. *J. Steroid Biochem. Mol. Biol.* **37**: 31–37
76. Lelongt, B, Trugnan, G, Murphy, G, and Ronco, PM (24-3-1997) Matrix metalloproteinases MMP2 and MMP9 are produced in early stages of kidney morphogenesis but only MMP9 is required for renal organogenesis in vitro. *J. Cell Biol.* **136**: 1363–1373
77. Michael, L and Davies, JA (2004) Pattern and regulation of cell proliferation during murine ureteric bud development. *J. Anat.* **204**: 241–255
78. Sakai, T, Larsen, M, and Yamada, KM (19-6-2003) Fibronectin requirement in branching morphogenesis. *Nature.* **423**: 876–881
79. Hieda, Y, Iwai, K, Morita, T, and Nakanishi, Y (1996) Mouse embryonic submandibular gland epithelium loses its tissue integrity during early branching morphogenesis. *Dev. Dyn.* **207**: 395–403
80. Qiao, J, Cohen, D, and Herzlinger, D (1995) The metanephric blastema differentiates into collecting system and nephron epithelia in vitro. *Development.* **121**: 3207–3214
81. Davies, J, Lyon, M, Gallagher, J, and Garrod, D (1995) Sulphated proteoglycan is required for collecting duct growth and branching but not nephron formation during kidney development. *Development.* **121**: 1507–1517
82. Michael, L (2003) PhD Thesis. (University of Edinburgh)
83. Davies, JA, Ladomery, M, Hohenstein, P, Michael, L, Shafe, A, Spraggon, L, and Hastie, N (15-1-2004) Development of an siRNA-based method for repressing specific genes in renal organ culture and its use to show that the Wt1 tumour suppressor is required for nephron differentiation. *Hum. Mol. Genet.* **13**: 235–246
84. Yang, Y, Spitzer, E, Meyer, D, Sachs, M, Niemann, C, Hartmann, G, Weidner, KM, Birchmeier, C, and Birchmeier, W (1995) Sequential requirement of hepatocyte growth factor and neuregulin in the morphogenesis and differentiation of the mammary gland. *J. Cell Biol.* **131**: 215–226
85. Niemann, C, Brinkmann, V, Spitzer, E, Hartmann, G, Sachs, M, Naundorf, H, and Birchmeier, W (19-10-1998) Reconstitution of mammary gland development in vitro: Requirement of c-met and c-erbB2 signalling for branching and alveolar morphogenesis. *J. Cell Biol.* **143**: 533–545
86. Sainio, K, Suvanto, P, Davies, J, Wartiovaara, J, Wartiovaara, K, Saarma, M, Arumae, U, Meng, X, Lindahl, M, Pachnis, V, and Sariola, H (1997) Glial-cell-line-derived neurotrophic factor is required for bud initiation from ureteric epithelium. *Development.* **124**: 4077–4087
87. Cramer, T, Schipani, E, Johnson, RS, Swoboda, B, and Pfander, D (2004) Expression of VEGF isoforms by epiphyseal chondrocytes during low-oxygen tension is HIF-1 alpha dependent. *Osteoarthritis. Cartilage.* **12**: 433–439
88. Shweiki, D, Itin, A, Soffer, D, and Keshet, E (29-10-1992) Vascular endothelial growth factor induced by hypoxia may mediate hypoxia-initiated angiogenesis. *Nature.* **359**: 843–845

89. Ladoux, A and Frelin, C (15-9-1993) Hypoxia is a strong inducer of vascular endothelial growth factor mRNA expression in the heart. *Biochem. Biophys. Res. Commun.* **195**: 1005–1010
90. Kearney, JB, Kappas, NC, Ellerstrom, C, DiPaola, FW, and Bautch, VL (15-6-2004) The VEGF receptor flt-1 (VEGFR-1) is a positive modulator of vascular sprout formation and branching morphogenesis. *Blood.* **103**: 4527–4535
91. Wilting, J, Christ, B, Bokeloh, M, and Weich, HA (1993) In vivo effects of vascular endothelial growth factor on the chicken chorioallantoic membrane. *Cell Tissue Res.* **274**: 163–172
92. Sainio, K, Saarma, M, Nonclercq, D, Paulin, L, and Sariola, H (1994) Antisense inhibition of low-affinity nerve growth factor receptor in kidney cultures: Power and pitfalls. *Cell Mol. Neurobiol.* **14**: 439–457
93. Davies, JA (1996) Mesenchyme to epithelium transition during development of the mammalian kidney tubule. *Acta. Anat. (Basel).* **156**: 187–201
94. Davies, J (2001) Intracellular and extracellular regulation of ureteric bud morphogenesis. *J. Anat.* **198**: 257–264
95. Perantoni, AO, Williams, CL, and Lewellyn, AL (1991) Growth and branching morphogenesis of rat collecting duct anlagen in the absence of metanephrogenic mesenchyme. *Differentiation.* **48**: 107–113
96. Sakurai, H, Bush, KT, and Nigam, SK (2001) Identification of pleiotrophin as a mesenchymal factor involved in ureteric bud branching morphogenesis. *Development.* **128**: 3283–3293
97. Hacohen, N, Kramer, S, Sutherland, D, Hiromi, Y, and Krasnow, MA (23-1-1998) Sprouty encodes a novel antagonist of FGF signalling that patterns apical branching of the *Drosophila* airways. *Cell.* **92**: 253–263
98. Casci, T, Vinos, J, and Freeman, M (5-3-1999) Sprouty, an intracellular inhibitor of Ras signalling. *Cell.* **96**: 655–665
99. Mailleux, AA, Tefft, D, Ndiaye, D, Itoh, N, Thiery, JP, Warburton, D, and Bellusci, S (2001) Evidence that SPROUTY2 functions as an inhibitor of mouse embryonic lung growth and morphogenesis. *Mech. Dev.* **102**: 81–94
100. Jaskoll, T and Melnick, M (2004) Embryonic salivary gland branching morphogenesis. In Davies, JA, Branching morphogenesis (Landes bioscience), pp. 156–171
101. Ohuchi, H, Hori, Y, Yamasaki, M, Harada, H, Sekine, K, Kato, S, and Itoh, N (2-11-2000) FGF10 acts as a major ligand for FGF receptor 2 IIIb in mouse multi-organ development. *Biochem. Biophys. Res. Commun.* **277**: 643–649
102. Nogawa, H and Takahashi, Y (1991) Substitution for mesenchyme by basement-membrane-like substratum and epidermal growth factor in inducing branching morphogenesis of mouse salivary epithelium. *Development.* **112**: 855–861
103. Nakanishi, Y, Morita, T, and Nogawa, H (1987) Cell proliferation is not required for the initiation of early cleft formation in mouse embryonic submandibular epithelium in vitro. *Development.* **99**: 429–437
104. Nakanishi, Y, Sugiura, F, Kishi, J, and Hayakawa, T (1986) Scanning electron microscopic observation of mouse embryonic submandibular glands during initial branching: Preferential localization of fibrillar structures at the mesenchymal ridges participating in cleft formation. *J. Embryol. Exp. Morphol.* **96**: 65–77
105. Nakanishi, Y, Nogawa, H, Hashimoto, Y, Kishi, J, and Hayakawa, T (1988) Accumulation of collagen III at the cleft points of developing mouse submandibular epithelium. *Development.* **104**: 51–59
106. Fukuda, Y, Masuda, Y, Kishi, J, Hashimoto, Y, Hayakawa, T, Nogawa, H, and Nakanishi, Y (1988) The role of interstitial collagens in cleft formation of mouse embryonic submandibular gland during initial branching. *Development.* **103**: 259–267
107. Hayakawa, T, Kishi, J, and Nakanishi, Y (1992) Salivary gland morphogenesis: Possible involvement of collagenase. *Matrix Suppl.* **1**: 344–351

108. Spooner, BS and Wessells, NK (1970) Effects of cytochalasin B upon microfilaments involved in morphogenesis of salivary epithelium. *Proc. Natl. Acad. Sci. U.S.A.* **66**: 360–361
109. Patan, S (2004) How is branching of animal blood vessels implemented? In Davies, JA, Branching morphogenesis (Landes Bioscience)
110. Patan, S, Alvarez, MJ, Schittny, JC, and Burri, PH (1992) Intussusceptive microvascular growth: A common alternative to capillary sprouting. *Arch. Histol. Cytol.* **55 Suppl**: 65–75
111. Patan, S, Haenni, B, and Burri, PH (1996) Implementation of intussusceptive microvascular growth in the chicken chorioallantoic membrane (CAM): 1. Pillar formation by folding of the capillary wall. *Microvasc. Res.* **51**: 80–98
112. Bjorndahl, M, Cao, R, Eriksson, A, and Cao, Y (11-6-2004) Blockage of VEGF-induced angiogenesis by preventing VEGF secretion. *Circ. Res.* **94**: 1443–1450
113. Dor, Y, Porat, R, and Keshet, E (2001) Vascular endothelial growth factor and vascular adjustments to perturbations in oxygen homeostasis. *Am. J. Physiol. Cell Physiol.* **280**: C1367–C1374
114. Yang, S, Xin, X, Zlot, C, Ingle, G, Fuh, G, Li, B, Moffat, B, de Vos, AM, and Gerritsen, ME (2001) Vascular endothelial cell growth factor-driven endothelial tube formation is mediated by vascular endothelial cell growth factor receptor-2, a kinase insert domain-containing receptor. *Arterioscler. Thromb. Vasc. Biol.* **21**: 1934–1940
115. Parsons-Wingerter, P, Elliott, KE, Clark, JI, and Farr, AG (2000) Fibroblast growth factor-2 selectively stimulates angiogenesis of small vessels in arterial tree. *Arterioscler. Thromb. Vasc. Biol.* **20**: 1250–1256
116. Boo, YC and Jo, H (2003) Flow-dependent regulation of endothelial nitric oxide synthase: Role of protein kinases. *Am. J. Physiol. Cell Physiol.* **285**: C499–C508
117. Yao, X, Kwan, HY, and Huang, Y (2002) A mechanosensitive cation channel in endothelial cells. *J. Card. Surg.* **17**: 340–341
118. Chen, CH and Chen, SC (1987) Evidence of the presence of a specific vascular endothelial growth factor in fetal bovine retina. *Exp. Cell Res.* **169**: 287–295
119. Gospodarowicz, D, Brown, KD, Birdwell, CR, and Zetter, BR (1978) Control of proliferation of human vascular endothelial cells. Characterization of the response of human umbilical vein endothelial cells to fibroblast growth factor, epidermal growth factor, and thrombin. *J. Cell Biol.* **77**: 774–788
120. Ferrara, N, Houck, KA, Jakeman, LB, Winer, J, and Leung, DW (1991) The vascular endothelial growth factor family of polypeptides. *J. Cell Biochem.* **47**: 211–218
121. Shoji, W, Isogai, S, Sato-Maeda, M, Obinata, M, and Kuwada, JY (2003) Semaphorin3a1 regulates angioblast migration and vascular development in zebrafish embryos. *Development.* **130**: 3227–3236
122. Fouquet, B, Weinstein, BM, Serluca, FC, and Fishman, MC (1-3-1997) Vessel patterning in the embryo of the zebrafish: Guidance by notochord. *Dev. Biol.* **183**: 37–48
123. Metzger, RJ and Krasnow, MA (4-6-1999) Genetic control of branching morphogenesis. *Science.* **284**: 1635–1639

5

Morphogenesis by Cell Proliferation and Death

CHAPTER 5.1

GROWTH, PROLIFERATION, AND DEATH: A Brief Overview

Growth and cell proliferation feature in the development of every multicellular organism. They are frequently linked so that cell proliferation drives organismal growth, but they do not have to be: early embryos can divide by cleavage without growth, tissues can grow by cell expansion in the absence of cell proliferation, and the volume of an embryo can increase by fluid uptake without an increase in the total volume of living cytoplasm (see Figure 5.1.1).

Because cells can alter in both volume and number, one has to know the value of both of these variables in order to calculate the total volume of a tissue. In most cases, the volume is determined by the product of the number and the size of the cells (exceptions to this simple rule occur where the extracellular matrix occupies a significant fraction of the volume as it does, for example, in cartilage). A number of classical experiments have suggested that cell number and cell size are controlled by separate mechanisms and that either can change without altering the other. What is more, these two variables can compensate for one another in achieving the optimum size for a tissue.

Mutual compensation by cell size and volume is illustrated well by mutants that affect cells in the developing wings of *Drosophila melanogaster*. Like other adult structures, these develop as imaginal discs in the embryo that then evert. Cells in the wing disc can be forced to proliferate more than usual by constitutive expression of the cell cycle regulator E2F, the function of which will be discussed in more detail later. If E2F is overexpressed, wings have abnormally many cells, but the size of the whole wing remains normal because each cell is proportionately smaller.[1] If, on the hand, proliferation is inhibited by temperature-sensitive inactivation of a molecule required for proliferation, the wing achieves a normal size through cell enlargement.[2] Plant mutants, too, show compensation between cell size and number in organs such as leaves.[3] The ability of an organ (and an organism) to alter cell number to compensate for changes in cell volume that are beyond its control can also be demonstrated by comparison of diploid and tetraploid salamanders. The cells of the tetraploid animals are larger than those of diploids, yet the animals themselves are the same size because the tissues of the tetraploid contain correspondingly fewer cells.[4] The size-regulation system therefore manipulates either or both variables (cell number and cell size) to achieve a set product of the two (see Figure 5.1.2).

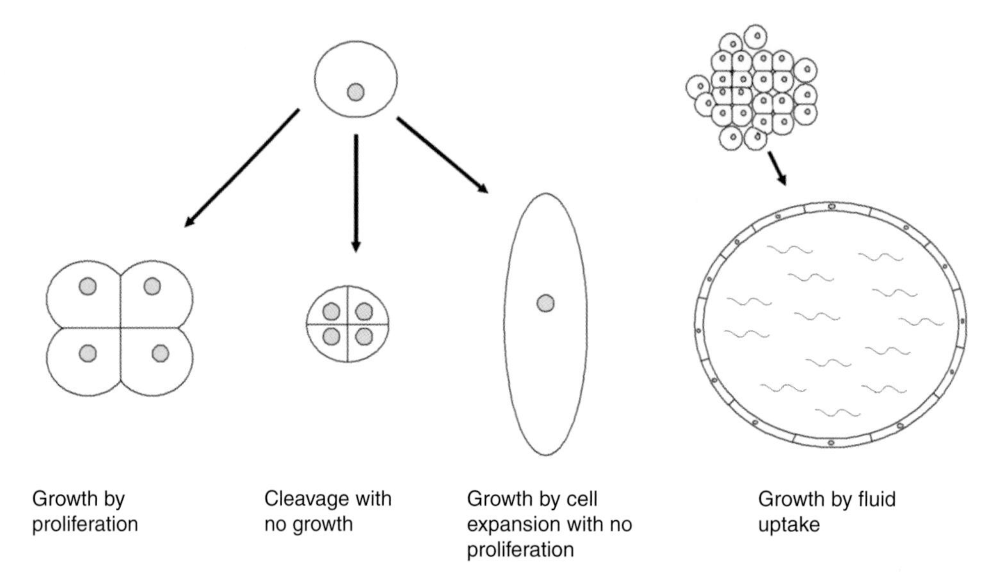

FIGURE 5.1.1 Growth and proliferation during development. Growth is commonly achieved through cell proliferation, particularly in the development of "higher" animals. The two processes do not have to be linked, however, as cleavage can take place without growth (as it does in embryos), growth can take place with no cell division, and growth of the whole embryo can take place by fluid uptake with no change in cell volume.

Mechanisms of cell expansion, particularly the anisotropic expansion that is so important in plant morphogenesis, have been discussed in Chapters 2.2 and 2.4. For this reason, this section of the book will focus mainly on the role of cell proliferation in driving tissue growth and on the role of cell death in limiting the size of cell populations. This chapter will introduce the control of cell proliferation itself; Chapter 5.2 will discuss how spatial control of cell division can be used to drive morphogenesis directly, and Chapter 5.3 will discuss the role of cell death in morphogenesis.

CELL PROLIFERATION IS CONTROLLED AT SEVERAL SCALES

Rates of cell proliferation are controlled by mechanisms that operate on scales ranging from that of the whole body to that of a cell's immediate neighbours. In the animal kingdom, at least, an individual's body tends to grow to a species- and sex-specific size, and there is only a modest variation between individuals. Domestic cats do not grow to the size of leopards, adders do not grow to the size of pythons, and dogfish do not grow to the size of basking sharks, even though the existence of these large relatives shows that there is nothing about the basic body plan that makes large size impossible. The maximum size an animal reaches is not simply a function of available resources, as it would be, for example, in the case of a hyphal fungus. A well-nourished mammal such as a human will reach a maximum height by adulthood, and supplying excess nutrients

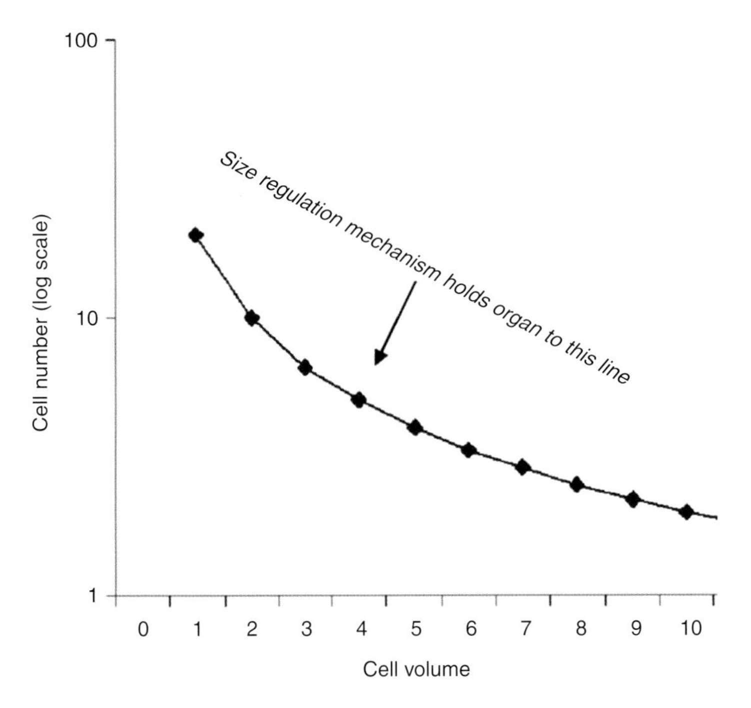

FIGURE 5.1.2 The size regulation system of organs can compensate for changes in either cell size (volume) or cell number by altering the other one, in order to achieve a set final organ size. In this diagram, cell volume is expressed in an arbitrary scale in which 1.0 happens to be 1/20th of the final organ volume.

will not cause the person to grow taller (though the person will—alas!—grow broader). Developing animals must therefore possess a global size-sensing and regulating system.

Within the body, there is strict size regulation of component parts (and this is true even for plants; leaves, for example, have a typical maximum size however large their tree is). In animals, paired organs are the same size on each side of the body and show much less variation than do the same organs between unrelated individuals of the same species. This can be verified from the outside of the body for some accessible organs, such as eyes, teeth, and testes, and also for appendages such as fingers. The similarity of size of paired internal organs can be seen clearly on dissection. More subtle than the balancing of the two copies of one organ is the need for different organs and tissues to be the correct size for each other. The liver has to be large enough to meet the metabolic needs of the rest of the body, the heart needs to be large enough to pump blood around that body, pools of motor neurons need to be large enough to innervate the muscles, and countless other examples of systems have to be balanced properly. What is more, for a developmental program that is able to evolve in response to a changing environment, the rest of the body has to be able to accommodate, automatically, a change in the size of one component that may have arisen through mutation and may offer a selective advantage.

Within these organs are many different cell types, and the ratio of their populations has to be correct within close limits if the organ is to function optimally. Because these different cell types form mixed communities (for example, epithelia and their supportive stroma), this correct proportion cannot just be regulated as an average proportion across an organ but must be maintained even at micro scales. This demands regulation mechanisms that can operate at these scales as well as at scales that balance the sizes of paired organs across the body.

The control of proliferation therefore takes place at multiple levels that interact with one another. All end up, though, affecting the machinery of cell replication so that it is a logical place to begin a more detailed description.

CONTROL OF CELL PROLIFERATION

The control of cell proliferation is a large and complex subject to which entire textbooks[5,6,7] and journals[8] are devoted. It is therefore not possible to do the whole field justice here, and neither is it desirable, since this book is intended to focus closely on morphogenesis. This chapter will therefore use a small number of examples to illustrate how regulatory signals are connected to the machinery that drives cell proliferation and hence to morphogenetic events. Given the large variety of systems that exert control over cell proliferation, it is important to exercise great caution in extrapolating from the examples discussed here to unrelated cell types and developmental events; the general principles will probably be the same, but the detailed identities of molecules and pathways may well not be.

A BRIEF INTRODUCTION TO THE CELL CYCLE

In unicellular organisms, cell proliferation is the default behaviour, and its rate is limited only by the availability of energy and raw materials.[9] The cell cycle of a eukaryotic unicell, such as the fission yeast *Schizosaccharomyces pombe*, consists of a growth♣ phase (G1), a phase of DNA synthesis (S), a second growth phase (G2), and mitosis itself (M), repeated *ad infinitum* (see Figure 5.1.3). Some of these phases are themselves divided into distinct sub-phases that are separated by "checkpoints." Progression from one stage to the next, and from one phase to the next, is controlled by complexes of signalling proteins that are specific to each stage. The system is elaborate and, although this book is not the place to describe it in detail (excellent reviews can be found elsewhere[10,11,12,13,14]), it is useful to sketch the general features of cell cycle control in unicells before considering the complications found in metazoa.

The general principle of cell cycle control in *S. pombe* is that each checkpoint is controlled by a feedback mechanism that confirms the completion of the previous stage. Progression from the metaphase to anaphase stages of mitosis is dependent, for example, on a signal that verifies that all kinetochores have become properly attached to the mitotic spindle (see Chapter 2.2). Unattached kinetochores activate the diffusible protein Mad2, which inhibits the Anaphase Promoting Complex of proteins.[15] Once all kinetochores are bound to the spindle, Mad2 activation ceases and, as already-activated Mad2 remains active for only a short time, the Anaphase Promoting Complex is freed of inhibition. It can therefore ubiquitinate and hence destroy proteins that cause sister

♣G1 and G2 originally stood for "gap" 1&2, but in the light of more information about the cell cycle it is now convenient to associate "G" with the word "growth" instead.

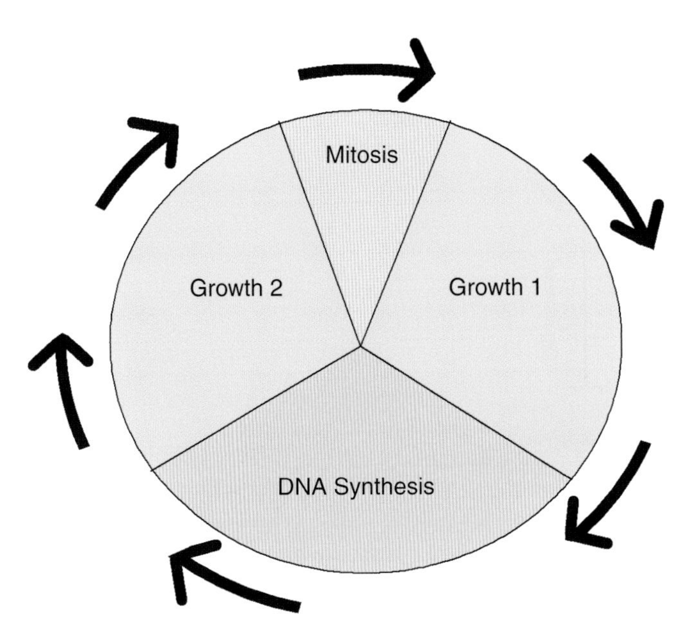

FIGURE 5.1.3 The basic cell cycle of *S. pombe*, in which the cell enters four stages in sequence and then begins again.

chromatids to be stuck together. The chromatids can therefore separate, and anaphase can begin (see Figure 5.1.4).

Progression from G1 to S phase must be delayed until the cell has grown large enough, something that depends on the availability of nutrients as well as on elapsed time. Activation of DNA synthesis requires the cyclin-dependent kinase Cdc2 to be active. Cdc2 is present at all stages of the cell cycle but is active only when bound to a suitable cyclin. The cyclin Cig2 activates cdc2, and Cig1 is produced in G1 some time before the G1-S transition takes place.[16] Precocious activation of cdc2 is prevented by the protein Rum1, which is made during the preceding mitosis and remains in the cell.[17] Progression from G1 to S therefore depends on inactivation of Rum1. Rum1 can be deactivated by the cyclin Puc1, which forms a complex with cdc2 that phosphorylates Rum1 and targets it for destruction. Puc1 is itself up-regulated in response to an increase in cell size,[18] although the manner in which cell size modulates Puc1 expression is not yet understood. The net result of this pathway is that progress from G1 to S is blocked until a cell has reached sufficient size for this transition to be reasonable. There are plenty more checkpoints, many of which use a variation on the cyclin/cdk system described above. What they have in common, in *S. pombe* and similar unicells, is that each depends only on sensors that determine whether the previous stage has been completed.

The cell cycles of multicellular animals operate according to broadly similar principles. The idea of there being many checkpoints is conserved, as is the implementation of them by cyclins and cyclin-dependent kinases. Indeed, the machinery of specific checkpoints, such as the metaphase-anaphase transition, is remarkably similar in animals and fission yeast. What is different, though, is that the cell cycle machinery of animal cells is not self-sufficient in the way it is in unicells, and progress through

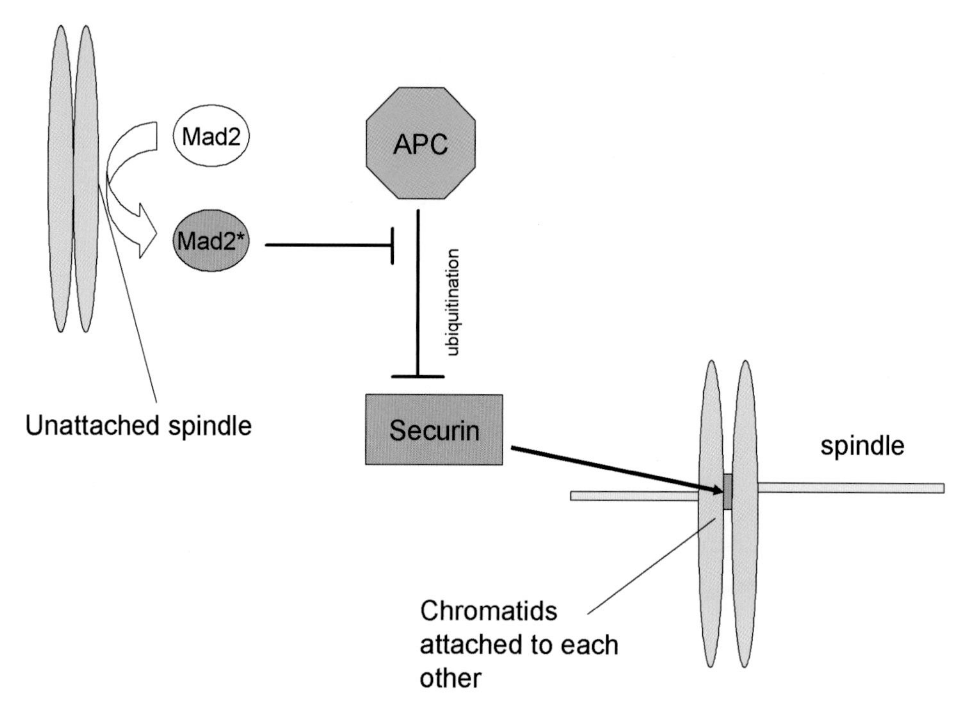

FIGURE 5.1.4 Essential features of the spindle assembly check in the fission yeast *S. pombe*. Signals generated by unattached kinetochores of one chromatic pair can block anaphase separation even of another chromatid pair that is correctly bound to the spindle, so the whole cell has to wait until spindle attachment is complete. The * symbol denotes activation of Mad. Facilitators of anaphase progression are coloured green and inhibitors are coloured pink.

one critical checkpoint, usually called the "restriction point," depends on a combination of signals confirming execution of the last stage, with signals that derive from the outside of the cell.♣ Without both signals being present, cells enter a quiescent state. For most cells, the critical checkpoint that depends on external signals lies in G1, and the quiescent state is called G_0 (see Figure 5.1.5). Once cells have passed the restriction point, they complete their cycle even in the absence of external signals, and only after reaching the decision point again do they enter G_0. The extracellular signals that control progress through the restriction point vary between different types of cells and stages of development and include both positive and negative regulators.

The G1 restriction point of mammalian cells is mediated in part by the actions of an intracellular anti-proliferation protein, Rb. Rb blocks the transcription of genes encoding proteins that are essential for progression to S phase, such as cyclin A and the enzymes thymidine kinase and dihydrofolate reductase (both of which are required to make raw ingredients for DNA synthesis).[19] Rb blocks the transcription of these genes by inhibiting a transcription factor, E2F, that their transcription requires.[20,21] Rb can itself be inhibited by phosphorylation by the cyclin-dependent kinases cdk2 and cdk4/6 when they are complexed with cyclin E or cyclin D1, respectively (see Figure 5.1.6).

♣Cells in embryos that are still at a stage too early to have shown differentiation into different cell types may be an exception to this rule.

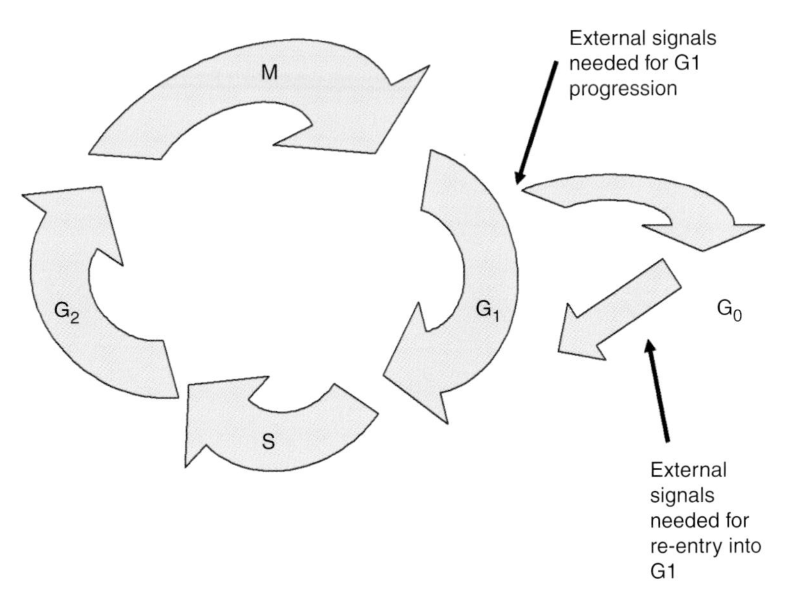

FIGURE 5.1.5 The basic arrangement of the cell cycle in metazoa. External signals are required at a critical point of the cell cycle, usually in early G1, in order to ensure cell cycle progression. In their absence, the cell enters the quiescent state G_0 or dies (see Chapter 5.3). Some cells have their decision point in G_2 instead.

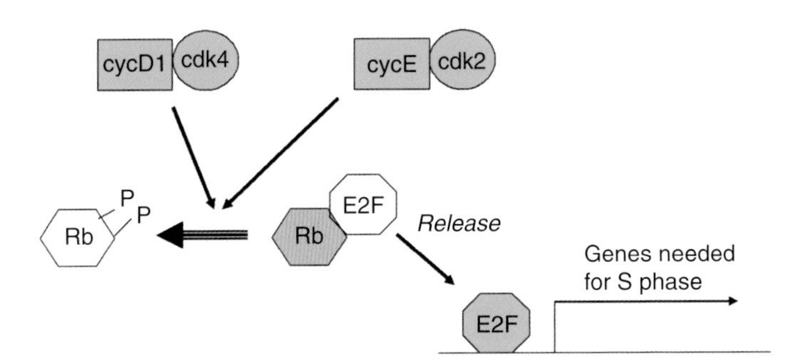

FIGURE 5.1.6 Rb normally blocks the transcription of genes needed for progress to S phase (cyclin A, dihydrofolate reductase, and so on) by inhibiting the activity of the transcription factor E2F. Phosphorylation of Rb renders it incapable of inhibiting E2F, so the cell can progress. This diagram is based on information in an article by C.F. Walsh.[22] Positive regulators of cycle progression are shown in green and inhibitors in pink.

The cdk2/cyclin E and cdk4/cyclin D1 complexes are the indirect target of extracellular signals in a large number of cell types, and a brief illustration of how they are tied to positive and negative extracellular signalling will serve to illustrate the general features of proliferation control.

Positive regulation of cell proliferation often uses the Erk MAP-kinase signal transduction pathway, as the full name of this pathway, “mitogen-activated protein kinase,” suggests. Erk MAP-kinase can be activated by a variety of pathways, but one

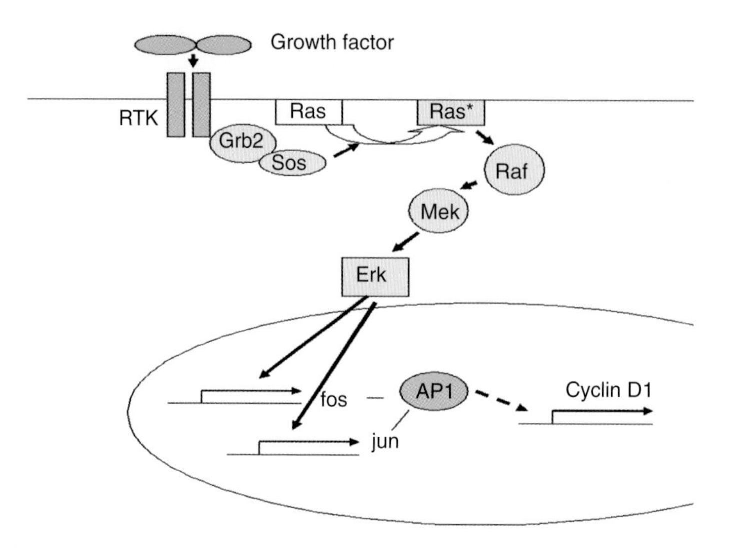

FIGURE 5.1.7 The connection (or at least, one of the connections) between growth factor signalling and cell cycle regulation. The diagram is based on information in an article by K. Roovers and R.K. Assoian.[25]

of the most common proceeds from receptor tyrosine kinases, via Grb2 and Sos, to activation of the small GTPase Ras by exchanging the GDP of inactive Ras for the GTP of active Ras. Activated Ras then activates Raf, which phosphorylates Mek, which in turn phosphorylates Erk (see Figure 5.1.7). Once activated, Erk enters the nucleus and stimulates transcription of the *fos* and *jun* genes, the products of which combine to form the transcription regulator AP1. There is a consensus AP1 site in the promoter of the cyclin D1 gene, and sustained Erk activation leads to transcription of cyclin.[23,24]

The activation of Erk is usually driven by signals that arise from growth factor receptors and from integrin-type matrix receptors. The pathways from these receptors converge in such a way that they effectively perform the Boolean "AND" operation, at least for physiological levels of ligand. In at least some cells, integrin-derived signals operate via small GTPases of the Rho family, which synergize with the signalling pathway upstream of Erk to ensure sustained Erk activation.[22] Cells will therefore respond to physiological levels of growth factors only if they are sitting on the correct matrix, which is an important error-checking mechanism in development and in adult life.

As proliferative signals can act by increasing production of cyclin-cdk complexes capable of inactivating Rb, thus ensuring progression of the cell cycle, anti-proliferative signals can act by inhibiting these cyclin-cdk complexes. For many cell types, TGFβ is an example of such an anti-proliferative signal. The TGFβ signalling pathway activates transcription of the $p15^{INK4B}$ gene, which encodes a protein that can bind the cyclin-dependent kinase cdk4. The $p15^{\mathrm{INK4B}}$ protein competes for cdk4 with a complex of cdk4, cyclin D1, and $p27^{\mathrm{kip1}}$; the effect of increasing the amount of $p15^{\mathrm{INK4B}}$ in the cell is therefore to liberate $p27^{\mathrm{kip1}}$ from this complex. Once free, the $p27^{\mathrm{kip1}}$ inhibits the activation of cdk2 by cyclin E and therefore block progress through the restriction point[26,27,28,29] (see Figure 5.1.8).

In practice, more than one pathway will converge on the restriction point, and cells inside developing embryos will use all of them to integrate information about the concentration of mitogenic and anti-proliferative signalling molecules in their vicinity

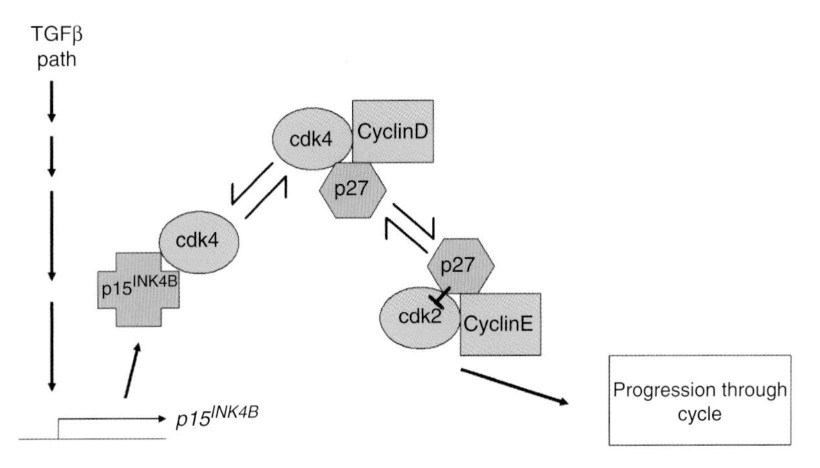

FIGURE 5.1.8 Coupling between extracellular signals and the restriction point of animal cell cycles. In this example, the anti-mitotic signalling molecule TGFβ causes the transcription of a protein that titrates cdk4 away from a cdk4-cyclin D-p27^{kip1} complex, and this liberates p27^{kip1} from that complex so that it can inhibit activation of cdk2/cyclin E.

with information about cell-cell contacts and cell-matrix contacts and about their own size and differentiation state. Detailed discussion about these pathways and how they converge can be found in a number of excellent reviews that go far beyond what can be described here.[22,25] The key point, as far as morphogenesis is concerned, is that cell proliferation is a regulated process and that each cell decides for itself, when it is at the restriction point, whether to divide or not. That decision depends on the signals that are present in its immediate environment.

LOCAL CONTROL OF CELL PROLIFERATION

Cells' ability to integrate diffusible and local signals to determine whether they should cycle allows them to regulate their proliferation according to a balance of local and "global" controls. A local reason to proliferate might be the need to replace a dead cell or to seal up a hole, whereas a global reason may be to increase the size of a complete organ or embryo.

An example of strictly local control that has been studied for many years gives rise to the phenomenon of contact inhibition of proliferation. Contact inhibition of proliferation prevents cells that are completely surrounded by neighbours from proliferating but allows those that make contact with free space proliferate to fill that space. The phenomenon is reminiscent, in its simplicity, of the "Game of Life" introduced in Chapter 1.2, and it is an efficient way to target proliferation so that spaces are filled but cells do not pile up where there is no more space. Its failure is one of the hallmarks of neoplastic transformation.

Contact inhibition of proliferation is still not completely understood, but the components that have been identified point to a link between adhesion systems and the cell cycle. In some cells, contact inhibition seems to act by modulating the activation of Erk MAP-kinase. For example, the activity of Erk, and hence the expression of cyclin D, in the MDCK renal epithelial cell line is higher when cells have free space than when they are crowded. Destruction of cell-cell junctions, using trypsin, restores the Erk activity to that of uncrowded cells, but it falls away again when junctions are re-established.[30] Treatment of the cells with sufficiently high concentrations of the growth factor HGF does succeed in activating Erk and cyclin D expression. Contact inhibition is therefore not an absolute state in these cells, and it can be overridden. This probably reflects the situation in a real embryo; intact epithelia do, after all, grow by cell proliferation, so it is clear that contact inhibition must be capable of being overruled in a controlled manner.

The link between cadherins and Erk is not yet clear, but in at least some cells, the ability of E-cadherin to inhibit Erk depends on the cells having a functional PI-3-kinase pathway, and constitutively active mutants of Akt, a component of the PI-3-kinase pathway, result in the constitutive suppression of Erk.[31] There is also the potential for a link via the signal transduction protein IQGAP1, which binds to Erk and inhibits it.[32] IQGAP1 also binds E-cadherin, and it is possible that the state of E-cadherin modulates competition between these two proteins and thus regulates Erk activity.

Contact inhibition can also occur via the proliferation inhibitor $p27^{kip1}$, the levels of which are elevated when cells make contact with each other in culture.[33,34] In at least some epithelial cells, this effect is mediated by the E-cadherin cell-cell adhesion molecule itself, and if the adhesive function of E-cadherin is inhibited using a monoclonal antibody, intracellular levels of $p27^{kip1}$ rise.[35] Similarly, transfection of breast tumour cells with dominant negative E-cadherin releases them from contact inhibition of proliferation.[36]

The connection between cadherins and $p27^{kip1}$ is not yet understood, but recent work has identified several candidate components.[35,36,37,38,39] One possibility is that the signal proceeds via β-catenin, which is distributed through the volume of isolated cells but is localized to the cytoplasmic domains of cell-cell junctions in cells that make contact with other cells.[39] The observation that antibody-mediated inhibition of E-cadherin function succeeds in releasing contact inhibition of proliferation without, apparently, releasing β-catenin from the junctions casts doubt on this, however.[35] The experiments were performed on different cell types, and the discrepancy in the results may simply be a warning that different cell types can process contact signals differently. In any case, it is very important to note that the above paragraphs described one or two examples just to illustrate how local signals may be used to control proliferation (see Figure 5.1.9). It would be unwise to extrapolate from these to all cells, and it is likely that many more mechanisms exist in the embryo as a whole. It is also important to remember that a cell's propensity to proliferate also changes with differentiation, an aspect of developmental biology that is deliberately omitted from this book (because it is covered so well elsewhere[9,40,41,42]).

TISSUE-SCALE CONTROL OF PROLIFERATION: A Mechanism for Keeping Different Cell Populations in Balance

Tissues consist of more than one type of cell, and the proportions of each type of cell generally have to be controlled within close limits if the tissue is to function properly (so do the spatial arrangements of the cells, but this chapter focuses only on control of

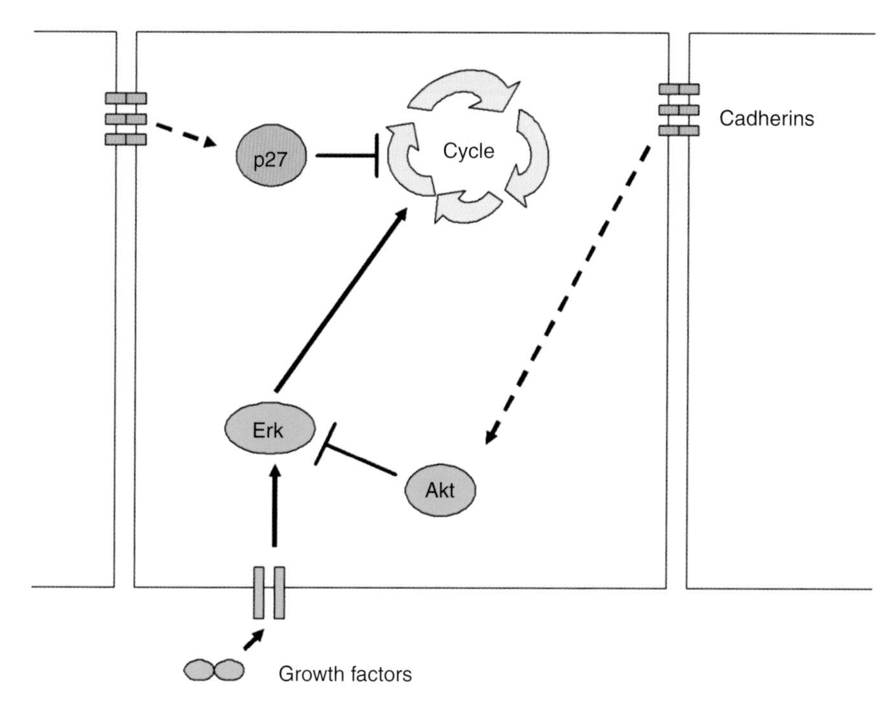

FIGURE 5.1.9 Summary of two pathways by which cell adhesion antagonizes proliferation. It is important to note that these pathways were characterized, as much as they have been, in different cells, and they may not both operate at once. There are probably many more (see main text).

cell number). There are, in principle, two main ways in which the proportions might be controlled. One method would be to rely on cells maintaining an intrinsic proliferation rate, characteristic of their state of differentiation, that is correct for the final tissue. This method would have the disadvantage that small errors may have serious consequences for the final ratio, because growth is exponential. The other, better, method would be for the proliferation of each cell type to be controlled by population-dependent signals that emanate from the other cell type(s) so that growth of each population is balanced. Signals that emanate from one cell type and act on another are described as paracrine.

The tissues of most organs consist of a mixture of epithelial and mesenchymal cells that perform the main functions of the tissue, and vascular endothelia that brings blood to the tissue to sustain it. The ratio of endothelia to the other cells is important; too little, and the other cells will be hypoxic, undernourished, and poisoned by their own waste products; too much, and the tissue becomes dominated by endothelia that get in the way of other functions. At least some of the signalling pathways that balance the growth of endothelia and other cells have been identified. Proliferation of tissue cells such as fibroblasts is controlled in part by the availability of oxygen. Cells cultured in hypoxic conditions fail to pass the G1 restriction point.[43,44] Oxygen is sensed by the protein Hypoxia Inducible Factor 1α (HIF1α). Under normal oxygen conditions, this molecule is rapidly ubiquitinated and destroyed so that cellular concentrations remain low. In low oxygen, ubiquitination fails, and HIF1α levels rise.[45,46] HIF1α is a transcription factor that increases the expression of the cycle inhibitor $p27^{kip1}$ and thus blocks the activity of cyclin E/cdk2 complexes.[47] Without cyclin E/cdk2 function, Rb remains unphosphorylated, and it will inhibit the E2F-dependent transcription of genes required for progression into S phase (see Figure 5.1.8). In monkey kidney cells, at least, this repression is deepened by an up-regulation of a phosphatase that returns any Rb that has been phosphorylated to its unphosphorylated, E2F-inhibiting, state.[48]

Hypoxia caused by a shortage of functional vascular endothelium therefore arrests proliferation in tissues so that they will not outgrow their blood supply. Among the genes regulated by HIF1α is one that encodes the extracellular signalling molecule, vascular endothelial growth factor (VEGF).[49,50] Cells that are therefore arrested by hypoxia produce VEGF and secrete it into their environment. Vascular endothelia bear the VEGF receptor tyrosine kinase, Flk-1, and VEGF acts as a powerful mitogen for endothelial cells, acting via Erk and also other signal transduction pathways such as PI-3-kinase.[51,52] The proliferation of endothelial cells is therefore limited by the extent of VEGF production by the cells of the tissue, which is in turn limited by the amount of oxygen made available by the endothelial system (see Figure 5.1.10).

The system described above is especially well studied because developing an ability to modulate the expansion of blood vessels pharmacologically is a promising strategy for restricting the growth of tumors.[53] It is, however, unusual in using oxygen

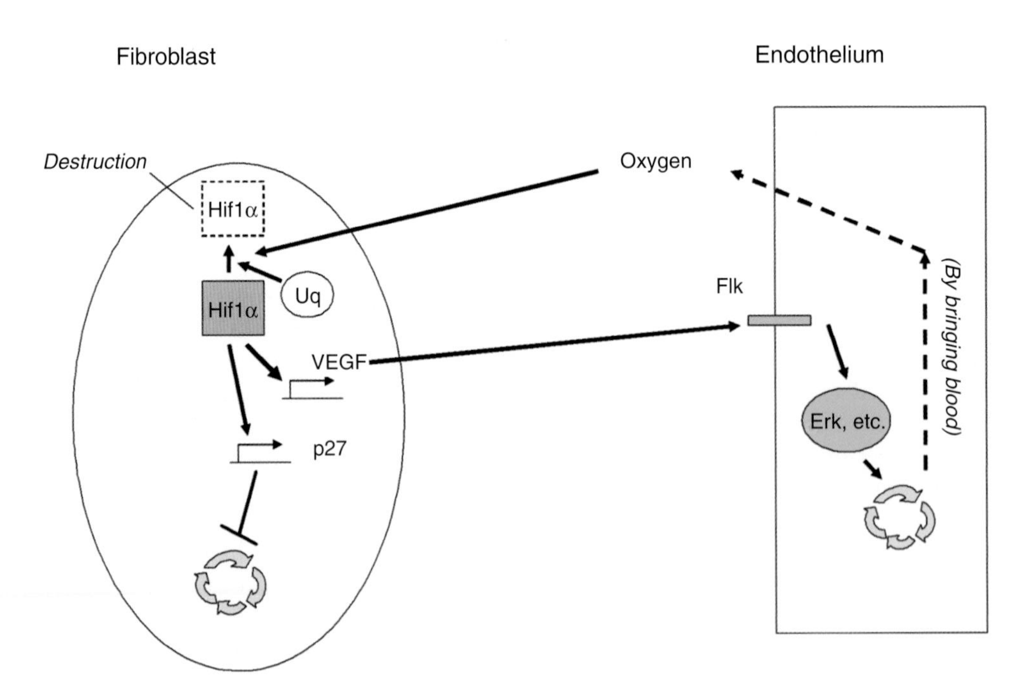

FIGURE 5.1.10 The feedback loop that balances the rates of proliferation of tissue cells and the endothelia that construct their blood supply (see main text for details).

as one of the signals, and in most cases, peptide growth factors mediate communication in both directions. In the developing kidney, for example, peptide growth factors are used to balance the growth of the epithelial collecting duct tree with that of the mesenchyme into which it expands (the development of the kidney is described further in Chapter 4.6). Expansion of the collecting duct system depends on mesenchyme-derived growth factors such as GDNF and HGF, whereas survival and expansion of the mesenchyme depends on epithelium-derived growth factors such as FGF2 and TGFα. The proliferation of the two types of cells cannot, therefore, become too unbalanced. Similar feedback can be found between epithelial and mesenchymal compartments of other organs.[54]

The type of feedback outlined in this section can be used either to ensure that two (or more) interdependent cell types grow proportionately during development, or to ensure tissue homeostasis in adulthood, depending on the "set point" of the feedback system. In growing tissues, the default behaviour of the organ tissue would be to grow, and it is held back by lack of oxygen. In static tissues, the default behaviour would not be to grow, and proliferation in excess of that required to replace cells lost through death would be invoked only for special purposes such as wound healing. The "set point" of the feedback is set up by other, long-range signals that are integrated with the local signals that have been described here. These are the signals that determine the size and proportions of the body as a whole.

LARGE-SCALE CONTROL OF CELL PROLIFERATION

As well as being controlled on the spatial scale of individual cells and tissues, the rate of cell proliferation is subject to controls that operate on a larger scale to organize the growth of entire organs and of the body as a whole. Organ growth seems to be regulated by a combination of controls internal and external to the organ. Body-wide control is achieved largely by the use of diffusible hormones. In mammalian foetal and infantile development, for example, the hormones IGFI, IGFII, and growth hormone (GH) are important regulators of cell proliferation, the IGFs playing a primary role in early development and GH playing a primary role once the pituitary gland has matured enough to make it. GH encourages the growth of a variety of tissues and is particularly important in the growth plates of long bones, which are zones at the ends of the shaft of each bone at which new bone formation takes place (see Figure 5.1.11). Events in the growth plates are complex, involving the proliferation, differentiation, and death of cells,[55] but only proliferation will be considered here.

The global signals provided by GH are "translated" into local signals carried by IGFI and IGFII at the growth plate itself.[56,57,58] IGFI and IGFII are both potent stimulators of chondrocyte proliferation,[59] so that the level of circulating GH controls the proliferation of cells within the growth place. Other local signals, carried by FGFs, BMPs and Ihh, also regulate proliferation.[55] GH may also have a direct effect as well as an indirect one via other growth factors.[60]

The IGF signalling pathway is very important in the size regulation of mammalian tissues in general. Mice lacking IGF1 or IGF2 are born very small and cannot be rescued by injections of GH.[61,62,63] Those lacking the receptor for IGF1 show an even greater defect of growth and generally die around birth.[63,64] The use of IGFs to regulate organ size is also highly conserved across the animal kingdom. Mutants of *Drosophila*

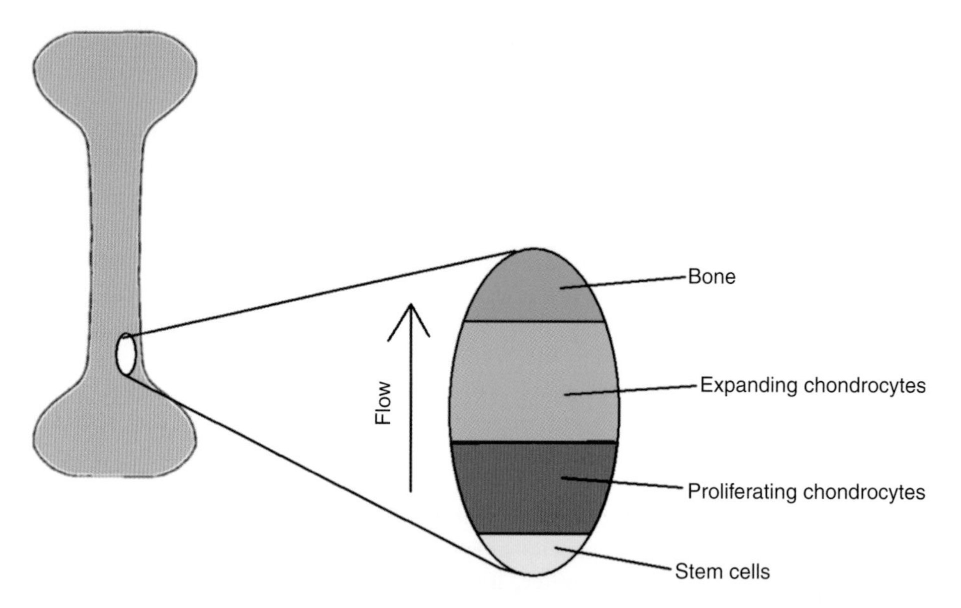

FIGURE 5.1.11 The growth plates of developing human long bones. Most proliferation takes place in a zone just proximal to the stem cells (which proliferate only enough to maintain their own numbers); this zone is coloured dark blue in the diagram. Cells within the growth plate are arranged in longitudinal columns so that they "flow" towards the shaft itself as they expand and differentiate. The contribution of new cells to the shaft drives elongation of the bone.

melanogaster that have reduced function of dINR, the insect's homologue of the IGF receptor, are half the weight of normal flies, and this reflects a reduction in both cell size and cell number. Conversely, mutants that show overexpression of ligands for dINR are significantly larger than normal flies.[65]

The fact that both cell size *and* cell number are affected by mutation of the IGF signalling pathway in *D. melanogaster* is especially interesting in light of the ability of embryos to compensate for proliferation defects using cell size and vice versa. The size regulation mechanism that can use cell size to compensate for defective proliferation must itself be liable to mutation, and the expected phenotype would be failure of this compensation. Such mutants exist, and include the genes *PTEN*, *tuberous sclerosis*, *InR*, *Akt*, and *chico*.[66,67,68] All of these are components of the IGF signal-transduction pathway (see Figure 5.1.12). Loss-of-function mutants of the positive regulators of this pathway, such as InR, PI-3-kinase, and s6k, all lead to smaller cells and organs, while loss-of-function mutations of repressors of the pathway, such as PTEN, result in abnormally large organs. (Some of these data come from studies in which mutants were generated only in defined areas of the embryo to avoid embryonic lethal phenotypes; only those parts of the embryos carrying the loss-of-function cells show the phenotype.) Conversely, overexpression of the inhibitor of the pathway PTEN results in small organs. This PTEN is the same molecule discussed in the context of cell guidance in Chapter 3.3.

The conclusion from these observations is that interfering with the insulin pathway alters cell size *and* removes an organ's ability to compensate for this. IGF signalling

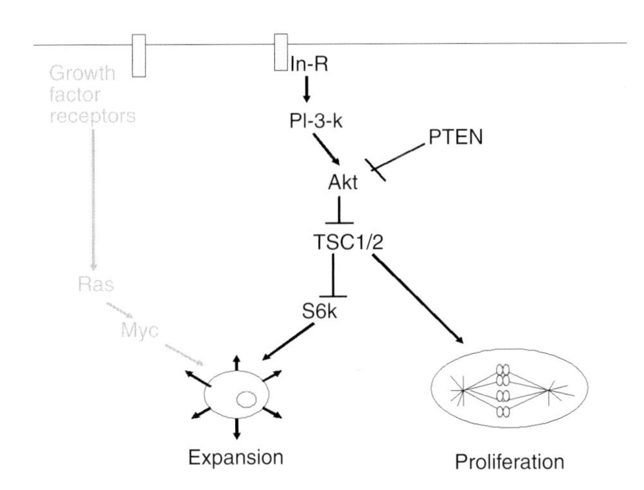

FIGURE 5.1.12 The role of the insulin receptor signal transduction pathway in controlling expansion and proliferation of cells. This diagram is based on Potter and Xu[69] and Potter, Pedraza, and Xu.[70]

is therefore probably the pathway that determines the "set point" of tissue size that can be reached through compensated regulation of cell size and number.

INTERPLAY BETWEEN GLOBAL AND ORGAN-SPECIFIC SIGNALS

The use of "global" signals allows the body to coordinate the growth of all of the components of complete systems. For example, hormones such as GH enable the coordination of the growth rates of different bones in the skeleton. In development of animals such as humans, this global coordination is important because there are distinct phases in long-bone growth. Rates of elongation are very high in foetal life but then fall away rapidly until the age of about 3 and fall away more slowly thereafter. At puberty, growth rates are again increased and then fall away again until the growth plates cease to exist and further growth is impossible. It is obviously important that different bones (for example, in the left and right legs) switch between different growth rates together.

The rate of growth in a growth plate is not, however, set purely by hormones derived from the pituitary gland, etc., but rather by a combination of external signals and regulatory systems internal to the growth plate. The existence of these internal regulatory systems can be demonstrated by an experiment in which the growth of a long bone in one rabbit limb is inhibited experimentally during part of the growth period. The contra-lateral leg grows normally, but when the treated leg is released from inhibition it increases its own rate of growth and catches up with the contra-lateral leg.[71] It is clear that for this to happen the treated leg must be able to detect its own size. It is not, however, clear how that leg "knows" how much it has to grow in order to catch up. One hypothesis is that the proliferative ability of a growth plate declines with proliferation so that the intrinsic falling away of proliferation rates in the early years of life is a reflection of the declining ability of GH to elicit proliferation. In this case, a growth plate that has been proliferating too little will have retained its capacity for elongation and, as long as there is enough GH around to sustain proliferation at all, it will be able to catch up with a bone that has been elongating normally.[71] Whatever the explanation, the ability of growth plates to catch up is useful in the treatment of human disorders that are characterized by abnormally low long-bone extension, such as Cushing's syndrome.

Other organs seem to have local controls as well. If foetal hearts are grafted into adult rat hosts, they continue to grow and develop, but the organs of the host do not start developing again.[72] Similarly, foetal kidneys grafted into adult hosts can grow so much that they are capable of sustaining life in the absence of host kidneys, at least over the short term, though there is no sign that the rest of the adult body has started to develop again.[73] In both of these cases, the foetal organs interpret the environment in which they find themselves to be conducive to growth, while adult organs do not. This is true even when the adult kidneys are present along with grafted fetal ones, which shows that the different response to the environment may be to do with either age or size but cannot just be a reflection of organ type. Some organs, such as human liver, show an impressive facility for "regeneration" in terms of creating new functional biomass even if not according to precisely the original anatomy. If a significant proportion of a rat liver is removed surgically, the remaining organ regenerates to the original mass but not beyond that point.[74,75] The size of the normal liver cannot therefore reflect a limit to cell proliferation—for example, because the ability of cells to undergo division was exhausted but must be imposed by a regulatory mechanism that holds back the potential for proliferation and releases this potential, temporarily, during regeneration.[69]

The details of the feedback mechanisms that regulate organ and body size are still not well understood, partly because they can be investigated properly only in the complicated context of a complete animal rather than in tissue-culture abstractions. Nevertheless, genetic approaches are beginning to uncover promising leads, and it may not be long before firm links are established among controls that operate at the scale of a whole animal, those that operate at the scale of tissues, and those that operate at the level of the proteins that control cell cycling itself.

Reference List

1. Neufeld, TP, de la Cruz, AF, Johnston, LA, and Edgar, BA (26-6-1998) Coordination of growth and cell division in the *Drosophila* wing. *Cell.* **93**: 1183–1193
2. Weigmann, K, Cohen, SM, and Lehner, CF (1997) Cell cycle progression, growth and patterning in imaginal discs despite inhibition of cell division after inactivation of *Drosophila* Cdc2 kinase. *Development.* **124**: 3555–3563
3. Tsukaya, H (2003) Organ shape and size: A lesson from studies of leaf morphogenesis. *Curr. Opin. Plant Biol.* **6**: 57–62
4. Frankenhauser, G (1945) The effects of changes in chromosome number on amphibian development. *Q. Rev. Biol.* **20**: 20–78
5. Whitaker, M (1996) Cell cycle. (JAI Press)
6. Fantes, P and Brooks, R (1994) The cell cycle: A practical approach. (Oxford University Press)
7. Sonnenschein, C and Soto, AM (1998) The society of cells. (Springer-Verlag)
8. Blagosklonny, M, Editor (2002) *Cell Cycle* (Landes Bioscience)
9. Greaves, M (2000) Cancer, the evolutionary legacy. 3–30 (Oxford University Press)

10. Lygerou, Z and Nurse, P (2000) Controlling S-phase onset in fission yeast. *Cold Spring Harb. Symp. Quant. Biol.* **65**: 323–332
11. Latif, C, Harvey, SH, and O'Connell, MJ (20-11-2001) Ensuring the stability of the genome: DNA damage checkpoints. *Sci. World J.* **1**: 684–702
12. Chang, F (2001) Studies in fission yeast on mechanisms of cell division site placement. *Cell Struct. Funct.* **26**: 539–544
13. Moser, BA and Russell, P (2000) Cell cycle regulation in *Schizosaccharomyces pombe*. *Curr. Opin. Microbiol.* **3**: 631–636
14. Russell, P (1998) Checkpoints on the road to mitosis. *Trends Biochem. Sci.* **23**: 399–402
15. Shah, JV and Cleveland, DW (22-12-2000) Waiting for anaphase: Mad2 and the spindle assembly checkpoint. *Cell.* **103**: 997–1000
16. Mondesert, O, McGowan, CH, and Russell, P (1996) Cig2, a B-type cyclin, promotes the onset of S in *Schizosaccharomyces pombe*. *Mol. Cell Biol.* **16**: 1527–1533
17. Benito, J, Martin-Castellanos, C, and Moreno, S (15-1-1998) Regulation of the G1 phase of the cell cycle by periodic stabilization and degradation of the p25rum1 CDK inhibitor. *Embo. J.* **17**: 482–497
18. Martin-Castellanos, C, Blanco, MA, de Prada, JM, and Moreno, S (2000) The puc1 cyclin regulates the G1 phase of the fission yeast cell cycle in response to cell size. *Mol. Biol. Cell.* **11**: 543–554
19. Weinberg, RA (1995) The retinoblastoma protein and cell cycle control. *Cell.* **81**: 323–330
20. DeGregori, J, Kowalik, T, and Nevins, JR (1995) Cellular targets for activation by the E2F1 transcription factor include DNA synthesis- and G1/S-regulatory genes. *Mol. Cell Biol.* **15**: 4215–4224
21. Angus, SP, Fribourg, AF, Markey, MP, Williams, SL, Horn, HF, DeGregori, J, Kowalik, TF, Fukasawa, K, and Knudsen, ES (10-6-2002) Active RB elicits late G1/S inhibition. *Exp. Cell Res.* **276**: 201–213
22. Welsh, CF (2004) Rho GTPases as key transducers of proliferative signals in g1 cell cycle regulation. *Breast Cancer Res. Treat.* **84**: 33–42
23. Lavoie, JN, L'Allemain, G, Brunet, A, Muller, R, and Pouyssegur, J (23-8-1996) Cyclin D1 expression is regulated positively by the p42/p44MAPK and negatively by the p38/HOGMAPK pathway. *J. Biol. Chem.* **271**: 20608–20616
24. Cheng, M, Sexl, V, Sherr, CJ, and Roussel, MF (3-2-1998) Assembly of cyclin D-dependent kinase and titration of p27Kip1 regulated by mitogen-activated protein kinase (MEK1). *Proc. Natl. Acad. Sci. U.S.A.* **95**: 1091–1096
25. Roovers, K and Assoian, RK (2000) Integrating the MAP kinase signal into the G1 phase cell cycle machinery. *Bioessays.* **22**: 818–826
26. Polyak, K, Kato, JY, Solomon, MJ, Sherr, CJ, Massague, J, Roberts, JM, and Koff, A (1994) p27Kip1, a cyclin-Cdk inhibitor, links transforming growth factor-beta and contact inhibition to cell cycle arrest. *Genes Dev.* **8**: 9–22
27. Hannon, GJ and Beach, D (15-9-1994) p15INK4B is a potential effector of TGF-beta-induced cell cycle arrest. *Nature.* **371**: 257–261
28. Reynisdottir, I, Polyak, K, Iavarone, A, and Massague, J (1-8-1995) Kip/Cip and Ink4 Cdk inhibitors cooperate to induce cell cycle arrest in response to TGF-beta. *Genes Dev.* **9**: 1831–1845
29. Liang, J, Zubovitz, J, Petrocelli, T, Kotchetkov, R, Connor, MK, Han, K, Lee, JH, Ciarallo, S, Catzavelos, C, Beniston, R, Franssen, E, and Slingerland, JM (2002) PKB/Akt phosphorylates p27, impairs nuclear import of p27 and opposes p27-mediated G1 arrest. *Nat. Med.* **8**: 1153–1160
30. Li, S, Gerrard, ER, and Balkovetz, DF (7-4-2004) Evidence for ERK 1/2 phosphorylation controlling contact inhibition of proliferation in Madin-Darby canine kidney epithelial cells. *Am. J. Physiol. Cell Physiol.* **287**: C432–439
31. Laprise, P, Langlois, MJ, Boucher, MJ, Jobin, C, and Rivard, N (2004) Down-regulation of MEK/ERK signalling by E-cadherin-dependent PI3K/Akt pathway in differentiating intestinal epithelial cells. *J. Cell Physiol.* **199**: 32–39

32. Roy, M, Li, Z, and Sacks, DB (23-4-2004) IQGAP1 binds ERK2 and modulates its activity. *J. Biol. Chem.* **279**: 17329–17337
33. Kato, A, Takahashi, H, Takahashi, Y, and Matsushime, H (21-3-1997a) Inactivation of the cyclin D-dependent kinase in the rat fibroblast cell line, 3Y1, induced by contact inhibition. *J. Biol. Chem.* **272**: 8065–8070
34. Kato, A, Takahashi, H, Takahashi, Y, and Matsushime, H (1997b) Contact inhibition-induced inactivation of the cyclin D-dependent kinase in rat fibroblast cell line, 3Y1. *Leukemia.* **11** (**Suppl** 3): 361–362
35. St. Croix, B, Sheehan, C, Rak, JW, Florenes, VA, Slingerland, JM, and Kerbel, RS (27-7-1998) E-Cadherin-dependent growth suppression is mediated by the cyclin-dependent kinase inhibitor p27(KIP1). *J. Cell Biol.* **142**: 557–571
36. Vizirianakis, IS, Chen, YQ, Kantak, SS, Tsiftsoglou, AS, and Kramer, RH (2002) Dominant-negative E-cadherin alters adhesion and reverses contact inhibition of growth in breast carcinoma cells. *Int. J. Oncol.* **21**: 135–144
37. Iurlaro, M, Demontis, F, Corada, M, Zanetta, L, Drake, C, Gariboldi, M, Peiro, S, Cano, A, Navarro, P, Cattelino, A, Tognin, S, Marchisio, PC, and Dejana, E (2004) VE-cadherin expression and clustering maintain low levels of survivin in endothelial cells. *Am. J. Pathol.* **165**: 181–189
38. Sasaki, CY, Lin, H, Morin, PJ, and Longo, DL (15-12-2000) Truncation of the extracellular region abrogrates cell contact but retains the growth-suppressive activity of E-cadherin. *Cancer Res.* **60**: 7057–7065
39. Dietrich, C, Scherwat, J, Faust, D, and Oesch, F (22-3-2002) Subcellular localization of beta-catenin is regulated by cell density. *Biochem. Biophys. Res. Commun.* **292**: 195–199
40. Gilbert, SF (2003) Developmental biology. **7th**: (Sinauer)
41. Alberts, B, Bray, D, Lewis, J, Raff, M, Roberts, K, and Watson, JD (1994) Molecular biology of the cell. **3rd**: (Garland)
42. Pollard, TD and Earnshaw, WC (2002) Cell biology. 725–765 (Saunders)
43. Gardner, LB, Li, F, Yang, X, and Dang, CV (2003) Anoxic fibroblasts activate a replication checkpoint that is bypassed by E1a. *Mol. Cell Biol.* **23**: 9032–9045
44. Schmaltz, C, Hardenbergh, PH, Wells, A, and Fisher, DE (1998) Regulation of proliferation-survival decisions during tumor cell hypoxia. *Mol. Cell Biol.* **18**: 2845–2854
45. Cockman, ME, Masson, N, Mole, DR, Jaakkola, P, Chang, GW, Clifford, SC, Maher, ER, Pugh, CW, Ratcliffe, PJ, and Maxwell, PH (18-8-2000) Hypoxia inducible factor-alpha binding and ubiquitylation by the von Hippel-Lindau tumor suppressor protein. *J. Biol. Chem.* **275**: 25733–25741
46. Maxwell, PH, Wiesener, MS, Chang, GW, Clifford, SC, Vaux, EC, Cockman, ME, Wykoff, CC, Pugh, CW, Maher, ER, and Ratcliffe, PJ (20-5-1999) The tumour suppressor protein VHL targets hypoxia-inducible factors for oxygen-dependent proteolysis. *Nature.* **399**: 271–275
47. Goda, N, Ryan, HE, Khadivi, B, McNulty, W, Rickert, RC, and Johnson, RS (2003) Hypoxia-inducible factor 1alpha is essential for cell cycle arrest during hypoxia. *Mol. Cell Biol.* **23**: 359–369
48. Krtolica, A, Krucher, NA, and Ludlow, JW (5-11-1998) Hypoxia-induced pRB hypophosphorylation results from downregulation of CDK and upregulation of PP1 activities. *Oncogene.* **17**: 2295–2304
49. Liu, Y, Cox, SR, Morita, T, and Kourembanas, S (1995) Hypoxia regulates vascular endothelial growth factor gene expression in endothelial cells. Identification of a 5′ enhancer. *Circ. Res.* **77**: 638–643
50. Forsythe, JA, Jiang, BH, Iyer, NV, Agani, F, Leung, SW, Koos, RD, and Semenza, GL (1996) Activation of vascular endothelial growth factor gene transcription by hypoxia-inducible factor 1. *Mol. Cell Biol.* **16**: 4604–4613
51. Millauer, B, Wizigmann-Voos, S, Schnurch, H, Martinez, R, Moller, NP, Risau, W, and Ullrich, A (26-3-1993) High affinity VEGF binding and developmental expression suggest Flk-1 as a major regulator of vasculogenesis and angiogenesis. *Cell.* **72**: 835–846

52. Yu, Y and Sato, JD (1999) MAP kinases, phosphatidylinositol 3-kinase, and p70 S6 kinase mediate the mitogenic response of human endothelial cells to vascular endothelial growth factor. *J. Cell Physiol.* **178**: 235–246
53. Davidoff, AM and Kandel, JJ (2004) Antiangiogenic therapy for the treatment of pediatric solid malignancies. *Semin. Pediatr. Surg.* **13**: 53–60
54. Davies, JA (2004) Branching morphogenesis. (Landes Biomedical)
55. van der Eerden, BC, Karperien, M, and Wit, JM (2003) Systemic and local regulation of the growth plate. *Endocr. Rev.* **24**: 782–801
56. Nilsson, A, Carlsson, B, Isgaard, J, Isaksson, OG, and Rymo, L (1990) Regulation by GH of insulin-like growth factor-I mRNA expression in rat epiphyseal growth plate as studied with in-situ hybridization. *J. Endocrinol.* **125**: 67–74
57. Smink, JJ, Koster, JG, Gresnigt, MG, Rooman, R, Koedam, JA, and Buul-Offers, SC (2002) IGF and IGF-binding protein expression in the growth plate of normal, dexamethasone-treated and human IGF-II transgenic mice. *J. Endocrinol.* **175**: 143–153
58. Nilsson, A, Isgaard, J, Lindahl, A, Dahlstrom, A, Skottner, A, and Isaksson, OG (1-8-1986) Regulation by growth hormone of number of chondrocytes containing IGF-I in rat growth plate. *Science.* **233**: 571–574
59. Olney, RC, Wang, J, Sylvester, JE, and Mougey, EB (14-5-2004) Growth factor regulation of human growth plate chondrocyte proliferation in vitro. *Biochem. Biophys. Res. Commun.* **317**: 1171–1182
60. Wang, J, Zhou, J, Cheng, CM, Kopchick, JJ, and Bondy, CA (2004) Evidence supporting dual, IGF-I-independent and IGF-I-dependent, roles for GH in promoting longitudinal bone growth. *J. Endocrinol.* **180**: 247–255
61. Liu, JL and LeRoith, D (1999) Insulin-like growth factor I is essential for postnatal growth in response to growth hormone. *Endocrinology.* **140**: 5178–5184
62. Powell-Braxton, L, Hollingshead, P, Warburton, C, Dowd, M, Pitts-Meek, S, Dalton, D, Gillett, N, and Stewart, TA (1993) IGF-I is required for normal embryonic growth in mice. *Genes Dev.* **7**: 2609–2617
63. Liu, JP, Baker, J, Perkins, AS, Robertson, EJ, and Efstratiadis, A (8-10-1993) Mice carrying null mutations of the genes encoding insulin-like growth factor I (Igf-1) and type 1 IGF receptor (Igf1r). *Cell.* **75**: 59–72
64. Baker, J, Liu, JP, Robertson, EJ, and Efstratiadis, A (8-10-1993) Role of insulin-like growth factors in embryonic and postnatal growth. *Cell.* **75**: 73–82
65. Brogiolo, W, Stocker, H, Ikeya, T, Rintelen, F, Fernandez, R, and Hafen, E (20-2-2001) An evolutionarily conserved function of the *Drosophila* insulin receptor and insulin-like peptides in growth control. *Curr. Biol.* **11**: 213–221
66. Huang, H, Potter, CJ, Tao, W, Li, DM, Brogiolo, W, Hafen, E, Sun, H, and Xu, T (1999) PTEN affects cell size, cell proliferation and apoptosis during *Drosophila* eye development. *Development.* **126**: 5365–5372
67. Potter, CJ, Huang, H, and Xu, T (4-5-2001) *Drosophila* Tsc1 functions with Tsc2 to antagonize insulin signalling in regulating cell growth, cell proliferation, and organ size. *Cell.* **105**: 357–368
68. Weinkove, D and Leevers, SJ (2000) The genetic control of organ growth: Insights from *Drosophila. Curr. Opin. Genet. Dev.* **10**: 75–80
69. Potter, CJ and Xu, T (2001) Mechanisms of size control. *Curr. Opin. Genet. Dev.* **11**: 279–286
70. Potter, CJ, Pedraza, LG, and Xu, T (2002) Akt regulates growth by directly phosphorylating Tsc2. *Nat. Cell Biol.* **4**: 658–665
71. Baron, J, Klein, KO, Colli, MJ, Yanovski, JA, Novosad, JA, Bacher, JD, and Cutler, GB, Jr. (1994) Catch-up growth after glucocorticoid excess: A mechanism intrinsic to the growth plate. *Endocrinology.* **135**: 1367–1371
72. Dittmer, JE, Goss, RJ, and Dinsmore, CE (1974) The growth of infant hearts grafted to young and adult rats. *Am. J. Anat.* **141**: 155–160

73. Rogers, SA and Hammerman, MR (2004) Prolongation of life in anephric rats following de novo renal organogenesis. *Organogenesis.* **1**: 22–25
74. Fausto, N (2004) Liver regeneration and repair: Hepatocytes, progenitor cells, and stem cells. *Hepatology.* **39**: 1477–1487
75. Black, D, Lyman, S, Heider, TR, and Behrns, KE (2004) Molecular and cellular features of hepatic regeneration. *J. Surg. Res.* **117**: 306–315

CHAPTER 5.2

MORPHOGENESIS BY ORIENTATED CELL DIVISION

Mitotic division can do more than simply expand the cell number in, and the volume of, a tissue. When cleavage planes have a specific orientation, mitosis can be used to expand a tissue in a particular direction and thus to create new form. The very earliest stages of animal development tend to be where the importance of accurately orientated cell division is most obvious. Following fertilization, most animals enter a phase of cleavage in which the large, unicellular zygote is divided into increasing numbers of smaller cells with no change in overall volume of the embryo. In some phyla, the precise spatial pattern of cleavage seems to be important in setting up the body plan. Most animals use one of two basic patterns of cleavage: radial and spiral (see Figure 5.2.1).

In radial cleavage, mitotic spindles are aligned precisely along some body axes and perpendicular to others so that cleavage planes are always parallel to or transverse to the meridian plane. In spiral cleavage, mitotic spindles are at an oblique angle to the animal-vegetal axis and also oblique to the meridian plane, and they alternate between being oblique-left and oblique-right to this plane; the result is a spiral arrangement of cells. In at least some animals with spiral cleavage, whether the alternation begins left or begins right is critical to the body plan. For example, the third cleavage division is right-handed in most snails, and the spiral in the shell of the adult animal is also right-handed. However, if the cleavage of the third division is left-handed then so is the adult body. This correlation is shown most dramatically by the snail *Lymnaea peregra*, which makes either right-handed or left-handed shells under the control of a single gene. The right-handed allele is dominant, and cytoplasm from right-handed eggs can convert left-handed ones to right-handed cleavage and adult body plan. This strongly suggests a causal relationship between direction of spiral cleavage in the early embryo and the chirality of the adult body,[1] but unless it can be shown that the product of the right-promoting allele affects *only* the orientation of cell division, the observations cannot prove causality. The cleavage planes of radially cleaving embryos may be less important; for example, experimental manipulations of the radially cleaving embryo of the frog *Xenopus laevis* can alter the primary cleavage plane without altering the future axis of the embryo.[2]

Unfortunately, very little is known about how spindle orientation is controlled during cleavage. Some theoretical work has been done on models of radial cleavage in

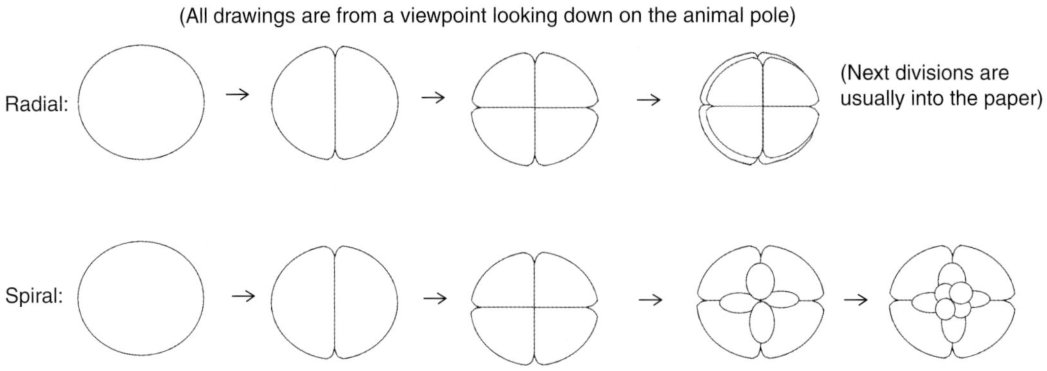

FIGURE 5.2.1 The two most common modes of cleavage in animal embryos. In radial cleavage, mitoses are arranged at right angles to each other in the meridian or transverse planes of the embryo; they may produce equally sized daughters, as shown here, or unequal ones. In spiral cleavage, mitoses proceed at an angle to the meridian and transverse planes, and daughter cells are placed in the "valleys" between other cells.

which furrows are placed to minimize free energy in the system.[3,4] Some of these models produce realistic arrangements of cells, but they concentrate more on the energetics of the final cell shape than on the earlier placement of the spindle, which is, in most systems, the determinant of the cleavage plane. I am therefore inclined to view them with some skepticism. It is possible—indeed probable—that cleaving embryos use the same mechanisms used in later mitoses, as described below.

The orientation of mitosis in epithelial tissues can have a profound effect on morphogenesis (the general structure of epithelial tissues is discussed further in Chapter 4.1). An example of this is seen in the first epithelium formed by a mouse embryo. The cells of this epithelium, the trophoblast, can divide either in the plane of the epithelium to produce two equal epithelial daughter cells, or they can divide perpendicularly so that one daughter remains within the plane of the epithelium while the other is driven in toward the centre of the embryo[5] (see Figure 5.2.2). This has a profound effect on the cells, for only those that leave the plane of the epithelium to form the inner cell mass will contribute directly to the mouse (the epithelium itself forms the placenta).

Changes in the orientation of mitosis can also affect the morphogenesis of mature epithelia. The morphology of the lining of mammalian uteri is sensitive to hormones, particularly oestrogens and progesterones, which together prepare the uterus for implantation of an embryo. In the absence of these hormones, the endometrium (uterine lining) is thin, but it thickens and acquires more layers of cells following treatment with oestrogen. In ovariectomized rats, which have extremely low levels of estrogens, the few mitoses that are present are orientated so that the cleavage plane is parallel to the basement membrane, and daughter cells remain in the same layer (see Figure 5.2.3). If those rats are treated with oestradiol, however, up to 40 percent of mitoses are orientated so that the cleavage plane is parallel to the basement membrane and one of the daughter cells is therefore pushed upward to become a new layer.[6] The molecular connection between hormones and mitotic orientation has not been established in this

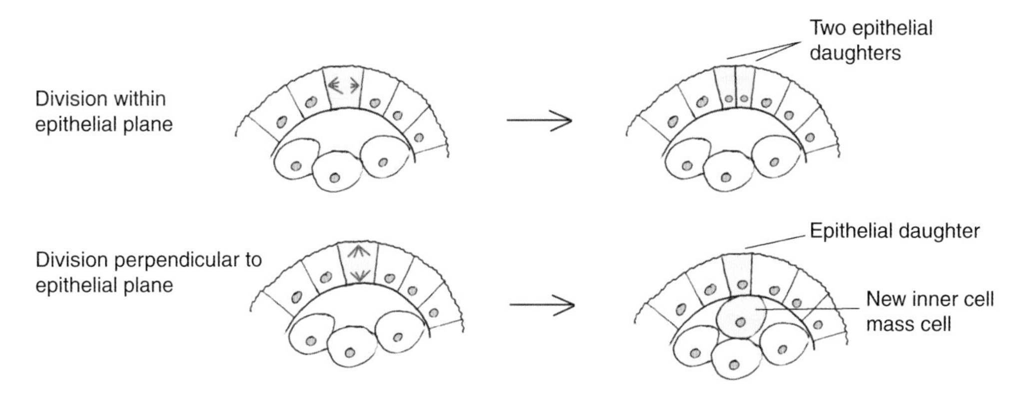

FIGURE 5.2.2 Orientation of mitoses in the mouse trophoblast determines whether both daughters of a division remain in the trophoblast or whether one joins the inner cell mass, which will give rise to the "embryo proper." This diagram is based on the data in an article by R. A. Pedersen, K. Wu, and H. Balakier.[5]

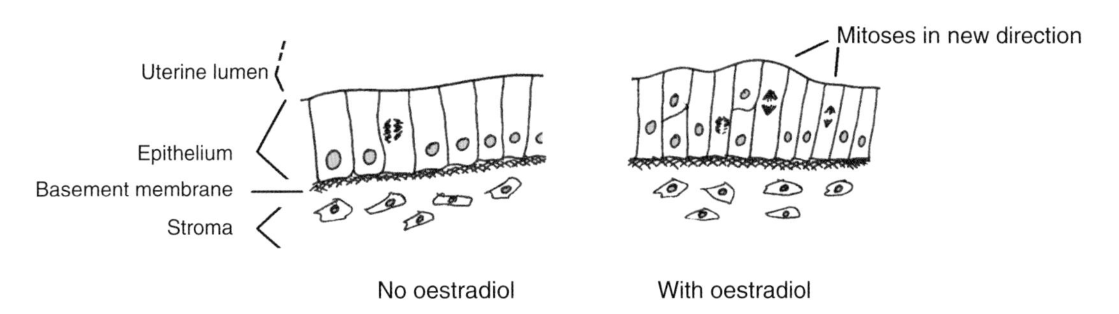

FIGURE 5.2.3 The increase in number of, and change in orientation of, mitoses in rat endometrium when treated with oestradiol. The diagram is based on micrographs in an article by A. G. Gunin.[6]

system, but this effect of oestrogen is specific to uterine epithelium and is not seen in irrelevant epithelia such as those of the gut.

The molecular mechanisms that control the positioning of the cleavage plane are still under investigation, but in the systems that have been studied so far it seems that the position and orientation of cytokinesis is determined by the position of the mitotic spindle. Cytokinesis does, after all, always run transversely across the line that was the axis of the spindle, as it must if it is to separate daughter nuclei. The relationship between spindle orientation and the cleavage plane holds even in cells from which the chromosomes have been removed,[7] demonstrating clearly that it is the spindle apparatus, in its widest sense, and not the chromosomes that it separates, that gives the cue. There is still some debate about whether spatial information for guiding cytokinesis comes from the interpolar parts of the spindle or from the astral microtubules (construction of the spindle was discussed in Chapter 2.2; Figure 5.2.4 is a reminder of its main features).

The astral model is supported mainly by an experiment in which the central portion of the large egg of the echinoderm *Echinarachnius parma* ("sand dollar") is

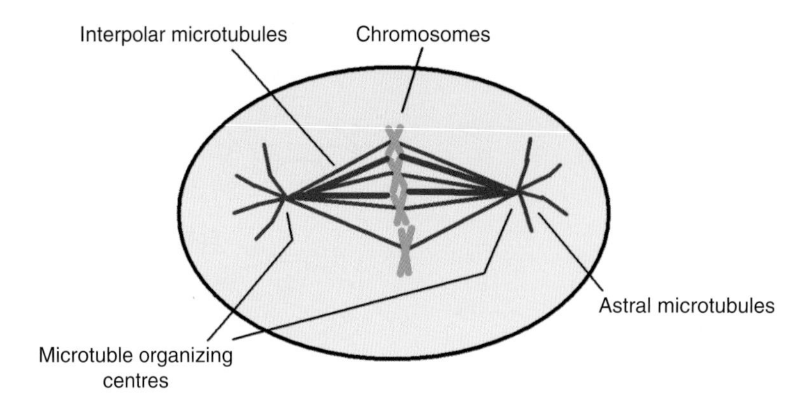

FIGURE 5.2.4 Nomenclature of spindle mitotic apparatus used in this chapter.

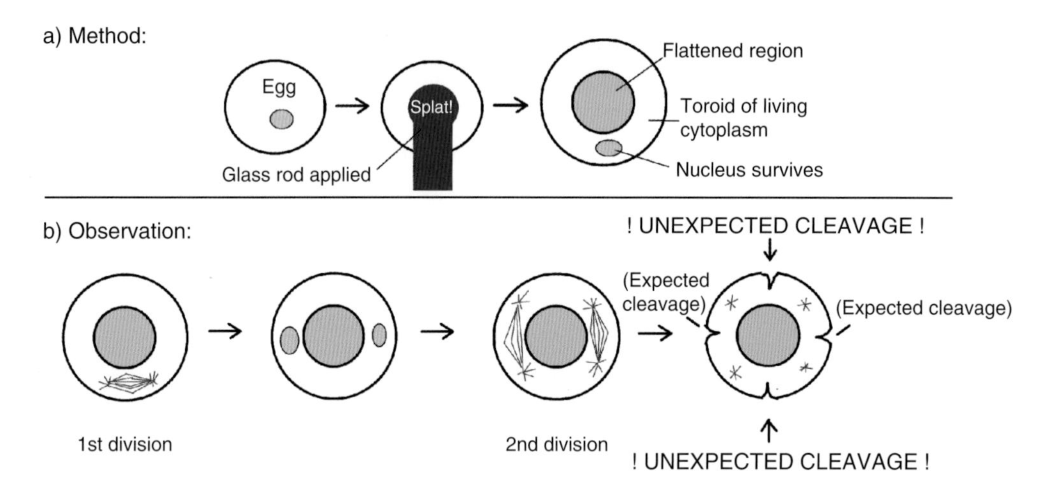

FIGURE 5.2.5 Evidence from manipulation of the sand dollar *Echinarachnius parma* that cleavage planes form between asters. Please see the main text for a rival point of view.

crushed with a glass rod so that the whole cell becomes a biconcave disk similar in shape to a mammalian red blood cell (see Figure 5.2.5). The first mitosis of these crushed eggs takes place normally except that cytokinesis fails to cross the crushed area and the daughter nuclei separate to opposite sides of the remaining ring of cytoplasm. The next mitosis also begins normally, and the effect is to produce asters that are spaced approximately equally around the cytoplasm. When cytokinesis begins, *four* furrows form, including a furrow between adjacent asters that have no spindle between them.[8] This observation is difficult to explain by a model in which interpolar tubules specify cytokinesis but easy to explain by a model in which cleavage furrows form between asters.

Evidence against the aster-driven model comes mainly from observations of the meiotic divisions a mutant of *Drosophila melanogaster*, called *asterless*. As its name suggests, the *asterless* mutant cannot form asters in its meiotic divisions, apparently because it lacks pericentriolar material, but cytokinesis takes place in the correct place anyway.[9]

Support for the cleavage plane being specified by interpolar microtubules comes from experiments in which communication between the spindle and the cell cortex is disrupted. If a blunt glass needle is inserted into a cell to act as a barrier between the spindle and the nearby cortex, it blocks cytokinesis when present from metaphase but fails to block it when only present after the first minute of anaphase.[10] This suggests that a transient signal travels from the spindle to the cell cortex during metaphase or the very beginning of anaphase. Components of this signal have now been identified, mainly through the genetics of *D. melanogaster*. They include chromosomal passenger proteins, such as the inner centromere protein INCENP and serine-threonine kinases of the Aurora family. These proteins bind chromosomes during prometaphase and metaphase but fall away at anaphase,[11,12] probably being released by the anaphase-promoting complex that was mentioned in Chapter 5.1. Once free of chromosomes, these proteins bind the microtubules at the equator of the spindle and assist in their stabilization as the chromosomes part. Mutations that block the ability of INCENP or Aurora proteins to associate with microtubules block the onset of cytokinesis.[13,14] It has been suggested that the high concentration of these proteins at the equator signals to the overlying cortex to initiate ingress of a cleavage furrow; some, such as INCENP, may interact directly with the furrow.[15] This theory does not, however, provide a ready explanation of how chromosome-free cells can still orientate cytokinesis properly.

Whether the astral model, the equatorial model, or some combination of both wins the day, orientation of cell division is controlled by the orientation of the mitotic spindle.

ORIENTATION OF THE MITOTIC SPINDLE

There are two broad ways in which correct orientation of the mitotic spindle might be achieved: MTOCs could be aligned correctly before spindle assembly begins, or the spindle could be allowed to form in any orientation and then brought into alignment afterwards. At least some cell types, such as mammalian neuroepithelial cells of the cerebral cortex, provide strong support for alignment taking place after spindle formation. In these cells, mitotic spindles rotate several times before taking up their final positions,[16] and, even while they are rotating, spindles often "stall" for a while at what will be their final orientation. This stalling suggests that something is trying to trap the spindles on this alignment but that it only succeeds after a few rotations have been made. Studies on mutants that randomize orientation of mitotic spindles provide strong hints about the nature of the trapping mechanism. Mutants that cause the astral microtubules to be unusually short and therefore not to reach into the cortical regions of the cell randomize the orientation of the spindle, suggesting that the trapping mechanism normally works by pulling the astral microtubules to specific regions of the cortex. Mutants of budding yeast that have lost the function of the microtubule binding and motor protein dynein,[17] and also mutants of the nematode *Caenorhabditis elegans* that have lost the function of the dynein-activating proteins Dnc1 and Dnc2,[18] all show randomization of spindle orientation. Also, if a laser is used to sever the central portion of the mitotic

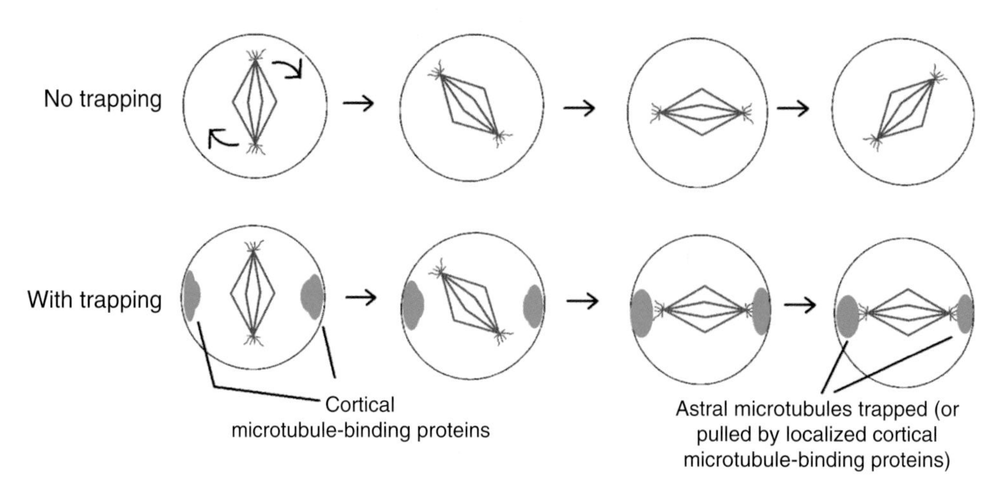

FIGURE 5.2.6 A model for orientation of the mitotic spindle through trapping by the cortical cytoplasm.

spindle of a one-celled embryo of *C. elegans*, the astral microtubules spring apart from each other as if they were being pulled toward the periphery of the cell.[19] Together, these observations suggest a mechanism in which specific regions of the cell cortex are (especially) able to trap astral microtubules using microtubule-binding proteins and/or to pull astral microtubules toward them using dynein (see Figure 5.2.6). The problem of orientating mitoses therefore becomes one of attracting astral microtubules to the correct parts of a cell.

The system that ensures that cells of simple epithelia normally divide only within the plane of the epithelium provides a good example of localized cortical microtubule-binding proteins. In epithelia, cadherin-rich adhesion sites such as adherens junctions are found on the lateral domains of cells but not on the basal or apical domains. (Epithelial structure in general, and the structure of these junctions in particular, is described in more detail in Chapter 4.1.) In the ectoderm epithelium of *D. melanogaster*, cell divisions are normally orientated so that the cleavage plane is perpendicular to the basement membrane, and each daughter cell therefore maintains contact with it. If formation of adherens junctions is disrupted by transfecting embryos with interfering RNA that targets junctional components, the normal orientation is lost. The position of the mitotic spindle in normal epithelial cells is such that its poles are close to adherens junctions, suggesting that there may possibly be a direct interaction between the two. This possibility is supported by the fact that the inner face of adherens junctions contains a large number of proteins, among which are some that can bind microtubules. For example, the protein APC is a component of adherens junctions, binding to β-catenin, and APC can in turn bind the microtubule-binding protein EB1, which could conceivably bind to the spindle. Mutation of either APC or EB1 causes the same loss of orientation control as loss of the junctions themselves, supporting the idea that the adherens-APC-EB1-spindle link is real and important (see Figure 5.2.7). There is also evidence from other systems that β-catenin can bind dynein,[20] and this may also act to keep astral microtubules in the region of adherens junctions.

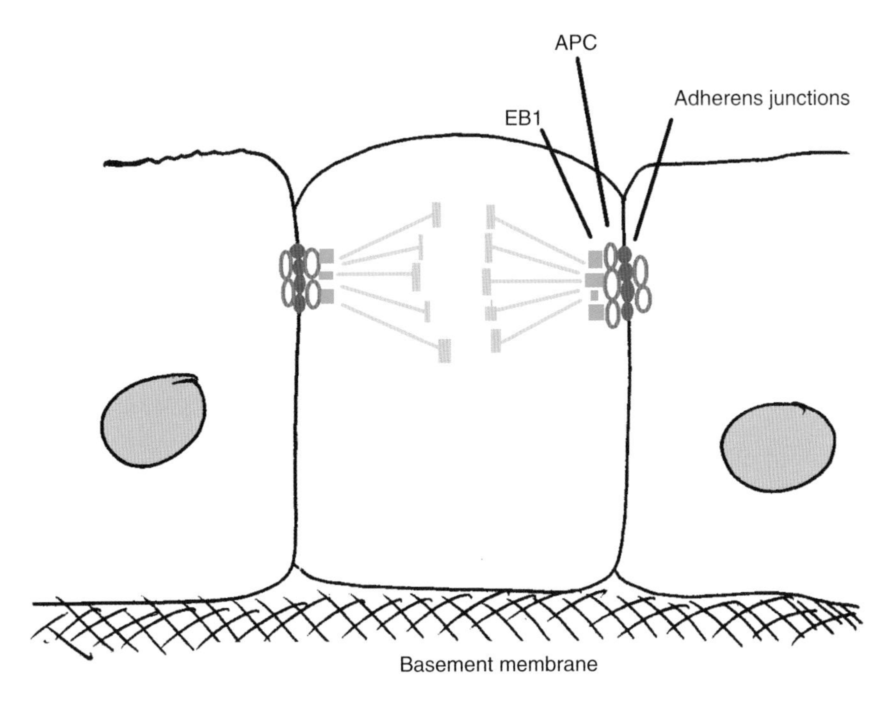

FIGURE 5.2.7 Possible means by which the positions of adherens junctions control the orientation of the mitotic spindle in epithelial cells to ensure that daughters each retain contact with the basement membrane. Clearly, it must be possible for this system to be overridden when division in the other direction is required (as, for example, in the endometrium example).

This adhesion-dependent system may control spindle orientation in other cell types too. In the testis of *D. melanogaster*, there are stem cells that maintain their own numbers and give rise to daughter cells that will eventually produce sperm. The stem cells bind to a central condensate of somatic cells called the "hub." Adhesion between stem cells and the hub is mediated, at least in part, by E-cadherin, which forms adherens junction–like complexes that include APC on their inner faces. Normally, the mitotic spindle of a dividing germ cell is orientated so that one pole is near the adhesion site; this will leave one daughter cell still contacting the hub directly and the other not (see Figure 5.2.8). The daughter still in contact with the hub retains stem cell character due to short-range signals coming from the hub, whereas the other cell becomes a gonioblast that undergoes further rounds of mitosis and then enters meiosis. If APC function is disrupted, mitotic orientation of the stem cell becomes randomized.[21]

It is interesting to note, in passing, that the protein APC acquired its name from the human disease adenomatous polyposis coli, which is associated with heterozygous APC mutation. Loss of heterozygosity (to an effectively homozygous mutant condition) by clones of cells in the colon causes them to develop into a disorganized overgrowth of the gut epithelium; this may later transform to become a carcinoma. Most theories about the aetiology of adenomatous polyposis coli concentrate on the ability of the APC protein to inhibit the Wnt signalling pathway,[22,23] but it is possible that a disorientation of cell division plays a major role in allowing cells to pile up on each other.

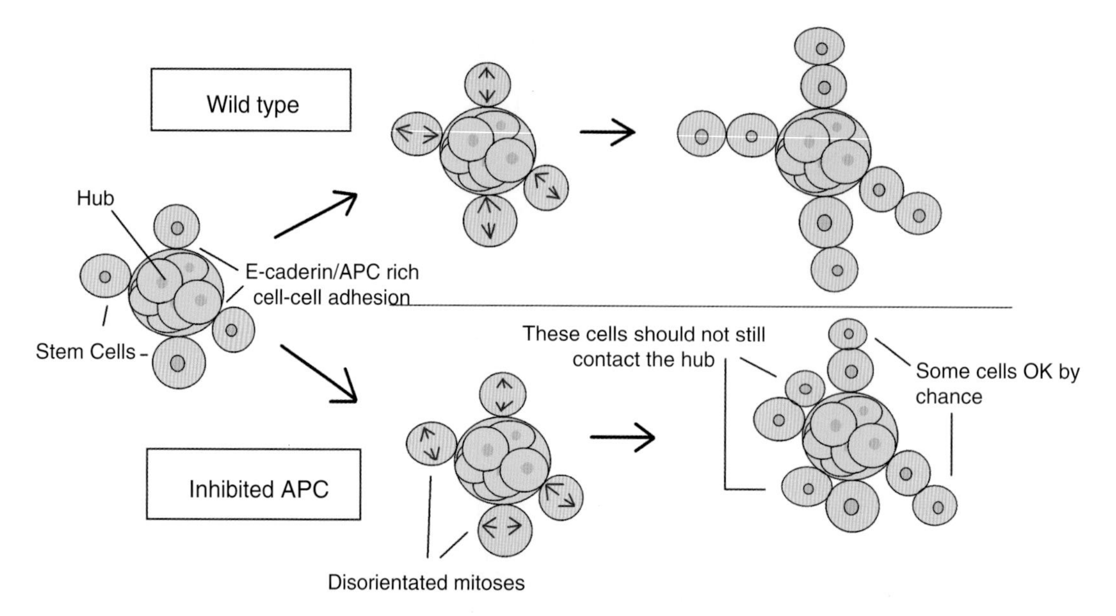

FIGURE 5.2.8 Schematic of stem cell mitosis in the testis of *D. melanogaster* showing the disorientating effects of loss of APC function.

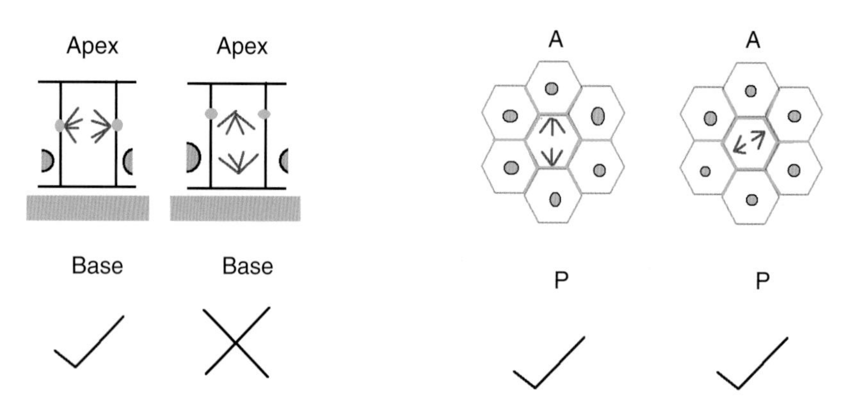

FIGURE 5.2.9 Adherens junctions (green) can orientate the plane of division in the apico-basal axis but cannot specify orientation within the plane of the epithelium itself.

The orientation system based on cell adhesion cannot be the only mechanism for controlling the orientation of division, partly because many cells do not have the correct junctions and partly because, even when they do, the circumferential arrangement of cell junctions in a typical epithelium cannot specify polarity within the plane of the epithelium (see Figure 5.2.9).

Orientation of cell division within the plane of an epithelium (and other tissues) is important in morphogenesis. It is, for example, one of the driving forces of vertebrate neurulation (described in more detail in Chapter 4.4). The mitoses in the epithelium on

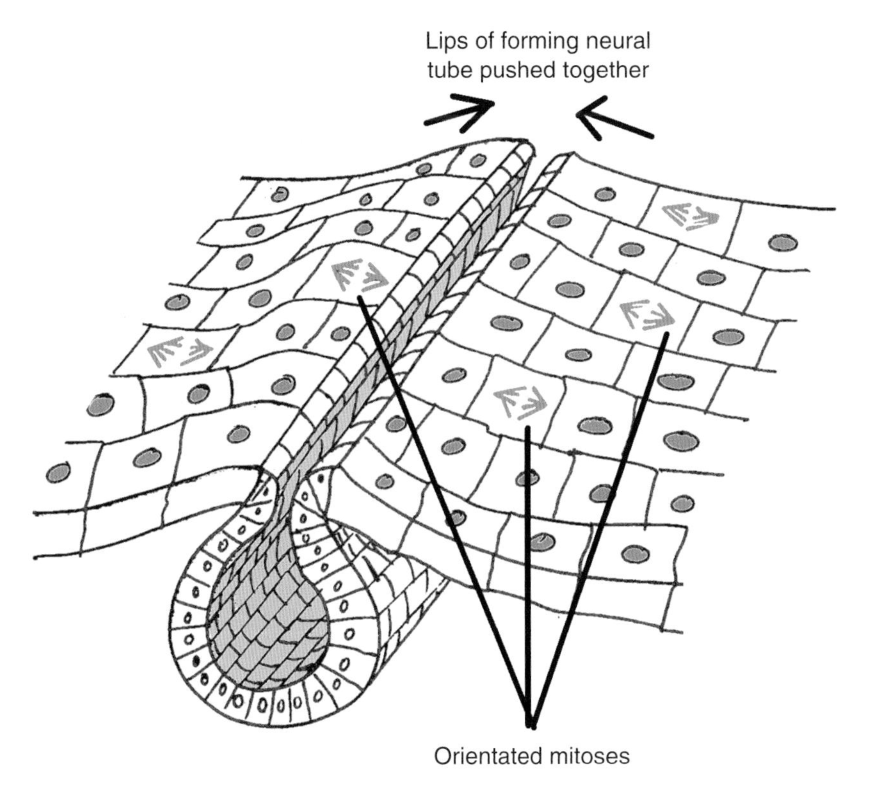

FIGURE 5.2.10 Orientated cell divisions in vertebrate neurulation.

each side of the neural tube are orientated so that daughter cells are produced towards the longitudinal axis of the embryo[24] (see Figure 5.2.10).

The mechanism of orientation is not yet well understood in the neural tube, but substantial progress has been made in understanding the coupling of mitotic orientation to planar polarity in *D. melanogaster*. The process is illustrated well by the development of the fly's mechanosensory organs. These organs consist of a bristle that projects from the notum (the dorsal part of the thorax) into the air-stream and of associated cells that support the bristle and relay information about its bending moments to the central nervous system. This information is used to control flight movements. Mechanosensory organs also detect direct contacts between the fly and anything else. The organs consist of five cells, all of which originate from a single progenitor called pI ("p" for "progenitor"). This cell and its daughters undergo a stereotyped series of orientated cell divisions to give rise to the five cells of the mature organ (see Figure 5.2.11).[25,26] The morphogenesis of the mechanosensory organ does not arise simply from orientated cell division but instead requires considerable cell rearrangement.[27] It is, nevertheless, useful for this discussion because it features orientated cell divisions in all three axes.

The first division of pI is within the plane of the epithelium and aligned with the antero-posterior axis of the body. Subsequent divisions take place perpendicularly to each of these axes, and this fact alone indicates that there must be more than one

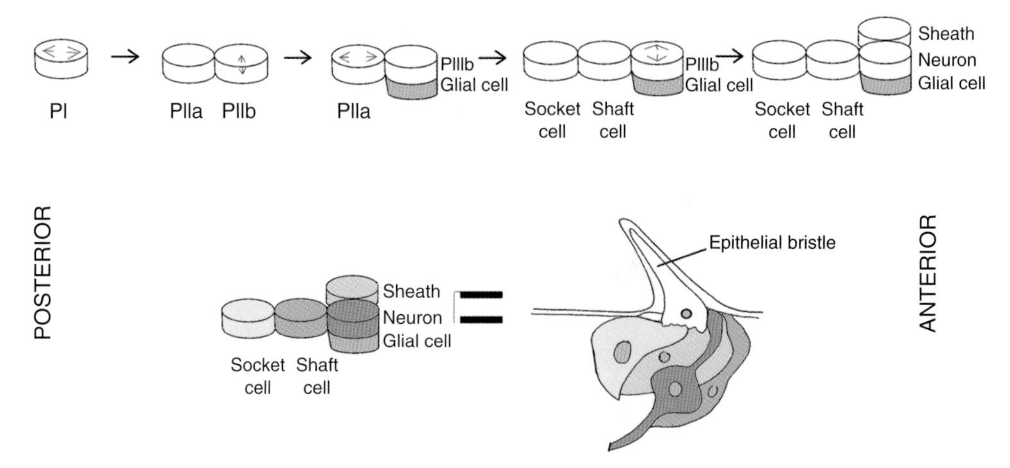

FIGURE 5.2.11 The sequence of orientated cell divisions during development of mechanoreceptors. The upper diagram shows, schematically, the sequence and orientation of divisions based on data in an article by G. V. Reddy and V. Rodriques.[27] The glial cell, shaded grey, is below the plane of the epithelium. The actual cell shapes are more complicated than the simple discs shown in the top figure, and considerable cell rearrangement is involved (particularly of the glial cell, which moves right away from the place of its birth[27]). The arrangement of cells in the mature mechanosensory organ is shown in the lower figure, based on the description in a recent review.[28]

orientation system available to cells since whatever causes one cell to divide along the body axis can be ignored by its daughter cell, which follows a quite different cue. Observation of forming spindles suggests that they arise with a random orientation and later rotate to be locked into the correct direction.[29] The orientation of pI's division depends on the same "planar polarity" system that guides the convergent extension discussed in Chapter 4.2, and which signals via the receptor Frizzled and its effector protein Dishevelled. Mutation of either of these proteins causes the division of pI to become disorientated.[26] (The randomization of bristle direction in patches of mutant epithelium is, of course, what gives flies their "frizzled" and "dishevelled" appearance.) It is not yet clear how planar polarity signals work at the scale of a whole tissue, but it does appear that their effect is to produce a higher level of Frizzled activation on one side of a cell than on the other. In the case of pI, Frizzled and Dishevelled are activated and located mainly at the posterior pole of the cell, whereas the proteins Strabismus and Prickled, also involved in planar cell polarity, are located mainly at the anterior (these proteins tend to segregate to areas at which Dishevelled is not present).[30] Strabismus recruits further proteins, such as Partner of Inscuteable and Disks Large, to the anterior end of the cell.[30]

The link between the planar cell polarity signalling system orientation of the mitotic spindle is not yet understood, but some very recently obtained data from vertebrate systems suggests how the connections may be made at both the posterior and anterior of the cell. At the posterior of the cell, Dishevelled may play the key role. Overexpression of vertebrate Dishevelled1 in neurons greatly increases the number of microtubules in their growing axons.[31] Dishevelled1 directly associates with microtubules in a complex that includes GSK3β and axin, and stabilizes them. This suggests

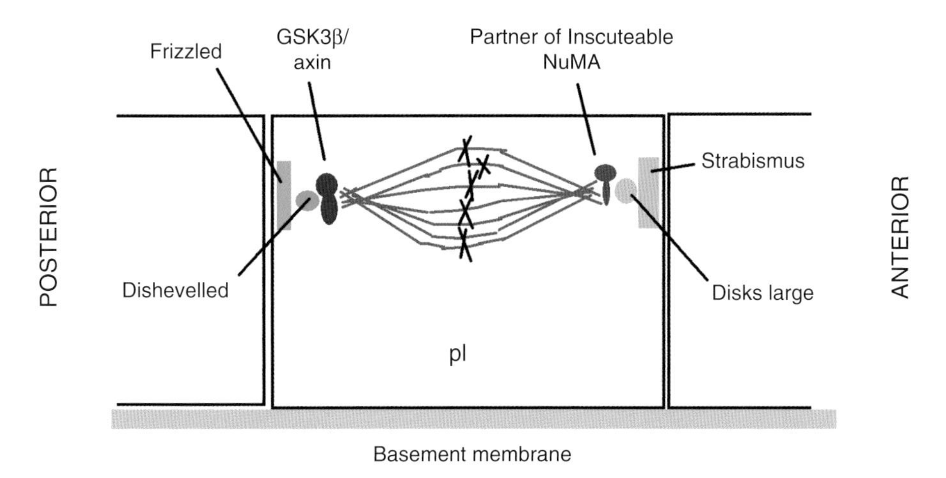

FIGURE 5.2.12 Schematic diagram to summarize the hypothetical links between the planar polarity signalling system and the mitotic spindle. This diagram is based on a mixture of data and guesswork, as explained in the text, and should be treated with caution.

that Dishevelled activation might act directly on the microtubule system associated with centrioles and with the spindle. It is important to note, however, that it is the "canonical" Wnt signalling pathway that proceeds through GSK3β, and the "planar polarity" pathway is supposed, at least at the time of writing, to proceed through Jun kinase *instead* (not "as well"). If this is indeed true, the data on Dishevelled/GSK3β/axin complexes are a red herring in terms of understanding orientation of mitoses. At the anterior end of the cell, Partner of Inscuteable may play a similar role. The vertebrate homologue of Partner of Inscuteable, LGN,[32] can bind with high affinity to the microtubule-organizing protein NuMA,[33] and may therefore act on the spindle (see Figure 5.2.12).

Control of the plane of mitosis in gastrulating zebrafish is also under the control of the Wnt-Frizzled-Dishevelled system[34]; such conservation across phyla suggests that the connection may be very widespread. In addition to being controlled by the Frizzled/Dishevelled system, the division of pI is probably also kept in the plane of the epithelium by the adherens junction–mediated system described above.

Once the pI cell has divided, the next cell division, that of pIIb, is at right angles to the division of pI, being aligned with the apico-basal axis of the cell. One of its daughters, the glial cell, is therefore pushed underneath the plane of the epithelium. Presumably, the spindle of pIIb is either disconnected biochemically from the influence of adherens junctions or made sensitive to an even stronger apico-basal influence. The divisions of pIIa and pIIIb seem to be controlled by local influences because if the mitosis of pI is disorientated by mutation of the Frizzled/Dishevelled system, the subsequent divisions of its daughters proceed normally with respect to the axis set up by the division of pI rather than with respect to the axis of the whole body.[26,35]

The changing sensitivity of cells to the various directional cues is an interesting unsolved problem. There are two possible explanations, which may operate together. One is that cells make their choices of orientation cue based on signals that they receive from their neighbours: each division will change the neighbour relationships of cells.

The other is that the choice is set by changes of cell state as differentiation proceeds. In the mechanosensory system, orientated cell division causes daughter cells to inherit unequal amounts of specific proteins that are distributed unequally in the cells.[27] It is possible that the inheritance of these cytoplasmic determinants affects how cells interpret the gradients, although this cannot be in a straightforward way since pI, which has determinants such as Numb, divides along the A/P axis, whereas pIIb, which inherits Numb, divides another way, and pIIa, which does not inherit Numb, divides in the same direction that pI did.

ADAPTIVE SELF-ORGANIZATION OF MITOTIC ORIENTATION AND HERTWIG'S RULE

So far, this chapter has concentrated on "programmed" orientation of cell division to achieve specific morphogenetic change. For much of embryogenesis, the orientation of mitosis is a reaction to other processes rather than a driving force and is important in allowing the production of cells by a tissue to align appropriately with changes in shape driven by other mechanisms such as invagination.

More than a century ago, Oscar Hertwig proposed a simple rule of self-organization that would cause mitoses in a distorted tissue to orientate themselves automatically in a direction that would reduce the forces associated with that distortion: he suggested that the spindle axis always aligns with the long axis of the cell. An updated formulation of Hertwig's rule, which would prefix the rule with a qualification such as "in the absence of overriding cues," seems to hold in many situations, in culture and in embryos. If a tissue—for example, an epithelial sheet—is stretched in one direction, its cells will also be stretched in that direction (assuming no rearrangement). Mitoses will therefore take place mainly in that same direction so that one stretched mother cell is replaced by two daughter cells. A physical model of the stability of the mitotic spindle has suggested a simple explanation for Hertwig's rule based on interactions between the spindle and the microtubules under the plasma membrane; repulsive forces are least when the spindle poles are in the least cramped parts of the cell.[36]

Valuable as it is for ensuring that daughter cells are produced in an orientation that minimizes the distortion of cells in an epithelium subjected to external forces, Hertwig's rule can create positive feedback that exacerbates the instability of shape of an unconstrained tissue in which mitosis is taking place. If the orientation of mitoses in a tissue were to be completely random and there were no compensating processes, the resultant expansion of the tissue would not be isometric but would instead show a cumulative distortion. There are two reasons for this. First, randomly orientated mitosis of cells in a random sequence would create a series of small steps of expansion in random directions. This is equivalent to a statistician's "drunken walk"; the distance from the origin that is travelled by a drunkard who has made n steps, each of unit length but in a random direction, is on average $\sqrt{n}$, *not* zero. Second, Hertwig's rule would mean that once one cell divides and its daughters begin to grow, their neighbours will be stretched along the direction of the first cell's division. They would therefore divide the same way, and by this positive feedback the tissue would be expected to be subject to runaway distortions. Embryos probably defend themselves from this by three methods: their cells can rearrange to relieve tissue stresses, their tissues are really bounded by other tissues so that growth is not unconstrained, and a combination of competition for mitogens and contact inhibition of proliferation will prevent too many mitoses occurring in the same place.

In case this section of the book leaves a false impression, it must be stressed that many morphogenetic events of animal development take place without orientated cell division and, indeed, without cell division at all. Even where mitoses do seem to be orientated consistently, their orientation may not actually be helpful to morphogenesis, and cell movement may be required to abrogate the deleterious morphological effect that the orientation of mitoses would otherwise have. An example of this is seen in the cnidarian *Hydra*, in which the directions of mitoses in the two layers of the body wall are different and would be expected to generate an awful mess if cell rearrangement were not able to sort everything out.[37] It should be borne in mind, therefore, that orientated cell division is just one of many morphogenetic mechanisms available to animals and that, even if present, it may work with or against other processes that are taking place.

ORIENTATED CELL DIVISION IN PLANTS

Orientated cell division is also seen in plants; indeed it is particularly important for organisms in which cell rearrangement is impossible. From the earliest stages of plant development, the form of the embryo emerges directly from the pattern of orientated cell divisions, which creates the characteristic dumbbell-shaped, octant and suspensor morphology (see Figure 5.2.13).

Orientated cell division is also critical to the setting up of the characteristic anatomy of growing roots and shoots, a fact that can be demonstrated by mutants

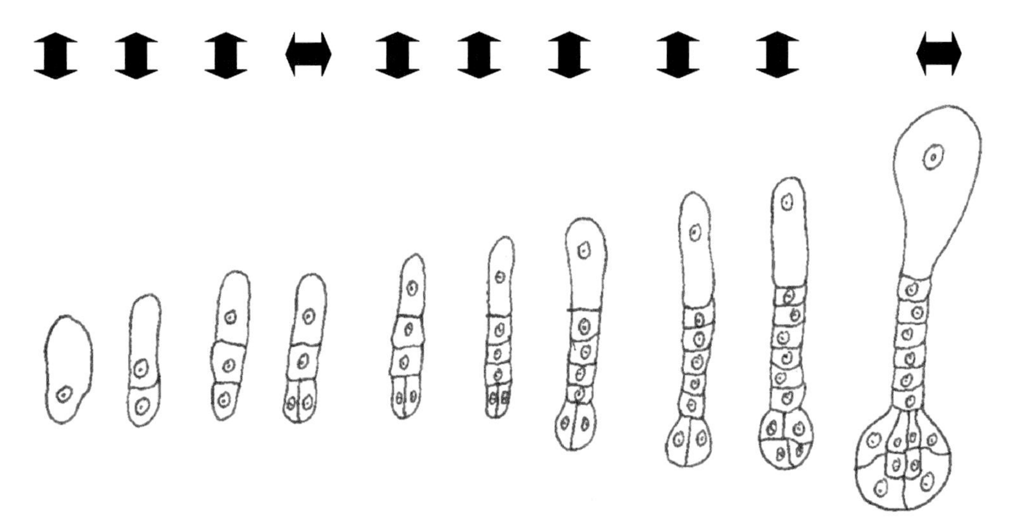

FIGURE 5.2.13 The succession of orientated cell divisions in the early development of the embryo of *Capsella bursa-pastoris* ("Shepherd's purse"): the orientation of each cell division is shown by the arrow above it, at the top of the diagram. Based on *Plant Anatomy* by A. Fahn.[38]

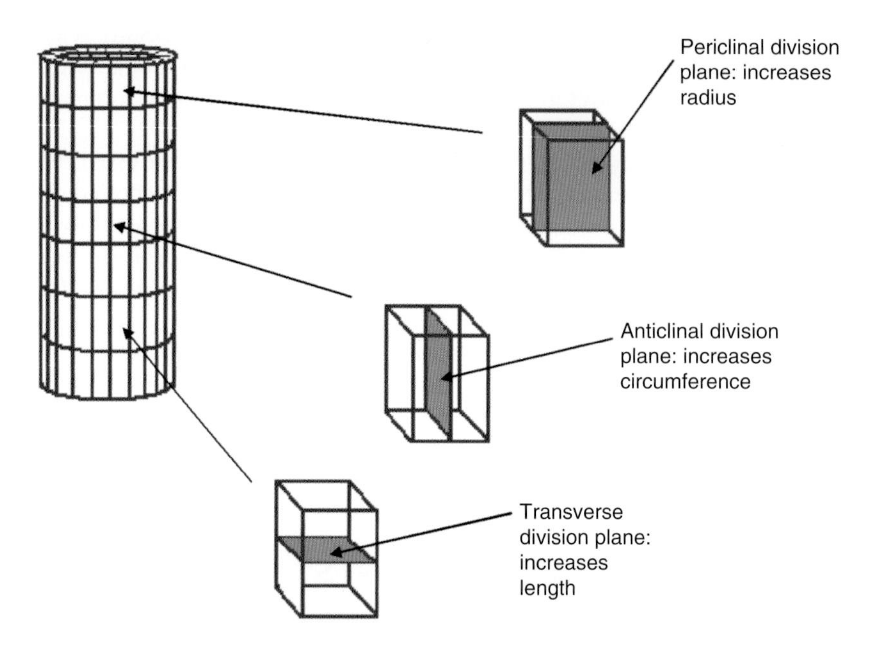

FIGURE 5.2.14 Nomenclature for division planes in the root tip.

of *Arabidopsis thaliana*. Roots grow by the addition of cells that derive from a population of "initials" in the meristem just under the root cap; initials have the characteristics that would earn them the name "stem cells" in animals, but that phrase is not used in botany because it would be an obvious source of confusion. There are four sets of initials,[39] each of which produces cells that give rise to a different concentric layer of the root; one set gives rise to pericycle and vascular tissue, one to columnella, one to the cortex/endodermis endoderm, and the other to the epidermis. The cortex and endodermis derive from the same daughter of an initial cell; this daughter divides to produce one cell destined to be cortex and one destined to be endoderm. This division is periclinal so that it produces one outer cell, destined to contribute to cortex, and one inner cell, destined to contribute to the endodermis (the nomenclature of division planes is illustrated in Figure 5.2.14). Each of these cells then undergoes a series of anticlinal divisions to produce a series of progenitor cells that occupy a one-cell-thick arc of the root tip. These then go on to divide and elongate to produce the root.

In the *schizoriza* mutant of *A. thaliana*, the pattern of division is disrupted so that there is more than one periclinal division before anticlinal division begins. Because each periclinal division produces daughter cells aligned along a radius, the effect of this disrupted division pattern is to produce an arc of cells more than one cell thick (see Figure 5.2.15). The genetics and cell biology of the mutant are complicated (differentiation is affected as well),[40] but the existence of a single mutant that has the effect of disrupting orientation of cell division underlines the basic fact that orientation of cell division is under strict control.

The mechanisms that control the orientation of mitoses in plants are not as well understood as they are in animals, but the question is an important one because of the

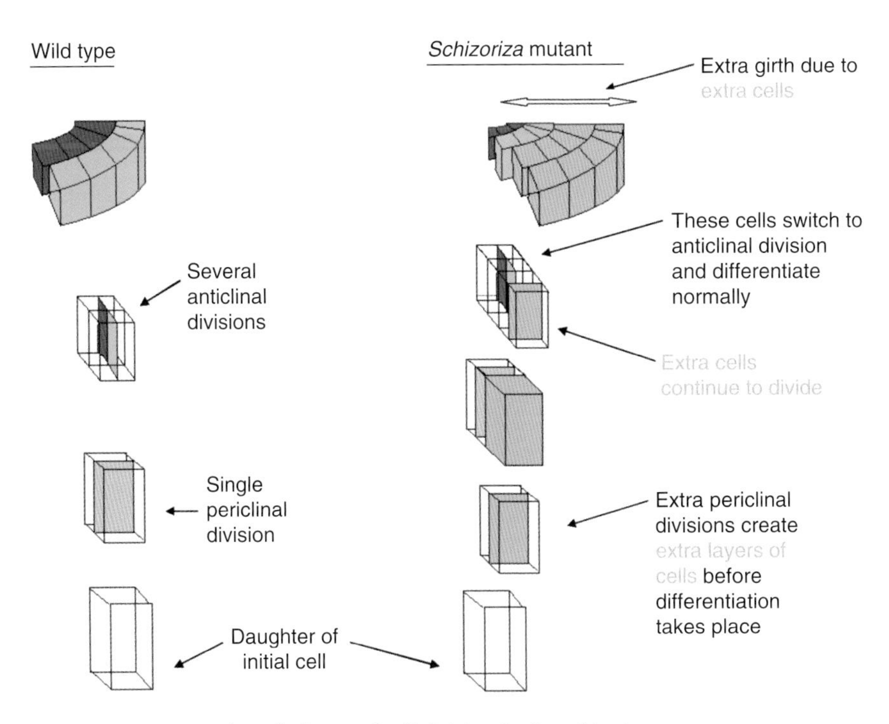

FIGURE 5.2.15 Altered planes of cell division in the *schizoriza* mutant.

very close relationship between division and morphogenesis in any organism in which cell rearrangement cannot override a cell's place of birth.

Reference List

1. Collier, JR (1997) Gastropods, the snails. 189–217. In Gilbert, SF and Raunio, AM, Embryology: Constructing the organism (Sinauer)
2. Black, SD and Vincent, JP (1988) The first cleavage plane and the embryonic axis are determined by separate mechanisms in Xenopus laevis. II. Experimental dissociation by lateral compression of the egg. *Dev. Biol.* **128**: 65–71
3. Goodwin, BC and Lacroix, NH (7-7-1984) A further study of the holoblastic cleavage field. *J. Theor. Biol.* **109**: 41–58
4. Goodwin, BC and Trainor, LE (21-8-1980) A field description of the cleavage process in embryogenesis. *J. Theor. Biol.* **85**: 757–770
5. Pedersen, RA, Wu, K, and Balakier, H (1986) Origin of the inner cell mass in mouse embryos: Cell lineage analysis by microinjection. *Dev. Biol.* **117**: 581–595
6. Gunin, AG (2001) Estrogen changes mitosis orientation in the uterine epithelia. *Eur. J. Obstet. Gynecol. Reprod. Biol.* **97**: 85–89
7. Zhang, D and Nicklas, RB (1-8-1996) 'Anaphase' and cytokinesis in the absence of chromosomes. *Nature.* **382**: 466–468
8. Rappaport, R (2003) Experiments concerning the cleavage stimulus in sand dollar eggs. *J. Exp. Zool.* **148**: 181–89
9. Bonaccorsi, S, Giansanti, MG, and Gatti, M (10-8-1998) Spindle self-organization and cytokinesis during male meiosis in asterless mutants of *Drosophila melanogaster*. *J. Cell Biol.* **142**: 751–761

10. Cao, LG and Wang, YL (1996) Signals from the spindle midzone are required for the stimulation of cytokinesis in cultured epithelial cells. *Mol. Biol. Cell.* **7**: 225–232
11. Cooke, CA, Heck, MM, and Earnshaw, WC (1987) The inner centromere protein (INCENP) antigens: Movement from inner centromere to midbody during mitosis. *J. Cell Biol.* **105**: 2053–2067
12. Murata-Hori, M, Tatsuka, M, and Wang, YL (2002) Probing the dynamics and functions of aurora B kinase in living cells during mitosis and cytokinesis. *Mol. Biol. Cell.* **13**: 1099–1108
13. Mackay, AM, Ainsztein, AM, Eckley, DM, and Earnshaw, WC (9-3-1998) A dominant mutant of inner centromere protein (INCENP), a chromosomal protein, disrupts prometaphase congression and cytokinesis. *J. Cell Biol.* **140**: 991–1002
14. Adams, RR, Maiato, H, Earnshaw, WC, and Carmena, M (14-5-2001) Essential roles of *Drosophila* inner centromere protein (INCENP) and aurora B in histone H3 phosphorylation, metaphase chromosome alignment, kinetochore disjunction, and chromosome segregation. *J. Cell Biol.* **153**: 865–880
15. Earnshaw, WC and Cooke, CA (1991) Analysis of the distribution of the INCENPs throughout mitosis reveals the existence of a pathway of structural changes in the chromosomes during metaphase and early events in cleavage furrow formation. *J. Cell Sci.* **98 (Pt 4)**: 443–461
16. Adams, RJ (1-12-1996) Metaphase spindles rotate in the neuroepithelium of rat cerebral cortex. *J. Neurosci.* **16**: 7610–7618
17. Li, YY, Yeh, E, Hays, T, and Bloom, K (1-11-1993) Disruption of mitotic spindle orientation in a yeast dynein mutant. *Proc. Natl. Acad. Sci. U.S.A.* **90**: 10096–10100
18. Skop, AR and White, JG (8-10-1998) The dynactin complex is required for cleavage plane specification in early Caenorhabditis elegans embryos. *Curr. Biol.* **8**: 1110–1116
19. Grill, SW, Gonczy, P, Stelzer, EH, and Hyman, AA (1-2-2001) Polarity controls forces governing asymmetric spindle positioning in the Caenorhabditis elegans embryo. *Nature.* **409**: 630–633
20. Ligon, LA, Karki, S, Tokito, M, and Holzbaur, EL (2001) Dynein binds to beta-catenin and may tether microtubules at adherens junctions. *Nat. Cell Biol.* **3**: 913–917
21. Yamashita, YM, Jones, DL, and Fuller, MT (12-9-2003) Orientation of asymmetric stem cell division by the APC tumor suppressor and centrosome. *Science.* **301**: 1547–1550
22. Bienz, M (18-3-2003) APC. *Curr. Biol.* **13**: R215–R216
23. Bright-Thomas, RM and Hargest, R (2003) APC, beta-Catenin and hTCF-4: An unholy trinity in the genesis of colorectal cancer. *Eur. J. Surg. Oncol.* **29**: 107–117
24. Sausedo, RA, Smith, JL, and Schoenwolf, GC (19-5-1997) Role of nonrandomly oriented cell division in shaping and bending of the neural plate. *J. Comp. Neurol.* **381**: 473–488
25. Le Borgne, R, Bellaiche, Y, and Schweisguth, F (22-1-2002) *Drosophila* E-cadherin regulates the orientation of asymmetric cell division in the sensory organ lineage. *Curr. Biol.* **12**: 95–104
26. Gho, M and Schweisguth, F (14-5-1998) Frizzled signalling controls orientation of asymmetric sense organ precursor cell divisions in *Drosophila*. *Nature.* **393**: 178–181
27. Reddy, GV and Rodrigues, V (1999) A glial cell arises from an additional division within the mechanosensory lineage during development of the microchaete on the *Drosophila* notum. *Development.* **126**: 4617–4622
28. Muller, U and Littlewood-Evans, A (2001) Mechanisms that regulate mechanosensory hair cell differentiation. *Trends Cell Biol.* **11**: 334–342
29. Bellaiche, Y, Gho, M, Kaltschmidt, JA, Brand, AH, and Schweisguth, F (2001) Frizzled regulates localization of cell-fate determinants and mitotic spindle rotation during asymmetric cell division. *Nat. Cell Biol.* **3**: 50–57
30. Bellaiche, Y, Beaudoin-Massiani, O, Stuttem, I, and Schweisguth, F (2004) The planar cell polarity protein Strabismus promotes Pins anterior localization during asymmetric division of sensory organ precursor cells in *Drosophila*. *Development.* **131**: 469–478

31. Ciani, L, Krylova, O, Smalley, MJ, Dale, TC, and Salinas, PC (19-1-2004) A divergent canonical WNT-signalling pathway regulates microtubule dynamics: Dishevelled signals locally to stabilize microtubules. *J. Cell Biol.* **164**: 243–253
32. Yu, F, Morin, X, Kaushik, R, Bahri, S, Yang, X, and Chia, W (1-3-2003) A mouse homologue of *Drosophila* pins can asymmetrically localize and substitute for pins function in *Drosophila* neuroblasts. *J. Cell Sci.* **116**: 887–896
33. Du, Q, Taylor, L, Compton, DA, and Macara, IG (19-11-2002) LGN blocks the ability of NuMA to bind and stabilize microtubules. A mechanism for mitotic spindle assembly regulation. *Curr. Biol.* **12**: 1928–1933
34. Gong, Y, Mo, C, and Fraser, SE (14-7-2004) Planar cell polarity signalling controls cell division orientation during zebrafish gastrulation. *Nature* **430**: 689–693
35. Le Borgne, R, Bellaiche, Y, and Schweisguth, F (22-1-2002) *Drosophila* E-cadherin regulates the orientation of asymmetric cell division in the sensory organ lineage. *Curr. Biol.* **12**: 95–104
36. Bjerknes, M (12-12-1986) Physical theory of the orientation of astral mitotic spindles. *Science.* **234**: 1413–1416
37. Shimizu, H, Bode, PM, and Bode, HR (1995) Patterns of oriented cell division during the steady-state morphogenesis of the body column in hydra. *Dev. Dyn.* **204**: 349–357
38. Fahn, A (1967) Plant anatomy. (Pergamon Press)
39. Dolan, L, Janmaat, K, Willemsen, V, Linstead, P, Poethig, S, Roberts, K, and Scheres, B (1993) Cellular organisation of the Arabidopsis thaliana root. *Development.* **119**: 71–84
40. Mylona, P, Linstead, P, Martienssen, R, and Dolan, L (2002) SCHIZORIZA controls an asymmetric cell division and restricts epidermal identity in the Arabidopsis root. *Development.* **129**: 4327–4334

CHAPTER 5.3

MORPHOGENESIS BY ELECTIVE CELL DEATH

By the time I was born, more of me had died than survived.
—LEWIS THOMAS[1]

As cells can be added to a system by proliferation, so can they be removed by death.[2] This process is important for controlling the size of cell populations, for clearing up mistakes, and also for directly driving some morphogenetic processes. The type of cell death that is a normal feature of development is achieved by active processes that are invoked by cells that have become committed to die. Some authors use the phrase "programmed cell death" to refer to any process by which cells die as a normal part of embryogenesis because cells die by running a specific internal "suicide program." Other authors use the phrase "programmed cell death" strictly to imply that the fate of an identifiable cell is reproducibly specified as death by the general developmental program of an embryo, in a manner similar to the specification of differentiated states. (This type of programmed death is epitomized by the development of the nematode *Caenorhabditis elegans*.[3,4]) This ambiguous usage of the phrase is causing problems in the literature and has inspired pleas for clearer thinking.[5] Some authors also use the term "apoptosis"[2] to refer to elective cell death in general, whereas others restrict the use of that term to a specific method of self-destruction,[6,7] creating confusion around that term also. This book will therefore use the general phrase "elective cell death" to refer to all processes by which a cell does away with itself, whatever the details of the actual mechanism used, and will use the more specific phrases in their strict sense and only when appropriate.

Elective cell death can accomplish a variety of tasks during morphogenesis, including clearing away temporary structures, removing lost cells, removing cells that are in the way of another process, adjusting the size of a cell population relative to another part of the embryo, and removing damaged cells. Cells can undergo elective death using one of a variety of possible mechanisms.[8] The best-characterized of these is that of apoptosis, in which a cascade of signalling events results in the activation of intracellular proteases of the caspase family.[9] These proteases activate enzymes that cleave

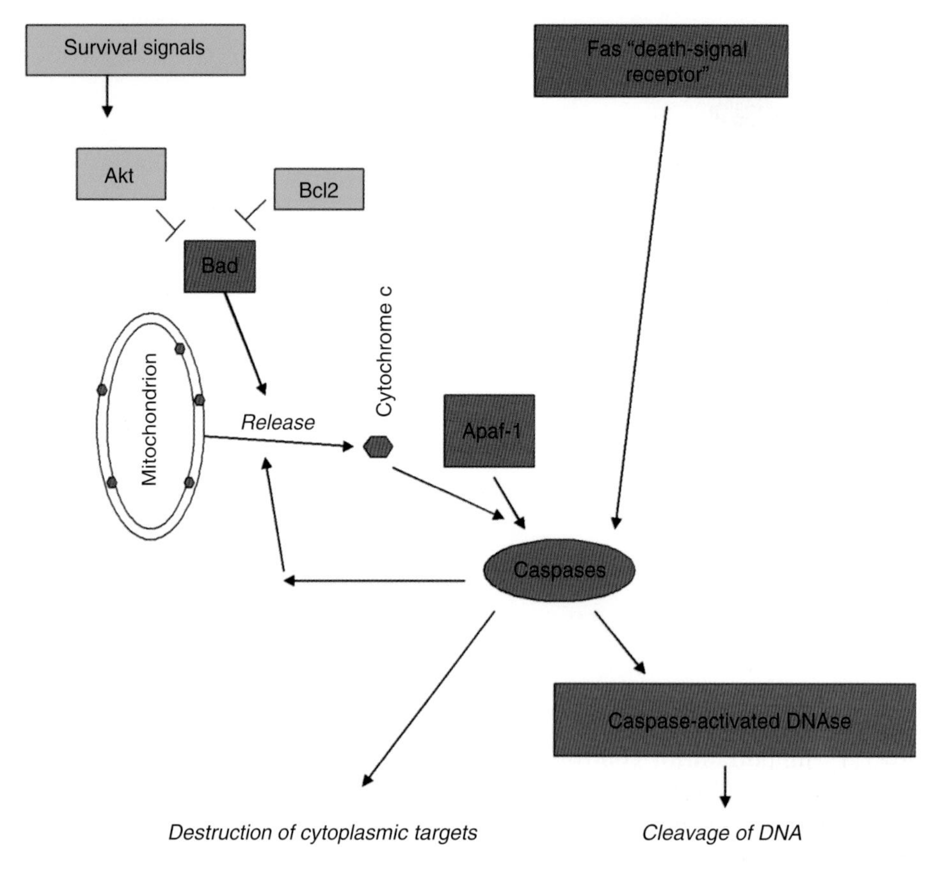

FIGURE 5.3.1 A summary of a typical pathway that controls elective cell death by caspase-mediated apoptosis. Pro-apoptotic influences are coloured red and anti-apoptotic ones are coloured green. Apoptosis can be triggered either by failure of survival signals or by activation of death receptors.

cellular DNA, destroy other cellular systems, and render the cell susceptible to phagocytosis (see Figure 5.3.1). An alternative mechanism, one that seems to be particularly common in development, is autophagy. This also involves caspase activation and DNA cleavage[10] but, in autophagy, cells digest themselves rather than relying on phagocytosis by a neighbour. Alternative internal mechanisms for elective cell death have been reviewed extensively elsewhere,[8,11,12,13] and detailed discussion of the intracellular cascades involved does not belong in a book that is about the generation of shape rather than about biochemical signalling. It is important to note, however, that the variety of possible internal mechanisms means that popular assays for elective cell death, which detect caspase activation or the ends of cleaved DNA, may suggest an absence of death when it is actually taking place but by a different mechanism. All studies on the role of elective cell death in morphogenesis, and particularly those reporting its absence, should therefore be interpreted with appropriate caution.

Elective cell death can be involved in morphogenesis in two main ways: as a direct means of the formation of a new shape or as part of a feedback system that clears

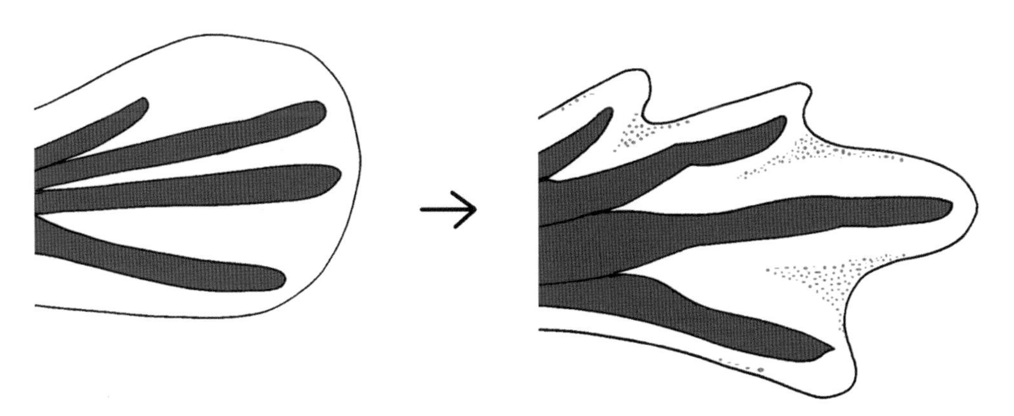

FIGURE 5.3.2 The role of elective cell death in separation of the digits. Mesenchymal condensates that lay down the cartilaginous models for the bones of fingers and toes develop inside the paddle-shaped end of the limb bud. Zones of elective cell death, shown as red dots, then eliminate tissue between the digits and thus separate them. This figure is sketched from micrographs in an article by V. Zuzarte-Luis and J.M. Hurle.[16]

up surplus and lost cells involved in a morphogenetic event that is driven by different mechanisms. The development of the mammalian hand and foot provide a well-studied example in which elective cell death plays a direct role in morphogenesis. During early development, the limb forms as a paddle-like outgrowth from the side of the trunk. Later, cells inside it undergo condensation to form the cartilaginous predecessors of limb bones, including the phalanges that will make up the fingers. The fingers do not, therefore, form as processes that push out from a paddle-like palm but rather form inside the paddle itself (see Figure 5.3.2). Once the condensations that herald each phalanx of a digit have formed, there is a large amount of elective cell death in the zones between the digits, the dead cells being cleared up by macrophages.[14] This cell death appears to take place by both caspase-dependent and caspase-independent mechanisms.[15]

Elective cell death in the interdigital zones seems to be induced specifically by proteins of the BMP family, which are involved with the patterning of the limb. This specific induction makes this type of elective cell death distinct from that seen in the population balancing and tidying-up applications described below, for in those applications cells die because they are not persuaded to live, whereas in the interdigital zone they die because they are told to die. BMP2, BMP4, and BMP7 all induce massive elective cell death in the interdigital zone, although they encourage growth and differentiation in nearby cells that are already committed to forming cartilage.[17,18,19,20] The importance of elective cell death in separating the digits of the hands and feet is emphasized by the mouse *hammertoe* mutant, in which interdigital death does not take place, and the toes fail to separate properly.[14] The development of webbed feet by this mutant can be prevented by inducing elective cell death in the interdigital zone through treatment of the embryos with exogenous retinoic acid, which happens to induce elective cell death in this system,[21] probably via BMPs.[22] An apparently similar pattern of elective cell death is also required to separate the paired bones in the lower arm (radius/ulna) and in the lower leg (tibia/fibula).[23,24] In the chick mutation *talpid3*, this zone of elective cell death fails, and the paired bones fuse.[25]

In birds in which digits are fully separated in the mature foot, such as chickens, elective cell death extends all the way down the digit. In webbed-footed species, such as ducks, it is restricted to the cells between the tips of the digits.[16] Ducks express the protein Gremlin, an antagonist of BMP action in their interdigital zones throughout what would be expected to be the period of cell death, and it probably saves the lives of the cells there; chickens do not do this.[26] This observation provides a rare insight into the molecular mechanisms that can underlie an evolutionary change in morphogenetic processes.

It should be noted that the death of cells is not in itself sufficient to obliterate a tissue because tissues include large amounts of extracellular matrix. This is destroyed in the interdigital zones by the action of proteases. The destruction of matrix is accompanied by the sloughing off of the ectoderm that covered dying mesoderm (in chicks, at least, elective cell death affects the mesodermal layer only).[16]

Another example of morphogenesis through elimination of surplus structures is seen in the development of the mammalian reproductive system. The early embryo develops the progenitors for the internal and external reproductive organs of both sexes at the same time (except for the gonads themselves, of which there are always just two that develop either as testes or ovaries). The external genitalia of both sexes develop from a common set of progenitors and achieve different morphologies through differences in their growth, folding, and fusion. The internal reproductive systems are based on different progenitors; the fallopian tubes, uterus, and upper vagina of the female derived from a pair of Mullerian ducts, whereas the vasa deferentia and ejaculatory ducts of the male derive from a pair of Wolffian ducts (also called nephric ducts; in the early embryo, these also form the drainage system for temporary kidneys). During normal development, the Mullerian duct is eliminated from males and the Wolffian duct from females. If this elimination does not take place, the result is one body that contains both male-type and female-type reproductive structures, usually in some disarray.[27,28,29,30,31,32,33] The choice of which duct is eliminated is determined by hormones produced by the testis; the survival of the Wolffian duct is promoted by testosterone from the Leydig cells of the testis, whereas the death of the Mullerian duct is promoted by anti-Mullerian hormone (AMH) from the Sertoli cells.[34] The Mullerian duct is sensitive to the presence of AMH only during a rather short interval of embryogenesis, between 13.5 and 14.5 days post-coitum.[35] During and after this period, the Mullerian ducts of both males and females show elective cell death, but the proportion of cells that undergo this death is much greater in males.[36] "Males" in which AMH function has been knocked out show the lower levels of elective cell death typical of females. There is also a difference in the form of cell death seen in the two sexes; both show a small number of cells that become phagocytosed by their neighbours, but in males a significant number of cells leave the epithelium altogether to become engulfed by phagocytes in the surrounding mesenchyme.[37]

Elective cell death can also be used to create hollows in tissues that were once solid. An example of this is seen in the development of the very early mammalian embryo. Between days 5 and 6 of mouse development, the packed inner cell mass of the blastocyst hollows out to form a cup-shaped "egg-cylinder" that consists of a single-layered, columnar epithelium. This hollowing out is achieved primarily by cells in the putative cavity removing themselves by elective cell death.[38] A similar process of cavitation by elective cell death also takes place in the hollowing out of glands that form by epithelial invagination, such as the mammary gland.

COMPETITION, DEATH, AND THE TROPHIC THEORY

Chapter 5.1 emphasized that, although proliferation is a default behaviour for unicells, the cells of multicellular organisms will proliferate only when they receive specific signals that permit them to do so. The difference is in fact even more extreme: in animals, at least, the default fate of cells seems to be to elect to die,♣ and they will survive only if provided with specific signals.[39] This view of the control of cell populations is often known as the "trophic theory."

The trophic theory was first developed to explain remarkable properties of adaptive self-organization that are displayed by the vertebrate nervous system. The limbs of tetrapods receive their motor and sensory innervation from the ventral and dorsal roots, respectively, of the spinal cord at axial levels corresponding to those of the limb. There are dorsal and ventral roots between each vertebra along the animal's back, but those that have to serve the limb as well as the trunk are significantly larger than, and contain more cells than, those that serve only the trunk. The size of pools of motor and sensory neurons that serve the limbs cannot be set by an intrinsic property of the spinal cord itself because, if a limb bud is removed from a developing embryo, the pool of neurons on that side of the body shrinks to a size characteristic of pools that do not serve limbs at all.[40]

Observation of the spinal cord during development reveals that all axial levels show an initial over-production of neurons compared to the numbers that will be required in the adult animal. The excess cells are removed by elective cell death, which takes place to a greater extent where there is no limb than where a limb exists. For example, in developing chicks, 40 to 50 percent of cells are lost even at the levels of developing limbs, and this loss exceeds 85 percent when the limbs are absent.[41,42] Observations such as this suggest that the limb itself, the target of the axons of these neurons, may be the source of signals that promote their survival. Experiments in which alterations in the precise size of limb muscles to be innervated by the motor pool cause a corresponding alteration in the percentage of motor neurons that survive, with a correlation coefficient better than 99 percent, support this hypothesis.[43] A long series of experiments (which have been reviewed elsewhere[44,45]) has demonstrated that targets for innervation, such as muscles of the developing limb, are sources of limited amounts of neurotrophic factors that promote the survival of neurons. Direct application of neurotrophins such as NGF to the chick spinal cord greatly reduces the amount of elective cell death that takes place there.[46]

Many neurotrophins signal to cells via the Trk family of receptor tyrosine kinases. (Other neurotrophic molecules, such as those in the GDNF family, use different receptor tyrosine kinases.) The details of how the signalling systems that are triggered by these extracellular molecules inhibit mechanisms of elective cell death are complex and vary with cell type; they are not completely understood, but more details can be found from a number of excellent recent reviews.[47,48,49,50] What is important from the point of view of regulation of neuronal number is that the amounts of survival factors produced by the neuronal targets are very small, and only just enough to support the correct number of innervating neurons. Some time after innervation of targets has begun, neurons become highly dependent on target-derived neurotrophins and, if they do not receive enough,

---

♣As they were in the chapter on proliferation, the cells of embryos too young to have shown any differentiation may be exceptions to this rule, as zygotes obviously have to be.

they undergo elective cell death. Neurons whose axons have failed to make contact with a suitable target will certainly die, and those that have reached a potential target will compete with each other for survival. The size of the neuronal pool is therefore adjusted automatically for the size of its target.

The dependence of the survival of one tissue type on the presence of another seems to be a general feature of development. In the kidney, for example, the survival of the mesenchyme depends on growth factors such as FGFs that are secreted by the epithelial ureteric bud.[51,52] In the foetal liver, the survival of red blood cell precursors depends on circulating levels of erythropoietin,[53] which is made mainly in the developing kidneys and in the liver itself.[54] Erythropoietin is made in response to an increased demand for blood cells, so the apoptotic mechanism is used as an alternative means of population control (alternative to proliferation, described in Chapter 5.1).

The competition between like cells for escape from death and hence for a contribution to the embryo, first noticed in the nervous system, seems to be a general phenomenon that selects for clones of healthy cells and against those that may be damaged in some way. This competition is illustrated particularly clearly in the development of the wing of *Drosophila melanogaster*. As explained in Chapter 4.4, wing discs develop by growth and subsequent evagination of imaginal discs that grow inside the larva. Their size has to be controlled accurately, with respect to the body as a whole and also with respect to each other (a fly with wings of different sizes would, presumably, have problems when it tried to take to the air). Wing discs grow by conventional mitotic proliferation (and by cell enlargement, as described in Chapter 4.4), but experiments using mosaic embryos have revealed that different clones of cells compete with each other for space.

There are various genotypes of *D. melanogaster* in which cells grow more slowly or more quickly than usual. For example, the spontaneous mutant *Minute*,[55] which has defective ribosomes, has slow-growing cells. Clones of wing cells bearing mutations in the receptor system for TGFβ (called Decapentaplegic or Dpp in *D. melanogaster*) also grow abnormally slowly,[56] as do those that have activity-reducing mutants in the growth-promoting small GTPase Ras[57] or in its effector dMyc.[58] Conversely, activating mutations in Ras[57] and dMyc[58] increase cell size and growth rates. For at least some of the slow-growing mutants—for example, *Minute* heterozygotes♣—slow growth of the cells does not prevent the normal development of a fly; it simply makes the development take place abnormally slowly. This makes an important point essential to the interpretation of the experiments described below: *Minute* cells can contribute normally to all structures of the fly when there are no wild-type cells available.

What happens when both slow-growing and wild-type cells are available in the same embryonic structure? This question can be answered by mosaic analysis in which mitotic cell recombination creates clones of one genotype in the background of another. One possible outcome might be that, at an early stage of disc development, each cell is somehow fated to have its progeny contribute to a defined proportion of the adult structure. In this case, one would expect the fast-growing clones of the mosaic embryos to finish producing their area of the disc and then to wait for the slow-growing cells to catch up. This is *not* what happens. Instead, a normal disc forms rapidly and it consists, when mature, mostly or exclusively of wild-type cells (subject to the proviso

♣*Minute* is dominant, as indicated by the capital letter of the mutant name.

that clones of fast-growing cells cannot cross the boundary between the anterior and posterior compartments of the wing).

This result has several implications. The first is that the mechanism for setting the size of the wing disc cannot be a simple counting of cell divisions by the progenitor cells, because if slow-growing clones contribute fewer cells to a mosaic wing then, given that the disc turns out to be a normal size, the wild-type founder cells must contribute more progeny to a mosaic wing than they would in a wild-type wing. There is a second implication: there must be a mechanism that regulates proliferation by cells of the disc according to how large the disc has actually become. This mechanism presumably calls for cell multiplication, and whichever cells are most able to respond are the ones to produce the progeny. The clones "compete" with each other for the area of the disc, in a manner analogous to that of ecological rivals.[59]

There are two main reasons for the dwindling contribution from slow-growing clones. The first arises simply from the exponential mathematics of cell multiplication. Consider, for example, a disc in which cells are either fast-growing type A, which divide once every 10 hours, or slow-growing type B, which divide once every 20 hours, and assume that the cells are initially present at a ratio of 1:1. After 60 hours, one might expect that the ratio would be 7:1 (see Figure 5.3.3a,b). This is not, however, what is observed, and in real mosaic wings the slow-growing clones shrink and are eliminated (see Figure 5.3.3c). Elimination of the slower-growing population can take place only if this population has a greater probability of undergoing elective cell death than the faster-growing population.

Staining of *Minute*/wild-type mosaic discs for the DNA ends characteristic of some pathways of elective cell death, using nick-end-labelling (TUNEL), shows that most of the elective cell death takes place in clones of *Minute* cells.[60] Once dead, the dead cells are eliminated by extrusion from the epithelium rather than by engulfment by other cells. Blockade of this pathway, using an intracellular survival-promoting protein derived from an insect virus, prevents elimination of slow-growing clones. The elimination of the clones can also be blocked by mutation of components of the Jun kinase (JNK) pathway, suggesting that this pathway is responsible for activating elective cell death in the disc. A series of further experiments using pathway mutants[60] have suggested that signalling by Dpp, the same molecule that controls proliferation in the disc, is critical to the control of death, too (see Figure 5.3.4). Mutation of the Dpp receptor (Thick-veins) causes an increase in the expression of Brinker (Brk), and this in turn activates the JNK pathway and hence cell death.

These observations suggest that *Minute* and wild-type cells compete for the Dpp signal and that *Minute* cells are for some reason weaker competitors for this signal. Work on other types of slow-growing clones shows that these, like *Minute*s, show a depression of the intracellular Dpp signalling pathway, although these clones have no direct genetic lesion in that pathway.[61] For example, when clones that carry function-inhibiting mutants of dMyc are grown in the context of wild-type cells, they show reduced activation of the Dpp pathway and consequent up-regulation of the pro-death JNK pathway.[61] This linkage between slow growth and low Dpp-transducing activity cannot, however, be explained simply by proposing a positive feedback loop in which the sensitivity of the Dpp pathway correlates with the growth rate of the cells. If this were true, slow-growing cells would die even in embryos in which there are no wild-type cells, contradicting the observation that such embryos survive. The survival of slow-growing cells in the absence of faster-growing cells and their death in the presence

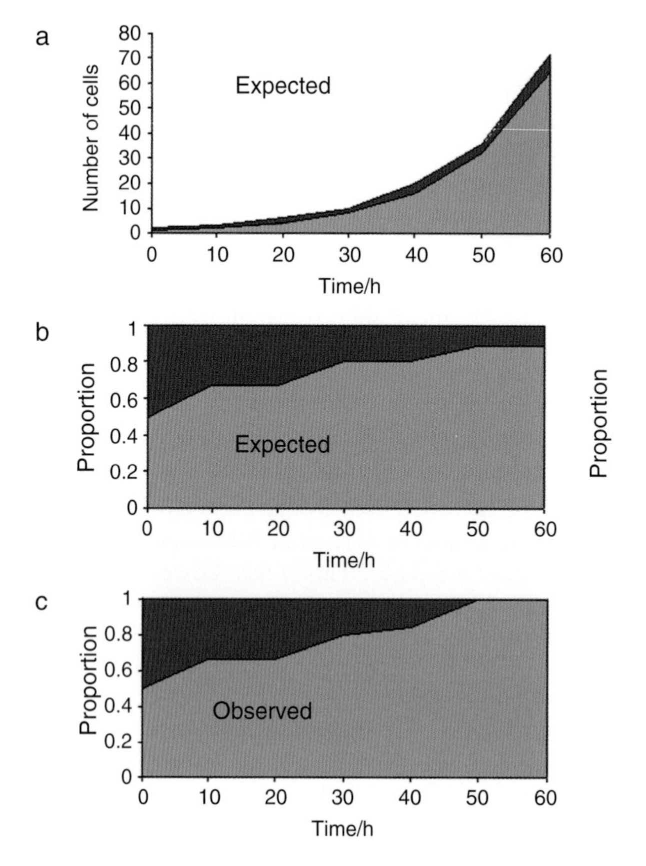

FIGURE 5.3.3 A comparison of the naively expected outcome of mixing fast-growing and slow-growing cells together in a growing structure with the observed outcome. For the purposes of illustration, the fast-growing cells (blue) are assumed to divide once every 10 hours and the slow-growing ones (red) once every 20 hours, and they are equally prevalent at the start of the experiment. Graph (a) shows the changing populations of cells as actual numbers, starting with one cell of each type, whereas graph (b) shows the changing proportions in the naïve model that assumes no cell competition. Graph (c) sketched to be representative of real experiments but *not* made using actual data from any experiment (they are not available in the right form), shows what is observed; the slow-growing population dwindles to nothing.

of faster-growing cells can be explained only if faster-growing cells act in a way that actively suppresses their slower-growing competitors.

The Dpp signal transduction pathway involves internalization of bound Dpp by an endosomal pathway so that it is removed from the extracellular fluid♣; internalization is required for Dpp signalling to take place.[64] If the rate at which this pathway operates were to correlate with proliferation rate, faster-growing clones would depress the local concentration of Dpp and therefore reduce Dpp signalling in adjacent cells (whether these cells were fast- or slow-growing). Engineering an overactive endocytic

♣Although, in the wing disc, the role of endocytosis is complicated by the fact that it may also be involved in passing Dpp from cell to cell to form a gradient across the tissue.[62,63]

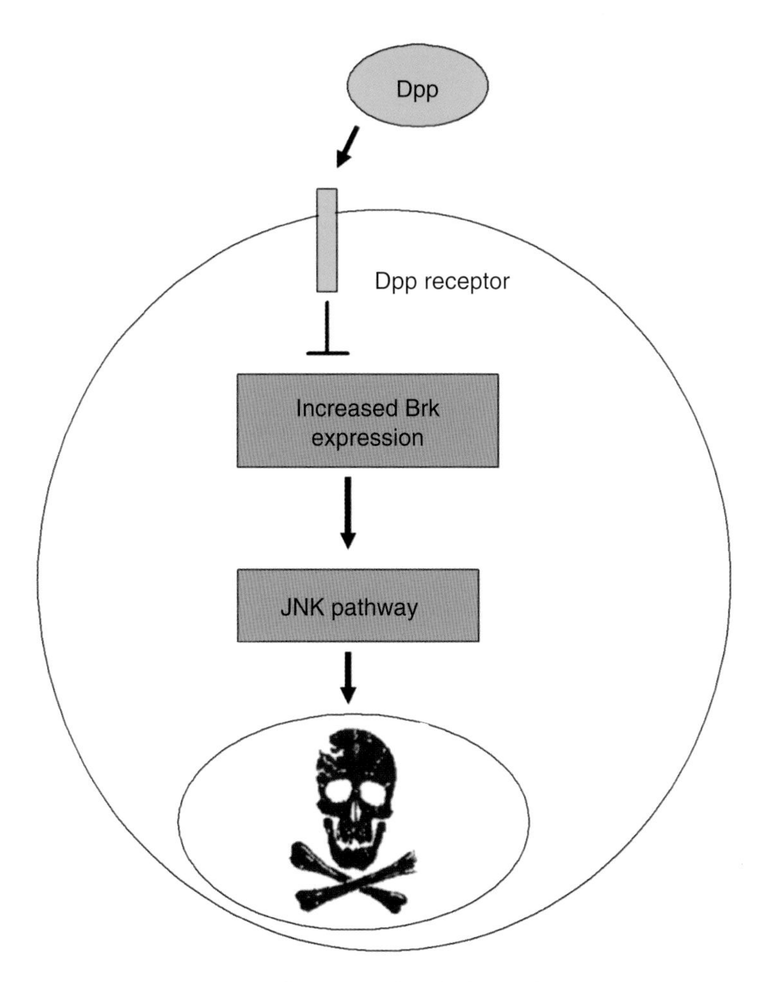

FIGURE 5.3.4 A possible pathway by which Dpp levels control elective cell death in the developing wing disc. The diagram is based on the discussion in E. Moreno, K. Basler, and G. Morata.[60]

pathway in the slower-growing clones of a mosaic wing (actually, a mosaic between wild-type and super-fast-growing mutants with hyperactive Myc) rescues the slower-growing cells so that they now thrive and are not swept aside by fast-growers. As well as showing an enhanced endocytosis, the cells show an up-regulation of Dpp pathway activation (although increased uptake of Dpp itself has not yet been demonstrated directly).

A regulatory pathway therefore emerges in which fast-growing cells also have the highest rates of Dpp endocytosis, which both stimulates their own Dpp signal transduction pathways and deprives their neighbours of Dpp (see Figure 5.3.5). If all clones in the same wing have the same rates of growth and therefore of endocytosis, they will survive whether they happen to be fast- or slow-growers, and it is only when clones of different growth rates are apposed that the slow-growing clones are out-competed and die.[60] The details of the link between rates of growth and of endocytosis have not yet

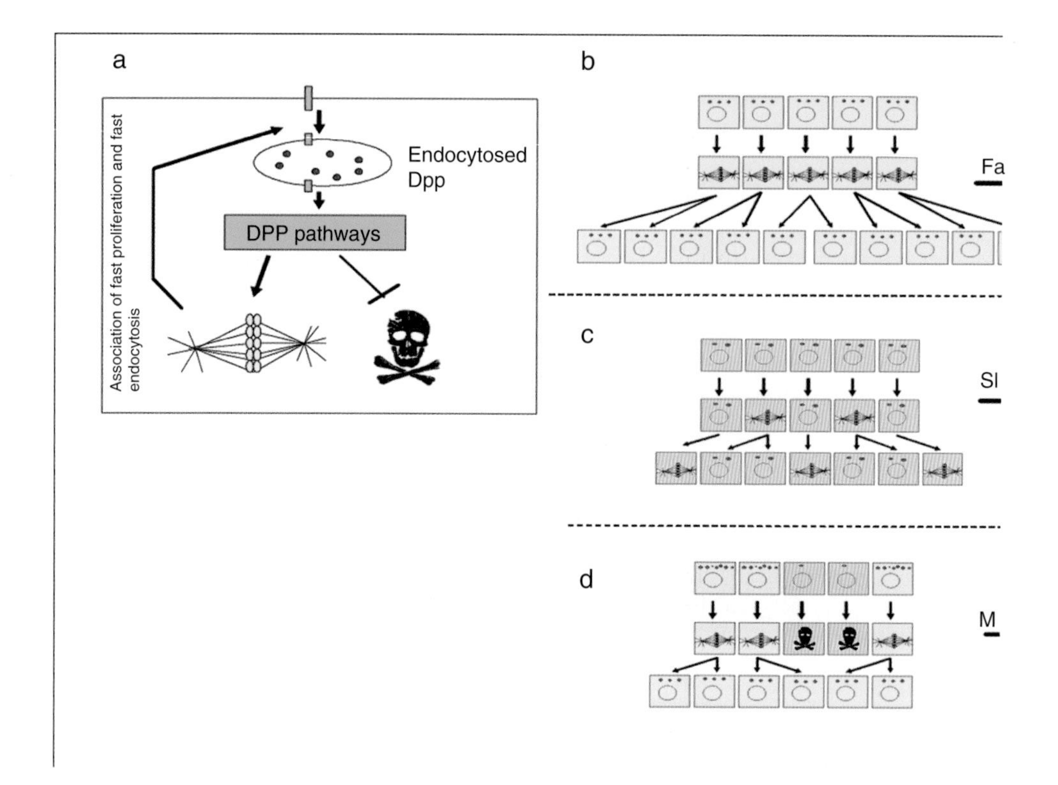

FIGURE 5.3.5 Dpp-mediated competition between clones in the wing disc. (a) Dpp is endocytosed by cells for which it acts as both a mitogen and a survival factor. Cells that proliferate quickly endocytose Dpp more quickly and thus sustain their own proliferation and survival. (b) Fast-growing cells (blue) all take up a moderate and approximately equal amount of Dpp (symbolized by three red dots) and all multiply; (c) Slow-growing cells (pink) take up less Dpp but still enough to survive and proliferate, albeit more slowly; (d) In a mixture of cell types, the high efficiency of Dpp uptake by fast-proliferating cells means that they can take up more than their "fair share" of Dpp, leaving the low-proliferating cells so starved of it that they die.

been established, although they may simply reflect the fact that high rates of protein synthesis that are associated with fast growth are also associated with high rates of membrane turnover.

This competition between clones for the ability to contribute to the wing disc—and, by extrapolation, the similar competition between clones of cells in other tissues and organs of other animals—provides yet another opportunity for adaptive self-organization at the scale of tissues and organs. Construction of a tissue does not depend on all founder cells being present and unmutated; any absences or mutations that limit the rate of growth will be made up for by the clones that arise from other founder cells. Such competition is, however, potentially very dangerous to long-lived animals such as ourselves because it means that clones of cells that carry mutations predisposing them to very rapid growth will be able to dominate a tissue. This feature is probably highly relevant to the development of human cancers.[65]

ANOIKIS AND ERROR CORRECTION

No morphogenetic process is perfect, and accumulated errors in any of the processes described in this book can result in a developmental mistake, typically a cell finding itself where it ought not to be. Simple probability makes this particularly relevant to the development of organisms such as mammals because they have many cells and their morphogenesis involves a great deal of cell migration and movement. As well as relying for their survival on diffusible signals such as growth factors, many cells also rely on survival signals transduced by specific matrix receptors and connect the pathways so that they effectively perform a Boolean AND operation. This allows cells to detect, at high spatial resolution, whether they are placed correctly, and to kill themselves if they are not. Elective cell death caused by displacement from a normal tissue environment is called "anoikis."♣

Anoikis was first discovered when the interaction between renal epithelial cells and their basement membrane was interrupted, using competitive peptides. The cells underwent elective cell death unless they were "transformed" beforehand by transfection with constitutively active forms of enzymes, such as Src or focal adhesion kinase (FAK), that normally transduce signals from cell-matrix adhesion complexes.[66,67] Subsequent work has revealed a number of different intracellular pathways that are used to link the cell matrix with survival, including PI-3-kinase, Erk, Rac/Rho/cdc42, and Jun kinase.[68]

All of the cells of a simple epithelium (see Chapter 4.1) maintain contact with the basement membrane. If any cell becomes displaced, either through damage or a misalignment of mitosis, it will lose contact with the membrane and will therefore undergo elective cell death. This facility is probably an important mechanism by which animals guard against the possibility of epithelial cells piling up in a way that would block tubes, and so on. It is striking how many cancer pathways are associated with the mutation of signalling pathways so that a cell can no longer realize it has lost contact with its basement membrane. Non-epithelial cells such as osteoclasts show a similar dependence on contact with the extracellular matrix and, because cells each express different types of matrix receptors, only the correct type of matrix will do to promote cell survival.[69] The system is ancient and, in primitive metazoa such as the cnidarian *Hydra*, grafts of cells from an unrelated individual do not adhere properly to the matrix of the host, and this causes their elimination by anoikis.[70]

Given the great age of the anoikis mechanism, it is not surprising that it has been co-opted by evolution to play an essential part in several morphogenetic mechanisms in higher animals. If cells elect to die if they lose contact with their matrix, one way in which they can be killed off is to arrange for the matrix to be destroyed. This order of events is seen in the involution of mammary glands, which is the process that returns them to a non-lactating state when a mother's latest offspring have weaned and no longer require milk. Before pregnancy, mammary glands consist of branched milk duct epithelia surrounded my mesenchyme and fat deposits. During pregnancy, the epithelia sprout alveoli (see Chapter 4.6) that make milk but, after weaning, these alveoli disappear by elective cell death in a process historically called "involution." Alveolar epithelial cells depend for their survival on making contact with components of the alveolar basement membrane, notably laminins that are bound by β1-integrin-containing receptors.[71] This basement membrane is stable during lactation, but the

♣From the Greek, meaning "homelessness."

transition to involution is heralded by the expression of proteases (e.g., gelatinase A and stromelysin 1) that destroy the proteins in this basement membrane. If the action of these proteases is inhibited by implanting into the mammary gland slow-release pellets of TIMPs (tissue inhibitors of metalloproteases), involution is delayed and the alveolar cells retain their healthy, differentiated state.[72,73] Conversely, precocious expression of a metalloproteinase during lactation resulted in a marked diminution of alveoli and milk production.[74] The propensity of cells to undergo anoikis can therefore be used to clear them away during morphogenetic change. Careful observation of mammary gland involution suggests that the control of elective cell death is a little more complicated, however. There are two waves of cell death, the first of which takes place even if the matrix is intact and the second of which requires matrix destruction.[75] The first wave is associated not with a disappearance of the matrix, but rather an altered conformation of β1-integrin that renders it unable to bind its matrix.[71] The anoikis mechanism can therefore be exploited from both sides, from the outside by the removal of matrix and from the inside by the switching off of matrix receptor activity.

Elimination of epithelial cells that are marooned in mesenchyme after palate fusion (see Chapter 4.5) may be another example of the use of anoikis by morphogenetic mechanisms. Indeed, the hollowing out of embryos and tubes, described earlier in the chapter, is arguably a related phenomenon because it is contact with the basement membrane that spares the lives of cells that surround their dying neighbours in the cavity. As more is discovered about elective cell death, it may be that anoikis merges back into the generality of death mechanisms and no longer seems to stand out as a separate mechanism.

The overall importance of elective cell death of all kinds to animal morphogenesis can be illustrated by reducing its extent, for example by knocking out or using inhibitors of caspases: the result is widespread and lethal dysmorphogenesis.[76]

ELECTIVE CELL DEATH IN PLANTS

Plants also show elective cell death during their (life-long) development. Elective cell death is seen as a response to pathogenic infection[77] and also as a means to eliminate embryonic structures that are not necessary to the future development of the plant. Elective cell death in plants is sometimes referred to as "senescence," although that term can create confusion with quite different processes in animals.♣ It also refers to tissue-scale phenomena that may not map well onto deaths of individual cells.

An example of a temporary structure that has to be cleared away is the endosperm, a triploid food-storing organ. The endosperm of the castor bean *Ricinus communis* stores oil and proteins and surrounds the cotyledons of the growing embryo. During development, these resources are transferred to the cotyledons and, once the stores in the endosperm are exhausted, the endosperm undergoes elective cell death, apparently by the release of massive quantities of endopeptidases from specific storage organs.[78] Another example of elective death in early development is the inner integument of rapeseed plant, *Brassica rapa*. The integuments cover the ovule of a flower and turn into the tough outer coat of the seed. The thick inner integument remains as protection

♣And since the word derives from the Latin *senex*, meaning "old man," the animal world seems to have a more just claim on the term than does the plant world.

for the seed, but shortly after embryonic growth begins, the inner integument destroys itself by elective cell death, and this avoids obstructing embryonic development.[79]

The process of development in plants is unending, and remodelling takes place throughout life. This means that complete organs such as leaves and shoots can be eliminated by programmed cell death in response to various triggers, including prolonged darkness. The process is local, and one shaded leaf of a plant can be degenerating while others, better lit, thrive, and it is active, requiring changes of expression in many genes.[80,81]

At the time of writing, very little is known about the internal mechanisms that control elective cell death in plants. Attempts to identify in plants processes that are typical of elective cell death in animals are also meeting with success. For example, dying broccoli cells show severing of DNA very similar to that seen in animal apoptosis, and the cells express homologues of regulators of animal apoptosis such as *bax*.[82] Genomic and proteomic techniques are now identifying groups of proteins that are regularly associated with elective cell death, and these may offer a promising direction for future research.[83,84]

DEATH FOR LIFE

That the default state of an animal cell (and perhaps a plant cell) should be to die rather than to proliferate and to thrive often strikes students, brought up on the Darwin-Wallace theory of evolution, as paradoxical. If it were a feature of truly solitary unicells, it would indeed be paradoxical, but in "higher" animals, cell proliferation is not the same thing as reproduction. A somatic cell may reproduce in a limited way by proliferating, but none of the daughter cells that it founds will outlive the body of which it forms a part. Only the germ cells, each of which carries half of the genome carried by the somatic cells, can give rise to the next generation. Since the selection pressure that drives evolution selects whole organisms that reproduce most successfully, the genotypes that encourage somatic cell death for the good of making a body that reproduces most efficiently will triumph over those that protect somatic cells and, by doing so, make a less efficient body. Occasionally, mutations arise that do allow somatic cells to survive and proliferate for their own sakes rather than in a way that is regulated by the rest of the body; indeed, such cells often end up regulating other body systems, for example blood supplies, to serve them. The neoplasias that this behaviour produces are a major cause of mortality in the developed world. The only reason that these mutations are not less common than they are is probably that they tend to happen after the reproductive years so would not be selected against.

In general, we owe our lives to our cells' willingness to die.

Reference List

1. Thomas, L (2004) The fragile species. (Macmillan)
2. Kerr, JF, Wyllie, AH, and Currie, AR (1972) Apoptosis: A basic biological phenomenon with wide-ranging implications in tissue kinetics. *Br. J. Cancer.* **26**: 239–57
3. Dong, KJ and Qian, JX (27-8-2004) Quantum algorithm for programmed cell death of *Caenorhabditis elegans. Biochem. Biophys. Res. Commun.* **321**: 515–516
4. Driscoll, M (1992) Molecular genetics of cell death in the nematode *Caenorhabditis elegans. J. Neurobiol.* **23**: 1327–1351

5. Ratel, D, Boisseau, S, Nasser, V, Berger, F, and Wion, D (2001) Programmed cell death or cell death programme? That is the question. *J. Theor. Biol.* **208**: 385–386
6. Sloviter, RS (2002) Apoptosis: A guide for the perplexed. *Trends Pharmacol. Sci.* **23**: 19–24
7. Alles, A, Alley, K, Barrett, JC, Buttyan, R, Columbano, A, Cope, FO, Copelan, EA, Duke, RC, Farel, PB, Gershenson, LE, Goldgaber, D, Green, DR, Honn, KV, Hully, J, Isaacs, JT, Kerr, JFR, Krammer, PH, Lockshin, RA, Martin, DP, McConkey, DJ, Michaelson, J, Schulte-Hermann, R, Server, AC, Szende, B, Tomei, LD, Tritton, TR, Umansky, SR, Valerie, K, and Warner, HR (1991) Apoptosis: A general comment. *Faseb J.* **5**: 2127–2128
8. Assuncao Guimaraes, C, and Linden, R (2004) Programmed cell deaths. Apoptosis and alternative deathstyles. *Eur. J. Biochem.* **271**: 1638-1650
9. Thornberry, NA and Lazebnik, Y (28-8-1998) Caspases: Enemies within. *Science.* **281**: 1312–1316
10. Martin, DN and Baehrecke, EH (2004) Caspases function in autophagic programmed cell death in *Drosophila. Development.* **131**: 275–284
11. Baehrecke, EH (2002) How death shapes life during development. *Nat. Rev. Mol. Cell Biol.* **3**: 779–787
12. Marino, G and Lopez-Otin, C (2004) Autophagy: Molecular mechanisms, physiological functions and relevance in human pathology. *Cell Mol. Life Sci.* **61**: 1439–1454
13. Danial, NN and Korsmeyer, SJ (23-1-2004) Cell death: Critical control points. *Cell.* **116**: 205–219
14. Zakeri, Z, Quaglino, D, and Ahuja, HS (1994) Apoptotic cell death in the mouse limb and its suppression in the hammertoe mutant. *Dev. Biol.* **165**: 294–297
15. Chautan, M, Chazal, G, Cecconi, F, Gruss, P, and Golstein, P (9-9-1999) Interdigital cell death can occur through a necrotic and caspase-independent pathway. *Curr. Biol.* **9**: 967–970
16. Zuzarte-Luis, V and Hurle, JM (2002) Programmed cell death in the developing limb. *Int. J. Dev. Biol.* **46**: 871–876
17. Macias, D, Ganan, Y, Sampath, TK, Piedra, ME, Ros, MA, and Hurle, JM (1997) Role of BMP-2 and OP-1 (BMP-7) in programmed cell death and skeletogenesis during chick limb development. *Development.* **124**: 1109–1117
18. Merino, R, Ganan, Y, Macias, D, Economides, AN, Sampath, KT, and Hurle, JM (1-8-1998) Morphogenesis of digits in the avian limb is controlled by FGFs, TGFbetas, and noggin through BMP signalling. *Dev. Biol.* **200**: 35–45
19. Macias, D, Ganan, Y, Rodriguez-Leon, J, Merino, R, and Hurle, JM (1999) Regulation by members of the transforming growth factor beta superfamily of the digital and interdigital fates of the autopodial limb mesoderm. *Cell Tissue Res.* **296**: 95–102
20. Merino, R, Ganan, Y, Macias, D, Rodriguez-Leon, J, and Hurle, JM (1999a) Bone morphogenetic proteins regulate interdigital cell death in the avian embryo. *Ann. N.Y. Acad. Sci.* **887**: 120–132
21. Ahuja, HS, James, W, and Zakeri, Z (1997) Rescue of the limb deformity in hammertoe mutant mice by retinoic acid-induced cell death. *Dev. Dyn.* **208**: 466–481
22. Rodriguez-Leon, J, Merino, R, Macias, D, Ganan, Y, Santesteban, E, and Hurle, JM (1999) Retinoic acid regulates programmed cell death through BMP signalling. *Nat. Cell Biol.* **1**: 125–126
23. Alles, AJ and Sulik, KK (1989) Retinoic-acid-induced limb-reduction defects: Perturbation of zones of programmed cell death as a pathogenetic mechanism. *Teratology.* **40**: 163–171
24. Dawd, DS and Hinchliffe, JR (1971) Cell death in the "opaque patch" in the central mesenchyme of the developing chick limb: A cytological, cytochemical and electron microscopic analysis. *J. Embryol. Exp. Morphol.* **26**: 401–424
25. Hinchliffe, JR and Thorogood, PV (1974) Genetic inhibition of mesenchymal cell death and the development of form and skeletal pattern in the limbs of talpid3 (ta3) mutant chick embryos. *J. Embryol. Exp. Morphol.* **31**: 747–760

26. Merino, R, Rodriguez-Leon, J, Macias, D, Ganan, Y, Economides, AN, and Hurle, JM (1999b) The BMP antagonist Gremlin regulates outgrowth, chondrogenesis and programmed cell death in the developing limb. *Development.* **126**: 5515–5522
27. Nishio, R, Fuse, H, Akashi, T, and Furuya, Y (2003) Persistent Mullerian duct syndrome: A surgical approach. *Arch. Androl.* **49**: 479–482
28. Avolio, L, Belville, C, and Bragheri, R (2003) Persistent mullerian duct syndrome with crossed testicular ectopia. *Urology.* **62**: 350
29. Okur, H and Gough, DC (2003) Management of mullerian duct remnants. *Urology.* **61**: 634–637
30. Ilias, I, Kallipolitis, GK, Sotiropoulou, M, Sofokleous, Ch, Loukari, E, and Souvatzoglou, A (2002) An XY female with Mullerian duct development and persistent Wolffian duct structures. *Clin. Exp. Obstet. Gynecol.* **29**: 103–104
31. Padmore, DE and Norman, RW (1998) Paratesticular tumor of Mullerian duct origin. *Can. J. Urol.* **5**: 564–565
32. Saharan, SP and Parulekar, SV (1993) Hemihaematometra with persistent undeveloped Wolffian duct. *J. Postgrad. Med.* **39**: 98–99
33. Malayaman, D, Armiger, G, D'Arcangues, C, and Lawrence, GD (1984) Male pseudohermaphrodism with persistent mullerian and wolffian structures complicated by intraabdominal seminoma. *Urology.* **24**: 67–69
34. Lee, MM and Donahoe, PK (1993) Mullerian inhibiting substance: A gonadal hormone with multiple functions. *Endocr. Rev.* **14**: 152–164
35. Dyche, WJ (1979) A comparative study of the differentiation and involution of the Mullerian duct and Wolffian duct in the male and female fetal mouse. *J. Morphol.* **162**: 175–209
36. Allard, S, Adin, P, Gouedard, L, di Clemente, N, Josso, N, Orgebin-Crist, MC, Picard, JY, and Xavier, F (2000) Molecular mechanisms of hormone-mediated Mullerian duct regression: Involvement of beta-catenin. *Development.* **127**: 3349–3360
37. Xavier, F and Allard, S (15-12-2003) Anti-Mullerian hormone, beta-catenin and Mullerian duct regression. *Mol. Cell Endocrinol.* **211**: 115–121
38. Coucouvanis, E and Martin, GR (20-10-1995) Signals for death and survival: A two-step mechanism for cavitation in the vertebrate embryo. *Cell.* **83**: 279–287
39. Raff, MC (2-4-1992) Social controls on cell survival and cell death. *Nature.* **356**: 397–400
40. Lamb, AH (1-12-1981) Target dependency of developing motoneurons in Xenopus laevis. *J. Comp. Neurol.* **203**: 157–171
41. Lanser, ME and Fallon, JF (1984) Development of the lateral motor column in the limbless mutant chick embryo. *J. Neurosci.* **4**: 2043–2050
42. Hamburger, V (1934) The effects of wing bud extirpation on the development of the central nervous system in chick embryos. *J. Exp. Zool.* **68**: 449–494
43. Tanaka, H and Landmesser, LT (1986) Cell death of lumbosacral motoneurons in chick, quail, and chick-quail chimera embryos: A test of the quantitative matching hypothesis of neuronal cell death. *J. Neurosci.* **6**: 2889–2899
44. Hutchins, JB and Barger, SW (1998) Why neurons die: Cell death in the nervous system. *Anat. Rec.* **253**: 79–90
45. Brown, K, Keynes, R, and Lumsden, A (2001) The developing brain. (Oxford University Press)
46. Hamburger, V and Yip, JW (1984) Reduction of experimentally induced neuronal death in spinal ganglia of the chick embryo by nerve growth factor. *J. Neurosci.* **4**: 767–774
47. Miller, FD and Kaplan, DR (2001) Neurotrophin signalling pathways regulating neuronal apoptosis. *Cell Mol. Life Sci.* **58**: 1045–1053
48. Patapoutian, A and Reichardt, LF (2001) Trk receptors: Mediators of neurotrophin action. *Curr. Opin. Neurobiol.* **11**: 272–280
49. Freeman, RS, Burch, RL, Crowder, RJ, Lomb, DJ, Schoell, MC, Straub, JA, and Xie, L (2004) NGF deprivation-induced gene expression: After ten years, where do we stand? *Prog. Brain Res.* **146**: 111–126

50. Barrett, GL (2000) The p75 neurotrophin receptor and neuronal apoptosis. *Prog. Neurobiol.* **61**: 205–229
51. Koseki, C, Herzlinger, D, and al Awqati, Q (1992) Apoptosis in metanephric development. *J. Cell Biol.* **119**: 1327–1333
52. Barasch, J, Qiao, J, McWilliams, G, Chen, D, Oliver, JA, and Herzlinger, D (1997) Ureteric bud cells secrete multiple factors, including bFGF, which rescue renal progenitors from apoptosis. *Am. J. Physiol.* **273**: F757–F767
53. Yu, H, Bauer, B, Lipke, GK, Phillips, RL, and Van Zant, G (15-1-1993) Apoptosis and hematopoiesis in murine fetal liver. *Blood.* **81**: 373–384
54. Wintour, EM, Butkus, A, Earnest, L, and Pompolo, S (1-11-1996) The erythropoietin gene is expressed strongly in the mammalian mesonephric kidney. *Blood.* **88**: 3349–3353
55. Morata, G and Ripoll, P (1975) Minutes: Mutants of *Drosophila* autonomously affecting cell division rate. *Dev. Biol.* **42**: 211–221
56. Burke, R and Basler, K (1996) Dpp receptors are autonomously required for cell proliferation in the entire developing *Drosophila* wing. *Development.* **122**: 2261–2269
57. Prober, DA and Edgar, BA (18-2-2000) Ras1 promotes cellular growth in the *Drosophila* wing. *Cell.* **100**: 435–446
58. Johnston, LA, Prober, DA, Edgar, BA, Eisenman, RN, and Gallant, P (17-9-1999) *Drosophila* myc regulates cellular growth during development. *Cell.* **98**: 779–790
59. Lawrence PA (1992) The making of a fly. (Blackwell)
60. Moreno, E, Basler, K, and Morata, G (18-4-2002) Cells compete for decapentaplegic survival factor to prevent apoptosis in *Drosophila* wing development. *Nature.* **416**: 755–759
61. Moreno, E and Basler, K (2-4-2004) dMyc transforms cells into super-competitors. *Cell.* **117**: 117–129
62. Gonzalez-Gaitan, M (2003) Endocytic trafficking during *Drosophila* development. *Mech. Dev.* **120**: 1265–1282
63. Gonzalez-Gaitan, M (2003) Signal dispersal and transduction through the endocytic pathway. *Nat. Rev. Mol. Cell Biol.* **4**: 213–224
64. Gonzalez-Gaitan, M and Jackle, H (1999) The range of spalt-activating Dpp signalling is reduced in endocytosis-defective *Drosophila* wing discs. *Mech. Dev.* **87**: 143–151
65. Abrams, JM (23-8-2002) Competition and compensation: Coupled to death in development and cancer. *Cell.* **110**: 403–406
66. Frisch, SM and Francis, H (1994) Disruption of epithelial cell-matrix interactions induces apoptosis. *J. Cell Biol.* **124**: 619–626
67. Frisch, SM, Vuori, K, Ruoslahti, E, and Chan-Hui, PY (1996) Control of adhesion-dependent cell survival by focal adhesion kinase. *J. Cell Biol.* **134**: 793–799
68. Frisch, SM and Screaton, RA (2001) Anoikis mechanisms. *Curr. Opin. Cell Biol.* **13**: 555–562
69. Sakai, H, Kobayashi, Y, Sakai, E, Shibata, M, and Kato, Y (13-4-2000) Cell adhesion is a prerequisite for osteoclast survival. *Biochem. Biophys. Res. Commun.* **270**: 550–556
70. Kuznetsov, SG, Anton-Erxleben, F, and Bosch, TC (2002) Epithelial interactions in Hydra: Apoptosis in interspecies grafts is induced by detachment from the extracellular matrix. *J. Exp. Biol.* **205**: 3809–3817
71. Prince, JM, Klinowska, TC, Marshman, E, Lowe, ET, Mayer, U, Miner, J, Aberdam, D, Vestweber, D, Gusterson, B, and Streuli, CH (2002) Cell-matrix interactions during development and apoptosis of the mouse mammary gland in vivo. *Dev. Dyn.* **223**: 497–516
72. Talhouk, RS, Bissell, MJ, and Werb, Z (1992) Coordinated expression of extracellular matrix-degrading proteinases and their inhibitors regulates mammary epithelial function during involution. *J. Cell Biol.* **118**: 1271–1282
73. Couch, FJ, Rommens, JM, Neuhausen, SL, Belanger, C, Dumont, M, Abel, K, Bell, R, Berry, S, Bogden, R, Cannonabright, L, Farid, L, Frye, C, Hattier, T, Janecki, T, Jiang, P, Kehrer, R, Leblanc, JF, Mcarthurmorrison, J, Mcsweeney, D, Miki, Y, Peng, Y, Samson, C, Schroeder, M, Snyder, SC, Stringfellow, M, Stroup, C, Swedlund, B, Swensen, J, Teng, D,

Thakur, S, Tran, T, Tranchant, I, Welverfeldhaus, J, Wong, AKC, Shizuya, H, Labrie, F, Skolnick, MH, Goldgar, DE, Kamb, A, Weber, BL, Tavtigian, SV, and Simard, J (1996) Generation of an integrated transcription map of the BRCA2 region on chromosome 13q12 q13. *Genomics.* **36**: 86–99

74. Sympson, CJ, Talhouk, RS, Alexander, CM, Chin, JR, Clift, SM, Bissell, MJ, and Werb, Z (1994) Targeted expression of stromelysin-1 in mammary gland provides evidence for a role of proteinases in branching morphogenesis and the requirement for an intact basement membrane for tissue-specific gene expression. *J. Cell Biol.* **125**: 681–693
75. Lund, LR, Romer, J, Thomasset, N, Solberg, H, Pyke, C, Bissell, MJ, Dano, K, and Werb, Z (1996) Two distinct phases of apoptosis in mammary gland involution: Proteinase-independent and -dependent pathways. *Development.* **122**: 181–193
76. Kuida, K, Haydar, TF, Kuan, CY, Gu, Y, Taya, C, Karasuyama, H, Su, MS, Rakic, P, and Flavell, RA (7-8-1998) Reduced apoptosis and cytochrome c-mediated caspase activation in mice lacking caspase 9. *Cell.* **94**: 325–337
77. Lam, E, Kato, N, and Lawton, M (14-6-2001) Programmed cell death, mitochondria and the plant hypersensitive response. *Nature.* **411**: 848–853
78. Gietl, C and Schmid, M (2001) Ricinosomes: An organelle for developmentally regulated programmed cell death in senescing plant tissues. *Naturwissenschaften.* **88**: 49–58
79. Wan, L, Xia, Q, Qiu, X, and Selvaraj, G (2002) Early stages of seed development in Brassica napus: A seed coat-specific cysteine proteinase associated with programmed cell death of the inner integument. *Plant J.* **30**: 1–10
80. Lin, JF and Wu, SH (2004) Molecular events in senescing Arabidopsis leaves. *Plant J.* **39**: 612–628
81. Gepstein, S (2004) Leaf senescence–not just a 'wear and tear' phenomenon. *Genome Biol.* **5**: 212
82. Coupe, SA, Watson, LM, Ryan, DJ, Pinkney, TT, and Eason, JR (2004) Molecular analysis of programmed cell death during senescence in Arabidopsis thaliana and Brassica oleracea: Cloning broccoli LSD1, Bax inhibitor and serine palmitoyltransferase homologues. *J. Exp. Bot.* **55**: 59–68
83. Swidzinski, JA, Leaver, CJ, and Sweetlove, LJ (2004) A proteomic analysis of plant programmed cell death. *Phytochemistry.* **65**: 1829–1838
84. Gechev, TS, Gadjev, IZ, and Hille, J (2004) An extensive microarray analysis of AAL-toxin-induced cell death in Arabidopsis thaliana brings new insights into the complexity of programmed cell death in plants. *Cell Mol. Life Sci.* **61**: 1185–1197

6 Conclusions and Perspectives

CHAPTER 6.1

CONCLUSIONS AND PERSPECTIVES

In Chapter 1.2, I argued that a field of research does not become a true science unless it has a theoretical framework that can be used to organize and to explain empirically derived facts. After sketching out a few key concepts that were needed for discussion of the experimental data in the core of this book, that chapter made the promise that the issue of principles would be revisited briefly in a chapter later in the book. This seems a sensible point to look back over what has been presented in an attempt to draw provisional conclusions from it and also to look forward to what remains to be discovered.

PROVISIONAL CONCLUSIONS FROM THE MECHANISMS DESCRIBED IN FOREGOING CHAPTERS

Emergence as a Link Between the Molecular-Biological and Systems-Biological Viewpoints

One of the key principles introduced in Chapter 1.2 was that of emergence—the way in which systems that involve only simple interactions can give rise to complex behaviour. In Chapter 1.2, this principle was illustrated with a computer model, the "Game of Life," because no real morphogenetic models had been described in detail by that stage of the book.

Hopefully, the real embryonic morphogenetic mechanisms that have been described, particularly those that could be described in molecular detail, have provided convincing biological examples of the principles of emergence. The simple interactions that govern microtubule polymerization and stability, for example, can "know" nothing of the shape of a mitotic spindle, yet the system of interacting molecules succeeds in assembling a spindle perfectly adapted for centreing and separating chromosomal kinetochores no matter where they happened to be when spindle assembly began (see Chapter 2.2). The local processes that regulate the assembly of a leading-edge cytoskeleton can "know" nothing of the large-scale migration pathways set down in the embryo, yet the feedback loops inherent in these processes ensure that the leading edge is assembled in such a way that it follows, automatically, these pathways (see Chapter 3.5). The molecular interactions that regulate the cytoskeleton of an epithelial cell near a wound can "know" nothing about the size and shape of a wound, yet the effect of these interactions, all following their own simple rules in all of the cells involved, is to close the wound perfectly (see Chapter 4.3). The simple, ignorant interactions involved in all of these systems can therefore generate a structure that behaves *as if* it has a large scale "knowledge" of itself and its surroundings.

The concept of emergence is still rather ill-defined, at least by the usual standards of scientific ideas, but it already fulfils a critical role in linking two very different views of biology. One view, epitomized by biochemistry and molecular biology, is essentially reductionist and focuses on the smallest components of living systems, the molecules themselves. The other view, the name of which changes with fashion but seems currently to be "systems biology," is holistic and focuses on the behaviour of whole organisms or physiological systems. These two views, both valid and both useful, have long been difficult to reconcile, and the dialog between their proponents has often been strained.[1,2,3] Emergence is a concept that can unite these views naturally because it focuses on the way in which interaction of agents studied by reductionists can give rise to phenomena studied by holists. Emergence is what frees researchers from looking for traces of high-level concepts in low-level components. That a leading edge can navigate does not mean that any of its interaction proteins somehow have the quality of navigation built into their structure. Similarly, the fact that humans show love does not have to mean, as the evolutionary biologist and theologian Teilhard de Chardin maintained, that love has to be a property even of the atoms from which humans and the rest of the world is made.[1,4]

The Multilayered Organization of Morphogenetic Processes

A key feature of the morphogenetic mechanisms described in this book is that they involve many different layers of organization and that, in the progression upward from the molecular level, features emerge at each layer that were absent from the layer below (see Figure 6.1.1). Furthermore, different morphogenetic processes can share the same bottom layers, or at least substantially overlapping sets of bottom layers, yet achieve quite different final outcomes because they use different upper layers of control.

The lowest layer of most morphogenetic processes is the simple self-assembly of macromolecular complexes.♣ Enzyme subunits come together to form a functional enzyme, actin monomers form complexes with nucleating proteins and with each other, adhesion molecules and junctional proteins form cross-linked adhesive complexes, and so on. Given the right mix of components, self-assembly of these macromolecular complexes is automatic, and to a first approximation (which will be discussed in more detail later) they can be taken for granted by higher layers.

Above the layer of self-assembly lie layers of control that dictate where and when the self-assembly takes place. For example, at the leading edge of a migrating cell, the lowest layer is the self-assembly of branched actin filaments nucleated by Arp2/3. However, the many instances of this layer are subject to a local control loop that uses

♣Self-assembly itself rests on layers such as chemistry, which determines the shapes and binding potentials of protein subunits, and chemistry rests on atomic physics, and so on. These are outside the scope of this book, which is why self-assembly is regarded here as the lowest layer relevant to the discussion.

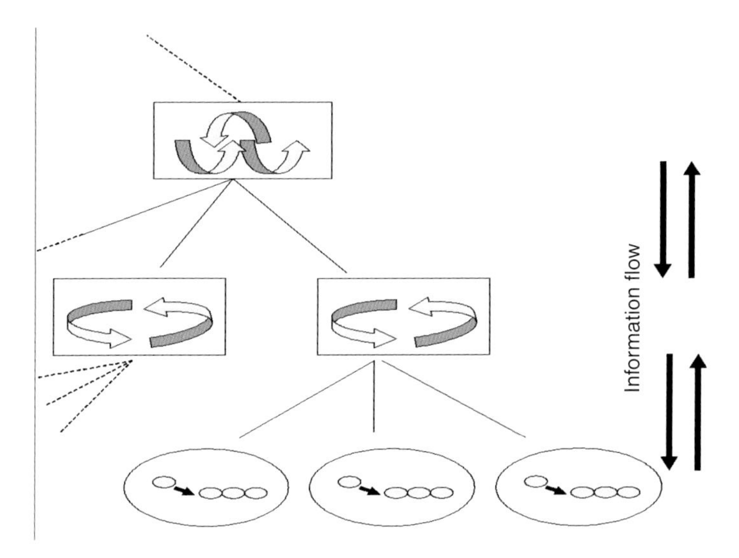

FIGURE 6.1.1 A general view of multilayered morphogenetic mechanisms. Simple, basic processes such as self-assembly (for example of cytoskeletal components) are combined with regulatory and feedback controls to form local morphogenetic systems such as an active part of the sub-membrane actin network in a motile cell. These parts are then further integrated into larger scale morphogenetic systems (for example, the leading edge of a cell, and so on). Each of these levels would correspond to an integron in the vocabulary discussed in the paragraphs on modularity, in the main text.

proteins such as WASP and SCAR and capping proteins so that assembly of the actin network is restricted to the very front of the leading edge (see Chapter 3.2). The "make a new branch of actin" layers are therefore brought together, with control elements, to form a "make a leading edge" layer of mechanism. The "make a leading edge" system is in turn a component, along with a "make contractile actin/myosin filament" system and others, of a higher level "make this cell move" system. This level imposes control on the lower-level systems to generate a cell that has a leading edge at one end and contractile filaments elsewhere and, as recounted in Chapter 3.2, it is capable of polarizing a cell even in the absence of external cues. Precisely where the boundaries between layers are drawn will always be somewhat arbitrary, but the point is that multiple layers exist and that some lower-level systems can be used by quite different high-level systems. For example, the low-level Rho-mediated assembly of actin/myosin tension fibres is invoked by the high-level process in charge of mesenchymal movement (see Chapter 3.2) and also by the high-level process in charge of closure of epithelial holes (see Chapter 4.3 and Figure 6.1.2).

This multilayered nesting of morphogenetic mechanisms extends over a wide variety of scales, and the development of complex tissues and organs will include most of the morphogenetic mechanisms described in this book. For an example, consider the development of the mammalian kidney, aspects of which were described briefly in Chapters 3.7, 4.5, and 4.6. At a morphological level, renal development involves growth and branching of an epithelial tube (the ureteric bud, which becomes

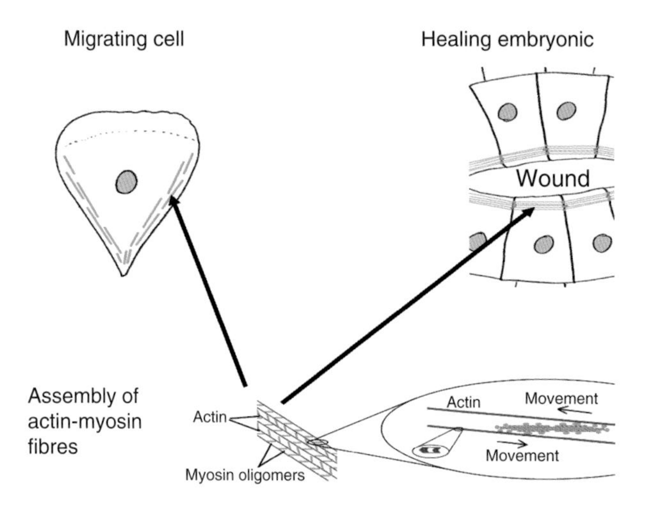

FIGURE 6.1.2 Use of the same low-level morphogenetic mechanism for different purposes by different higher-level mechanisms of different cells. An enlargement of the stress fibre diagram at the bottom of the figure appears in Chapter 2.1.

the branched urine collecting duct system),[5] proliferation of mesenchyme,[6] condensation of groups of mesenchymal cells,[7] differentiation of mesenchymal condensates to become long epithelial tubes called "nephrons,"[8] fusion of nephrons with the collecting duct system, differentiation of other mesenchymal cells to become supporting stroma, formation of blood vessels so that every nephron ends up with a blood supply, and invasion of the kidney by the axons of nerves.[9] Some low-level processes are called on in more than one of these events. For example, the low-level self-assembly of actin/myosin tension filaments is invoked in the branching epithelium, in the condensing mesenchyme cells, and in the migrating blood-cell progenitors. The formation of motile leading edges, itself a multilayered process as described above, is invoked in condensing mesenchyme, in blood-cell progenitors, and in the growth cones of invading neurons. Similar lists can be made for most of the low-level processes mentioned earlier in this book. The take-home message is that the different developmental programs followed by different cell types can invoke the same basic process many times (see Figure 6.1.3).

Higher-level mechanisms that operate beyond the level of macromolecular assemblies are also invoked by different cell types. Cell proliferation is an obvious example, affecting the epithelial and mesenchymal populations.[10] Elective cell death is another; it seems to be used to balance the rather large number of mesenchyme cells that are usually available to form nephrons with the smaller number actually needed,[11] and it clears away progenitor cells that have not been induced to do anything useful. At the highest level, whole-embryo mechanisms seem to ensure that the size of kidney matches the size of the developing body.

Feedback systems link all of these processes and enable the developing kidney (like any other organ) to organize itself properly, despite variability in precisely where components are placed and in the layout and metabolism in the rest of the embryo. Again, feedback as a concept operates at and between many different levels. At the finest level

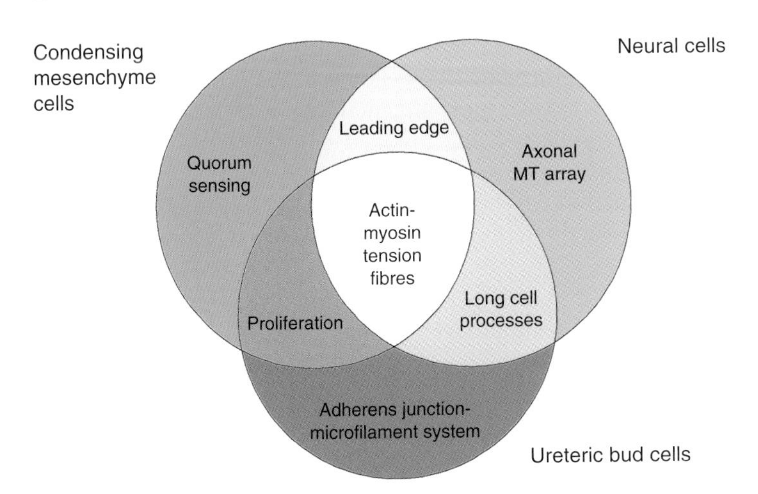

FIGURE 6.1.3 Morphogenetic modules can be used in different combinations by different cells. This figure depicts three cell types of the developing mammalian kidney as a Venn diagram. Some mechanisms are (as far as we know at the moment) invoked by only one of these cell types, some by two, and some by all three. There are, of course, many more processes involved, and this diagram just chooses seven, of different "levels," to illustrate the general point.

(10 to 1,000 nm spatial resolution), it is used to ensure the adaptive self-organization of cytoskeletal components and adhesion systems in stationary and moving cells (and to do other things). At coarser scales (10 to 100 μm spatial resolution), feedback is used to regulate the sizes of cell condensations and the spacing between branches of the growing collecting duct tree (and many other things). Feedback mechanisms do not operate only within a scale: the sensing of cell density will require the function of molecular-scale machines, and the effectors of the feedback will be molecular-scale machines (e.g., for cell movement or cell death).

The Useful-Yet-Dangerous Metaphor of Modularity

When seeking a word to describe the lower-level systems that are grouped to produce higher-level ones, which are in turn grouped, the word "module" springs naturally to mind. By analogy with engineering, each level consists of a "module" that can be called by higher-level processes in the developmental program, which will themselves form a module of another level. In some contexts, the self-assembly of a single macromolecular complex may be treated as a module, whereas in other contexts the assembly of a leading edge, or perhaps of a whole population of migrating cells, may be considered as a module. The language is very helpful in providing an easy way to discuss the many-layered organization of morphogenetic mechanisms, and the concept of a module is a very useful aid in the interpretation of whole-genome microarray studies of tissues that are undergoing a morphogenetic event. Mechanisms that may be considered to be "morphogenetic modules," such as the assembly of motile leading edges, or convergent extension, or activation of elective cell death, will be associated with specific patterns of gene expression.[12,13,14,15] When these patterns have been characterized in systems known to use these morphogenetic mechanisms, they can be sought bioinformatically in microarray data obtained from systems undergoing morphogenetic events that have not yet been characterized. Changes in the gene expression of the uncharacterized event that are similar to those seen when a known morphogenetic module is activated would naturally suggest that the same module is being invoked during the uncharacterized event. This type of approach promises a new and rapid way to study morphogenesis, at least at the level of gaining hints that will help in the design of rigorous experiments. It is not, however, foolproof because the whole idea of modularity is only a metaphor, and life is not really so simple.

True multilevel, modular systems, such as those used in engineering and particularly in software engineering, have a rule: high-level systems may use low-level modules in a variety of ways, but they do not alter processes that are local to the low-level module. Computing hardware is a typical example of a multilevel, modular system. Low-level systems, such as electromechanical relays or semiconductor logic gates, are modules that can perform primitive Boolean operations such as AND or OR. When connected together the right way, these modules can produce a higher-level module that can, for example, add two binary numbers, and modules such as this can be combined together to create higher-level machines right up to a complete computer. At each new level, properties emerge (such as addition) that were not present in the basic modules. Nevertheless, creation of different types of high-level machine is done solely by modifying the connections between basic modules and never by altering the workings of the modules themselves. Creation of a different high-level computing circuit in this way does not involve, for example, the addition of an extra contact to a certain relay or of an extra transistor to a certain logic gate, but only the use of the same basic level modules in a different way. Modern computer languages force programmers to use similar rules. Low-level processes tend to be written as "functions," and higher level processes call these self-contained functions by passing a defined set or parameters (e.g., numbers of be factorized) to them and accepting the results back. The writing of the high-level process determines the order in which basic functions are called, but there is no means for a high-level process to interfere directly with the way that a basic function works, to swap a "−" for a "+" in a line of code in the function, for example.

Biology breaks this rule. New cell and tissue shapes and behaviours *can* evolve simply by connecting existing basic morphogenetic modules in new ways, and that is the predominant model for evolution at the moment.[15,16] It is clear, though, that the evolution of new forms can also happen by tinkering with the functions of basic modules themselves, with or without a change in the way that they are connected. Although the details of their evolution are not known, the actin-polymerization mechanisms described in Chapter 2.3 are probably an example of this. Addition of specific proteins such as fimbrin or espin to a "make actin filaments" module changes the effect of that module, and these proteins work as structural components at the lowest levels of that module, not as high-level "commands."

Therefore, although the module metaphor is natural and helpful, within limits, to discussion mechanisms of morphogenesis, it is essential to remember that it is a metaphor. An alternative term was suggested by François Jacob, who wrote with great prescience about the emergence of new properties in multilevel biological systems as early as 1973. In Jacob's nomenclature, the "modules" are referred to as "integrons" because they arise through the integrated activity of lower-level agents. Although using this word does add extra jargon to a field already oversupplied with it, the word "integron" has the merit that it does not come loaded with shades of meaning more appropriate to engineering than biology. Jacob's own explanation of the word "integron" is as follows:

> At each level, sub-units of relatively well-defined size and almost identical structure associate to form a unit of the level above. Each of these units formed by the integration of sub-units may be given the general name "integron." An integron is formed by assembling integrons of the level below it; it takes part in the construction of the integron of the level above.[17]

Are Multilayered Morphogenetic Systems Hierarchical?

One concept that is frequently mentioned in standard textbooks about the molecular cell biology of development is that of hierarchical control. Hierarchy is a term that originates in organized religion♥ and refers to a chain of command in which agents are assigned to levels, and each level is in command of the levels below it but commanded by the levels above it. In many religions, junior priests are subordinate to high priests, and other social structures such as the army have adopted a similar hierarchical structure in which people of low rank are subordinate to those of higher rank. Like modularity, the concept of hierarchy is both useful and dangerous in the context of biology, and in the context of morphogenesis its dangers probably outweigh usefulness and it should be dropped.

The idea of hierarchical control entered molecular biology mainly through research into the molecular genetics of simple prokaryotic systems such as bacteriophage λ and the *lac* operon.[18] In these systems, the activity of one gene, for example the cI repressor of bacteriophage λ, controls the expression of a number of other genes and is therefore regarded as a "master regulator," at a superior position in a hierarchy of control. The hierarchical model is useful for simple and relatively inflexible systems such as viruses, but is less so for more complicated and flexible mechanisms. The problem is that the feedback in these systems, the same feedback that gives them the property of adaptive self-organization, breaks a fundamental rule of hierarchy. In a true hierarchy, command flows only one way: Captains give orders to Sergeants but Sergeants do not give orders to Captains. In the morphogenetic mechanisms described in this book, the chain of command can loop back so that, to pursue the military analogy, a lowly Private at the end of a long chain of command ends up giving an order to the Brigadier General back at the top. Thus a "high-level" genetic module that invokes the creation of migratory machinery in a cell, including the expression of target receptors, can be switched off by signals initiated by the receptors once they have bound their targets (see Chapter 3.6).

It is therefore most sensible to view morphogenetic processes as being multilayered, without considering those layers to follow the rules of a hierarchy.

Why Are Morphogenetic Mechanisms Organized the Way They Are?

"Why" questions are always dangerous to ask in science because they run the risk of descent into teleology and metaphysics. Nevertheless, asked with appropriate

♥From the Greek *hierarkhês*, meaning "high priest."

caution, a question about whether one would expect something to be the way it is, from basic principles, can help to highlight any unexpected features it has and help to suggest future lines of enquiry. In biology, "why" questions usually translate as "what is there about this system that would confer a selective advantage over rival systems when they all compete during natural selection?" Such questions demand that one imagines some alternative system for accomplishing the same task, with which the attributes of real biological systems can be compared.

At the lowest level, real morphogenetic mechanisms rest on the self-assembly of macromolecular complexes, a process that is directed primarily by the information present in the subunits themselves, but with some other regulatory controls. It is difficult to imagine any mechanism that does not rest at bottom on self-assembly, but it is clear that self-assembly is not itself sufficient to produce complex organisms, for the reasons discussed in Chapter 1.3. An alternative to multilayered, shared-integron mechanisms for morphogenesis that are based on a foundation of self-assembly therefore requires that alternatives be found to either the multilayering or the sharing of integrons, because there are no realistic alternatives to the foundation being self-assembly.

The presence of many layers gives morphogenetic systems great potential for feedback and control. Low-level, local feedback can organize and optimize structures at the very fine (<100 nm) spatial scale and the short (<10 s) temporal scale. Higher-level, more diffuse feedback systems can integrate the activities of local systems and regulate their number, while feedback at even larger spatial scales can coordinate the activities of 10- to 100-nm scale structures with the embryo as a whole. Having only low-level, short-range feedback would allow an organism to make accurate structures, but these would be hopelessly uncoordinated once the organism reached a size much more than that of a single cell. Having only long-range feedback would allow large-scale coordination, but it would not allow the structures being coordinated to be made with enough accuracy to be of any actual use. Multiple layers of control therefore confer direct selective advantages on morphogenetic systems. They would therefore be expected to be favored during evolution by natural selection.

Sharing of integrons is also likely to be favoured, but for slightly convoluted reasons. Organisms that succeed in evolving complex developmental programs and body plans must be founded on mechanisms that are amenable to rapid evolution: an organism founded on mechanisms that are not amenable to rapid evolution cannot evolve quickly. Consider a hypothetical primitive organism, the development of which involves only very few morphogenetic events. Modification of the developmental program so that an existing morphogenetic event is replaced with a new one would probably be relatively straightforward; perhaps the mutation of a single protein would be enough to turn a convex curvature into a concave one, for example. Modification of the developmental program so that a new event is added, without removing an existing event would, however, be more or less difficult, depending on how it was done. If it were done by the activation at an additional time and place of a morphogenetic mechanism (integron) that already existed somewhere else in the developmental program of the organism, the modification may be relatively straightforward. If, however, it had to be done by the evolution of a brand new mechanism, it would probably be much less likely to occur in a reasonable length of time. For this reason, evolution of new morphogenetic events by invoking existing morphogenetic mechanisms at new times and places, and in new combinations, is more likely than the creation of new mechanisms *de novo*. Sharing of low-level integrons by different high-level ones would therefore be expected to be a feature of phyla that have played the evolution game successfully. This does not, however, preclude some *de novo* invention and, as explained in the section on modularity above, there is evidence for "new" proteins having been added to modify the function of low-level integrons in certain times and placed in the embryo. Mutation and selection just happen; they are not constrained to respect the boundaries scientists draw between processes.

LOOKING FORWARD: What Remains to Be Done?

I have said many times in this book that our understanding of the mechanisms of morphogenesis remains at a fairly primitive level, despite the quite brilliant work that has been done by leaders in the field. Although the great questions of the field are at least 2,000 years old, researchers have had good enough tools to make a serious attempt at answering them for fewer than 20. For this reason, the field still feels young and it has only recently become fashionable.♣ Where might it be heading, and what remains to be done before morphogenesis is properly understood?

Without wanting to delve too deeply into epistemology, the answer to that question depends very much on what is meant by "understanding." Each science has, at a particular time, some agreed feeling of what constitutes an understanding. For a physicist, something may be considered to be "understood" if it can be expressed in a mathematical equation. For an organic chemist, understanding a reaction may mean being able to express it as an equation with "curly arrows." For many molecular biologists, the current measure of understanding is an ability to summarize a process in blob-and-arrow diagrams in which signalling pathways and so on are shown as chains of interaction and causation. The best-studied morphogenetic mechanisms can already by summarized in this way (and some have been earlier in this book), and an intensification of research into developmental cell biology promises that more will join their ranks soon. Is that level of understanding enough?

The problem with blob-and-arrow diagrams, useful as they are, is that they present a one-dimensional view of what is a three-dimensional process (four-dimensional, if time is included). They provide a useful summary of key processes, but they have almost no predictive power except in the crude sense of predicting what knockouts would cause morphogenesis to fail. A model would have a much more convincing claim to be an "understanding" if it had the power to predict the actual movements or shapes that would result from the action of a morphogenetic system. This involves quantitative modelling, and quantitative modelling that is more than guesswork requires at least some quantitative measurements to be made from the system under scrutiny. Fortunately, though, the nature of the feedback loops that give self-organizing properties to morphogenetic mechanisms mean that, when modelling a system with 30 different types of molecule, it is probably not necessary to measure the binding and diffusion constants of every molecule: feedback controls the system in a way that is almost

♣In writing this book, I have been struck by how many important and perceptive papers that considerably advance the field of morphogenesis are published in journals with low impact factors. Either the field of morphogenesis has a gratifyingly high proportion of researchers who pay no attention to the currently fashionable idea that publishing in certain journals is somehow intrinsically better than publishing in others, or it has been very hard for them to have their excellent work taken seriously by journals that concentrate on publishing only work likely to be cited many times soon after publication.

independent (within limits) of the precise quantitative characteristics of individual components. Useful predictive modelling of morphogenesis ought to be achievable using only the characteristics of higher-level integrons (although a "complete understanding" will require more detailed information).

Quantitative modelling, which nowadays means computer modelling, of the morphogenetic processes in real embryos promises to be one of the most powerful means of understanding morphogenesis we will have in the near future. It does, however, require a great deal of work if it is to be done properly, which means done using mechanisms that have been identified by real molecular cell biology and using key quantitative parameters that have been measured from life. Guessing mechanisms and parameters and then showing that a life-like shape results on the computer screen is of very limited use unless it leads directly to the design of real experiments because it is too easy to tweak a large set of parameters until the right shape happens to result. As a theoretical physicist is alleged to have said, "Give me three variables and I can describe an elephant: give me four and I can get it to wag its tail."

There are other ways of using computer modelling. One, often called the "A-life" (artificial life) approach, is to model extremely simple morphogenetic processes in simple "organisms" in silico, to learn more about what different types of morphogenetic mechanism are capable of doing in principle.[19] Although many biologists dismiss such activity as irrelevant game-playing, the results that emerge may well be increasingly useful in suggesting what mechanisms may be, or what mechanisms can*not* be, involved in real morphogenetic events. Such A-life organisms can also be used to study general features of the evolution of morphogenetic systems in general.

When a combination of molecular cell biology, computer modelling, and whatever new "–omics" appear has resulted in what appears to be a real understanding of morphogenesis, it will be valuable to test this understanding by creating simple systems *de novo*. In the same way that engineers can test their understanding of aeronautics by building a paper plane, so researchers into morphogenesis can test their understanding of morphogenesis in general by creating their own morphogenetic mechanisms in cells that do not normally do anything morphologically interesting, and testing whether these cells now undergo morphogenesis as predicted. This has already been done in very small ways—for example, by transfecting cells with adhesion molecules so that they cluster together.[20] In future, perhaps entire integrons and multilayered systems of integrons can be designed to make these aggregates of cells do something really interesting and *predictable*. That would be a really impressive demonstration that the principles of morphogenesis, and not just what happens in a few instances, really were understood.

This book has focused very much on basic science and has not emphasized clinical and industrial applications of morphogenetic knowledge, but such applications undoubtedly exist. Medically, the whole field of tissue engineering is hampered by difficulties in producing artificial organs, whether for use *ex corporo* or *in vivo*, that have cells in the correct spatial arrangement. Historically, the problem has been addressed by developing shaped non-living supports that force a particular arrangement on cells,[21] but it would be a great improvement if cells could be programmed to do this for themselves. Industrial applications also exist; an interesting recent example has been the development of fungi (by conventional breeding and use of carefully controlled culture conditions, rather than by morphogenetic engineering) that have the optimum degree of branching to act as a convincing meat-substitute for converts to vegetarianism who miss their favourite dishes. There are many other possible applications, particularly in plants, where alterations in morphogenetic mechanisms may be used, for example, to create root systems that are particularly useful in stabilizing coasts from erosion or in modulating the speed of water flow in areas prone to flooding.

One of the most important advances in understanding is likely to come from outside the field of development itself. The general problem of emergence crosses all of the sciences, from physical to biological, and even enters fields outside the hard sciences such as sociology and linguistics. It may always remain a woolly concept, but on the other hand it may one day gain a more rigorous definition (momentum/inertia and magnetism were, after all, woolly concepts once). If it does, the whole of developmental biology will benefit from a rigorous concept that will provide an even more useful framework for understanding and evaluating different morphogenetic events.

Our understanding of morphogenesis is changing quickly. It will be a bittersweet experience to see this book soon rendered out of date by the pace of progress, as I am sure it will be: I hope that it plays some role in inspiring those who will consign it to history.

Reference List

1. Medawar, PB and Medawar, J (1983) Aristotle to zoos. (Oxford University Press)
2. Sheldrake, R (1981) A new science of life: The hypothesis of formative causation. (Anthony Blond)
3. Mayr, E (1997) This is biology. (Belknap/Harvard)
4. de Chardin, T (1959) The phenomenon of man. 264–265. Translated by Wall, B.
5. Davies, J (2001) Intracellular and extracellular regulation of ureteric bud morphogenesis. *J. Anat.* **198**: 257–264
6. Saxen, L (1987) Organogenesis of the kidney. (Cambridge University Press)
7. Stark, K, Vainio, S, Vassileva, G, and McMahon, AP (1994) Epithelial transformation of metanephric mesenchyme in the developing kidney regulated by Wnt-4. *Nature.* **372**: 679–683
8. Davies, JA (1996) Mesenchyme to epithelium transition during development of the mammalian kidney tubule. *Acta. Anat. (Basel).* **156**: 187–201
9. Karavanov, A, Sainio, K, Palgi, J, Saarma, M, Saxen, L, and Sariola, H (21-11-1995) Neurotrophin 3 rescues neuronal precursors from apoptosis and promotes neuronal differentiation in the embryonic metanephric kidney. *Proc. Natl. Acad. Sci. U.S.A.* **92**: 11279–11283
10. Michael, L and Davies, JA (2004) Pattern and regulation of cell proliferation during murine ureteric bud development. *J. Anat.* **204**: 241–255
11. Coles, HS, Burne, JF, and Raff, MC (1993) Large-scale normal cell death in the developing rat kidney and its reduction by epidermal growth factor. *Development.* **118**: 777–784
12. Baier, G (2003) The PKC gene module: Molecular biosystematics to resolve its T cell functions. *Immunol. Rev.* **192**: 64–79
13. Baguna, J and Garcia-Fernandez, J (2003) Evo-Devo: The long and winding road. *Int. J. Dev. Biol.* **47**: 705–713
14. Harden, N (2002) Signaling pathways directing the movement and fusion of epithelial sheets: lessons from dorsal closure in *Drosophila*. *Differentiation.* **70**: 181–203
15. Gerhart, J and Kirschner, M (1997) Cells, embryos and evolution. (Blackwell)
16. Raff, RA (1996) The shape of life. (Chicago University Press)
17. Jacob, F (2004) The logoc of life: A history of heredity. (Pantheon)
18. Judson, HF (1979) The eighth day of creation. (Jonathan Cape)

19. Ward, M (1999) Virtual organisms. (Pan)
20. Nagafuchi, A, Shirayoshi, Y, Okazaki, K, Yasuda, K, and Takeichi, M (24-9-1987) Transformation of cell adhesion properties by exogenously introduced E-cadherin cDNA. *Nature.* **329**: 341–343
21. Kim, SS, Park, HJ, Han, J, Choi, CY, and Kim, BS (2003) Renal tissue reconstitution by the implantation of renal segments on biodegradable polymer scaffolds. *Biotechnol. Lett.* **25**: 1505–1508

INDEX

A

Acetylcholine receptors (AchR), 139f
ACF7 protein, microtubule binding by, 53
AchR. *See* Acetylcholine receptors
Acrosomal actin, pre-assembled, 63
Acrosomal process
  Brownian ratchet in, 62f
  microfilaments constructing, 60
  sea cucumber sperm with, 61f
Acrosomal reaction, ion channels opening in, 61–62
Actin
  bristles with, 68–69
  cable system of, 226
  filopodia/lamellipodia based on, 108–109
  growing process with monomers of, 60–61
  growth cones containing, 109f
  microfilaments constituted of, 24
  networks controlling assembly of, 111f
  one-dimensional self-assembly for, 24–26
  plasma membrane with, 105f
  tension-generating structures by, 113
  trichome elongation with, 87
  wound healing and, 228
Actin bands, 223–224, 225f
Actin fibre bundles, 240–241
Actin filaments
  accumulation of, 237
  animal cells with, 44
  bending of, 104
  flexibility of, 106
Actin microfilaments, 84
Actin polymerization, 24
Actin-based model of invagination, 234f
Actin-depolymerizing factors (ADFs), 86
Actin-myosin contraction model, 238f
Actin-polymerization mechanisms, 354
Adaptive self-organization
  feedback, self-assembly with, 13–15
  microfilament tension with, 48–50
  structures arranged by, 14–15
  tissues/organs with, 338
ADFs. *See* Actin-depolymerizing factors
Adherens junctions
  apico-basal axis with, 318f
  cadherin molecules used by, 201
  cell cooperation for, 218–219
  epithelial cells with, 45, 48f
  imaginal disc epithelium linked by, 240
  linkage between, 46f
  mitotic spindle orientating, 317f
  rap1 activity determining, 238
Adhesion
  antibodies interfering with, 157
  cell cycle linked to, 299–300
  growth cones breaking, 153f
  haptotaxis and, 149
Adhesion molecules, 155–156
Adhesion-dependent system, 317
Adhesive cells, 187f
Adult skin, 229–230
Amnioserosa, 224–225
Amphipathic molecule, 21–22
Angioblasts, 281
Angiogenesis, 97
Animal cells
  actin microfilaments of, 44
  shapes of, 41

Anisotropic cells, 37f
plants using, 37
shapes of, 240
Anoikis, 339–340
Anterior half-somites, 160
Antibodies, 157
APC protein, 317
Apico-basal axis, 318f
Apoptosis
elective cell death controlled by, 330f
fusion from, 253
Arabidopsis thaliana, 324
Archenteron, 233f
Aristotle, 3
Arp2/3
as branch-initiating complexes, 110
filaments with, 107f
Artificial phospholipid bilayers, 24
Aster, 314f
Astral model, 313–314
Attraction, 155–156
Autocatalytic process, 22, 24
Autocrine signals, 271
Axial invagination, 231–235
Axon elongation, 161
Axon extension mechanism, 14
Axonal growth cone, 174
Axons
healing sprouting, 142
outgrowth restriction of, 160f

B

Bacon, Francis, 7
Baconian mode, 7
Basement membrane, 339–340
Basidiomycete fungus, 39f
Bilayered membranes, 18–24
Binding energy, 18
Biological membranes, 140
Biology, 3
Blastopore, 143–144, 144f
Blastula stage, 199
Blob and arrow diagram, 5f, 357–358
Brachydanio rerio, 280
Branched processes, 35
Branched tube, 260f
Branching
clefting with, 273–277
dipodial/monopodial patterns of, 261f
kidney with, 270f
modes switched for, 269–270
organs with, 268t
processes of, 88
renal cells and, 262
Smad proteins inhibiting, 264
sprouting with, 259–269
tracheae with, 274f
tubules with, 265f
Branching epithelia, 271
Branching systems, 280–281
Branching tips, 273
Branching tree, 271–273
Branchless signaling, 273
Bristle movement, 67
Bristles, 68–69, 69f
Brownian ratchet, 106
as filament elongation, 62–63
object moving by, 150f
Brownian ratchet, elastic
membrane advancement basis from, 104–105
peripheral actin fibres with, 105f
Brush border, 65
Buriden's ass, 132

C

Cadherin molecules
Adherens junctions using, 201
cells expressing, 186f
desmsomes using, 201
Erk linked to, 300
fat family with, 208
Caenorhabditis elegans, 315, 329
cAMP. *See* Cyclic adenosine monophosphate
Carrassius auratus, 175
Cascade, 123
Cell(s)
actin self-assembly controlled for, 24
advance of, 111–113
aligned fibres guiding, 157–159
anisotropic expansion by, 37f
basement membrane contacting, 339–340
cadherin expressed by, 186f
cavitation of, 36f
cell-cell adhesion molecules inducing, 185
coalescence of, 97–98
connections of, 320–321
cooperation of, 218–219
critical mass formed by, 192–193
directional cues to, 321–322
directional sense by, 137–138
dispersal of, 96–98, 100
electric fields influencing, 138

elongation of, 72–73, 73, 83
extracellular ligands binding with, 154
filopodia crawling in, 107–110
flattening, elongation of, 33–34
fusion of, 36f
galvanotaxis guiding, 137
gap junctions connecting, 203
group translocation of, 98–100
interfacial tensions with, 189f
leading edge activity by, 132
mitotic spindle construction by, 71f
morphogenesis deforming, 213
morphogenesis of, 34f, 35f
multiplication of, 280
procollagen molecules secreted by, 25
as self-reproducing, 21–22
shape command made by, 5
shapes of, in suspension, 42f
signaling for, 155–156
small microvilli in, 64
spindle construction problem in, 70
tension/compression structures in, 43f
Cell adhesion
cell proliferation antagonized by, 301f
condensation through, 185–190
condensations altering, 194
measuring, 153f
neurites stimulated by, 156
orientation system based on, 318
Cell boundaries
different types of, 214–215
myosin targeting, 217–218
Cell cycle
adhesion system linked to, 299–300
extracellular signals and, 299f
four stages of, 295f
growth factor signaling and regulating, 298f
metazoa arranged for, 297f
multicellular animals with, 295–296
phases of, 293–299
restriction point in, 299
Cell divisions
mechanoreceptors with, 320f
mechanosensory systems with, 322
vertebrate neurulation and, 319f
Cell locomotion
crawling as, 103
signal transduction system interfering with, 161
Cell membranes
artificial phospholipid bilayers studying, 24
phospholipid constituent of, 19–20
Cell migration
chemotaxis with directional, 119
morphogenesis with, 96
neural crest with, 100
sexual reproduction with, 95
Cell movements
electrical fields influencing, 137–140
hair follicle with, 191f
Cell populations, 333
balance mechanism for, 300–303
size controlled for, 329
Cell processes
filopodia as, 35
migration by, 101
production of, 34–35
Cell proliferation, 352
cell adhesion antagonizing, 301f
control local for, 299–300
controlling of, 292–294
growth achieved through, 292f
insulin receptor signal transduction pathway's role in, 305f
large-scale control of, 303–305
positive regulation of, 297–298
reproduction different from, 341
tissue-scale control of, 300–303
Cell shape
regulation ordinary for, 41
tensegrity model describing, 43
tissue morphogenesis driven by, 37–39
Cell tension, 224–225
Cell types, 189
morphogenetic modules used by, 353f
organs and, 293–294
Cell wedging, 237–238, 238f
Cell-cell adhesion molecules, 185
Cell-matrix junctions, 48f
Cell-scale feedback system, 14
Cell-substrate adhesion, 153–154
Cellular mechanisms
fold formation with, 280
interssusceptive branching with, 279
multicellular epithelial tubes and, 267–268
Cellulose
biosynthesis of, 83
as glucose polymer, 77
microfibrils of, 80
plant cells synthesizing, 78
synthase complexes of, 80f
Central nervous system
bristle movement information to, 67
migration pathways in, 170
structure developing ventral, 171f

Centripetal chronotopic relationship, 175
cGMP signaling system
  dynamics of, 128–129
  feedback system with, 128
  influence of, 127
Charged molecule, 139f
Chemical system, 138
Chemiosmotic gradients, 140
Chemoattractant, 120
Chemorepellant, 131–132
Chemorepulsion, 130–131
Chemotactic crawling migration, 130
Chemotactic fields, 120
Chemotactic gradient, 119–125
Chemotactic migration, 129–130
Chemotaxis
  cell migrating by, 126
  cell migrating directionally in, 119
  integration of, 132–133
  principles of, 129
  sperm attracted by, 95–96
Chromatids, 295
Cleavage
  aster with, 314f
  cells spiral arrangement in, 311
  communications disrupted in, 315
  embryo with, 312f
  mitotic division and, 311
Cleft palate, 253f
Clefting
  branching by, 273–277
  collagen fibers associated with, 276
  mesenchyme-derived signals required for, 277
Cloacal cavity, 256
Clones, 338f
Collagen, 25–26, 25f
Collagen fibers, 276
Collagen fibrils, 26
Commissural axons, 172f
Communication, 4
Complex structures, 13
Component parts, 293
Compression structures, 62
Computer hardware, 354
Condensates, 192–194
Condensation(s)
  cell adhesion system altered by, 194
  cell adhesion with, 185–190
  dermal cells with, 192f
  hyaluronic acid destruction to, 191–192
  interstitial matrix with, 190–192
  limb mesenchyme with, 189
  mesenchymal cells with, 185, 193
Contact inhibition of proliferation, 299
Contractile actin cables, 228
Contralateral projection, 178–179
Convergent extension
  corella inflata with, 215f
  dishevelled signaling regulating, 220
  elongated axis formation from, 213–214
  embryo with, 217f
  epithelial tubes with, 213–214
  gut primordium elongation with, 216f
  length-change processes in, 218f
  planar polarity system guiding, 320
Conway, John, 9–11
Corella inflata, 213, 215f
Cortical cytoplasm, 316f
Cortical cytoskeleton, 45f
Cortical microtubule
  cellulose synthase complexes with, 80f
  plant cells with, 81f
Crawling cells
  as cell locomotion, 103
  force exerted by, 111
Critical mass, 192–193
Cross-linking proteins
  fascin as, 68
  microfilaments linked by, 44
Current flow
  blastopore with, 144
  posterior intestinal portal with, 145f
Cyclic adenosine monophosphate (cAMP), 122
Cylinders, 206
Cysts, 262f
Cytokinesis, 313–314
Cytoskeleton
  accurate element placement for, 43–44
  animal cell shapes influenced by, 41
  depolymerization of, 113
  member system theories of, 42
  morphogenesis in, 14
  tension/compression systems of, 55
  wounds regulated by, 350

D

D. discoideum, 126, 129–130, 185, 316
Darwin-Wallace theory, 13
DE-cadherin, 250
Depolymerization, 84, 113
Dermal cells, 192f
Dermis, 190–191

Desmsomes, 201
Differentiation, 4
Diffuse expansion, 38
Diffusible chemorepellant system, 130
Diffusion
  coefficient/degradation in, 121f
  equation for, 120
  gradients sharpening with lower, 124f
  simple systems with, 120f
Digits, 331f
Dipodial branching, 259–261, 261f
Dishevelled locations, 220
Dishevelled signaling, 220
Division planes, 324f
Dorsal aorta, 98f
Dorsal closure
  accurate navigation for, 226–227
  Drosophila Melanogaster with, 223–227
  embryo with, 224f
  filopodial activity of, 227f
  laser ablation influencing, 225f
  leading edge actin bands in, 225f
Double-layered sheets, 20–21
Dpp endocytosis, 337–338
Dpp levels, 337f
Dpp signal, 335–336
Dpp signal transduction pathway, 336–337
Drosophila Melanogaster, 167, 190, 214, 217f, 219, 265, 291
  dorsal closure in, 223–227
  mechanosensory organs of, 319
  sensory bristles in, 67–69
  tracheal fusion in, 249–250
Durotaxis, 155
Dynamic instability, 53
Dynein, 53

E

E-cadherin, 228–229
Ecdysone reception, 242f
Echinarachnius parma, 313
Echinoderm egg, 59–60
Elective cell death
  anoikis and, 339
  apoptosis controlling, 330f
  digits separated with, 331f
  Dpp levels controlling, 337f
  hollows created by, 332
  plants with, 340–341
  as senescence, 340
  tasks accomplished by, 329–330
Electric currents, 142
Electric fields
  AchR biased by, 139f
  cell movement response to, 137–140
  cells influenced by, 138
  charged molecule influenced by, 139f
  living systems with, 140–145
  migrating cells responding to, 137
  posterior intestinal portal with, 144
  vertebrate limb bud with, 144–145
  vibrating probe measuring, 143f
  wound healing and, 142
Electrophoresis, 138
Embryo
  attractive molecules in, 156–157
  blastopore of, 143–144, 144f
  cell division oriented in, 323f
  cleavage modes in, 312f
  cloacal cavity in, 256
  convergent extension in, 217f
  direction finding process by, 167
  dorsal closure in, 224f
  electric fields in, 144
  epithelial structure forced on, 247
  morphogenetic mechanism of, 15f
  spaghetti-squash mutation in, 241–242
  tracheal system connections in, 249f
  uterus implanted with, 312–313
  wound healing in, 227–230
Embryogenesis, 69
Embryonic branching systems, 265
Embryonic development, 4–5
Embryonic epithelium, 229
Embryonic structure, 334–335
Embryonic wounds, 228
Emergence, 359
  simple operations in, 9
  trap-door algorithms analogous to, 11–12
  views united with, 350
Endometrium, 313f
Endosperm, 340–341
Endothelial cells, 36f
Eph/Ephrin gradients, 175–176
Ephrin A5, 161
Ephrin-A-EphA, 162f, 176f
Ephrins, 160
Epistemology, 357
Epithelia
  dominance of, 200f
  growth controlled by, 275
  paracrine signals controlling, 267

Epithelial cells
  adherens junctions with, 45, 48f
  alignment of, 218f
  brush border in, 65
  clefting initiated by, 276–277
  division of, 316
  flattening/elongation of, 33, 34f
  germ-band extension with, 215–217
  junctions of, 202f
  structure of, 201
  trophectoderm with, 204
Epithelial convergent extension, 220–221
  germ band elongation of, 214
  planar polarity pathway for, 221f
Epithelial folding, 239–240
Epithelial fusion
  distinction of, 248
  palate development with, 250–256
  tube formation from, 254
Epithelial structure, 247
Epithelial tissues
  internal anatomy dominated by, 199–200
  leading edge of, 225–226
  mitosis orientation in, 312
Epithelial tubes
  branching of, 249
  convergent extension in, 213–214
Epithelium
  Adherens junctions linking, 240
  forces shaping, 205–208
  formation of, 204f
  making of, 203–205
  plane of, 217–218
  self-organization by, 272–273
Erk, 298, 300
Error correction, 339–340
Espin, 66
Ethylene, 82
Evagination, 240–243
External regulation, 17
Extracellular ligands, 154
Extracellular matrix
  integrin-mediated focal adhesions to, 49
  mesenchymal cells separated by, 190
  protease destroying, 332
Extracellular signals, 299f

F

FAK. *See* Focal adhesion kinase
Fasciculation
  Growth cones choosing, 175
  nerves constructed by, 157
Fascin, 68
Fast-growing cells, 336f
Fat family, 208
Fear of Intimacy, 189
Feedback, 356
  cell cycle controlled by, 294–295
  output process control from, 14
  process organization through, 352–353
  proliferation rates balanced by, 302f
  self-assembly, adaptive self-organization with, 13–15
Feedback loops, 55
Feedback system, 273
  cGMP signaling system with, 128
  kidney with, 272f
FGF receptor (FGFR) tyrosine kinase, 156
FGFR. *See* FGF receptor tyrosine kinase
Fibrils, collagen, 25–26
Fibronectin, 158–159
Fick's law, 61f, 119
Filaments
  Arp2/3-mediated nucleation of, 107f
  branch half-life in, 107
  elongation of, 62–63
  lecithin derivatives forming, 21
  modulation controlling length of, 67f
  overlap knuckles in, 68
Filamin, 47
Filopodia
  actin basis of, 108–109
  activity of, 227f
  as cell processes, 35
  cells with crawling, 107–110
  DE-cadherin accumulated by, 250
  formation control of, 110–111
  tubules formed from, 267
Fimbrin, 47
Five-cell arrangement, 11f
Flowering plant
  pollen tube growth in, 85f
  root cell expansion in, 38f
  root hairs growing for, 38f
Focal adhesion kinase (FAK), 154, 155, 339
Focal adhesions
  destruction of, 114
  as integrin-containing junctions, 112–113
  tension promoting, 49f
Focused cell growth, 83–84
Fold formation, 280
Follicular dendritic cells, 35
Four-cell starting arrangement, 10f

Fraenkel-Conrat and Williams, 28
Free energy, 186, 187f
Frictionless wire, 206f
Frizzled locations, 220
Fuller, R. Buckminister, 42
Fusion
  apoptosis with, 253
  hole closure differences with, 248f
  human failure for, 248
  orthogonal invagination with, 255f
  secondary palate with, 254f
  vessel lumen division by, 278f

G

G1 restriction point, 296–297
Galvanotaxis
  cell guidance by, 137
  issue of, 145
  myxamoebae and, 140
Game of Life
  Conway providing, 9–11
  four-cell starting arrangement in, 10f
  neighbors in, 9f
  trap-door functions in, 12f
Gametes, 96f
Gap junctions, 203
Genes, 297f
Genital development, 255–256
Germ band, 214
Germ cells
  migration representation for, 168f
  migration timing in, 168
  waypoint navigation of, 167–170
Germ-band extension, 215–217
Giberellic acid, 82
Glial cells, 170–171
Glider, 11f
Glucose polymer, 77
Glycolipids, 20f
Glycosaminoglycan
  hyaluronic acid as uncharged, 191
  palate shelve adhesion from, 252–253
Gonads
  cell types of, 189
  formation of, 190f
  primordial germ cells colonizing, 169
Gradients
  diffusion constant lowered for, 124f
  diffusion/destruction creating, 122
  sharpening, 124, 125f
  as steep internal, 123
Green fluorescent protein (GTP), 124
Growth, cell proliferation achieving, 292f
Growth cones
  actin contained in, 109f
  adhesion broken for, 153f
  cell-substrate adhesion steering, 153–154
  collapse of, 161f
  Fasciculation chosen by, 175
  as migratory system, 101
  movement restrained for, 154f
  protein synthesis at, 173
  waypoint navigation by, 170–173
Growth factor signaling, 298f
Growth plates, 304f, 305–306
  human long bones developed by, 304f
Growth-associated hormone, 82
GTP. *See* Green fluorescent protein
GTPase
  microfilaments controlled by, 51f
  Rac member of, 110–111
GTP-tubulin molecules, 51
Guidance system, 175–176

H

Hair follicle, 191f
Haptotaxis
  adhesion gradient guidance for, 149–150
  method demonstrating, 151–152
  simple boundary for, 152f
Harris, K.G., 111, 187
Healing, 142
Helix, 82
Hemidesmosomes, 203
Hertwig, Oscar, 322
HGF, 264
Hierarchial control, 355
HIF1α. *See* Hypoxia Inducible Factor 1α
HMGCoA reductase, 169
Hole closure, 248f
Hollow sphere, 248f
Hollows, 332
Homophilic adhesion molecules, 189–190
Human congenital diseases, 4
Human long bones, 304f
Human mouth, 253f
Humans, 248
Hyaluronic acid, 191
Hydrogen bonds, 19–20
Hydrophilic head, 19–20
Hyphae, 39

Hypoxia, 302
Hypoxia Inducible Factor 1α (HIF1α), 301

I

IGF signaling pathway, 303–304
Imaginal discs, 240–243
Inductive signals, 72–73
Infolding, 277f, 278f
Initials, 324
Inner ear, 66
Inside-out polymerization, 59–63
Insulin pathway, 304–305
Insulin receptor signal transduction pathway, 305f
Integrin-containing adhesion complex, 154
Integrin-containing junctions, 203
Integrons
  modules as, 354
  sharing of, 356–357
Interfacial force, 188
Interfacial tensions, 189f
Internal anatomy, 199–200
Interssusceptive branching, 279
Interstitial matrix, 190–192
Intracellular vacuoles, 36
Intralumenal pillars, 278f
Intralumenal tissue fold, 279f
Intussusceptive branching, 277–280
Invagination
  actin-based model for, 234f
  archenteron with, 233f
  gel swelling model for, 234f
  mechanisms of, 233
  tracheal system from, 266
  tube formation without, 239–240
  tube produced by, 231
Ion channels, 61–62
Ion pumps, 141f

J

Jacob, Francois, 354
JNK. *See* Jun kinase pathway
Jun kinase (JNK) pathway, 335

K

Kidney
  branching ureteric bud epithelium of, 270f
  condensation in, 192f
  feedback system in, 272f
  mesenchyme survival in, 334
  stromal cells supporting, 193–194
  Wnt4 signal removed from, 194f
Kinetochores, 70

L

Lamellipodium, 103–107
  actin basis of, 108–109
  formation control of, 110–111
  internal structure of, 104f
  production of, 125–126
  rival parts of, 151f
Laser ablation, 225f
Lateral pseudopodia, 127f
Leading edge
  epithelial tissues with, 225–226
  integrin-containing cell-substrate adhesion complexes in, 151
  migrating cells with, 150–151, 350–351
  protrusion of, 103–107
Lecithin derivatives, 21
Lecithin smear, 22f
Leg disc, 241f
Limb mesenchyme, 189
Living systems, 140–145
Local interactions, 8
Locomotion, 159–162
Long cylinders, 206–207, 207f
Ltmnaea peregra, 311
Lumens, 36

M

Macromolecular complexes, 350, 356
Macroscopic models, 18
Major sperm protein (MSP), 103
Mammalian branching systems, 268f
Mammalian cells, 296–297
Mammalian erythrocyte, 45f
Mammalian hand, 331
Mammalian neuroepithelial cells, 315–316
Mammalian tissue, 303–304
Mammalian wound-healing response, 130
Mammary epithelium, 269–270
Mammary glands, 271f
Mantis religiosa, 71–72
MAP-kinase activation, 274f
MAP-kinase pathway, 264
MDCK cells
  active cells sprouting in, 264–265
  branching tubules produced by, 265f
  cysts in, 263
  HGF influencing, 264

Mechanical force, 49
Mechanical movement, 35
Mechanisms
embryonic epithelium with similar, 229
idea of, 8–11
identification requirements for, 11
invagination with, 233
shapes generated by, 5, 6, 8–9
Mechanoreceptors, 320f
Mechanosensory organs, 319
Mechanosensory systems, 322
Medawar, Peter, 3
Meiotic spindles, 71–72
Membrane advancement, 104–105
Membrane-located molecules, 108f
Mesenchymal cells
condensations of, 185, 193
extracellular matrix separating, 190
kidney survival of, 334
Mesenchyme-derived signals, 277
Mesoderm cells, 158f
Metazoa, 297f
Metazoan development, 129
Microfibrils, 79f
Microfilament bundles
myosin II generating tension in, 47f
outside-in polymerization of, 63–64
Microfilament-based structures, 59–63
Microfilament-cell adhesion system, 50
Microfilaments
acrosomal process constructed from, 60
actin constituent of, 24
as compression structures, 62
core of, 65–66
cross-linking proteins linking, 44
destruction of, 50
elongation of, 60f, 88
GTPase/Rho controlling, 51f
mechanical force and, 49
small microvillus with, 63
stability of, 48
tension of, 48–50, 54f
treadmilling of, 66
trichome with, 87f
Microtubule organizing centers (MTOCs), 52, 70
Microtubule-associated proteins, 53
Microtubule-based structures, 69–70
Microtubules
ACF7 protein binding with, 53
as actin microfilaments, 84
branching processes with, 88
cellulose microfibrils orientation determined by, 80
dynamic instability of, 53
flexural rigidity of, 52
GTP-tubulin molecules forming, 51
Reorientation of, 82
system for, 51–55
Microvillus
bundles formed in, 68
microfilament core of, 65–66
Microvillus, large, 65–66
Microvillus, small
barbed ends of, 65f
basic structure of, 64f
cells with, 64
cross-linked microfilaments in, 63
regression of, 64
Midline
attraction to, 171
optic chiasm and, 178f
proteins regulating crossing of, 172
Migrating cells
electric fields and, 137
leading edge of, 150–151, 350–351
noise influencing decision making by, 132
substrate interaction with, 161–162
tension exerted by, 113f
Migration
cell processes with, 101
central nervous system with, 170
Migratory grex, 97
Migratory system, 101
Mill, John Stuart, 7
Minute cells
Dpp signal for, 335–336
structures contributed by, 334
Mitosis, 324–325
endometrium with, 313f
epithelial tissues with, 312
mouse trophoblast oriented by, 313f
Wnt-Frizzled-Dishevelled system controlling, 321
Mitotic division, 311
Mitotic orientation, 322–323
Mitotic spindles
adherens junctions orientation of, 317f
cortical cytoplasm with, 316f
cytokinesis determining position of, 313
embryogenesis with, 69
orientation of, 315–322
planar polarity system linked to, 321f

Mitotic spindles (*continued*)
  purpose of, 70
  radial cleavage with aligned, 311
Mitral cells, 173
Modularity, 353–355
Modules, 354
Molecular biological techniques, 4–5
Molecular biology, 349–350
Molecular mechanisms, 8
Molecular revolution, 3
Molecular ruler, 28
Molecules, 156–157
Monopodial, 261f
Monopodial branching, 259–261
Monopolar mechanism, 219
Morphogenesis
  animal cell cytoskeleton with, 14
  cell migration/locomotion contributing to, 96
  different cells and, 352f
  dorsal aorta with, 98f
  elective cell death involved in, 330–331
  as four-dimensional, 3–4
  individual cell deformation in, 213
  multilayered organization of, 350–353, 351f
  predictive modeling of, 358
  as shape production, 3
Morphogenetic mechanisms
  abstract principles of, 5–6
  embryo with, 15f
Morphogenetic modules, 353f
Morphological event, 279–280
Motility, 125–129
MSP. *See* Major sperm protein
MTOCs. *See* Microtubule organizing centers
Mullerian duct, 332
Mullins-Sekerka instability, 261
Multicellular animals, 295–296
Multicellular epithelial tubes, 267–268
Multicellular fungi, 39
Multilayered morphogenetic systems, 355
Multilayered organization, 350–353, 351f
Mutual antagonism, 128
Mutual compensation, 291
Myosin
  cell boundaries targeted by, 217–218
  oligomers formed by, 46–47
  tension generated by, 113
  vertebrate cells with, 114
Myosin II
  lateral pseudopodia suppressed by, 127f
  microfilament bundles and, 47f
  tension generated by, 46
Myotube, 36
Myxamoebae
  chemotactic migrating, 122
  galvanotaxis with, 140
Myxamoebal cell, 97

N

Nasal retina, 178
Natural selection, 13
Negative feedback
  branching tree organized by, 271–272
  microfilament stability regulated by, 48
  multiple layers of, 14
Neighbor exchange, 213
Neoplasias, 341
Nephric duct, 205
Nervous system
  development of, 101
  fasciculation constructing, 157
Netrin, 177
Networks, 111f
Neural crest
  cell migration in, 100, 156–157
  migration routes of, 100f
  outgrowth of, 159f
Neural plate
  curvature formation in, 235
  hinge points of, 237f
Neural tube
  cell elongation in, 73
  dorsal parts meeting for, 239f
  orthogonal invagination and, 235–239
  vertebrate with, 236
Neurites
  cell adhesion molecules stimulating, 156
  galvanotactic response of, 138f
Neurons, 333
Neuropepithelial cells, 177
Neurotrophins, 333–334
Neutrophils, 129–130
Notochord, 280

O

Olfactory bulb, 173
Oligomers, 46–47
Optic chiasm
  contralateral projection at, 178–179
  midline crossing decision at, 178f
  optic nerves joining at, 177–178
Optic nerves, 177–178

Optic tectum
guidance molecules expressed by, 160–161
target cell axons in, 179
Optic vesicle, 72–73
Organs
cell type populations for, 293–294
dipodial branching in, 259–261
local controls for, 306
size regulation system for, 293f
Organ-specific signals, 305–306
Orthogonal invagination
fusion in, 255f
neural tube formation and, 235–239
Osmotic turgor pressure, 78–79
Outside-in polymerization, 63–64

P

Palate development, 250–256
Palate, secondary, 252f, 254f
Palate shelves
glycosaminoglycan adhesion for, 252–253
growth of, 252
Palisading, 72
Paracrine signals
cell types for, 301
epithelia controlled by, 267
self-organization by, 271
Pathways, 173
Penis, 255f, 256f
Phase space, 18
Phospholipid
bilayer in, 20f
as cell membrane constituent, 19–20
double-layered sheets in, 20–21
general structure of, 19f
Phospholipid geometry, 21
Phospholipid vesicle, 21f
Physical process, 13–14
Physical structures, 13
Physical systems, 13
PI-3-kinase, 263–264
Placodes, 190–191
Planar polarity
convergent extension guided by, 320
epithelial convergent extension and, 221f
epithelium plane as, 217–218
mitotic spindle linked to, 321f
proteins asymmetrical localization in, 226
signaling system in, 220f
Plant cells
cell wall rigidity in, 77
cellulose synthesized by, 78
cortical microtubule production in, 81f
elongation direction setting for, 79
focal growth in, 84
osmotic turgor pressure expanding, 78–79
shapes of, 77–78
structure of, 78f
Plants
anisotropic cell expansion used by, 37
cell division in, 323–325
diffuse cell elongation in, 78–83
elective cell death in, 340–341
mitosis orientation controlled in, 324–325
Plasma membrane, 105f
Plexins, 131f
Polarity, 219
Pollen tubes
components of, 86
flowering plant with, 85f
gradient concentration in, 86
root hair similar to, 84
Polymerization
as inside-out, 59–63
as outside-in, 63–64
self-assembly by, 17
Popper, Karl, 7
Popperian mode, 7
Positive feedback loop, 124
Posterior intestinal portal, 144, 145f
Post-hoc reasoning
conclusions from, 12–13
dangers of, 11–13
Predictive modeling, 358
Primordial cells, 98
gonads colonized by, 169
intermediate navigation systems missing for, 169–170
migration of, 99f
Principles, 7
Processes
categories of, 3
cell morphogenesis examples of, 34f, 35f
Procollagen molecules, 25
Protease, 332
Proteins
capping of, 110
FAK as signaling, 155
midline crossed by, 172
stemmed loop interacting with, 27–28
synthesis of, 173

Protoplasts, 83
Protrusion
leading edge with, 103–107
plexins inhibiting, 131f

Q

Quality control, 28–29
Quantitative modeling, 358
Quorum sensing, 193–194

R

Rac, 110–111
Radial cleavage, 311
Ramogens
branching epithelia and, 271
branching organs with, 268t
function of, 262–263
Rap1 activity, 238
Rb protein, 296, 297f
Receptor system, 131
Renal cells, 262
Repulsion, 179
Repulsive molecules, 130–131
Restriction point, 299
Retinal ganglion cells
centrifugal sequence of, 175
cues guiding, 177f
growth pattern of, 174
Retinotectal mapping, 179
Retinotectal projection system, 180f
Retraction, 113
Rho pathway
activation of, 242
microfilaments controlled by, 51f
stress fiber formation encouraged by, 113–114
RNA (Ribonucleic acid)
as molecular ruler, 28
TMV coat protein mixed invitro with, 27
Root hairs
components of, 86
depolymerizing microtubules of, 84
microtubule cytoskeleton alignment for, 83–84

S

Salivary gland, 275f
Sand dunes, 13–14
Sb-sbd. *See* Stubble-stubbloid
SCAR (suppressor of cAR), 106, 351
Schizoriza mutant, 325f
Schizosaccharomyes pombe, 294
Sea cucumber sperm, 61f
Sea urchin archenteron, 231–235
Self-assembly
actin with one-dimensional, 24–26
adaptive self-organization, feedback with, 13–15
bilayered membranes with, 18–24
collagen fibrils requiring regulation of, 26
collagen with one-dimensional, 25–26
external regulation not required in, 17
macromolecular complexes with, 350, 356
macroscopic models with, 18
magnetic tiles floating for, 19f
physical structures built with, 13
polymerization with, 17
process energy requirements for, 17–18
simple viruses with three dimensional, 26–28
small viruses building themselves by, 26–27
small-scale structures using, 29
structures of, 28–29
subunit binding ability in, 17
uncomplicated structures for, 18
Semaphorins, 130–131
Senescence, 340
Sensory bristles, 67–69, 68f
Sexual reproduction, 95
Shapes
mechanisms generating, 5, 6, 8–9
morphogenesis producing, 3
Sheet, 214f
Sheldrake, Rupert, 8
Shh. *See* Sonic Hedgehog
Shroom
cell wedging induced by, 237–238
Shh regulating, 238–239
Sigmoidal curve, 22
Signal transduction pathways, 154
Signaling events, 4, 5f
Signaling pathway, 128f
Simple systems, 120f
Single-layered structure, 21f
Size-regulation system, 291
Slit proteins, 130
Slow-growing cells, 336f
Smad proteins, 264
Small-scale structures, 29
Snelson, Kenneth, 42
Social amoeba, 96–97
Sodium oleate
as amphipathic molecure, 21–22
micelles forming by, 22f

vesicle formation kinetics by, 23f
vesicle reproductive cycle of, 23f
Somatic cells, 341
Sonic Hedgehog (Shh)
Shroom regulated by, 238–239
signaling by, 275
Spaghetti-squash (sqh), 241–242
Sperm, 95–96
Spherical vesicles, 21
Spinal cord/hindbrain, 101f
Spindle (assembly), 72, 296f
Sprout, 262f
Sprouting cells
branching by, 259–269
branching tips with, 273
viscous fingering providing, 261
Sprouty protein, 273, 277
Sqh. *See* Spaghetti-squash
Steinberg hypothesis, 186, 187
Stem cell mitosis, 318f
Stem cells, 324
Stemmed loop, 27–28
Stereocilia, 66
Stress fibers
regulators of, 50
Rho encouraging formation of, 113–114
tension promoting, 49f
Stromal cells, 193–194
Structures
adaptive self-organization arranging, 14–15
self-assembly and, 18
Stubble-stubbloid (sb-sbd), 242
Submandibular salivary glands, 275
Subunit binding, 17
Surface area
epithelial folding increasing, 239–240
tension minimizing, 205–206
Surface tension, 206–207
Systems-biology, 349–350

T

Tangential forces, 188f
Target cell axons, 179
Temporal retina, 178
Tensegrity, 42–43, 43f
Tensile microfilaments, 44–47
Tension
as myosin-generated, 113
surface area minimized by, 205–206
Thermodynamics
free energy minimized in, 41
model for, 187
Thompson, D'Arcy, 206
Ti1 axon, 170f
TIMPs (tissue inhibitors of metalloproteases), 340
Tissue engineering, 358–359
Tissues
morphogenesis of, 37–39
proportions controlled for, 300–301
whole-genome microarray studies of, 353–354
TMV. *See* Tobacco Mosaic Virus
TMV coat protein, 27
Tobacco Mosaic Virus (TMV), 26, 27f
Tracheae, 274f
Tracheal branching, 269f
Tracheal fusion, 249–250, 251f
Tracheal system
embryo with, 249f
embryonic branching systems similar to, 265
tubule structure in, 266f
Transepithelial potential, 141, 141f, 142f
Transmember protein, 189
Trap-door algorithms, 11–12
Trap-door functions, 12f
Tre, signaling pathway of, 169
Treadmilling, 107
Trichome, 87, 87f, 88
Trichoplax adhaerens, 199
Trimer, 44–45
Trophectoderm, 204
Trophic theory, 333
Trophoblast, mouse, 313f
Tubes
creation methods for, 232f
epithelial fusion for, 254
formation of, 239–240
invagination producing, 231
Tubules
filopodia sprouts forming, 267
functions of, 70
tracheal system with, 266f
Tubulin, 52f

U

Ureteric bud, 269f
Urethral groove, 255–256, 256f, 257
Uterus, 312–313

V

Vascular endothelial growth factor (VEGF), 302
Vasculogenesis, 97

Vegetal plate, 233
VEGF. *See* Vascular endothelial growth factor
Vertebrate cells
  myosin activity in, 114
  posterior half-somite of, 159
Vertebrates
  lens of, 73f
  limb buds of, 144–145
  neural tube formation in, 236
  neurulation of, 319f
  skeletal muscles of, 36f
  visual system connections in, 174
  visual system of, 174–181
Vesicles, 22–24, 23f
Vibrating probe, 143f
Viruses
  organizing itself by, 28
  self-assembly building, 26–27
  three-dimensional self-assembly for, 26–28
Viscous fingering, 261
Visual system
  simultaneous cues for, 181
  vertebrates with, 174
V-shaped dimer, 47

W

WASP (Wiscott-Aldrich syndrom protein), 106, 351
Waypoint navigation, 154
  germ cells with, 167–170
  growth cones with, 170–173
  midline crossing with, 172
  protein synthesis in, 173
Waypoints
  glial cells as, 170–171
  vertebrate visual system with, 174–181
Whole-genome microarray studies, 353–354
Wild-type cells, 335–336
Wnt signaling pathway, 317
Wnt4 signal, 194f
Wnt-Frizzled-Dishevelled system, 321
Wounds
  actin filament bands in, 228
  adult skin responding to, 229–230
  cytoskeleton regulating, 350
  electric field and, 142
  embryo healing of, 227–230
  transepithelial potential influenced by, 142f

X

Xenopus laevis, 311

Y

Yeast, 296f
Young' modulus, 155

Z

Zipper/Spaghetti squash complex, 219